The Mycota

Edited by
K. Esser and P.A. Lemke

Springer
Berlin
Heidelberg
New York
Barcelona
Budapest
Hong Kong
London
Milan
Paris
Santa Clara
Singapore
Tokyo

The Mycota

The Mycota

A Comprehensive Treatise
on Fungi as Experimental Systems
for Basic and Applied Research

Edited by K. Esser and P.A. Lemke†

IV Environmental and Microbial Relationships

Volume Editors:
D.T. Wicklow and B.E. Söderström

With 82 Figures and 31 Tables

 Springer

Series Editors

Professor Dr. Dr. h.c. mult. Karl Esser
Allgemeine Botanik
Ruhr-Universität
D-44780 Bochum
Germany

Professor Dr. Paul A. Lemke†, Auburn, USA

Volume Editors

Professor Dr. Donald T. Wicklow
Mycotoxin Research Unit
United States Department of Agriculture
Agricultural Research Service
National Center for Agricultural Utilization Research
1815 N. University St.
Peoria, IL 61604
USA

Professor Dr. Bengt E. Söderström
Department of Microbial Ecology
Lund University
Ecology Building
22362 Lund
Sweden

ISBN 3-540-58005-0 Springer-Verlag Berlin Heidelberg New York

Library of Congress Cataloging-in-Publication Data. (Revised for vol. 2) The Mycota. Includes bibliographical references and index. Contents: 1. Growth, differentiation, and sexuality/editors, J.G.H. Wessels and F. Meinhardt – 2. Genetics and biotechnologh. 1. Mycology. 2. Fungi. 3. Mycology – Research. 4. Research. I. Esser, Karl, 1924– . II. Lemke, Paul A., 1937– . OK603.M87 1994 589.2 ISBN 3-540-57781-5 (v. 1: Berlin: alk. paper) ISBN 0-387-57781-5 (v. 1: New York: alk. paper) ISBN 3-540-58003-4 (v. 2: Berlin) ISBN 0-387-58003-4 (v. 2: New York)

Production Editor: P. Venkateswara Rao

Cover design: E. Kirchner, Springer-Verlag

Typesetting by Best-set Typesetter Ltd., Hong Kong

SPIN: 10043385   31/3137/SPS – 5 4 3 2 1 0 – Printed on acid-free paper

Series Preface

Mycology, the study of fungi, originated as a subdiscipline of botany and was a descriptive discipline, largely neglected as an experimental science until the early years of this century. A seminal paper by Blakeslee in 1904 provided evidence for self-incompatibility, termed "heterothallism", and stimulated interest in studies related to the control of sexual reproduction in fungi by mating-type specificities. Soon to follow was the demonstration that sexually reproducing fungi exhibit Mendelian inheritance and that it was possible to conduct formal genetic analysis with fungi. The names Burgeff, Kniep and Lindegren are all associated with this early period of fungal genetics research.

These studies and the discovery of penicillin by Fleming, who shared a Nobel Prize in 1945, provided further impetus for experimental research with fungi. Thus began a period of interest in mutation induction and analysis of mutants for biochemical traits. Such fundamental research, conducted largely with *Neurospora crassa*, led to the one gene: one enzyme hypothesis and to a second Nobel Prize for fungal research awarded to Beadle and Tatum in 1958. Fundamental research in biochemical genetics was extended to other fungi, especially to *Saccharomyces cerevisiae*, and by the mid-1960s fungal systems were much favored for studies in eukaryotic molecular biology and were soon able to compete with bacterial systems in the molecular arena.

The experimental achievements in research on the genetics and molecular biology of fungi have benefited more generally studies in the related fields of fungal biochemistry, plant pathology, medical mycology, and systematics. Today, there is much interest in the genetic manipulation of fungi for applied research. This current interest in biotechnical genetics has been augmented by the development of DNA-mediated transformation systems in fungi and by an understanding of gene expression and regulation at the molecular level. Applied research initiatives involving fungi extend broadly to areas of interest not only to industry but to agricultural and environmental sciences as well.

It is this burgeoning interest in fungi as experimental systems for applied as well as basic research that has prompted publication of this series of books under the title *The Mycota*. This title knowingly relegates fungi into a separate realm, distinct from that of either plants, animals, or protozoa. For consistency throughout this Series of Volumes the names adopted for major groups of fungi (representative genera in parentheses) are as follows:

Pseudomycota

Division:	Oomycota (*Achlya, Phytophthora, Pythium*)
Division:	Hyphochytriomycota

Eumycota

Division:	Chytridiomycota (*Allomyces*)
Division:	Zygomycota (*Mucor, Phycomyces, Blakeslea*)
Division:	Dikaryomycota

Subdivision: Ascomycotina
    Class: Saccharomycetes (*Saccharomyces, Schizosaccharomyces*)
    Class: Ascomycetes (*Neurospora, Podospora, Aspergillus*)
Subdivision: Basidiomycotina
    Class: Heterobasidiomycetes (*Ustilago, Tremella*)
    Class: Homobasidiomycetes (*Schizophyllum, Coprinus*)

We have made the decision to exclude from *The Mycota* the slime molds which, although they have traditional and strong ties to mycology, truly represent nonfungal forms insofar as they ingest nutrients by phagocytosis, lack a cell wall during the assimilative phase, and clearly show affinities with certain protozoan taxa.

The Series throughout will address three basic questions: what are the fungi, what do they do, and what is their relevance to human affairs? Such a focused and comprehensive treatment of the fungi is long overdue in the opinion of the editors.

A volume devoted to systematics would ordinarily have been the first to appear in this Series. However, the scope of such a volume, coupled with the need to give serious and sustained consideration to any reclassification of major fungal groups, has delayed early publication. We wish, however, to provide a preamble on the nature of fungi, to acquaint readers who are unfamiliar with fungi with certain characteristics that are representative of these organisms and which make them attractive subjects for experimentation.

The fungi represent a heterogeneous assemblage of eukaryotic microorganisms. Fungal metabolism is characteristically heterotrophic or assimilative for organic carbon and some nonelemental source of nitrogen. Fungal cells characteristically imbibe or absorb, rather than ingest, nutrients and they have rigid cell walls. The vast majority of fungi are haploid organisms reproducing either sexually or asexually through spores. The spore forms and details on their method of production have been used to delineate most fungal taxa. Although there is a multitude of spore forms, fungal spores are basically only of two types: (i) asexual spores are formed following mitosis (mitospores) and culminate vegetative growth, and (ii) sexual spores are formed following meiosis (meiospores) and are borne in or upon specialized generative structures, the latter frequently clustered in a fruit body. The vegetative forms of fungi are either unicellular, yeasts are an example, or hyphal; the latter may be branched to form an extensive mycelium.

Regardless of these details, it is the accessibility of spores, especially the direct recovery of meiospores coupled with extended vegetative haploidy, that have made fungi especially attractive as objects for experimental research.

The ability of fungi, especially the saprobic fungi, to absorb and grow on rather simple and defined substrates and to convert these substances, not only into essential metabolites but into important secondary metabolites, is also noteworthy. The metabolic capacities of fungi have attracted much interest in natural products chemistry and in the production of antibiotics and other bioactive compounds. Fungi, especially yeasts, are important in fermentation processes. Other fungi are important in the production of enzymes, citric acid and other organic compounds as well as in the fermentation of foods.

Fungi have invaded every conceivable ecological niche. Saprobic forms abound, especially in the decay of organic debris. Pathogenic forms exist with both plant and animal hosts. Fungi even grow on other fungi. They are found in aquatic as well as soil environments, and their spores may pollute the air. Some are edible; others are poisonous. Many are variously associated with plants as copartners in the formation of lichens and mycorrhizae, as symbiotic endophytes or as overt pathogens. Association with animal systems varies; examples include the predaceous fungi that trap nematodes, the microfungi that grow in the anaerobic environment of the rumen, the many insect-associated fungi and the medically important pathogens afflicting humans. Yes, fungi are ubiquitous and important.

There are many fungi, conservative estimates are in the order of 100 000 species, and there are many ways to study them, from descriptive accounts of organisms found in nature to laboratory experimentation at the cellular and molecular level. All such studies expand our knowledge of fungi and of fungal processes and improve our ability to utilize and to control fungi for the benefit of humankind.

We have invited leading research specialists in the field of mycology to contribute to this Series. We are especially indebted and grateful for the initiative and leadership shown by the Volume Editors in selecting topics and assembling the experts. We have all been a bit ambitious in producing these Volumes on a timely basis and therein lies the possibility of mistakes and oversights in this first edition. We encourage the readership to draw our attention to any error, omission or inconsistency in this Series in order that improvements can be made in any subsequent edition.

Finally, we wish to acknowledge the willingness of Springer-Verlag to host this project, which is envisioned to require more than 5 years of effort and the publication of at least nine Volumes.

Bochum, Germany
Auburn, AL, USA
April 1994

KARL ESSER
PAUL A. LEMKE
Series Editors

Addendum to the Series Preface

In early 1989, encouraged by Dieter Czeschlik, Springer-Verlag, Paul A. Lemke and I began to plan *The Mycota*. The first volume was released in 1994, three other volumes followed in the years 1995 and 1996. Also on behalf of Paul A. Lemke, I would like to take this opportunity to thank Dieter Czeschlik, his colleague Andrea Schlitzberger, and Springer-Verlag for their help in realizing the enterprise and for their excellent cooperation for many years.

Unfortunately, after a long and serious illness, *Paul A. Lemke* died in November 1995. Without his expertise, his talent for organization and his capability to grasp the essentials, we would not have been able to work out a concept for the volumes of the series and to acquire the current team of competent volume editors. He also knew how to cope with unexpected problems which occurred after the completion of the manuscripts. His particular concern was directed at Volume VII; in this volume, a posthumous publication of his will be included.

Paul A. Lemke was an outstanding scientist interested in many fields. He was extremely wise, dedicated to his profession and a preeminent teacher and researcher. Together with the volume editors, authors, and Springer-Verlag, I mourn the loss of a very good and reliable friend and colleague.

Bochum, Germany
January 1997

KARL ESSER

Volume Preface

In their concept of *The Mycota*, Karl Esser and Paul Lemke (Series Editors) determined that a volume was needed to examine research on fungal populations and communities. We were invited to organize Volume IV, *Environmental and Microbial Relationships*, and were instructed to concentrate on fungal responses to the physical environment, interactions with other fungi, microorganisms and invertebrates, and the role of fungi in ecosystem processes such as decomposition and nutrient cycling. Individual chapter authors were asked to concentrate on ecological themes that could be supported by selected studies in depth and to judge fungal systems for their promise as research tools. We were advised not to solicit exhaustive reviews to cover all ecological groups of fungi.

Several authors were asked to emphasize the technology transfer of ecological information, showing how specific knowledge of the ecology and biology of fungi has application in biological control, in enzymatic conversions of plant biomass, biodegradation of toxic organic pollutants, and the discovery of natural products. Here, we asked chapter authors to take into account the basic ecological underpinnings of such technologies. Initially, we intended to place chapters emphasizing biological control strategies or industrial mycology in a section entitled "Technology transfer of ecological information." However, it was found more desirable to integrate these specific chapters within the ecological framework of the volume. We hope that this approach will better unite aspects of ecological research classified as fundamental vs. applied mycology. The immediate challenge for mycological ecology is to identify those examples where ecological studies of fungi in nature provide basic information leading to the development of a particular technology.

The book begins with a section entitled "Fungal life history and genetic strategies." Here, J.H. Andrews and R.F. Harris examine the interconversion of different growth forms in a fungal life cycle, noting that the precise nature of the environmental signals which trigger these phases and how they are transduced by the organism are unknown. Fungal ecological genetics has become the most dynamic area of mycological research in the 1990s, largely due to the efforts of A.D.M. Rayner and colleagues at the University of Bath. In the present volume, M. Ramsdale and A.D.M. Rayner outline some of the highlights of progress made in recent years and also present their vision for the future of fungal ecological genetics.

The second section is concerned with the role of selected environmental factors and ecosystem disturbance in determining the species structure of fungal communities. J.C. Zak and S.C. Rabatin begin by examining various experimental designs, approaches and methodologies for analyzing and describing fungal communities. They emphasize the importance of scale in community ecology and argue that disturbance may be the single most important process regulating the structure and functioning of fungal communities in nature. C.F. Friese, S.J. Morris and M.F. Allen evaluate disturbance dynamics over a wide array of scales and interacting factors, highlighted by a case linking site disturbance by harvester ants and the renewal of plant communities every 100 to 1000 years. R.M. Miller and D.J. Lodge identify research needed to understand how fungi respond to disturbance created by management practices in agriculture and forestry. Disruption of mycorrhizal hyphal networks and the response of saprophytic hyphae have impacts on soil structure and nutrient pools associated with the fungal hyphae. M. Wainwright and G.M. Gadd show how a more accurate assessment of the effects of pollutants on the growth and activity of fungi in the environment has been made possible by new methods for determining active fungal biomass and enlightened approaches to in vitro experimentation. In reviewing examples of fungi living in extreme environments, N. Magan argues a critical need for experiments which better simulate fluctuating abiotic parameters and their impact on the ecophysiology of these fungi. A knowledge of global, regional and endemic patterns of fungal distribution is important for the understanding of evolutionary processes and biodiversity patterns and E.J.M. Arnolds observes that such information may be applied to the control and spread of crop pests as well as in fungal conservation.

The interactions of fungi with one another, with other microbes, nematodes and arthropods have produced a wealth of ecological information with the potential for development of biological control strategies. This is considered in the third section, "Fungal interactions and biological control strategies." In reviewing fungal competition, P. Widden relates this knowledge to what is known about plant competition and the predictive value of competition theory. Because many decomposer fungi have to replace an existing microflora in order to colonize a substrate, interference competition can have an important impact on decomposition. P. Jeffries recognizes mycoparasitism as a widespread phenomenon and presents numerous examples of mycoparasitic interactions. Mycoparasitism is believed to have an important influence on competitive interactions involving fungi in nature, and in the biological control of phytoparasitic fungi. *Trichoderma harzianum* is presented as a model system of a typical mycoparasite. I. Chet, J. Inbar and Y. Hadar also examine research to improve plant resistance to fungal pathogens by integrating cloned fungal chitinase with antifungal polypeptides. Again, *Trichoderma* is shown to be a useful model system. While recognizing that mycoinsecticides have had little impact on insect pest control to date, A.K. Charnley suggests a promising future for these entomopathogenic fungi. This optimism is based on the current rate of progress in research on epizootology, mass production, formulation, application and mechanisms of pathogenesis. Fungal agents for controlling plant pathogenic nematodes often perform poorly or inconsistently because they have been released prematurely, without sufficient basic knowledge of their biology and ecology. B. Kerry and B. Jaffee explain the importance relating basic information on the mode of action and epidemiology of selected fungal biocontrol agents to methods of mass production, formulation and application. In his examination of the potential of mycoherbicides, D.A. Shisler emphasizes the importance of understanding phylloplane microbial dynamics in order to obtain consistent field efficacy of a mycoherbicide, including strategies for bolstering the pathogen at this same weak point. T. McGonigle shows how the selective grazing of soil and litter fungi by arthropods can have an impact on community structure. Fungi have responded in different ways to reduce the negative effects of predation. J.B. Gloer offers numerous examples showing how observations in fungal ecology have generated hypotheses about fungal antagonism and defense which, in turn, have led to the discovery of novel bioactive fungal metabolites.

The final section recognizes fungus-mediated decomposition and the nutrient mobilizing potentials of fungi in both aquatic (marine and freshwater) and terrestrial ecosystems. J. Dighton considers aspects of metal ion accumulation and enyzmatic competence of saprotrophic fungi. Such basic information can be applied to the degradation of toxic organic compounds and the uptake and accumulation of metal ions in contaminated soils. J.R. Leake and D.J. Read review evidence that mycorrhizae in decomposing litter have a direct role in recycling of organic nutrients. There is surprisingly little information on the extent to which processes of nutrient mobilization occur in nature. M.O. Gessner, K. Suberkropp and E. Chauvet examine the role of fungi in plant litter decomposition in aquatic environments (e.g. salt marshes, mangrove swamps and streams), and they observe that mechanisms controlling the allocation of resources between mycelium and reproductive structures are not yet understood. In an effort to encourage lateral thinking, M.J.R. Nout and co-authors examine different approaches for estimating fungal biomass in food fermentations and consider the appropriateness of various approaches for fungal ecologists investigating litter decomposition. Likewise, the bioconversion of plant fibres to fuel, feed or precursors for chemical syntheses is examined by R. Sinsabaugh and M.A. Liptak, scientists whose principal research interest is in the biochemistry of plant litter decomposition and the ecology of the decomposers in nature.

We hope that not only professional biologists who wish to learn research directions and opportunities in mycological ecology will find this volume interesting, but also that graduate students in mycology and microbial ecology will find it useful. Thanks are especially due to those individual contributors who produced outstanding original figures and to all for promptly responding to our various editorial requests.

Peoria, Illinois, USA
Lund, Sweden
January 1997

DONALD WICKLOW
BENGT SÖDERSTRÖM
Volume Editors

Contents

List of Conributors

ALLEN, M.F., Soil Ecology and Restoration Group, Department of Biology, San Diego State University, San Diego, California 92182-4614, USA

ANDREWS, J.H., Department of Plant Pathology, University of Wisconsin-Madison, Madison, Wisconsin 53706, USA

ARNOLDS, E.J.M., Biological Station, Centre for Soil Ecology, Agricultural University Wageningen, Kampsweg 27, 9418 PD Wijster, The Netherlands

CHARNLEY, A.K., School of Biology and Biochemistry, University of Bath, Claverton Down, Bath, BA2 7AY, UK

CHAUVET, E., Centre d'Ecologie des Systèmes Aquatiques Continentaux, CNRS-UPS, 29 rue Jeanne Marvig, 31055 Toulouse Cedex, France

CHET, I., Department of Plant Pathology and Microbiology, The Hebrew University of Jerusalem, Faculty of Agriculture, Rehovot 76100, Israel

DIGHTON, J., Division of Pinelands Research, Institute of Marine & Coastal Sciences, Department of Biology, Rutgers University, Camden, New Jersey 08102, USA

FRIESE, C.F., Microbial and Environmental Ecology Research Group, Department of Biology, University of Dayton, 300 College Park, Dayton, Ohio 45469–2320, USA

GADD, G.M., Department of Biological Sciences, University of Dundee, Dundee, DD1 4HN, UK

GESSNER, M.O., Forschungszentrum für Limnologie, EAWAG, 6047 Kastanienbaum, Switzerland

GLOER, J.B., Department of Chemistry, University of Iowa, Iowa City, Iowa 52242, USA

HADAR, Y., Department of Plant Pathology and Microbiology, The Hebrew University of Jerusalem, Faculty of Agriculture, Rehovot 76100, Israel

HARRIS, R.F., Department of Soil Science, University of Wisconsin-Madison, Madison, Wisconsin 53706, USA

INBAR, J., Department of Plant Pathology and Microbiology, The Hebrew University of Jerusalem, Faculty of Agriculture, Rehovot 76100, Israel

Present address: Department of Biochemistry and Molecular Biology, Israel Institute of Biological Research, POB 19, Ness Ziona 70450, Israel

JAFFEE, B.A., Department of Nematology, University of California at Davis, Davis, California 95616-8668, USA

JEFFRIES, P., Research School of Biosciences, University of Kent, Canterbury, Kent, CT2 7NJ, UK

KERRY, B.R., Department of Entomology and Nematology, IACR-Rothamsted, Harpenden, Herts AL5 2JQ, UK

LEAKE, J.R., Department of Animal and Plant Sciences, The University of Sheffield, Western Bank, Sheffield S10 2TN, UK

LIPTAK, M.A., Biology Department, University of Toledo, Toledo, Ohio 43606, USA

LODGE, D.J., Center for Forest Mycology, USDA-Forest Service, Forest Products Laboratory, Box 1377, Luquillo, Puerto Rico 00773-1377, USA

MAGAN, N., Applied Mycology Group, Cranfield Biotechnology Centre, Cranfield University, Cranfield, Bedford MK43 0AL, UK

McGONIGLE, T.P., Department of Land Resource Science, University of Guelph, Guelph, Ontario, Canada N1G 2W1

MILLER, R.M., Terrestrial Ecology Group, Environmental Research Division, Argonne National Laboratory, 9700 S. Cass Avenue, Argonne, Illinois 60439, USA

MORRIS, S.J., Department of Plant Biology, The Ohio State University, 1735 Neil Ave., Columbus, Ohio 43210, USA

NOUT, M.J.R., Department of Food Science, Agricultural University, Bomenweg 2, 6703 HD Wageningen, The Netherlands

RABATIN, S.C., Plant Disease Control/Biocontrol, Ricera, Inc., Painesville, Ohio 44077, USA

RAMSDALE, M., Department of Zoology, University of Cambridge, Downing Street, Cambridge CB2 3EJ, UK

RAYNER, A.D.M., School of Biology and Biochemistry, University of Bath, Claverton Down, Bath BA2 7AY, UK

READ, D.J., Department of Animal and Plant Sciences, University of Sheffield, Western Bank, Sheffield S10 2TN, UK

RINZEMA, A., Department of Food Science, Agricultural University, Bomenweg 2, 6703 HD Wageningen, The Netherlands

SCHISLER, D.A., Fermentation Biochemistry Research, National Center for Agricultural Utilization Research, USDA-ARS, 1815 N. University St., Peoria, Illinois 61604, USA

SINSABAUGH, R.L., Biology Department, University of Toledo, Toledo, Ohio 43606, USA

SMITS, J.P., Division Agrotechnology and Microbiology, TNO Nutrition and Food Research Institute, P.O. Box 36, 3700 AJ Zeist, The Netherlands

SUBERKROPP, K., Department of Biological Sciences, University of Alabama, Tuscaloosa, Alabama 35487-0344, USA

WAINWRIGHT, M., Department of Molecular Biology and Biotechnology, University of Sheffield, Sheffield, S10 6UH, UK

WIDDEN, P., Department of Biology, Concordia University, 1455 de Maisonneuve Boulevard West, Montreal, Quebec, Canada H3G 1M8

ZAK, J.C., Ecology Program, Department of Biological Sciences, Texas Tech University, Lubbock, Texas 79409-3131, USA

Life History and Genetic Strategies

1 Dormancy, Germination, Growth, Sporulation, and Dispersal

J.H. ANDREWS[1] and R.F. HARRIS[2]

CONTENTS

I. Introduction

To place this topic within the broad context of fungal biology, we note as a point of departure that the body plan of the fungi allies them with the modular rather than the unitary class of organisms (Table 1, Fig. 1). The modular lifestyle of the fungi has been discussed elsewhere (Andrews 1991, 1992, 1994; Carlile 1995); here, it need only be emphasized that many of their key ecological attributes emerge from the property of modularity. Among these are the following features directly relevant to the issues of dormancy, growth, sporulation, and dispersal. First, the fungi, perhaps more so than any other group of organisms, show pronounced phenotypic and genotypic plasticity. The ability of many species to interconvert between yeast and mycelial forms (Vanden Bossche et al. 1993) and the various nuclear states ranging from haploidy through heterokaryosis to diploidy (Raper and Flexer 1970) are but two examples. Second, fungi grow as colonies in relatively fixed positions. Sooner or later, resources will become limiting (rate of uptake exceeds rate of resupply), creating resource depletion zones (Andrews 1991). The "choice" for the organism then becomes one of either switching to an inactive state to conserve energy until the local environment changes, or growth to acquire new resources. The latter may take the form of unspecialized (e.g., hyphal strands) or specialized (e.g., rhizomorphs) colonization as the fungal clone expands its feeding range. Alternatively, it may form propagules for dispersal to habitats beyond the reach of hyphal spread.

II. Life-Cycle Components

Figure 2 presents a generalized schematic of fungal life cycles. The basic functional components are dormant survival, growth (unrestricted, and restricted by nutrient limitation or other factors), and sporulation (sexual and asexual), linked by the processes of activation and germination, and dispersal. Specific morphotypes used by different fungi to meet these life-cycle needs are summarized in Table 2. As noted by Friday and Ingram (1985), structural composition needs are different for different functional parts of the fungal life cycle. The nature of dormant survival forms reflects dispersal strategy (wind, water, living vectors) and the mechanism of remaining dormant under unfavorable competitive growth conditions. Commonly recognized dormant forms (Fig. 2) are sclerotia, rhizomorphs, chlamydospores, and thick-walled oospores, but fungi that do not form such specialized resting structures must rely on other morphotypes to ensure survival under conditions unfavorable for growth. Activation of resting, undifferentiated vegetative hyphae, sclerotia, rhizomorphs, and chlamydospores, and germination of sexual spores (such as oospores, zygospores, ascospores, and basidiospores) and asexual spores (such as sporangiospores, arthrospores, blastospores, and conidia) are suppressed by unfavorable physical (water and temperature) and chemical (pH, toxic or inhibitory chemicals and biochemicals, nutrient deficiency) conditions. Suppression of activation and

---

[1] Department of Plant Pathology and
[2] Department of Soil Science, University of Wisconsin-Madison, Madison, Wisconsin 53706, USA

The Mycota IV
Environmental and Microbial Relationships
Wicklow/Söderström (Eds.)
© Springer-Verlag Berlin Heidelberg 1997

Table 1. Major attributes of unitary and modular organisms[a]

Attribute	Unitary organisms	Modular organisms
Branching	Generally nonbranched	Generally branched
Mobility	Mobile; active	Nonmotile[b]; passive
Germ plasm	Segregated from soma	Not segregated
Development[c]	Typically preformistic	Typically somatic embryogenesis
Growth pattern	Noniterative; determinate	Iterative; indeterminate
Internal age structure	Absent	Present
Reproductive value	Increases with age, then decreases; generalized senescence	Increases; senescence delayed or absent; directed at module
Role of environment in development	Relatively minor	Relatively major especially among sessile forms
Examples	Mobile animals generally, especially the vertebrates	Many of the sessile invertebrates such as hydroids; corals; colonial ascidians; also plants; fungi; bacteria

[a] Modified from Andrews (1991). These are generalizations; there are exceptions (see text for examples).
[b] Juvenile or dispersal phases mobile. Many bacteria and protists are somewhat mobile.
[c] Pertains to degree to which embryonic cells are irreversibly determined. Preformistic = all cell lineages so determined in early ontogeny; somatic embryogenesis = organisms capable of regenerating a new individual from some cells at any life-cycle stage (cells totipotent).

Fig. 1. Merging colonies of the bryozoan *Membranipora* on the frond of a seaweed. Each chamber, called a zooid, is an asexual unit capable of independent existence. Resemblance to the colonial, modular growth pattern of fungi, especially yeasts, is apparent. *Bar* 1 mm. (Photo courtesy D. Padilla)

byproducts accumulate, resulting in growth-limiting conditions. Fungal strategies under growth-restricted conditions include: (1) physiological retrenchment to a scavenging maintenance rather than growth mode (essentially a *K*-strategy; Andrews and Harris 1986), or morphotype conversion to (2) foraging rather than localized colonization by hyphae or blastospores (to explore for new substrate sources and/or escape toxic byproducts, e.g., some dimorphic yeasts; Gimeno and Fink 1992; Gimeno et al. 1992), (3) sexual sporogenous hyphae, (4) asexual sporogenous hyphae, or (5) resting, differentiated hyphal structure such as sclerotia or chlamydospores. The shift to sporogenous hyphae may involve specific sporulation inducers in place of or in addition to growth-restricting nutritional or environmental conditions.

Sexual or asexual reproduction involves the differentiation of sporogenous hyphae into fruiting bodies and ultimately results in the production of sexual or asexual spores (Fig. 2, Table 1). Sexual reproduction involves exchange of genetic information between individuals and exposure of the new genotype to selection. Asexual sporulation commonly involves maximized conversion of biomass to massive numbers of small propagules and is, for example, a typical escape mechanism used by *r*-strategists (Andrews and Harris 1986). The size and structural properties (e.g., wall thickness, pigmentation, storage materials) of sexual and asexual spores commonly reflect adaptation for dispersal.

germination may also require the absence of specific assimilatory (body building) or dissimilatory (energy-generating) substrates, biochemical inducers, and/or the absence of a temperature or other physical shock.

The life cycle is presumably adjusted through natural selection such that "success" (genes contributed to gene pool) is maximized by activation of resting forms under nutritional and environmental conditions favoring unrestricted vegetative growth and competitive exploitation. With continued growth of the fungus (and competing organisms), substrates become scarce and/or toxic

Table 2. Functional morphotypes of fungi

Phyla/class[a]	Functional morphotype							Genus example
	Dormancy[b]	Growth		Reproduction				
		Unrestricted	Restricted	Sexual sporulation		Asexual sporulation		
				Fruiting bodies	Spores	Fruiting bodies	Spores	
Basidiomycota	Basidiospores	Hyphae	Hyphae	Basidia	Basidiospores	Conidiophores	Conidiospores	*Puccinia*
	Rhizomorphs/basidiospores	**Rhizomorphs**/hyphae	**Rhizomorphs**/hyphae	Basidia	Basidiospores	—	—	*Armillaria*
	Chlamydospores/basidiospores	Hyphae	Hyphae	Basidiocarp, basidia	Basidiospores	—	—	*Agaricus*
Ascomycota	Conidiospores/ascospores	Hyphae	Hyphae	Ascocarp, asci	Ascospores	Conidiophores	Conidiospores	*Venturia*
	Conidiospores	Hyphae	Hyphae	—	—	Conidiophores	Conidiospores	*Aspergillus*
	Conidiospores	Hyphae	Hyphae	—	—	Conidiophores	Conidiospores/arthrospores/fission spores	*Colletotrichum*
	Chlamydospores/blastospores	Blastospores/hyphae	Blastospores/swollen cells/hyphae	—	—	—	Blastospores	*Aureobasidium*
	Blastospores	Blastospores	Blastospores/pseudohyphae	—	—	—	Blastospores	*Saccharomyces*
	Sclerotia	Hyphae	Hyphae	—	—	—	—	*Sclerotium*
Zygomycota	**Zygospores/**sporangiospores	Hyphae	Hyphae	Fused gametangia	Zygospores	Sporangiophores, sporangia	Sporangiospores	*Rhizopus*
Chytridiomycota	**Zygospores/resistant sporangia**	Hyphae	Hyphae	Fused gametangia	Zygospores	Sporangiophores, sporangia	Zoospores	*Allomyces*
Heterokonta Oomycetes	**Oospores/encysted zoospores**	Hyphae	Hyphae	Fused gametangia	Oospores	Sporangiophores, sporangia	Zoospores	*Pythium*
Hypochytrio-mycetes	Hyphae	Hyphae	Hyphae	—	—	Sporangia	Zoospores	*Rhizidiomyces*

a Classification system of Barr (1992).
b Bold print identifies classical, specialized resting structures.

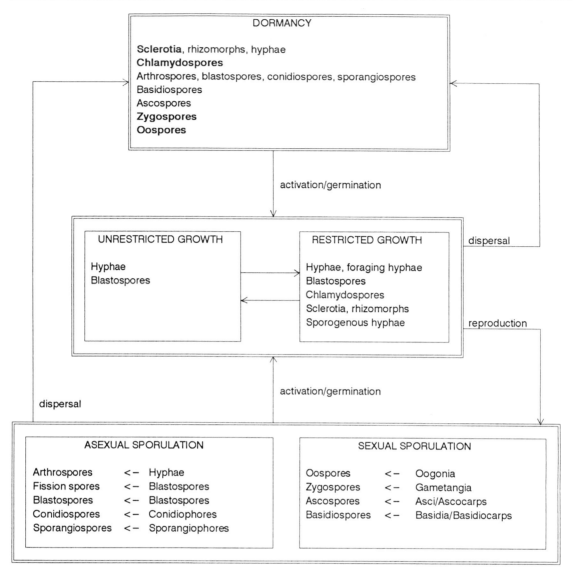

Fig. 2. Functional morphotype life cycles of fungi. *Bold print* identifies classical, specialized resting structures

Resting propagules may remain dormant for considerable time periods, depending on the nutritional and environmental conditions to which they are exposed. Under fluctuating nutrient and environmental conditions, shifts between unrestricted and restricted growth will occur (Fig. 2), and the commitment to sporulation or dormancy may be reversed, at least in the early stages. Under certain nutrient-limiting conditions, signals for germination and sporulation may coexist, resulting in a highly abbreviated life cycle, microcycle conidiation. This can occur, for example, when there is a finite nutrient flux through a nutrient pool of growth-limiting concentration due to nutrient uptake potential in excess of nutrient input rate (Cascino et al. 1990).

A. Dormancy

All organisms have periods when metabolism is slowed. For modular, sessile creatures, these take the form of distinctive life-cycle phases characterized by morphological as well as physiological change called dormancy. A compelling case can be made that dormancy in its various physical manifestations is adaptive because the phenomenon occurs in many organisms differing widely in ancestry (evolutionary convergence). The dormant condition appears to have arisen in response to two frequently related demands: (1) dispersing gametes, progeny, or the parental clone to new environments, and (2) surviving adverse conditions that would be lethal to the vegetative state. Packaging of the genome and necessary cell machinery in resistant units such as spores and related structures is one evidently successful response to these dual challenges.

In biochemical terms, the intensity of dormancy may range from complete metabolic inertia to varying degrees of endogenous and/or exogenous electron donor and electron acceptor metabolism. Endogenous electron donor substrates include sugar, fatty acid, amino acid, and purine and pyrimidine monomers derived from cell polymers such as polysaccharides, proteins, lipids, and nucleic acids. The oxidized electron donor products are largely carbon dioxide and ammonium. For facultatively anaerobic fungi such as yeasts, fermentation products such as alcohols and saturated fatty acids are released in addition to carbon dioxide. Use of endogenous electron donors implicitly involves cell decay. Exogenous electron donors include similar monomers available in the surrounding environment; they generally require active uptake systems for transportation into the cell. The exogenous electron acceptor, oxygen, is taken up by passive mechanisms. In the absence of oxygen, obligately aerobic organisms (most fungi) can only survive by complete metabolic shutdown.

Maintenance of a state of suspended animation achieved by completely shutting down metabolism is by definition characterized by lack of carbon dioxide production, oxygen uptake, and other indices of metabolic activity. Maintenance of integrity involving low-level endogenous electron donor metabolism results inevitably in increasing decay and death with time. This form of dormancy metabolism is characterized by varying degrees of oxygen uptake and carbon dioxide production. Maintenance of integrity involving active uptake and use of exogenous electron donors to attenuate endogenous electron donor metabolism, together with matching exogenous electron acceptor (oxygen) metabolism, may be invoked to attenuate decay and death. At some point, the maintenance of integrity involving osmoregulation of substrates into and products out of the cell,

merges from dormancy to active maintenance of the non-growing cell.

General mechanisms of achieving and maintaining dormancy in seeds apply also to fungi. According to Harper (1957): "some seeds are born dormant, some acquire dormancy and some have dormancy thrust upon them." Three similar kinds of dormancy have been recognized for fungi: (1) constitutional (constitutive) dormancy, involving an innate dormancy property; (2) induced dormancy, the imposition of constitutional dormancy on a spore that was born exogenously dormant, and (3) exogenous dormancy, the delay of development because of unfavorable environmental conditions (Sussman and Douthit 1973). The three types of mechanisms for achieving dormancy (innate, induced, and exogenous) give rise to two different states of dormancy (constitutional and exogenous). Constitutional dormancy is associated with pronounced changes in ultrastructure, chemical composition, respiration, water content, and cell permeability; it involves compartmentation of enzymes and substrates and production of self-inhibitors, and various metabolic blocks (Sussman and Douthit 1973; Van Etten et al. 1983). In physiological terms, constitutional dormancy most likely involves minimal, if any, osmoregulation, and endogenous rather than exogenous electron donor metabolism. The state of exogenous dormancy is characterized by cells showing dormancy because of unfavorable conditions rather than because of a change in the physiological structure of the cell; these cells potentially invoke a combination of exogenous and endogenous donor metabolism for active maintenance of viability. Exogenous dormancy of fungal propagules in soil typically results from energy deprivation and is manifested by the well-known phenomenon of fungistasis (Lockwood 1990).

The mechanisms that trigger dormancy and maintain the dormant state are apparently complex and poorly understood. The specific factors will vary depending on the nature of the resting structure (sexual or asexual spore, sclerotium, etc.). Sporulation is covered in a Section II.C, below. Sclerotia are induced by wounding of the mycelium, accumulation of staling products, spatial confinement, various environmental manipulations, including nutrients, light, temperature, pH, and aeration, and by the activity of phenols and polyphenoloxidases (Willetts and Bullock 1992).

There is not a clear distinction among the resting cell classes, and determining the operative mechanism of constitutional or exogenous dormancy is not straightforward. Addition of a nutrient, for example, may alleviate exogenous dormancy by providing a needed substrate to activate starved spores; alternatively, a metabolic block may be removed, in which case the added nutrient would be relieving constitutional dormancy (Van Etten et al. 1983). Hence, to categorize dormancy requires that the mode of action of the manipulation be known (Van Etten et al. 1983). Regardless, the ecological significance of these various controls is that they serve to prevent growth at the site of spore production and to signal onset of growth under environmental conditions conducive for competitive survival.

While diverse sexual and asexual spores (for examples, see Table 2) are typically the most resistant of the life-cycle phases (Sussman 1966), virtually any fungal morphotype can assume a more or less dormant state. The only exceptions are actively motile forms such as flagellated zoospores. Mycelium can differentiate into tissues such as sclerotia, cords or rhizomorphs, and fruiting bodies, all of which play a role in survival. Alternatively, hyphae may simply become thick-walled and pigmented (Dauermycelium or Dauerzellen of yeasts; Sussman 1966). In the life cycles of fungi as well as other organisms, dormancy is typically associated with sexual recombination (Bonner 1958) though it need not occur exclusively at this point in the life cycle. This linkage is advantageous because the shutdown phase serves either as a convenient vehicle, typically a spore, for dispersing the variable products of meiosis to new environments, or for survival in situ until favorable conditions return. Where sex is tied to the production of a resistant form, it may be either or both that are selected for. That the target could be the resistant spore per se is suggested for certain homothallic strains of *Neurospora* (Perkins and Turner 1988). In these strains, the ascospore is the only spore form produced but, because of inbreeding, it contributes essentially nothing in terms of genetic variation. However, it plays a major role as a survival structure since the long-lived ascospores of *Neurospora* can survive more than 20 years to be activated for germination by heat shock (R.L. Metzenberg, Univ Wisconsin, pers. comm.).

Next to spores, sclerotia (Fig. 3) are probably the most important survival form for many higher fungi in the Ascomycotina and Basidiomycotina.

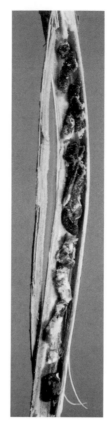

Fig. 3. Sclerotia of *Sclerotinia sclerotiorum* in the stem of a soybean plant. Not magnified. (Photo courtesy C. Grau)

Sclerotia are hard, resting structures composed of aggregated hyphae that may be more or less differentiated (Chet and Henis 1975; Willetts and Bullock 1992), spanning the range from "true sclerotia", composed of an external melanized rind and internal medulla (typical of *Sclerotinia* and *Botrytis*), to microsclerotia or bulbils of *Verticillium*, and the stromatal tissue or pseudosclerotia of *Monilinia* (Willetts and Bullock 1992). The structure, ontogeny, and general biology of sclerotia have been reviewed frequently (e.g., Chet and Henis 1975; Willetts 1978; Willetts and Bullock 1992).

Overall, with few exceptions, the mechanistic basis for the various forms of dormancy, and the different triggering factors identified by laboratory experiments, remain unknown. Likewise, the relative importance of the factors to the ecology of the fungi in nature is unclear.

B. Germination and Growth

Numerous environmental factors have been associated with activation from dormancy, among them rehydration of propagules, application of nutrients (or various chemicals acting as "nutrients" or by some other means), heat shock, or light (Sussman 1966; Van Etten et al. 1983). Prediction and understanding of the factors controlling activation and germination (spores) of dormant structures involves consideration of the state of dormancy (constitutional, exogenous) and the mechanisms of dormant cell activation (induction, derepression). Constitutional dormancy is commonly associated with complete metabolic shutdown of the dormant cells. Activation of such cells tends to require drastic environmental changes such as heat shock or high concentrations of electron donor/carbon substrates conducive to passive rather than requiring active uptake, and shows a distinct lag phase for physiological restructuring prior to germination. Exogenously dormant cells are responsive to relatively small environmental and nutrient changes and germinate with minimal lag after activation. For insurance against germination under marginal nutrient conditions, ecophysiological considerations predict that the substrate concentrations needed to trigger activation of exogenously and particularly constitutionally dormant cells would be greatly in excess of the half saturation concentrations for substrate-limited growth.

Environmental conditions triggering activation and germination are most likely conducive to unrestricted development following growth initiation. Mycelial fungi will characteristically grow vegetatively as hyphae, and yeasts as blastospores, as long as environmental conditions are nonlimiting. Under optimized environmental conditions, growth rates are maximized and are controlled by intrinsic intracellular rather than substrate uptake properties. This phase of growth corresponds to the establishment and early stage of exploitation within the continuum "arrival at, establishment in, exploitation of, and exit from a resource" (Rayner et al. 1985). With increasing biomass, substrate demand ultimately exceeds the rate of substrate supply by diffusion, and/or the rate of inhibitory product production exceeds the rate of product removal by diffusion. These conditions signal a need to complete the exploitation phase and prepare for exit from the resource (Rayner et al. 1985). The fungal response is to depart from such growth-limiting conditions (resource depletion zones, see Dispersal, Sect. II.D below), by colonization of new territory and/or by production of dispersal morphotypes. Interest-

ingly, conditions for unrestricted fungal growth are analogous to conditions conducive to habitat selection for spatial refuges by colonial benthic invertebrates (where the location of the refuge can be sensed by the recruiting larvae), and an exploratory restricted growth response by fungi corresponds to the directional growth mechanism of invertebrates (where the position of the refuge is spatially unpredictable) (Buss 1979). The various foraging modes (mycelial polymorphisms) and the important role of phenotypic plasticity in the fungi are discussed elsewhere (e.g., Rayner and Coates 1987; Andrews 1992). For dimorphic pathogenic yeasts, the minimal media conditions of saprophytic growth can induce the mycelial rather than the yeast form (Cole and Nozawa 1981). Similarly, yeasts such as *Saccharomyces cerevisiae* may shift from blastospore to foraging pseudohyphal growth under nutrient-limiting conditions (Gimeno and Falk 1992; Gimeno et al. 1992). However, this phenomenon is not universal for yeast-like fungi, since the distribution of blastospore versus hyphal morphotypes was not affected by nutrient status for pleomorphic *Aureobasidium pullulans* (Andrews et al. 1994). Under fungal growth-limiting conditions caused by something other than the electron donor/carbon substrate (e.g., N or P deficiency, pH shift, or toxic product accumulation), there is commonly a phase of intracellular storage carbon (e.g., glycogen) or extracellular polysaccharide (e.g., pullulan) accumulation. Ultimately, intracellular and/or extracellular inducer or derepressor signals promote a physiological shift to asexual or sexual sporulation or production of other forms of resting structures such as chlamydospores or sclerotia.

C. Sporulation

Diverse environmental factors influence sporulation, including nutrition, light, pH, aeration, injury, and moisture (Dahlberg and Van Etten 1982). In general, sporulation is promoted by nutrient conditions that restrict growth, but this is not invariably the case (e.g., Huang and Cappellini 1980). Sporulation triggers may also take the form of growth-inhibiting chemical and physical factors, and hormonal inducers. Sporulation initiated in response to nutrient limitation involves reorganization of endogenous resources as well as use of exogenous substrates, although the potential for such mycelial reorganization appears to decrease with age (Cascino et al. 1990).

Formation of thick-walled, often melanized, chlamydospores resistant to environmental extremes is ubiquitous across fungal groups (Table 2). Chlamydospore formation may occur in response to adverse environmental conditions, such as medium acidification (Andrews et al. 1994), or may be a gradual response, such as is commonly observed in fungal liquid and solid medium cultures subjected to prolonged incubation. For instance, in *Aureobasidium pullulans*, chlamydospore formation initiated by nitrogen starvation involves intermediate formation and accumulation of intracellular storage carbon and extracellular polysaccharide by swollen cells as they convert to chlamydospores (Andrews et al. 1994).

Asexual sporulation occurs in most fungi, and morphotypic differences are important in fungal classification (Table 2). For yeasts, asexual sporulation involves blastospore budding and ultimate release of fully formed blastospores from the mother cell, or occasionally from hyphae. For mycelial fungi, asexual sporulation is typically initiated by differentiation of hyphae into sporophores, which are borne naked or within fruiting bodies of varying complexity. Asexual sporulation by septate fungi (Basidiomycota and Ascomycota) ranges in complexity from simple fragmentation of hyphae to produce chains of arthrospores, to well-defined conidiophores of diverse ontogeny and morphology. Modes of conidial (blastic versus thallic) and conidiogenous (determinate, ampullate, percurrent, basauxic, tretic, etc.) cell development are reviewed by Cole (1986). Chytridiomycota and Zygomycota characteristically produce nonseptate hyphae that differentiate during asexual reproduction into sporangiophores producing membrane-enclosed sporangia which rupture when mature, releasing sporangiospores.

Sexual sporulation is the primary basis for subdivision of fungi into major classes (Table 2). Generalized terms for sexual fruiting bodies and sexual spores produced by basidiomycetes are, respectively, basidiocarps, and basidiospores. Similarly, ascomycetes produce ascocarps and ascospores; zygomycetes and chytridiomycetes produce zygospores; and oomyctes produce oospores (Table 2). Factors and mechanisms controlling initiation and development of sexual sporulation are much less well defined than for asexual sporulation.

As discussed below, spores provide an effective way to protect, package, and disseminate genes. The capacity of fungi to form propagules is truly prodigious. Buller (1922) estimated that a large fruiting body of the puffball *Calvatia gigantea* contained 7×10^{12} spores, while a single fruiting body of *Fomes applanatus* could liberate 3×10^{10} basidiospores per day over a 6-month fruiting period, or a total of 5.5×10^{12} propagules annually. Another advantage of the spore is that, as a single cell, it allows the fungus to start a new cycle without the load of accumulated somatic mutations that a multicellular reproductive unit would carry (Andrews 1992).

D. Dispersal

In general ecological terms, the "success" of a sessile organism such as a fungus, as measured by its distribution and abundance, will be determined in large part by its ability to reach sites conducive for colonization. In a patchy environment, such favorable or habitable sites (Fig. 4) are islands in space and time where the organism can become established and reproduce (Gadgil 1971; Harper 1981; Andrews 1991). Hence, a dispersal phase is an integral part of the life cycle and should be maintained by natural selection as long as the chances overall for "finding" a better site outweigh the hazards of dying en route or of colonizing a worse site (Gadgil 1971). The adaptive value of dispersal is attested to by the occurrence in such diverse lineages as the actinomycetes, myxomycetes, pteridophytes, bryophytes, fungi, and higher plants, of numerous and occasionally intricate devices to insure that it occurs.

More specifically, dispersal enables the sessile organism to escape from sites depleted of nutrients (so-called resource depletion zones; Harper 1985; Andrews 1991). For the fungi, dispersal can also play a role in the sexual process by bringing compatible genomes together (as in the dissemination of spermatia), and by distributing zygotes or the products of meiosis to new environments. This is not to suggest that dispersal, either for escape or for recombination, is a planned course of action by the organism. Rather, the patterns we see have been shaped by past events that have influenced the survival of genotypes.

Broadly speaking, the dispersal process includes three phases: the discharge of propagules, their transport by some agency, and deposition.

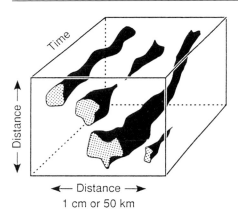

Fig. 4. The habitable site concept. Distance is arbitrary and relative to the size and activity of the organism. A given space may be occupied by more than one species which may increase, decrease, or become locally extinct temporarily or permanently. (Andrews 1991)

"Propagule" in this context is most commonly a spore, though other types of structures, including sporangia, sclerotia, and hyphal fragments can be dispersed. Discharge mechanisms may be active or passive; the several forms of each have been reviewed by Ingold (1965) and Gregory (1973). Water, raindrops, wind, animal vectors, and radial or directed growth are the major modes of dispersal. With respect to the latter, fairy rings expand centrifugally into new sources of organic matter, while rhizomorphs or cords can advance a fungus in linear fashion into new territory. A conservative estimate is 20 cm for the annual linear extension of rhizomorphs of *Armillaria bulbosa* in a northern Michigan hardwood forest where snow cover is prolonged and the growing season is short (Smith et al. 1992).

Detailed reviews of dispersal are numerous and include those by Ingold (1965), Gregory (1973), Pedgley (1991), and Malloch and Blackwell (1992). Here we have space to note only a few interesting examples. Because of the difference in composition between the air spora and the predominant soil mycoflora, Gregory (1973) felt that air spora originated primarily from fungi growing on vegetation. While acknowledging that much of this comes from wild plants, he compared the "pollution" of the air with pathogens and allergens by agriculture to air pollution with gases and particulates by industry. Indeed, he estimated that soil receives the equivalent of 5 kg N per hectare annually by fallout of air spora (including bacteria). The presence of a significant spora even over

oceans and polar regions is well documented (reviewed by Gregory 1973). There are many examples of relatively long-distance transport (Gregory 1973; Malloch and Blackwell 1992), though as a practical matter, dispersal gradients from sources of inoculum are typically steep (Gregory 1973). For discussion of the epidemiological implications, see Zadoks and Schein (1979) and Gilligan (1985).

There are many fascinating examples of dispersal of fungal propagules by animals. Among these are the transmission by insects of pycniospores of the black stem rust fungus among pycnia on barberry leaves. The insects are attracted by the viscous pycnial exudates; during the course of feeding, they transfer pycniospores to receptive hyphae of the compatible mating type (Buller 1938).

In the hypogeous gastromycetes, such as the truffle *Tuber melanosporum*, odor attracts animals that dig up and eat the fruiting bodies (Maser et al. 1978). A perhaps less known but interesting and complex relationship involves *Drosophila* and the yeast community of cacti (reviewed by Starmer et al. 1991). The yeasts include species of *Pichia*, *Cryptococcus*, and *Candida*. This is a geographically widespread, mutualistic association in which the yeasts rely on the insects for transport to new hosts and the drosophilids derive nutrients from the yeasts. A coevolutionary relationship is likely because different yeast communities, including members with distinct physiological properties, are associated with different species of *Drosophila*. Injury of the cacti sets a train of events in motion by exposing tissue to invasion by microorganisms. Odor from the decay attracts the drosophilid vectors, which lay eggs in the rotted tissue. The hatching larvae forage on microorganisms and rotted cactus tissue, pupate, and the emerging adults eventually disperse to newly formed rot pockets, completing the life cycle.

An intriguing instance of the role of animals in transmitting fungal gametes involves the rust fungus *Puccinia monoica*, which infects *Arabis* species (Roy 1993). The systemic mycelium dramatically alters in plant's morphology, inhibiting flowering and redirecting host growth into extraordinary structures that look strikingly like the flowers of other species. Indeed, these "pseudoflowers" appear so genuine that they have been mistaken at a distance as being real by professional botanists. Insects attracted to the "flowers" fertilize the rust. Moreover, because the behavior of insect pollinators is affected, pollination of other plants such as buttercups, blooming nearby at the same time, is affected. Thus, in addition to the host, many members of the *Arabis* community are influenced by this rust pathogen (Roy and Bierzychudek 1993).

III. Conclusions

The life cycle has been called "the central unit in biology" (Bonner 1965). For the fungi, this includes the phases we have addressed briefly here – dormancy, growth, and sporulation compartments connected by the processes of germination and dispersal. Natural selection acts on each of these individually as well as on the cycle as a whole.

Dormancy is an adaptation for growth in a changing environment but "opting out" represents time lost that could otherwise have been spent in colonization and contributing genes to the population gene pool. Entry into and emergence from the dormancy and growth compartments are triggered by environmental signals. The precise nature of these and how they are transduced by the organism are unknown. Evidently, many are tied directly or indirectly to nutrient levels. As noted earlier, dimorphic transition of *Saccharomyces cerevisiae* to a linear cell shape and filamentous growth form is induced by starvation for nitrogen and controlled by the *RAS* signal transduction pathway (Gimeno et al. 1992). Microcycle conidiation, involving germination of conidia which then form spores directly from the germ tube, occurs in many fungi. This phenomenon can be induced in the laboratory by various manipulations such as heat shock. However, because nutrient signals can be involved (Cascino et al. 1990), in nature the microcycle presumably has ecological significance as a simultaneously operating growth and dispersal strategy in response to growth-restricted environmental conditions.

Morphological plasticity, a hallmark of the fungi, is seen at every level of the life cycle. Indeed, the diversity in life-cycle variants within the fungi is attributable to their unsurpassed ability to interconvert among different growth forms; to vary the timing, extent, and mode of differentiation; to develop in various nuclear states (ploidy levels); and to decouple sexual from asexual reproduction.

References

Andrews JH (1991) Comparative ecology of microorganisms and macroorganisms. Springer, Berlin Heidelberg New York

Andrews JH (1992) Fungal life history strategies. In: Carroll GC, Wicklow DT (eds) The fungal community, 2nd edn. Dekker, New York, pp 119–145

Andrews JH (1994) All creatures unitary and modular. In: Blakeman JP, Williamson B (eds) Ecology of plant pathogens. CAB International, Wallingford, UK, pp 3–16

Andrews JH, Harris RF (1986) r- and K-selection and microbial ecology. Adv Microb Ecol 9:99–147

Andrews JH, Harris RF, Spear SH, Lau GW, Nordheim EV (1994) Morphogenesis and adhesion of *Aureobasidium pullulans*. Can J Microbiol 40:6–17

Barr DJS (1992) Evolution and kingdoms of organisms from the perspective of a mycologist. Mycologia 84:1–11

Bonner JT (1958) The relation of spore formation to recombination. Am Nat 92:193–200

Bonner JT (1965) Size and cycle. Princeton Univ Press, Princeton

Buller AHR (1922) Researches on fungi, vol II. Further investigations upon the production and liberation of spores in hymenomycetes. Longmans and Green, London

Buller AHR (1938) Fusions between flexuous hyphae and pycniospores in *Puccinia graminis*. Nature 141: 33–34

Buss LW (1979) Habitat selection, directional growth and spatial refuges: why colonial animals have hiding places. In: Larwood CR, Rosen BR (eds) Biology and systematics of colonial organisms. Academic Press, New York, pp 459–497

Carlile MJ (1995) The success of hypha and mycelium. In: Gow NA, Gadd CTM (eds) The growing fungus. Chapman and Hall, London, pp 3–19

Cascino JH, Harris RF, Smith CS, Andrews JH (1990) Spore yield and microcycle conidiation of *Colletotrichum gloeosporioides* in liquid culture. Appl Environ Microbiol 56:2303–2310

Chet I, Henis Y (1975) Sclerotia morphogenesis in fungi. Annu Rev Phytopathol 13:169–192

Cole GT (1986) Models of cell differentiation in conidial fungi. Microbiol Rev 50:95–132

Cole GT, Nozawa Y (1981) Dimorphism. In: Cole GT, Kendrick B (eds) Biology of conidial fungi, vol 1. Academic Press, New York, pp 97–133

Dahlberg KR, Van Etten JL (1982) Physiology and biochemistry of fungal sporulation. Annu Rev Phytopathol 20:281–301

Friday A, Ingram DS (1985) The Cambridge encyclopedia of life science. Cambridge University Press, Cambridge

Gadgil M (1971) Dispersal: population consequences to evolution. Ecology 52:253–261

Gilligan CA (ed) (1985) Mathematical modelling of crop disease, vol 3. Advances in Plant Pathology. Academic Press, New York

Gimeno CJ, Fink GR (1992) The logic of cell division in the life cycle of yeast. Science 257:626

Gimeno CJ, Ljungdahl PO, Styles CA, Fink GR (1992) Unipolar cell divisions in the yeast *S. cerevisiae* lead to filamentous growth: regulation by starvation and *RAS*. Cell 68:1077–1090

Gregory PH (1973) The microbiology of the atmosphere, 2nd edn. Leonard Hill, Aylesbury

Harper JL (1957) Biological flora of the British Isles, *Ranunculus acris* L., *Ranunculus repens* L., *Ranunculus bulbosa* L. J. Ecol 45:289–342

Harper JL (1977) Population biology of seed plants. Academic Press, London

Harper JL (1981) The meanings of rarity. In: Synge H (ed) The biological aspects of rare plant conservation. Wiley, New York, pp 189–203

Harper JL (1985) Modules, branches, and the capture of resources. In: Jackson JBC, Buss LW, Cook RE (eds) Population biology of clonal organisms. Yale Univ Press, New Haven, pp 1–33

Huang BL, Cappellini RA (1980) Sporulation of *Gibberella zeae*. VI. Sporulation and maximum mycelial growth occur simutaneously. Mycologia 72: 1231–1235

Ingold CT (1965) Spore liberation. Clarendon Press, Oxford

Lockwood JL (1990) Relation of energy stress to behaviour of soil-borne plant pathogens and to disease development. In: Hornby D (ed) Biological control of soil-borne plant pathogens. CAB International, Wallingford, UK, pp 197–214

Malloch D, Blackwell M (1992) Dispersal of fungal diaspores. In: Carroll GC, Wicklow DT (eds) The fungal community, 2nd edn. Dekker, New York, pp 147–171

Maser C, Trappe JM, Nussbaum RA (1978) Fungus – small mammal interrelationships, with emphasis on Oregon coniferous forests. Ecology 59:799–809

Pedgley DE (1991) Aerobiology: the atmosphere as a source and sink for microbes. In: Andrews JH, Hirano SS (eds) Microbial ecology of leaves. Springer, Berlin Heidelberg New York, pp 43–59

Perkins DD, Turner BC (1988) *Neurospora* from natural populations: toward the population biology of a haploid eukaryote. Exp Mycol 12:91–131

Raper JR, Flexer AS (1970) The road to diploidy with emphsis on a detour. In: Charles HP, Knight BCJG (eds) Organization and control in prokaryotic and eukaryotic cells. 20th Symp of the Society for General Microbiology. Cambridge Univ Press, Cambridge, pp 401–432

Rayner ADM, Coates D (1987) Regulation of mycelial organization and responses. In: Rayner ADM, Brasier CM, Moore D (eds) Evolutionary biology of the fungi. Cambridge Univ Press, Cambridge, pp 115–136

Rayner ADM, Watling R, Frankland JC (1985) Resource relations-an overview. In: Moore D, Casselton LA, Wood DA, Frankland JC (eds) Developmental biology of higher fungi. Cambridge Univ Press, Cambridge, pp 1–40

Roy BA (1993) Floral mimicry by a plant pathogen. Nature 362:56–58

Roy BA, Bierzychudek P (1993) The potential for rust infection to cause natural selection in apomictic *Arabis holboellii* (Brassicaceae). Oecologia 95:533–541

Smith ML, Bruhn JN, Anderson JB (1992) The fungus *Armillaria bulbosa* is the among the largest and oldest living organisms. Nature 356:428–431

Starmer WT, Fogleman JC, Lachance M-A (1991) The yeast community of cacti. In: Andrews JH, Hirano SS (eds) Microbial ecology of leaves. Springer, Berlin Heidelberg New York, pp 158–178

Sussman AS (1966) Dormancy and spore germination. In: Ainsworth GC, Sussman AS (eds) The fungi, vol 2. Academic Press, New York, pp 733–764

Sussman AS, Douthit H (1973) Dormancy in microbial spores. Annu Rev Plant Physiol 24:311–352

Vanden Bossche H, Odds FC, Kerridge D (1993) Dimorphic fungi in biology and medicine. Plenum Press, New York

Van Etten H, Dahlberg KR, Russo CM (1983) Fungal spore germination. In: Smith JE (ed) Fungal differentiation. Dekker, New York, pp 235–266

Willetts HJ (1978) Sclerotium formation. In: Smith JE, Berry DR (eds) The filamentous fungi, vol 3. Developmental mycology. Arnold, London, pp 197–213

Willetts HJ, Bullock S (1992) Developmental biology of sclerotia. Mycol Res 10:801–816

Zadoks JC, Schein RD (1979) Epidemiology and plant disease management. Oxford Univ Press, New York

2 Ecological Genetics

M. RAMSDALE[1] and A.D.M. RAYNER[2]

CONTENTS

I. Introduction

Ecological genetics is a synthesis of ecology and population genetics which integrates description and interpretation of the heterogeneous distribution patterns of individual organisms with assessments of genetic diversity in natural populations. It therefore involves characterizing the processes which underlie the short- and long-term dynamic relationships between environmental factors and the phenotypes and genotypes of organisms. In other words, it provides meaningful answers to fundamentally important questions concerning where organisms (and their offspring) are and what they do there, how they arrived, whether they will persist and how they are likely to change in character and/or distribution.

Such questions are not so readily addressed by traditional approaches to fungal ecology. The latter have focused on compiling species abundance and diversity indices (without reference to intraspecific variation), counting fungi (without defining individual reference boundaries), weighing fungi (without acknowledging heterogeneity) and assessing metabolic activity (rather than ecological role).

Part of the reason for the shortcomings of traditional approaches to fungal ecology, and at the same time a major opportunity for ecological genetic approaches, lies in the indeterminate nature of the mycelial body form. By maintaining the potential for growth over indefinite time periods within an interconnected system, fungal mycelia organize themselves into a form that is difficult to quantify meaningfully by conventional methods. On the other hand, this form demonstrates, perhaps more clearly than any other, the significance of a circumstance-driven organizational (epigenetic) programme and the feedback relationship between the functioning of a living system, its environment and its genetic information content. This relationship can be explored by mapping the distributional patterns of mycelia and ascertaining their degree of heterogeneity.

This chapter aims first to identify the fundamental issues that need to be resolved in order to circumscribe and understand the distribution of genetic and organizational variation in natural populations of mycelial fungi. The utility of a variety of practical approaches available to address these issues will then be assessed. Finally, the outcome of applying these approaches will be exam-

[1] Department of Zoology, University of Cambridge, Downing Street, Cambridge CB2 3EJ, UK
[2] School of Biology and Biochemistry, University of Bath, Claverton Down, Bath BA2 7AY, UK

The Mycota IV
Environmental and Microbial Relationships
Wicklow/Söderström (Eds.)
© Springer-Verlag Berlin Heidelberg 1997

ined in the context of presently available information bearing on the extent, sources and adaptive significance of genetic and epigenetic diversity within fungal populations.

II. Fundamental Issues

A. Spatial Distribution and Ecological Role

Before any conclusions can be reached about an organism's relationship with its environment, it is important to establish where the organism is located, i.e. to define its boundary limits. For the many organisms, including most animals, which have a determinate body form, this does not present an immediate problem. They cannot be in two places at once and so can be assigned spatial coordinates in simple, integral Euclidean dimensions. The problem for animal ecological geneticists is therefore not usually one of assigning position at one instant but rather of tracking mobile organisms as they change position in time (Ford 1971).

However, organizationally indeterminate body forms like fungal mycelia cannot readily be pinpointed in space. They can be in more than one place at a time – often in varied guises, they do not permeate space completely (see below) and are often wholly or partly hidden. On the other hand, their spatiotemporal trajectories are preserved in and to some extent can be forecast from their structural topography. In fact, the maps produced by keeping track of animal trajectories can bear a striking resemblance to the spatial patterns produced by mycelia (López et al. 1994; Rayner and Franks 1987).

The relation between coverage of domain (extent) and space-filling capacity (content) is critical to characterizing ecological roles, whether recording the spatially indeterminate development of mycelia, or the temporally indeterminate behaviour of animals. There is therefore a need to configure both regional boundaries around individual domains or territories and topographical boundaries defining the concentration of structure or effort within these domains.

Identification of regional and topographical boundaries is, however, beset by considerable practical as well as conceptual problems due to the varied operational scales and heterogeneity of indeterminate systems. With mycelial fungi, vari-

ations in operational scale have immediate repercussions with respect to deciding on suitable sampling patterns and what can be considered to be part of the same system or individual.

Correspondingly, a critical consideration in delimiting regional boundaries is the degree of genetic heterogeneity which can be tolerated within these boundaries, both by the observer and by the system itself. For reasons to be given later, regional boundaries are best defined, both practically and biologically, on the basis that somatic proliferation within these boundaries occurs from a single haploid, diploid or heterokaryotic genetic source. Systems whose regional boundaries are defined in terms of such self-proliferation can be described as genetic individuals or genets (Brasier and Rayner 1987; Dickman and Cook 1989; Rayner 1991a,b, 1992, 1994a,b).

There is already evidence that single mycelial genets may span domains measurable in anything from micrometres to kilometres (Smith et al. 1992) in different species, and, indeed, there may also be considerable heterogeneity in the size range of genets within a species. Since it is not possible in advance to know either the degree of heterogeneity in a population or the number of individuals present, there is considerable danger that any given sampling pattern may over- or underrepresent the amount of variation. This will particularly apply to population structures in which the disposition of regional boundaries corresponds with that of a fractal foam (Fig. 1). In order to characterize such population structures, it is essential to have some way of identifying the minimum scale (resolution) at which all samples are identical at a variety of locations. However, this may often be difficult to achieve in practice, and has rarely even been attempted in population studies with fungi to date.

Some insights are possible, however, into how to determine, a priori, the likely upper limits to the extent of regional boundaries. These insights may be gained by understanding the relationships between various modes of genetic self-proliferation and the heterogeneous and/or discontinuous distribution of resources in natural habitats.

Where resources are discontinuous, i.e. packaged within discrete (fully bounded) packages or units, then two extreme kinds of distribution pattern can arise. Non-unit restriction occurs via mycelial proliferation (growth) between units, resulting in systems that are or have been interconnected (Cooke and Rayner 1984; Rayner et al.

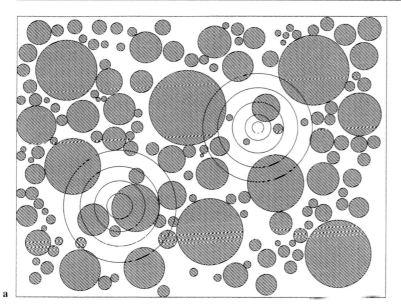

a

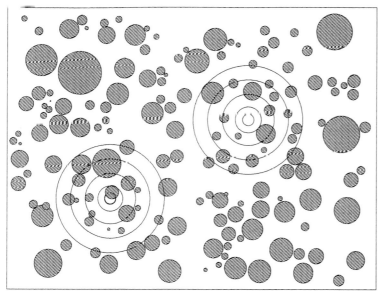

b

Fig. 1a,b. Fractal foams indicating the distribution of indeterminate genets in a plane. The fractal dimension is low in **a**, and high in **b**. *Shaded circles* represent individual genets, *nested circles* represent example quadrats for the determination of D

1985). The regional boundaries of such systems can come to cover enormous territories (Smith et al. 1992). Where mycelial growth between resource units is not possible, however, then proliferation necessitates dispersal and hence reproduction. If reproduction involves nonrecombinatorial mechanisms (homokaryotic fruiting or production of asexual spores), then a single genet may still become widespread, but will necessarily be discontinuous. Where reproduction necessitates sexual outcrossing, each meiospore will give rise to a unique genet. Either way, regional boundaries around physiologically unified systems will be confined within the physical boundaries of the resource units, and may indeed be reinforced by barriers such as pseudosclerotial plates. Such systems can therefore be described as unit-restricted, and sampling pattern can then be prescribed by the size of the resource units, with the smallest units requiring sampling on the finest scales (Cooke and Rayner 1984; Rayner 1992, 1994b).

Challenging questions regarding the scale at which to depict and quantify (as opposed to locate) topographical and regional boundaries are also posed by the heterogeneities or irregularities of indeterminate mycelial systems. Mycelia have fractal properties, in that their density varies with the scale at which they are observed (Obert et al. 1990, 1991; Ritz and Crawford 1990; Bolton and Boddy 1993; Matsuura and Miyazima 1993; Crawford and Ritz 1994; Mihail et al. 1994).

A further important property of indeterminate mycelial systems, of great significance to understanding ecological roles, is that their degree of irregularity may vary in different spatiotemporal locations, i.e. they can have multifractal organization (Benzi et al. 1984). This property is associated with the capacity to change organizational mode or state in correspondence with different circumstances and functional requirements (Gregory 1984; Rayner and Coates 1987; Stenlid and Rayner 1989; Andrews 1992; Rayner 1994a,b; Rayner et al. 1994a,b). Differences between these states, e.g. between yeast and mycelial, sporulating and non-sporulating, coenocytic and septate, profusely branched and sparsely branched, submerged and emerged, diffuse and aggregated, pigmented and unpigmented, can be considerable. An immediate practical problem may lie in even recognizing that they originate from the same organism.

The practical problems may be exacerbated in situations where certain modes are less readily detected than others, for example because they are not easily cultured. This may apply especially to fungi which exist for part of their life spans as biotrophic, endozoic or endophytic inhabitants of living host tissue, and then switch to more actively degradative growth when this tissue becomes dysfunctional (Carroll 1988; Boddy and Griffith 1989).

Of course, many fungi are not easily culturable at any stage in their life cycles, so that the distribution of their mycelia has either to be detected directly (which can be difficult) or inferred from the distribution of sporophores.

Production of sporophores, which can be observed easily and identified using conventional taxonomic procedures, has traditionally been a major basis for defining fungal distribution patterns. However, as determinate offshoots, sporophores do not constitute the whole fungus and may at least partly be subject to selection processes that differ from those operating on mycelia (but see below). Their absence does not therefore necessarily imply the absence of mycelium. They can, like mycelia, also be variable in form (pleomorphic), and without additional work it is not generally possible to establish whether separate collections emanate from the same genetic source.

Sporophores do not, therefore, in themselves, provide a reliable or even a legitimate basis for locating individual fungi. However, they can provide useful samples of genetic material and so facilitate the location of regional boundaries around and between mycelial systems, which is the most fundamental requirement of fungal ecological genetics. Providing that their genetic origin is known, sporophores are also of considerable importance to the fungal ecological geneticist for a variety of other reasons.

Firstly, the variability of sporophore and mycelial phenotypes can be characterized and compared. In this way, valuable information can be provided about the regulation of determinate and indeterminate developmental phases and the relative importance of reproductive and mycelial characteristics in the determination of ecological niche. Secondly, the modes of production and dispersal of spores have obvious significance with regard to the scale and pattern of genetic variation in natural populations. Thirdly, reproductive and mycelial phases of fungi are not as selectively independent of one another as might at first seem, and it is important to understand the kinds of feedback which may operate between these phases. The scale and, indeed, the existence of sporophores is dependent on resources supplied via the mycelium, and since sporophores are made up of hyphal components, the pattern-generating processes involved in their formation will inevitably reflect those in mycelia. By the same token, modes of reproduction, particularly with regard to whether they involve recombinatorial or clonal mechanisms, are likely to influence the interactive properties and developmental versatility of mycelia via their relationship to r- and K-selection (see Sect. II.C).

B. Past History

In order to account for current genetic distribution patterns, three distinctive but related questions, each requiring a different kind of approach, need to be answered. How much overall genetic similarity or dissimilarity is there in the population? How is this variation spatially distributed? To what extent is this distribution correlated with the degree of relatedness between neighbours in terms of their ancestral lineages?

Some means of calibrating the amount of genetic variation in local populations against spatial or temporal reference scales is therefore needed. The resulting information can then be used to assess the relative contributions of selection, long- and short-range dispersal mechanisms, and recombinatorial and clonal modes of reproduction, in giving rise to disparate, graduated or homogeneous patterns of genetic variation. Such knowledge can, in turn, provide insights into the nature and strength of selective influences operating along spatiotemporal transects and so allow the importance of adaptive processes in determining genetic distribution patterns to be evaluated. For example, if it can be shown that genetically similar members of a local population are not related to one another by recent descent, then this provides clear evidence for adaptation. This is an important issue because in assessing the origins and maintenance of diversity, it is necessary to understand the relationship between that which is **impelled**, a consequence of organizational drives, and that which is **compelled**, by selective conditions (Rayner 1994a,b; Rayner et al. 1994a,b). Moreover, when attempting to explain evolutionary success, it is necessary to gauge the contribution of serendipity – being in the right place at the right time rather than necessarily having an adaptive genotype.

Here, it should be noted that evolutionary theory has predominantly been concerned with the way in which adaptive adjustments of genetic composition are made via the operation of natural selection on generations of entities that are essentially particulate in space and time. This has probably resulted from emphasis on animals in which somatic alterations do not enter the germ line and which are also relatively unable to adjust their genetically prescribed developmental programming (as opposed to behaviour) to accord with variable circumstances. Even here, the question remains as to how much variation is generated endogenously and **accommodated**, as opposed to **instigated**, by selective conditions. Is there selective **vacuum** or selective **pressure**?

In developmentally indeterminate systems, such as those of mycelial fungi, the contributions of genetic prescription and environmental opportunities and constraints in the determination of phenotype become even more inseparable. Radical shifts in organizational pattern can and do occur as circumstances change, even within the life span of a single organism or genet (Slutsky et al.

1985; Rayner and Coates 1987; Stenlid and Rayner 1989; Rayner et al. 1994a,b). If history is to be used to understand present and predict future distribution, it is essential to know the extent to which self-organizing processes or drives are mediated by genetically prescribed mechanisms.

Here, it is important to appreciate the fact that, as energy-gathering and -distributing systems, mycelia require means of both gaining access to and retaining resources. As such, they can be thought of as non-linear hydrodynamic systems whose developmental pattern depends on counteractive feedback between processes that drive and constrain expansion of hyphal boundaries. According to a recent hypothesis, these processes are of four fundamental kinds: conversion, regeneration, distribution and recycling, and are regulated by three basic parameters (Fig. 2). The latter comprise the resistances to deformation and passage of water and solutes (i.e. the degree of "insulation", see Rayner et al. 1994b) across hyphal boundaries and the degree of protoplasmic partitioning within hyphae. Resistance to deformation dependes on the interplay between plasticization and rigidification processes due to generation and cross-linking of microfibrillar and amorphous wall components (Bartnicki-Garcia 1973; Wessels 1986). Resistance to passage of solutes and water can be varied by depositing and releasing, polymerizing and depolymerizing variably hydrophobic compounds, including polypeptides, phenolics and terpenoids. Some of these processes may be partly independent of genetic information in that once initiated, via the generation of free radical moieties, subsequent chemical reactions are determined by local environmental conditions. Protoplasmic continuity is determined by the presence of septa and anastomoses, by septal sealing and localized protoplasmic death (also the likely outcome of the generation of free radicals).

C. Future Status Under Routine Selection

If the degree to which observed patterns of genetic variation are determined by adaptive requirements is known, then it will be possible to forecast the extent to which these patterns will persist, providing that selective conditions are maintained or recur.

Selective conditions in environments that are either maintained in dynamic equilibrium or in which there are regular cyclical changes may be

I CONVERSION   II REGENERATION
   [CONSERVATION]   [ASSIMILATION]

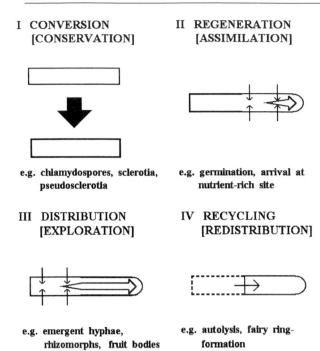

e.g. chlamydospores, sclerotia,   e.g. germination, arrival at
     pseudosclerotia   nutrient-rich site

III DISTRIBUTION   IV RECYCLING
     [EXPLORATION]   [REDISTRIBUTION]

e.g. emergent hyphae,   e.g. autolysis, fairy ring-
     rhizomorphs, fruit bodies   formation

Fig. 2. Four fundamental processes in elongated hydrodynamic systems, as determined by boundary deformability, permeability and internal partitioning. Rigid boundaries are shown by *straight lines*; deformable boundaries as *curves*; impermeable boundaries by *thicker lines*; degenerating boundaries as *broken lines*; protoplasmic disjunction by an *internal dividing line*; *simple arrows* indicate input across permeable boundaries into metabolically active protoplasm; *tapering arrows* represent throughput due to displacement

referred to as routine (Brasier 1987). Following Grime (1979), they can be divided into three primary categories based on the relative occurrence of "disturbance", "stress" and the incidence of competitors (e.g. Cooke and Rayner 1984). An important contribution to ecological genetics may be made by assessing observed distribution patterns to see how well they conform with theoretical expectations based on the effects of these different kinds of selection, as will now briefly be itemized.

Disturbance may be defined as any environmental event or process, which, by enrichment of living space or destruction of residents, makes new resources available for exploitation. Immediately following disturbance, conditions are liable to be at their most open and homogeneous. Disturbance therefore imposes R-selection (= r-selection), favouring opportunist or ruderal organisms, equipped for rapid arrival and exploitation of readily assimilable resources. As conditions aggravate or competitors establish, pioneers

unable to adapt to these changes will come under increasing pressure to disseminate their genes.

R-selected fungi (or organizational modes) are therefore likely to be biased towards conversional and regenerative processes (Fig. 2). Hence, they may be expected to have short individual life spans, to reproduce rapidly – without genetic diversification and with minimal resources – and to have a limited organizational repertoire.

Stress may be defined as any more or less continuously imposed environmental feature, other than competition, which limits the productivity of the majority of organisms (or modes) under consideration. The minority of S-selected entities that are able, by means of specialized attributes, to develop effectively under stressful conditions may therefore do so in the relative absence of competitors. They may generally be expected to emphasize conversional processes and so to have long individual life spans, either because of the maintenance of stressful conditions over extended time periods, or because of an ability to persist once these conditions are alleviated. Commitment to reproduction is therefore liable to be slow, though the ultimate investment of both assimilable and refractory resources in this process may be considerable. Whether reproduction should be recombinatorial or clonal may depend on the degree of heterogeneity of stressful habitats, and susceptibility of the population to spread of transmissible infections. Similarly, where conditions are likely to vary in space and/or time, a versatile developmental repertoire that enables suitable adjustment to changed circumstances would be advantageous.

In the relative absence of disturbance and selective stress, the potential incidence of competitors, of both the same and different species within a habitat, may be considerable, and C-selected or combative organisms (or modes) will be favoured. These may either retain resources captured during earlier stages of colonization under stressful or disturbed conditions, hence emphasizing conversional processes, or replace former inhabitants, hence emphasizing distributional and recycling processes. As K-selected inhabitants of closed communities, combative entities are prone to have long individual life spans, during which they encounter (and create) considerable biotic and abiotic heterogeneity. They may therefore be expected to possess means for both genetic and epigenetic diversification, the former being associated with a relatively low reproductive rate.

D. Capacity for Change Under Episodic Selection

Radical, non-repetitive shifts in the selective conditions that a population is exposed to constitute episodic selection (Brasier 1987). The exposure of a pathogen population to a new host, or vice versa, following intercontinental transfer, provides an example which illustrates the potentially explosive and fundamentally unpredictable (unstable) consequences that result from such selection (Gibbs 1978; Fry et al. 1992). These consequences depend critically on both the degree of genetic and epigenetic variation already present in the population or its founder, as well as the capacity of the latter to interconvert between recombinatorial and clonal modes of proliferation. A successful response to episodic selection depends on being able to produce the right variant at the right place and/or time, to proliferate in such a way as to exploit the opportunity maximally, and then to revert to those properties most apposite to routine selection conditions. It is perhaps in their capacity to respond to episodic selection that the indeterminate life forms of mycelial fungi have found some of their most spectacular successes, generating episodic selection pressures of their own as they invade communities and change their pattern of functioning irreversibly (Brasier 1988). Moreover, it is in understanding just what it is about the genetic and epigenetic systems of mycelial fungi that enables them to respond in this way that fungal ecological geneticists find their severest challenge and may gain their deepest insights into evolutionary processes.

III. Methodologies

A. Locating Regional Boundaries

The approach most appropriate to locating the regional boundaries of genets will be determined primarily by the operational scale, visibility and culturability of the mycelium.

Many fungal mycelia inhabit relatively large domains and are either directly visible or have visible effects on their substratum, as in fairy rings, lichen communities, decay columns in wood, cankers in bark, rhizomorphic and mycelial cord systems and diebacks caused by root-rotting fungi. Such mycelia can be allocated to unit-restricted and non-unit-restricted categories and their distri- bution circumscribed, at least provisionally, by direct observation (Thompson 1984; Mason et al. 1987). Moreover, because of characteristic physiological and morphological changes that produce demarcation zones at their interfaces, adjacent mycelial systems, both of the same and different species, commonly map their own regional boundaries (e.g. Rayner 1994b). However, it is important here to be able to distinguish demarcation zones due to interaction from other kinds of boundary zones, such as those delimiting pseudosclerotia (see Rayner and Todd 1979, 1982a,b).

Where the distribution of mycelia cannot be so readily visualized, or where confirmation or elaboration of provisional maps is necessary, then techniques which distinguish between samples on a more rigorous genetic basis are required. Here, it is particularly important to use a technique which discriminates at the level of resolution most appropriate to the issue being addressed. Whilst some techniques may detect fine-scale genetic variation even below the level of genet (Jacobson et al. 1993), others discriminate only at the sibling species level or higher (Forster et al. 1989, 1990; Gardes et al. 1991; Gardes and Bruns 1993). Moreover, it may be important to distinguish between biologically relevant and irrelevant variation.

In culturable ascomycetes and basidiomycetes, a simple test can be made by pairing isolates from different locations and recording whether they are somatically compatible, capable of complete physiological integration, or incompatible, producing rejection or demarcation zones (e.g. Childs 1963; Rayner and Todd 1982a,b). This test usually discriminates sensitively between genets, often even when these are sister-related. However, incompatibility reactions are very variable, may be obscured or enhanced in particular organizational modes or on certain media, and need to be carefully distinguished from other kinds of inhibition – such as may be evident in self-pairings (see also Chap. 12, Vol. I).

In sexually outcrossing basidiomycete populations, somatic incompatibility is expressed between heterokaryotic or exceptionally (as in *Armillaria*) allodiploid genets, whereas in ascomycetes, it typically occurs between homokaryotic genets. However, somatic incompatibility need not absolutely preclude heterokaryon formation in ascomycetes, and, indeed, unstable heterokaryons do form between somatically incompatible genets of, for example, members of the

Xylariaceae and Diatrypaceae (Sharland and Rayner 1986, 1989a). The fact that such apparently contradictory consequences of somatic non-self hyphal fusion as rejection and heterokaryon formation can occur in the same mycelial interaction is a feature of many fungi and a reflection of their organizational heterogeneity (Rayner 1991a).

This fact is important because vegetative (=somatic) incompatibility in ascomycetes has commonly been equated with the absolute inability to form a heterokaryon (i.e. heterokaryon incompatibility; e.g. Leslie 1993). Heterokaryon incompatibility tests, often based on some kind of nutritional complementation, have become widespread in studies of ascomycete population biology. However, it will be evident that, at least in some cases, they are liable to resolve differences only between groups of related genets, rather than between individual genets. They may even discriminate between genetically identical strains that are deficient in hyphal fusion (Correll et al. 1989).

Other cultural criteria for discrimination between genets include morphology and distribution of mating alleles. Morphology is unreliable because of epigenetic variation, but can be helpful when used in combination with other criteria. Mating alleles have been used widely in studies of basidiomycete population biology, being especially favoured because they are genetically well defined. However, they are rather time-consuming to identify, and cannot discriminate between sister-related genets.

An alternative or complementary approach to the use of cultural criteria involves the use of molecular markers (e.g. Michelmore and Hulbert 1987; Egger 1992). Such markers need to be chosen carefully, taking into account their selective neutrality, ease of scoring and abundance. The biological relevance of the differences that they detect needs considerable attention.

Isoenzyme analysis has been used successfully. However, it suffers in being dependent on detecting gene expression, which may vary with developmental mode and genetic dominance, and, partly, because the lack of selective neutrality shown by these markers is likely to resolve differences only between groups of genets rather than individuals (e.g. Stenlid 1985; Micales et al. 1986; Sen 1990).

More direct evidence of genetic difference can be obtained by examination of nuclear DNA (nDNA), mitochondrial DNA (mtDNA) and ribosomal DNA (rDNA) polymorphisms. There are two main strategies for detecting these polymorphisms: restriction fragment length polymorphism (RFLP)/DNA fingerprinting studies and approaches based upon polymerase chain reaction (PCR) technology.

Multilocus RFLP and fingerprinting studies (e.g. Anderson et al. 1987; Coddington et al. 1987) use total DNA, extracted from different organisms, as the target for analyses that detect variation throughout the genome. These approaches require the hybridization of radio-labelled or digoxigenin/biotin-labelled DNA probes to multiple sites within large quantities of restriction enzyme-digested DNA that has been electrophoresed and transferred to a supporting membrane. Not only do such procedures require a considerable investment in terms of time and money, but they may have restricted utility in many fungal ecological genetic studies, where often only small quantities of material are available and/or large numbers of samples need to be analyzed, e.g. when herbarium specimens or unculturable, small or cryptically colonizing fungi are being studied.

In such situations, random amplified polymorphic DNA (RAPD) and arbitrarily primed PCR (AP-PCR) analyses are potentially invaluable (Welsch and McClelland 1990; Williams et al. 1990; Foster et al. 1993). These require only small amounts of DNA from which segments of the genome are selectively amplified (using PCR-based technologies) between suitably chosen primer-annealing sites. These procedures generate populations of small DNA molecules that can be electrophoresed and directly visualized with ethidium bromide on agarose gels.

Difficulties may be encountered when interpreting DNA polymorphisms in the absence of detailed knowledge of the mechanisms that generate molecular variation in natural populations and in the laboratory. Such variation is largely produced by a combination of point mutations (which alter restriction enzyme or primer annealing recognition sites), and insertions/deletions (which alter the length of DNA fragments generated from restriction digests or PCR-based amplifications). However, in contrast to other better-characterized systems, the relative contribution of such events is largely unknown for filamentous fungi. Furthermore, in PCR-based studies, the choice of initial reaction conditions is vital to the efficient amplification of a product(s), and different reaction conditions can, for the same sample, yield different products. The question therefore arises as to whether the variation observed amongst RAPDs is present (or relevant) in a biological sense, or is an artefact. This is of particular concern, as few studies have examined the inheritance of RAPD markers, or demonstrated that they are, in fact, homologous (cf. Doudrick et al. 1993). Ideally, RAPD procedures can identify differences at the level of the individual and can provide strong evidence for identity between samples, as long as adequate controls are used and the well-known risks of amplifying contaminant DNA are avoided (especially when working with natural material).

B. Quantifying Structural Heterogeneity Within Regional Boundaries

Any differences in organizational mode between samples that are shown, by the methods just described, to have a common genetic identity will be evident as phenotypic variation within a regional boundary. The next problem is to find some way of describing the relation of such variation to structural disposition in a quantitative way. There are two basic requirements for this. The first is to have some means of assaying the material content of fungal biomass within regional boundaries. The second is to have some means of relating the irregular distribution of this content to its overall extent, thus arriving at an estimate of its space-filling capacity. Density is an inappropriate measure of this capacity because fractally organized systems cannot usefully be quantified in units of length, area or volume. Moreover, by averaging out the heterogeneities of mycelial systems, density estimates eliminate all kinds of functionally important local detail of potential value in comparisons between different taxa, genets and organizational states.

The most appropriate means of achieving the second end may therefore be to characterize the topographical distribution of mycelial systems in terms of their degree of irregularity or fractal (fractional) dimension (Mandelbrot 1982). This measure can be obtained by relating the content or mass (m) of material in a portion of a system to its extent, the radius of the field within which it is contained (R), according to the formula:

$$m = kR^D,$$

where k is a constant and D is the dimension. For homogeneous structures, D is an integer, but for heterogeneous, fractal structures it is fractional and can be found as the slope of the graph of $\ln m$ against $\ln R$. The higher the fractal dimension, the more thoroughly it permeates space.

The process of measuring values of m for particular values of R can, however, be problematic. If some kind of image of the system is available, then image analysis techniques can help to reduce the amount of labour involved (e.g. Bolton and Boddy 1993). For some systems, such as macroscopically visible mycelial cord systems, provision of such images is relatively easy. For more microscopic systems, or systems that are embedded in opaque material, there is greater difficulty, in both visualizing the systems themselves and being sure of their identity. Ideally, some technique that enables direct observation, using sectioned material or surfaces viewed using light or electron microscopy, combined with some means of diagnostic labelling, is required. Suitable labels might include antibodies (Miller and Martin 1988; Dewey 1990) or fluorescence/radio-labelled in-situ hybridization probes (Li et al. 1993). Where this is not possible, then existing methods for biomass estimation (Newell 1992) might be applied in a nested quadrat design at a variety of locations.

C. Calibrating Genetic Heterogeneity Within Populations

1. Assessment of Overall Genetic Variation

At least in theory, techniques which distinguish between individual genets may also provide indication of how genetically disparate they are. For example, somatic incompatibility reactions commonly vary in intensity and/or pattern according to the degree of genetic difference between isolates. Unfortunately, use of this criterion is hampered by the often complex expression of somatic incompatibility and by the lack of quantitative criteria for scoring variations in intensity. Variations in the number and character of polymorphisms in molecular markers may also be used to calibrate genetic diversity. The latter approach has found particular favour in the application of RAPD techniques. However, the validity of such applications is dubious in view of the uncertainties already mentioned about the inheritance and homology of the polymorphisms revealed by these techniques.

2. Geometric Distribution of Variation

In order to analyze the geometric distribution of genetic variation, the first requirement is to assess the extent of regional boundaries over a sufficient range of scales to avoid over- or underestimation of numbers. In situations where the boundaries are easily visible within a confined area or volume, for example in a section across a decaying log, this can be done relatively easily. In other cases, it is necessary to use a series of nested quadrat designs, whereby the population is sampled in successively smaller (or larger) fields, as indicated in Fig. 1. Assuming that maximum and minimum size ranges of regional boundaries have been identified by adequate sampling and that the measurements are made within an enclosed reference boundary, the numbers of genets detected in fields of different size can then be used to provide an estimate of the fractal dimension of the population (Maurer 1994). In a plane, D close to 0 implies genetic uniformity (one genet), D close to 1 implies extreme genetic predominance (widely spaced genets, very disproportionate in size), D close to 2 implies extreme genetic diversity (many genets of similar size). In this context, D therefore provides a direct index of population diversity that takes into account both the spatial dominance and genetic heterogeneity of an indeterminate system. Simultaneously, the measure also gives clues to the roles that organisms play or strategies they adopt within a community. Traditional diversity indices (Magurran 1988; Anagnostakis 1992) are limited to descriptions of straightforward genetic constitution and apply only to particulate, spatially determinate systems, not to those with an indeterminate mode of organization.

3. Assessment of Relatedness

In order to determine whether different genets arise from the same lineage, it is necessary to have genetic markers which are highly polymorphic in the population as a whole, selectively neutral and, ideally, either unable to segregate following meiosis or non-transferable from one genetic line to another. The latter criterion is best fulfilled by cytoplasmic markers, notably in mtDNA. In many fungi, mtDNA has been shown to be sufficiently polymorphic in this respect to be useful (Hintz et al. 1985; Taylor 1986; Smith et al. 1990; Bruns et al. 1991). However, far more information on variation of mtDNA in local populations is needed be-

fore its utility in determining lineage can be assessed.

Where mtDNA proves not to be polymorphic within populations, other criteria may be helpful. Highly multiallelic genes, notably those of basidiomycete mating type factors provide an excellent example; however, whereas the presence of common alleles between genets provides strong evidence that they are related, absence of common alleles need not imply that they are unrelated. At the molecular level, certain single locus RFLPs may likewise be helpful in tracing ancestry, as they have proved to be in determining parentage in human beings.

Where an array of polymorphisms at varied sites are available for comparison (for example, following RAPD analysis), then some kind of cluster analysis may allow dendrograms showing most likely relationships to be drawn up. Such analysis would follow the same kinds of procedures currently used to work out phylogenetic relationships (e.g. Bruns et al. 1991; Farris 1994; Felsenstein 1994). However, for reasons already described, it may be important to interpret the relationships with caution.

Overall, there has been very little work on tracing relationships within natural fungal populations in these ways, and further development of this field is vital.

D. Assessing Genetic and Epigenetic Changeability

Adaptive responses to environmental heterogeneity in space or time, under both routine and episodic selection, can be achieved either through genetic changes within populations or epigenetic changes within individual genets. There are several key elements in the assessment of the potential for such changes.

1. Assessing Genetic Flexibility

With respect to genetic mechanisms, of primary concern is whether the reproductive pathways result in recombinatorial or clonal modes of proliferation. First, it should be established whether the population produces anamorphic and/or teleomorphic life-cycle stages. Where teleomorphs can be identified, sibling single-spore progeny should be compared to determine whether they are genetically the same or different. In nonculturable forms, this may require the direct application of molecular methods. In culturable forms, it can be achieved by pairing isolates from the same fruit body in all combinations and scoring any rejection or mating reactions. Where there are no such reactions and the cultures appear identical, a non-outcrossing breeding strategy can be inferred. Where the progeny are variable and exhibit mating or rejection reactions, an outcrossing strategy is indicated (e.g. Ainsworth 1987; Sharland and Rayner 1989a,b). These conclusions can be confirmed by molecular studies of appropriate resolution (see above).

If anamorphic phases occur, there is the potential for extensive clonal proliferation either locally or, perhaps less usually, over long range. This possibility should be checked using pairing tests and/or molecular methods.

Finally, the mutability of an organism may be assessed by noting the extent to which spontaneous changes in phenotype can be correlated with somatic genetic alterations, detected either by molecular methods or classical segregation analysis (Anagnostakis 1987; Hintz et al. 1988; Samac and Leong 1988; Kistler and Miao 1992).

2. Assessing Capacity for Epigenetic Variation

If separate samples from a single source exhibit phenotypic differences and can yet be shown to be genetically homogenous, then the occurrence of epigenetic variation has been demonstrated. Several categories of epigenetic variation may occur, though not all of these have been widely recognized. First, there is that which is due to direct effects of the environment on physiological functioning; these can usually be identified as continuous changes in phenotype when an organism is grown over a range of conditions. Second, there are changes in organizational state which are often regarded as normal differentiation processes, characterized by the production of identifiable, recurrent, alternative phenotypes that vary discontinuously. They may arise spontaneously and/or be correlated with particular environmental circumstances. Third, there are inconsistent variations in organizational pattern whose exact form and location are unpredictable. The two former categories can result directly from alterations in gene expression. However, the third is supernumary, a kind of cultural or hyperepigenetic condition which results from processes that are driven by non-genetic feedback, such as free radi-

cal chain reactions initiated but not directed by phenol-oxidizing enzymes.

As has been implied, the main means of detecting epigenetic variation involves examining numerous samples of a fungus from the same genetic source under both uniform and varied conditions. A particularly useful approach is to make use of heterogeneous matrices of the kind shown in Fig. 3, which allow a mycelium to interconnect between discrete microenvironmental sites and so display the wide, if not full range of its epigenetic repertoire.

IV. Observations

Many of the issues and approaches identified in this chapter have not previously been widely recognized. Correspondingly, our primary intention has been to look towards the future development of fungal ecological genetics rather than to survey its limited past. Nonetheless, it is appropriate now to outline what progress has been made, both to provide a foundation for future work and to indicate where lacunae exist.

A. How Variable Are Natural Populations and Individuals?

The variability of fungi has long been recognized, often ruefully, and encapsulated in such dismissive descriptions of these organisms as a "mutable and treacherous tribe". However, understanding of the mechanistic origins of variation, much less its ecological and evolutionary significance, has been slow in coming.

Variation at the population level is implicit in the presence of recombinatorial mechanisms, and it has now become widely recognized that many species possessing such mechanisms consist of innumerable genets. These not only differ in their phenotype, but also, as a result of somatic incompatibility systems, remain to a large extent discrete, i.e. physiologically and genetically isolated from one another, except when participating in sexual exchange (e.g. see Rayner 1991a,b; Leslie 1993).

The extent to which recombinatorial mechanisms are responsible for dissemination is therefore a determinant of population diversity, in terms of how many individual types occur within

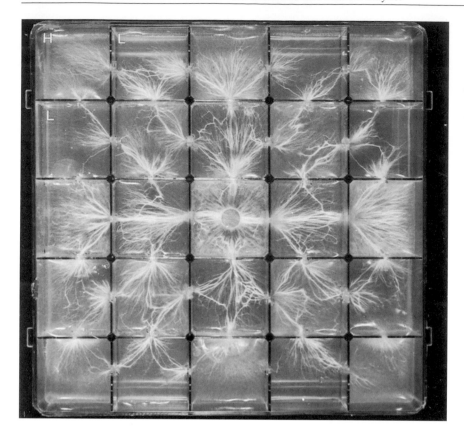

Fig. 3. Networking of *Coprinus picaceus* mycelium grown in a heterogeneous matrix of interconnecting chambers alternately containing high (*H*) low and (*L*) nutrient media (2% malt agar and tap water agar). (Rayner 1994b)

a reference boundary. Where clonal dissemination predominates over short range but recombinatorial mechanisms predominate over long range, then greater diversity may be expected between than within populations. Examples are provided by *Cryphonectria parasitica* and *Cryptostroma corticale* (Bevercombe and Rayner 1984; Milgroom et al. 1991). Also, different populations, or parts of the same population, may not have the same diversity if the relationship between clonal and recombinatorial modes varies (cf. Anagnostakis 1992). The ability to switch between clonal and recombinatorial modes provisions fungal systems with a remarkable responsiveness to environmental uncertainty.

In several *Stereum* species, both outcrossing and non-outcrossing populations, with consequently very different spatial structures, have been detected in different geographical locations (Ainsworth 1987). In *Ophiostoma novo-ulmi*, a clonal distribution pattern at epidemic fronts is quickly superseded by a highly diverse structure in the bark of diseased trees (Brasier 1988).

Even where recombinatorial mechanisms predominate, there may be variations in population diversity with respect to different reference boundaries due to the indeterminate nature of mycelial systems. For example, there may be high diversity in a forest stand, equivalent to a fractal D = 2 in a cross section or plane, because a different genet can be detected in each tree, but low diversity within each tree (i.e. D ≈ 0) because only one or a few genets are present. This situation has been found amongst several heartrot or specialized opportunist fungi, such as *Phaeolus schweinitzii*, *Piptoporus betulinus* and *Daldinia concentrica* (Barrett and Uscuplic 1971; Adams et al. 1981; Boddy et al. 1985), and indicates restricted establishment from individual spore sources. By contrast, a low D in both individual trees and whole stands, such as probably occurs in *Armillaria* sp. and *Phellinus weirii* (e.g. Childs 1963; Dickman and Cook 1989; Smith et al. 1992), as well as mycorrhizal *Suillus* species (Dahlberg and Stenlid 1990), indicates significant spread by means of migratory mycelium. Intermediate values for D (e.g. Fry et al. 1991) may occur as a result of selection pressures acting differentially upon members of a semiclonal population. On the other hand, a high D value in individual trees, as exhibited by such specialized opportunists as *Hypoxylon fragiforme* in beech (*Fagus* spp.; Chapela and Boddy 1988) and *H. fuscum* in hazel (*Corylus avellana*; Rayner 1993), is indicative of establishment from variable spores at numerous locations.

B. How Much of the Variability Can Be Accommodated on a Genetic or an Epigenetic Basis?

Although the fact that individual genets can be identified confirms that there is genetic variability in natural fungal populations, many other issues are unresolved. Some phenotypic variation, as has been widely assumed in the past, can be accounted for purely on a genetic basis (Anderson et al. 1992; Kistler and Miao 1992). However, other sources of variation should not be ignored. For example, as more work of the kind described by Sharland and Rayner (1989a,b) is done, it seems likely that alternative phenotypes and hyperepigenetic variation will be found to be widespread. Even when genetic variation is detected, it is difficult to quantify without a suitable reference scale against which to make comparisons. It may be possible, using molecular techniques to compare the amount of difference both within and between individual genets and also between populations (Horgen et al. 1984; Jahnke et al. 1987; Vilgalys and Johnson 1987; Smith and Anderson 1989). Generally, little work of this kind has been done, and when variation is detected at the molecular level, it is often difficult to interpret what the differences signify.

C. How Much Variation Is Adaptive or Non-Adaptive?

This question, too, remains largely unanswered at present. In some cases, it may be possible to infer the adaptive significance of variations in particular traits such as virulence and pathogenicity amongst biotrophic and necrotrophic parasites, since these traits clearly determine the outcome of infection. Likewise, if the expression of particular attributes of different genets or organization states can clearly be correlated with particular environmental circumstances, then these attributes can be regarded as having adaptive value. This would particularly apply if the attributes can be allocated

a functional role, for example through application of ecological strategies theory. The production of unicellular and mycelial modes of proliferation during different phases of colonization of a plant or animal may be interpreted in this way.

The problems of assessing adaptivity currently have two main sources. First, there is a dearth of information about the spatial distribution of genetic polymorphisms within natural populations of fungi. Even where such information exists, its significance is not necessarily clear. For example, the enormous number of mating alleles in basidiomycete populations (e.g. Williams and Todd 1985) is hugely in excess of what is needed to maintain a high outbreeding bias (which many assume to be its function). Such an adaptation needs to be demonstrated, not simply inferred. Second, there is the indeterminacy of mycelial systems, which makes it so difficult to allocate fungal individuals to particular environmental locations.

On the other hand, the ability of fungal individuals to grow in heterogeneous domains opens up prospects for new insights into the interplay between genetic, epigenetic and hyperepigenetic sources of variation as determinants of distribution. Hyperepigenetic variation, in particular, may prove to have general significance within serendipitous indeterminate systems. If so, fungal ecological genetics has as much to contribute to future evolutionary theory as it has to gain from past work with other organisms.

V. Conclusions

The natural distribution patterns and diversity of fungi are the outcome of a complex relationship between genotype, phenotype and environment. Understanding this relationship depends on being able both to assess the occurrence and to recognize the dynamic origins of fungal boundaries, and hence poses considerable practical and conceptual challenges. In particular, there is a need to come to terms with the fundamental properties of mycelia as versatile, indeterminate systems. These systems can only be closely approximated to additive assemblies of discrete selectable units under extreme (R-selective) circumstances, when resources are freely and uniformly (or randomly) available in the external environment. Under all other circumstances, counteraction between integration and differentiation causes variation in the scale, heterogeneity and interconnectedness of fungal boundaries. The future task of fungal ecological genetics will be to relate this organizational variation – and its underlying genetic, epigenetic and hyperepigenetic mechanisms – to the diverse environmental contexts in which fungi thrive and survive.

References

Adams TJH, Todd NK, Rayner ADMR (1981) Antagonism between dikaryons of *Piptoporus betulinus*. Trans Br Mycol Soc 76:510–513

Ainsworth AM (1987) Occurrence and interactions of outcrossing and non-outcrossing populations of *Stereum*, *Phanerochaete* and *Coniophora*. In: Rayner ADM, Brasier CM, Moore D (eds) Evolutionary biology of the fungi. Cambridge University Press, Cambridge, pp 285–299

Anagnostakis SL (1987) Chestnut blight: the classical problem of an introduced pathogen. Mycologia 79:23–37

Anagnostakis SL (1992) Diversity within populations of fungal pathogens on perennial parts of perennial plants. In: Carroll GC, Wicklow DT (eds) The fungal community – its organization and role in the ecosystem (2nd edn). Marcel Dekker, New York, pp 183–192

Anderson JB, Petsche DM, Smith ML (1987) Restriction fragment length polymorphisms in biological species of *Armillaria mellea*. Mycologia 79:69–76

Anderson JB, Kohn LM, Leslie JF (1991) Genetic mechanisms in fungal adaptation. In: Carroll GC, Wicklow DT (eds) The fungal community – its organization and role in the ecosystem (2nd edn). Marcel Dekker, New York, pp 73–98

Andrews JH (1992) Fungal life-history strategies. In: Carroll GC, Wicklow DT (eds) The fungal community – its organization and role in the ecosystem (2nd edn). Marcel Dekker, New York, pp 119–146

Barrett DK, Uscuplic M (1971) The field distribution of interacting strains of *Polyporus schweinitzii* and their origin. New Phytol 70:581–598

Bartnicki-Garcia S (1973) Fundamental aspects of hyphal morphogenesis. In: Ashworth JM, Smith JE (eds) Microbial differentiation, 23rd Symp Society of General Microbiology. Cambridge University Press, Cambridge, pp 245–267

Benzi R, Paladin G, Parisi G, Vulpiani A (1984) On the multifractal nature of fully developed turbulence and chaotic systems. J Phys A Math Gen 17:3521–3531

Bevercombe GP, Rayner ADM (1984) Population structure of *Cryptostroma corticale*, the causal fungus of sooty bark disease of sycamore. Plant Pathol 33:211–217

Boddy L, Griffith GS (1989) Role of endophytes and latent invasion in the development of decay communities in sapwood of angiospermous trees. Sydowia 41:41–73

Boddy L, Gibbon OM, Grundy MA (1985) Ecology of *Daldinia concentrica*: effect of abiotic variables on mycelial extension and interspecific interactions. Trans Br Mycol Soc 85:201–211

Bolton RG, Boddy L (1993) Characterization of the spatial aspects of foraging mycelial cord systems using fractal geometry. Mycol Res 97:762–768

Brasier CM (1987) The dynamics of fungal speciation. In: Rayner ADM, Brasier CM, Moore D (eds) Evolutionary biology of the fungi. Cambridge University Press, Cambridge, pp 231–260

Brasier CM (1988) Rapid changes in genetic structure of epidemic populations of *Ophiostoma ulmi*. Nature 332:538–541

Brasier CM, Rayner ADM (1987) Whither terminology below the species level in fungi? In: Rayner ADM, Brasier CM, Moore D (eds) Evolutionary biology of the fungi. Cambridge University Press, Cambridge, pp 379–388

Bruns TD, White TJ, Taylor JW (1991) Fungal molecular systematics. Annu Rev Evol Syst 22:525–564

Carroll GC (1988) Fungal endophytes in stems and leaves: from latent pathogen to mutualistic symbiont. Ecology 69:2–9

Chapela I, Boddy L (1988) Fungal colonization of attached beech branches. II. Spatial and temporal organization of communities arising from latent invaders in bark and functional sapwood, under different moisture regimes. New Phytol 110:47–57

Childs TW (1963) *Poria weirii* root rot. Phytopathology 53:1124–1127

Coddington A, Matthews PM, Cullis C, Smith KH (1987) Restriction digest patterns of total DNA from different races of *Fusarium oxysporum* f.sp. *pisi* – an improved method of race identification. J Phytopathol 118:9–20

Cooke RC, Rayner ADM (1984) Ecology of saprotrophic fungi. Longman, London

Correll JC, Klittich CJR, Leslie JF (1989) Heterokaryon self-incompatibility in *Gibberella fujikuroi* (*Fusarium moniliforme*). Mycol Res 93:21–27

Crawford J, Ritz K (1994) Origin and consequences of colony form in fungi: a reaction diffusion mechanism for morphogenesis. In: Ingram DS, Hudson A (eds) Shape and form in plants and fungi. Academic Press, London, pp 311–328

Dahlberg A, Stenlid J (1990) Population structure and dynamics in *Suillus bovinus* as indicated by spatial distribution of fungal clones. New Phytol 115:487–493

Dewey FM (1990) The use of monoclonal antibodies to detect plant invading fungi. In: Schots A (ed) Monoclonal antibodies in agriculture. Pudoc, Wageningen

Dickman A, Cook S (1989) Fire and fungus in a mountain hemlock forest. Can J Bot 67:2005–2016

Doudrick RL, Nelson CD, Nance WL (1993) Genetic analysis of a single urediniospore culture of *Cronartium quercuum* f.sp. *fusiforme*, using random amplified DNA markers. Mycologia 85:902–911

Egger KN (1992) Analysis of fungal population structure using molecular techniques. In: Carroll GC, Wicklow DT (eds) The fungal community – its organization and role in the ecosystem (2nd edn). Marcel Dekker, New York, pp 193–208

Farris J (1994) The logical basis of phylogenetic analysis. In: Sober E (ed) Conceptual issues in evlutionary biology. MIT Press, Cambridge, pp 333–362

Felsenstein J (1994) The detection of phylogeny. In: Sober E (ed) Conceptual issues in evolutionary biology. MIT Press, Cambridge, pp 363–376

Ford EB (1971) Ecological genetics, 3rd edn. Chapman & Hall, London

Forster H, Kinscher TGF, Leong SA, Maxwell DP (1989) Restriction fragment length polymorphisms of the mitochondrial DNA of *Phytophthora megasperma* isolated from soybean, alfalfa, and fruit trees. Can J Bot 67:529–537

Forster H, Oudemans P, Coffey MD (1990) Mitochondrial and nuclear DNA diversity within six species of *Phytophthora*. Exp Mycol 14:18–31

Foster LM, Kozak, KR, Loftus MG, Stevens JJ, Ross IK (1993) The polymerase chain reaction and its application to filamentous fungi. Mycol Res 97:769–781

Fry WE, Drenth A, Spielman LJ, Mantel BC, Davidse LC, Goodwin SB (1991) Population genetic structure of *Phytophthora infestans* in The Netherlands. Phytopathology 81:1330–1336.

Fry WE, Goodwin SB, Matuszak JM, Spielman LJ, Milgroom MG (1992) Population genetics and intercontinental migrations of *Phytophthora infestans*. Annu Rev Phytopathol 30:107–129

Gardes M, Bruns TD (1993) ITS primers with enhanced specificity for basidiomycetes – applications to the identification of mycorrhizae and rusts. Mol Ecol 2:113–118

Gardes M, White TJ, Fortin JA, Bruns TD, Taylor JW (1991) Identification of indigenous and introduced symbiotic fungi in ectomycorrhizae by amplification of nuclear and mitochondrial ribosomal DNA. Can J Bot 69:180–190

Gibbs JN (1978) Intercontinental epidemiology of Dutch elm disease. Annu Rev Phytopathol 16:287–307

Gregory PH (1984) The fungal mycelium: an historical perspective. Trans Br Mycol Soc 84:1–11

Grime JP (1979) Plant strategies and vegetation processes. John Wiley, Chichester

Hintz WEA, Mohan M, Anderson JB, Horgen PA (1985) The mitochondrial DNAs of *Agaricus*: heterogeneity in *A. bitorquis* and homogeneity in *A. brunnescens*. Curr Genet 9:127–132

Hintz WEA, Anderson JB, Horgen PA (1988) Nuclear migration and mitochondrial inheritance in the mushroom *Agaricus bitorquis*. Genetics 119:35–41

Horgen PA, Arthur R, Davy O, Moum A, Herr F, Straus N, Anderson J (1984) The nucleotide sequence homologies of unique DNAs of some cultivated and wild mushrooms. Can J Microbiol 30:587–593

Jacobson KM, Miller OK, Turner BJ (1993) Random amplified polymorphic DNA markers are superior to somatic incompatibility tests for discrimating genotypes in natural populations of the ectomycorrhizal fungus *Suillus granulatus*. Proc Natl Acad Sci USA 90:9159–9163

Jahnke KI, Bahnweg G, Worral JJ (1987) Species delimitation in the *Armillaria mellea* complex by analysis of nuclear and mitochondrial DNAs. Trans Br Mycol Soc 88:572–575

Kistler HC, Miao VPW (1992) New modes of genetic change in filamentous fungi. Annu Rev Phytopathol 30:131–200

Leslie JF (1993) Fungal vegetative incompatibility. Annu Rev Phytopathol 31:127–150

Li S, Harris CP, Leong SA (1993) Comparison of fluorescence in-situ hybridization and primed in-situ labeling methods for detection of single-copy genes in the fungus *Ustilago maydis*. Exp Mycol 17:301–308

López F, Serrano JM, Acosta FJ (1994) Parallels between the foraging strategies of ants and plants. Trends Ecol Evol 9:150–153

Magurran AE (1988) Ecological diversity and its measurement. Princeton University Press, Princeton

Mandelbrot BB (1982) The fractal geometry of nature. Freeman, New York

Mason PA, Last FT, Wilson J, Deacon JW, Fleming LV, Fox FM (1987) Fruiting and succession of ectomycorrhizal fungi. In: Pegg GF, Ayres PG (eds) Fungal infection of plants. Cambridge University Press, Cambridge, pp 253–268

Matsuura S, Miyazima S (1993) Colony of the fungus *Aspergillus oryzae* and self-affine fractal geometry of growth fronts. Fractals 1:11–19

Maurer BA (1994) Geographical population analysis: tools for the analysis of biodiversity. Blackwell, Oxford

Micales JA, Bonde MR, Peterson JL (1986) The use of isoenzyme analyses in fungal taxonomy and genetics. Mycotaxon 27:405–489

Michelmore RW, Hulbert SH (1987) Molecular markers for genetic analysis of phytopathogenic fungi. Annu Rev Phytopathol 25:383–404

Mihail JD, Obert M, Taylor SJ, Bruhn JN (1994) The fractal dimension of young colonies of *Macrophomina phaseolina* produced from microsclerotia. Mycologia 86:350–356

Milgroom MG, MacDonald WL, Double ML (1991) Spatial analysis of vegetative compatibility groups in the chestnut blight fungus, *Cryphonectria parasitica*. Can J Bot 69:1407–1413

Miller SA, Martin RR (1988) Molecular diagnosis of plant disease. Annu Rev Phytopathol 26:409–432

Newell SY (1992) Estimating fungal biomass and productivity in decomposing litter. In: Carroll GC, Wicklow DT (eds) The fungal community – its organization and role in the ecosystem (2nd edn). Marcel Dekker, New York, pp 521–562

Obert M, Pfeifer P, Sernetz M (1990) Microbial growth patterns described by fractal geometry. J Bacteriol 172:1180–1185

Obert M, Neuschulz U, Sernetz M (1991) Comparison of different microbial growth patterns by fractal geometry. In: Peitgen HO, Henriques JM, Penedo CF (eds) Fractals in the fundamental and applied sciences. Elsevier, New York, pp 293–306

Rayner ADM (1991a) The challenge of the individualistic mycelium. Mycologia 83:48–71

Rayner ADM (1991b) The phytopathological significance of mycelial individualism. Annu Rev Phytopathol 29:305–323

Rayner ADM (1992) Monitoring genetic interactions between fungi in terrestrial habitats. In: Wellington EMH, van Elsas JD (eds) Genetic interactions among microorganisms in the natural environment. Pergamon Press, Oxford, pp 267–285

Rayner ADM (1993) New avenues for understanding processes of tree decay. Arboricult J 17:171–189

Rayner ADM (1994a) Evolutionary processes affecting adaptation to saprotrophic life styles in ascomycete populations. In: Hawksworth DL (ed) Ascomycete systematics (lichenized and non-lichenized: problems and perspectives in the nineties). Plenum Press, New York, pp 261–271

Rayner ADM (1994b) Pattern-generating processes in fungal communities. In: Ritz K, Dighton J, Giller KE (eds) Beyond the biomass. Wiley-Sayce, Chichester, pp 247–258

Rayner ADM, Coates D (1987) Regulation of mycelial organization and responses. In: Rayner ADM, Brasier CM, Moore, D (eds) Evolutionary biology of the fungi. Cambridge University Press, Cambridge, pp 115–136

Rayner ADM, Franks NR (1987) Evolutionary and ecological parallels between ants and fungi. Trends Ecol Evol 2:127–132

Rayner ADM, Todd NK (1979) Population and community structure and dynamics of fungi in decaying wood. Adv Bot Res 7:333–420

Rayner ADM, Todd NK (1982a) Population structure in wood-decomposing basidiomycetes. In: Frankland JC, Hedger JN, Swift MJ (eds) Decomposer basidiomycetes: their biology and ecology. Cambridge University Press, Cambridge, pp 109–128

Rayner ADM, Todd NK (1982b) Ecological genetics of basidiomycete populations in decaying wood. In: Frankland JC, Hedger JN, Swift MJ (eds) Decomposer basidiomycetes: their biology and ecology. Cambridge University Press, Cambridge, pp 129–142

Rayner ADM, Watling R, Frankland JC (1985) Resource relations – an overview. In: Moore D, Casselton LA, Wood DA, Frankland JC (eds) Developmental biology of higher fungi. Cambridge University Press, Cambridge, pp 1–40

Rayner ADM, Griffith GS, Ainsworth AM (1994a) Mycelial interconnectedness. In: Gow NAR, Gadd GM (eds) The growing fungus. Chapman & Hall, London, pp 21–40

Rayner ADM, Griffith, Wildman HG (1994b) Differential insulation and the generation of mycelial patterns. In: Ingram DS, Hudson A (eds) Shape and form in plants and fungi. Academic Press, London, pp 291–310

Ritz K, Crawford J (1990) Quantification of the fractal nature of colonies of Trichoderma viride. Mycol Res 94:1138–1141

Samac DA, Leong SA (1988) Two linear plasmids in mitochondria of Fusarium solani f.sp. cucurbitae. Plasmid 19:57–67

Sen R (1990) Intraspecific variation in two species of Suillus from scots pine (Pinus sylvestris L.) forests based on somatic incompatibility and isoenzyme analyses. New Phytol 114:607–616

Sharland PR, Rayner ADM (1986) Mycelial interactions in Daldinia concentrica. Trans Br Mycol Soc 86:643–650

Sharland PR, Rayner ADM (1989a) Mycelial interactions in out-crossing populations of Hypoxylon. Mycol Res 93:187–198

Sharland PR, Rayner ADM (1989b) Mycelial ontogeny and interactions in non-outcrossing populations of Hypoxylon. Mycol Res 93:273–281

Slutsky B, Buffo J, Soll DR (1985) High frequency switching of colony morphology in Candida albicans. Science 230:666–669

Smith ML, Anderson JB (1989) Restriction fragment length polymorphisms in mitochondrial DNAs of Armillaria: identification of North American biological species. Mycol Res 93:247–256

Smith ML, Duchesne LC, Bruhn JN, Anderson JB (1990) Mitochondrial genetics in a natural population of the plant pathogen Armillaria. Genetics 126:575–582

Smith ML, Bruhn JN, Anderson JB (1992) The fungus Armillaria bulbosa is among the largest and oldest living organisms. Nature 356:428–431

Stenlid J (1985) Population structure of Heterobasidion annosum as determined by somatic incompatibility, sexual incompatibility and isoenzyme patterns. Can J Bot 63:2268–2273

Stenlid J, Rayner ADM (1989) Environmental and endogenous controls of developmental pathways: variation and its significance in the forest pathogen, Heterobasidion annosum. New Phytol 113:245–258

Taylor JW (1986) Fungal evolutionary biology and mitochondrial DNA. Exp Mycol 10:259–269

Thompson W (1984) Distribution, development and functioning of mycelial cord systems of decomposer basidiomycetes of the deciduous woodland floor. In: Jennings D, Rayner ADM (eds) The ecology and physiology of the fungal mycelium. Cambridge University Press, Cambridge, pp 185–214

Vilgalys R, Johnson JL (1987) Extensive genetic divergence associated with speciation in filamentous fungi. Proc Natl Acad Sci USA 84:2355–2358

Welsch J, McClelland M (1990) Fingerprinting genomes using PCR with arbitrary primers. Nucleic Acids Res 18:7213–7218

Wessels JGH (1986) Cell wall synthesis in apical hyphal growth. Int Rev Cytol 104:37–79

Williams END, Todd NK (1985) Numbers and distribution of individuals and mating type alleles in populations of Coriolus versicolor. Genet Res 46:251–262

Williams JGK, Kubelik AR, Livak KJ, Rafalski JA, Tingey SV (1990) DNA polymorphisms amplified by arbitrary primers are useful as genetic markers. Nucleic Acids Res 18:6531–6535

Determinants of Fungal Communities

3 Organization and Description of Fungal Communities

J.C. Zak[1] and S.C. Rabatin[2]

CONTENTS

I. Introduction

After more than a century of growth and sophistication in theory and methodology, biological communites are recognized as components of hierarchical systems whose patterns of species composition and abundance reflect the accommodations that occur among members of the community. The organisms comprising a community are delineated as such not only because they occupy a common space, but also because they have adapted to each other and the abiotic characteristics of the shared habitat (Wilson 1980). In any instant of time, a community reflects the prior processes of accommodation of one species to another (Allen and Hoekstra 1992). Viewed in this manner, communities are more than just the space occupied by particular species. Rather, they are defined by the relationships that occur among the component parts.

As with all living systems, communites are dynamic. The degree of heterogeneity that results from community fluctuations is observed and quantified on a level set by human observers and thus is arbitrary and dependent upon the scale chosen (Sousa 1984; Allen and Hoekstra 1992). The predominant goal of community ecology is to identify the spatial and temporal factors that underlie the observed community structure and function. Ultimately, ecologists seek to relate community patterns to ecosystem-level processes.

Fungal communites occupy environments which are a reflection of evolutionary consequences, and the resultant broad taxonomic diversity that typifies the fungi (Hawksworth 1991). Fungi usually colonize habitats that are ephemeral in time and heterogenous in space. Furthermore, habitat quality for most fungal communites may change rapidly over short intervals such that shifts in fungal species composition and abundances can be dramatic. The composition and densities of the component species in a fungal community are not constant, but vary over a time scale that is dependent on the life-history patterns of the species within the community (Frankland 1981; Andrews 1992).

Our goal in this chapter is to examine the current status of research in fungal community ecology. We begin with a review of the general concepts underpinning fungal community ecology, followed by a review of the general approaches and methodologies used by fungal ecologists. Finally, we discuss the approaches used to characterize fungal communities, and conclude with a discussion of the state of our knowledge

[1] Ecology Program, Department of Biological Sciences, Texas Tech University Lubbock, Texas 79409-3131, USA
[2] Plant Disease Control/Biocontrol Ricerca, Inc., Painesville, Ohio 44077, USA

The Mycota IV
Environmental and Microbial Relationships
Wicklow/Söderström (Eds.)
© Springer-Verlag Berlin Heidelberg 1997

with regard to community dynamics and bio-diversity patterns.

II. General Concepts of Fungal Community Ecology

Changes in fungal community composition associated with a specific substrate, when viewed over time, constitute the process that we call succession (Rayner and Todd 1979; Frankland 1992). The processes of fungal succession and substrate decomposition occur simultaneously. Decomposition results in the gradual disappearance of the habitat as carbon is mineralized to CO_2 during fungal respiration. The substrate is subsequently modified as fungi metabolize the various chemical constituents of the organic material. Concomittently, there are associated changes in fungal species composition and biomass in response to substrate (habitat) modification. It is important to note that fungal succession differs from the classic theory of plant succession. The concept of a seral stage defining a specific point in the development of a plant community on a specific habitat, usually leading to a "climax" community with a specific species composition that is self-perpetuating, does not apply to fungal community development. Fungal communities lose species over time as the resources of the substrate are depleted. The development of fungal communities on any substrate (Hudson 1968; Cooke and Rayner 1984; Frankland 1989) is only superficially similar to patterns of plant species replacement. The underlying mechanisms for changes in species composition of plant and fungal communities over time are fundamentally different, though we use the same term to describe the emergent processes.

Species replacements and changes in dominance structure in fungal communities during decomposition occur in part in response to seasonal patterns in either moisture and temperature of the habitat (Gochenaur 1984), changes in the chemical characteristics of the substrate (Kjøller and Struwe 1980), alteration in the physical structure of the substrate (Boddy 1992), predation of fungal mycelium by microarthropods (Lussenhop 1992) or by the synergistic interaction of these factors. Ultimately, the composition and abundances of the species comprising any fungal community are dependent upon the spatial and temporal variation in the impacts of these factors. Moreover, the relative importance of these factors and others is highly dependent upon the substrate type and the taxonomic groups of fungi that comprise the community of interest.

The history of the community concept for fungi has represented an attempt to equate patterns in the development of fungal communities to that of plant communities (Cooke and Rayner 1984), and an attempt to integrate fungal community ecology with main areas of ecological theory (Wicklow and Carroll 1981; Carroll and Wicklow 1992). Concepts initially developed for understanding plant community dynamics, i.e., life-history strategies (Grime 1977), have provided insights into the underlying population interactions that contribute to fungal community composition and dynamics (Pugh and Boddy 1988). The shift in thinking away from adherence to theory drawn from plant communities led to recognition that resource and scale could greatly influence our understanding of fungal community dynamics (Swift 1976).

Except for studies of wood-decomposing basidiomycetes and investigations that have examined the effects of substrate quality on patterns of fungal community development (Robinson et al. 1994), much of fungal community ecology is descriptive and/or correlative. The causal mechanisms underlying the dynamics of fungal communites have rarely been addressed experimentally. Swift (1976), in an early and influential theoretical treatment, envisioned fungal communities as having a hieracrchical structure, and specifically warned against the dangers of losing sight of scale differences. Swift (1976) identified the fundamental component of a decomposer community as a "unit community", which he defined as . . . "a species assemblage which inhabits a volume of resource that is delimited in some clear and unequivocal way such that, whereas the species within the unit community may be expected to interact in some way, interactions between organisms in different or even neighboring 'unit communities' will be minimized." Examples of "unit communities" include seperate twigs, isolated leaves, and fecal pellets (Swift 1984). He did acknowledge that boundaries could become uncertain when leaves blanket a forest floor or when a single fungal mycelium might colonize a number of resource units over a large area, as was shown by Thompson and Boddy (1983) for a basidiomycete colonizing tree boles. What was intrinsically appealing with this approach was that the "unit community" defined both the species composition of the community and the functional unit. We now recognize that unit communities are not self-contained functional unities, because fungi can obtain nutrients from sources beyond the substrates that we are examining, further complicating how we define a community. However, the recognition that fungal community development on a specific resource should manifest collective (i.e., species abundance distributions) and emergent properties (species diversity) that are also expressed at larger scales of resolution finds increasing relevance in the current discussions on the roles of scale and hierarchy theory in community ecology (Allen and Hoekstra 1992). A hierarchy is developed from the unit community level by grouping the communities inhabitating the same resource (e.g., beech leaves), the same habitat (litter layer), and finally the same ecosystem (beech-maple forest).

As conceptualized by Swift, the resource units colonized by fungi have been considered to be analogous to an "island" (Wildman 1992), with the difference being that the resource island is continually being decomposed by the inhabitants. Thus, the chemical makeup, physical features, and even the size of the resources are being changed as decomposition proceeds. These islands (new leaves, fecal pellets), while separated in time and space, are involved in cyclic episodes of colonization, dispersal, and extinction. Swift's conceptual model was a significant advance because it specifically addressed fungal community dynamics. Moreoever, having a hierarchical framework, fungal dynamics could be "scaled up" from community to ecosystem.

Important to understanding the dynamics of fungal communities, or any community for that matter, is the recognition that the ecological unit does not have a reality beyond our observations (Allen and Hoekstra 1992). Fungal communities, like those of higher plants, are ecological frameworks designed to focus observations such that they may be repeatable in space or time. What constitutes a community, then, is very much dependent upon our scale of observation (Hoekstra et al. 1991; Allen and Hoekstra 1992; Zak et al. 1995). Frankland (1992) alluded to the subjectivity in examining fungal communities when she made reference to the fact that fungal communities are not self-contained entities. Ignoring the human subjectivity in our ecological observations will not make it disappear; a more rational approach

which clearly delineates the criteria used for establishing the ecological framework, i.e., community, is required rather than being capricious.

Fungal ecologists not only approach the study of communities with much the same subjectivity in delineating communities as do plant and animal ecologists; mycologists are at a severe disadvantage in that they, for the most part, are enumerating organisms that they cannot directly observe. Garrett (1952), a pioneer in fungal ecology, succinctly summarized the difficulties inherent to fungal community ecology when he stated that . . . "for most fungi we see what we cannot identify, and identify what we cannot see."

III. Determining the Composition of Fungal Communities

A. Identification of Taxa and Enumeration by Direct Means

Studies of fungal community dynamics generally rely on isolation procedures for enumerating the composition of the attendant community and restrict the range of observations to specific substrates (e.g., leaves, root surfaces, wood). There are specific habitats, and taxonomic groups, however, where the direct observation of fruitbodies or conidial occurrence can be used to determine the presumptive composition of the fungal community. These special cases are described first, followed by a discussion of the approaches taken to determine species composition of soil and litter fungal communities.

1. Aquatic Hyphomycotina

Species occurrences of conidial fungi responsible for litter decomposition in streams can be obtained simply by examining the edges of leaf pieces for the presence of conidia (Suberkropp 1992). Scanning subsamples of leaf surfaces following staining for conidia has also been used to determine the percentage frequency of occurrence of each species in the community (Suberkropp 1984). Species lists for stream habitats can readily be determined by filtering water, staining the filters, and enumerating the species occurrence from the conidia present on the filter (Iqbal and Webster 1973). Filtering techniques, however, provide only a list of species and do not represent a single community because the conidia are produced from multiple substrates (Shearer and Webster 1985). Moreover, densities, and thus the ability to detect conidia, are influenced by losses from entrapment in debris and filtering from invertebrates (Shearer and Lane 1983).

2. Coprophilous Fungi

The fungal community that exists on herbivore dung provides an excellent model system for examining various aspects of fungal community dynamics, as well as for testing community theories (Yocom and Wicklow 1980; Wicklow 1981, 1992; Webster 1988). The dung community is enumerated by directly scanning the dung surface for the occurrence of fungal fruitbodies. Studies employing this model system for fungal community analysis have shown that characteristic communities of fungi develop on the dung of different herbivores. Despite the correlations between fungal occurrence and specific type of dung, coprophilous fungal communities are also influenced by temperature, substrate moisture patterns, and dung age (Yocom and Wicklow 1980; Angel and Wicklow 1983; Kuthubutheen and Webster 1986).

3. Epigeous Basidiomycotina

Quantification of basidiomycete sporocarps in field plots can provide informtion on saprophytic and putative ectomycorrhizal communities, and the effects of small-scale soil disturbance (Sagara 1992). It is important, however, that certain factors, which are key determinants of fruiting patterns, are considered when designing a sampling regime. Sporocarp production varies considerably in time and space. Spatial distributions of sporocarps are highly dependent upon the distribution of woody litter and other substrates for decomposers, and on the distribution of roots for mycorrhizal fungi. Spropcarp numbers have been shown to range considerably within and between years (Bills et al. 1986; Villeneuve et al. 1989). Therefore, a well-considered sampling design should encompass sufficiently large areas with long-term monitoring (Vogt et al. 1992). Approaches for sampling vary from establishing large permanent plots that are subdivided into subplots to assist in mapping, to quantification using belt transects. The objectives of the study, and the scale relevant to the objectives will be major determinants of the sampling design: Table 1 lists examples.

Regardless of the sampling design, methods which involve trampling of the plots while sampling may bias results, and care should be taken to minimize or control for these effects. While the effects of sporcarp removal for identification or biomass measurements on subsequent production have not been extensively studied, several reports (Arnolds 1988) have indicated that fruitbody removal does not alter production. To minimize potential removal effects, Ammirati et al. (1987) used a nondestructive approach to mapping fruitbodies by placing a dot of enamel paint on each sporocarp.

The occurrence of epigeous fungi in forest soils can be related to disturbance events that alter soil chemical characteristics or open up new habitats for fungal colonization. Examples of such phenomena can be found in the altered patterns of sporocarp production associated with tree fall gaps. Other groups are the specific groups of Basidiomycotina, "ammonia fungi", that occur on forest soils after deposition of urine and/or dung of animals, or following the decomposition of dead animal carcasses (Sagara 1975, 1981). Sagara (1992) pointed out that, although dung has long been recognized as an important substrate for fungi, the role of urine deposits and locations of decomposed fecal material and bodies as specific habitat for epigeous forest fungi has been unappreciated.

4. Hypogeous Basidiomycotina

Enumerating the community composition of hypogeous fungi is complicated by their occurrence below ground, the necessity to destructively sample the forest floor, and the amount of time necessary to search plots for sporocarps. To estimate species occurrences and sporocarp production, studies have employed variously sized plots that may or may not be subdivided (Hunt and Trappe 1987), or that employed circular plots located along parallel transects (Luoma et al. 1991) through the forest stand (Table 1). In most instances, sampling was destructive and plots were not resampled. Because of the destructive sampling of the forest floor when searching for hypogeous fungi, the area sampled must be large enough to be representative of the forest type, but should not be so large that after a few sample periods most of the forest floor is disturbed (Vogt et al. 1992). In a Douglas-fir forest in Oregon, Luoma et al. (1991) reported that the estimation

Table 1. Examples of sampling designs and schedules used to enumerate epigeous and hypogeous fungal communities. (Vogt et al. 1992)

Study duration (years)	Experimental design	Sampling schedule	Reference
5 (Epigeous)	$12 - 40 \times 40\,m^2$ plots	2-week intervals	Ohtonen (1986)
4 (Epigeous)	$52 - 1000\,m^2$ plots	5 times per year	Ruhling et al. (1984)
3 (Epigeous)	$345 - (2 \times 50\,m)$ plots	1–2-week intervals	Ohenoja (1984)
3 (Hypogeous)	$12 - 4\,m^2$ plots without resampling	Monthly	Hunt and Trappe (1987)
3 (Hypogeous)	$50\,m^2$ plots, random	Monthly	Fogel (1976)
1 (Epigeous)	$2 \times 50\,m$ plots	Weekly	Vogt et al. (1981)
1 (Hypogeous)	$12 - 2 \times 2\,m$ plots	Weekly	Vogt et al. (1981)

of sporocarp biomass would be overestimated on a hectare basis if the total sample area were less than $800\,m^2$. Selection of a sampling design using large plots should always be balanced by a consideration of the time required to search during periods of maximum sporocarp production. Such constraints can severely restrict sampling efficiency and ultimately determine the size of the area sampled (Fogel 1981).

Sampling efficiency may also depend upon the year-to-year variation in sporocarp production. In the first year of sampling, only two thirds of the total number of species that were eventually detected over a 32-month period of a study conducted in a Douglas-fir stand in western Oregon were found (Hunt and Trappe 1987). Furthermore, the number of species detected at this site continued to increase over the 32-month period. These data would suggest that the inflection point of the species area curve was not reached even after sampling for almost 3 years at this location. The magnitude of fluctuations in community dynamics of hypogeous fungi is only increased by the knowledge that a diverse number of animals consume considerable biomass of these fungi (Fogel and Trappe 1978). How animal consumption might influence species detection and patterns of occurrence is unclear. Nevertheless, there is little doubt that the relationship between hypogeous fungi and their animal consumers can strongly influence hypogeous fungal species distribution and abundance.

B. Isolation Methods for Identification and Enumeration

1. Leaf Surfaces

Leaf surfaces are dynamic communities in time and space that quickly reflect the interactions be-

tween immigration of propagules, emigration due to physical loss or removal of propagules, biomass increases and propagule production, and local extinctions (Kinkel 1991). The composition of the fungal communities associated with leaf surfaces are determined in practice by the methods and media used for isolation, and incubation temperatures. Early work on the fungal communities of leaves focused on species occurrences (Preece and Dickinson 1971; Dickinson and Preece 1976) and changes in species composition (successional patterns) as a leaf develops from a bud to a mature structure (Wildman and Parkinson 1979). This descriptive research was followed by examinations of the interspecies interactions on leaves and the potential application of selected saprophytes as biocontrol agents (Fokkema 1981). To better understand the characteristics of leaf surface habitats as they influence fungal community dynamics, Andrews (1984) and Andrews and Harris (1986) examined the "life-history strategies" (patterns) of phylloplane fungi using the concepts of r- and K-selection. These investigations, along with an analysis of leaves as "ecological islands" by Andrews et al. (1987) and Wildman (1992), have moved the study of leaf-surface fungal communities from the descriptive phase to an examination of the underlying population processes that contribute to fungal community dynamics.

Kinkel (1991) has emphasized that the characteristics of phylloplane communities are determined by the degree of resource limitation and whether population growth is density-independent or density-dependent. Furthermore, understanding the relative influences of immigration, emigration, and death rates of fungi on leaf surfaces should provide the means to link community dynamics to population events directly (Kinkel 1991). Earlier, Kinkel et al. (1989a) had found that the size of fungal communities associated with apple leaves was consistently smaller than predicted based on immigration estimates alone. They concluded that fungal growth could not compensate for high emigration and death rates following the initial immigration event. However, depending upon the organism, single large immigration events could have large short-term consequences on phylloplane community size and composition (Kinkel et al. 1989b). These data indicate that immigration rates may regulate phylloplane community dynamics and may be important in maintaining community size (Kinkel 1991). Furthermore, during the initial stages of community assembly, phylloplane community development may indeed be immigration-limited. Therefore, studies of phylloplance community dynamics should not focus on within-community interactions, but should test the applicability of population dynamics models that emphasize the open nature of the phylloplane (Kinkel 1991).

2. Soil Fungi

The media used to isolate fungi depend largely upon the taxonomic group of interest, for no single medium will allow for the isolation of all fungi present in a particular habitat. Most studies of saprophytic fungi use a general-purpose medium such as malt extract agar, potato dextrose agar, or cornmeal agar to which antibiotic or other antibacterial compounds have been added (Frankland et al. 1990). Selective media have also been devised that allow for the isolation of particular fungal taxa that may not be able to be isolated on general-purpose media. For example, the incorporation of benomyl and dichloran into potato dextrose agar was found to be an effective medium for general isolation of hymenomycetes (Worral 1991), while modified Melin-Norkans medium is routinely used for isolation of ectomycorrhizal fungi (Molina and Palmer 1982). Recipes for a wide range of culture media can be found in Stevens (1974).

The method one chooses for isolating fungi from soil organic matter and from litter determines the composition of the particular fungal community, and will vary with the substrate to be examined. During the early period of the study of soil fungal communities, the soil dilution plate was the predominant method both for isolating fungi from soil and enumerating the composition of the fungal community (Parkinson 1981). Warcup (1955), however, demonstrated that the majority of the colonies that developed on a dilution plate originated from spores or other propagules and not from actively growing mycelium. Thus, the soil dilution procedure, while easy to employ, does not provide an accurate assessment of the composition of the fungal community at the time of sampling. The dilution method indicates only what fungi are present as spores. Moreover, the procedure is biased towards those fungi, such as *Penicillium* species, which produce an abundance of small propagules.

Recognizing the limitations of the soil dilution procedure, Warcup (1955) introduced the direct hyphal isolation method for determining soil fungal communities. The results from his studies of wheat field and pasture soil showed that a completely different fungal community was enumerated with the hyphal isolation procedure, including the isolation of basidiomycetes, than was obtained with the soil dilution procedure. One major drawback of the direct hyphal isolation procedure is that it is a labor-intensive method. A later study by Williams and Parkinson (1964) also reported low success with the direct hyphal isolation approach; hyphae associated with organic material could not be extracted. For those substrates such as wood, where macroscopic aggregations of mycelium (e.g., rhizomorphs or cords) are obvious, attempts can be made to transfer these materials to an appropriate medium either directly or following surface sterilization (Frankland et al. 1990) to enumerate species occurrences.

For the isolation and enumeration of fungal communities from the organic components of soil and from plant litter, various investigators have employed modifications of the root-washing procedure described by Harley and Waid (1955). Modifications of the root-washing procedure generally place the soil or plant litter on a series of sieves while bubbling air through the material to remove spores. The procedure can be automated (Bissett and Widden 1972) for various purposes. The significance of the washing procedure for soils is that organic and inorganic material of various sizes can be plated out onto an appropriate medium, thus retaining information on the origin of the species. Care should be taken when choosing the size of particles. Bååth (1988) advocated that small particles should be used since the size of the particles affected the number and composition of species obtained from each particle. With plant litter the material can be separated into different tissue types (Frankland 1966) before washing or left intact and subsequently cut into appropriate sizes for effective enumeration of the fungal community (Durall and Parkinson 1990).

3. Woody Litter

Wood that falls to the soil surface already contains established fungal communities that were initiated either from colonization of the living tissue, or from saprophytes that colonized the standing-dead resource (Boddy and Griffith 1989). In either case, the subsequent patterns of fungal community development will depend upon the competitive abilities of the soil fungi to replace the resident populations in the wood. The dynamics of fungal communities that develop during wood decay have been likened to a complex battleground where the interactions between individuals are marked by lines of demarcation (interaction zones) and whose persistence or impermanence reflects the degree of community stability (Rayner 1995).

Enumeration of fungal communities from wood, generally because of the size and structural complexity, requires alternative sampling approaches from other types of litters. Much of the older information on the fungal communities of wood was obtained from either random or ordered sampling of the substrate. While random sampling provides a species list and perhaps an indication of species abundance, the sampling strategy did not reveal the spatial structure of the community. Boddy et al. (1985) used three approaches to obtain as complete a picture as possible of the three-dimensional structure of the fungal community on branch wood. They suggest that the first approach to visualizing the three-dimensional structure of the fungal community associated with wood is to saw the wood into sections and map the internal patterns of decay or discolored wood along with lines demarcating genetically incompatible isolates. From these decay patterns, isolations can be made to determine the species that are responsible for the decay or discoloration. Finally, moist chamber incubation of wood pieces may allow mycelium to grow out that can then be used for isolation and identification using various culture-based keys (Staplers 1978; Rayner and Boddy 1988; Nakasone 1990).

IV. Fungal Community Analyses

A. Species Richness

The simplest approach for characterizing a fungal community has been to list the kinds and numbers of species (species richness) that one has observed or isolated. Though simple in design, even this first-level attempt at describing a community is fraught with problems that are in part methodological, as discussed above, and in part due the problem of scale in delimiting what constitutes a community. Prior to initiating a community study, a "species area curve" (number of subsamples versus the number of new species encountered) should be determined. Such a graphic representation allows the investigator to establish the level of sampling beyond which further efforts will not be as productive in detecting additional species. Unfortunately, species-area curves have rarely been generated prior to sampling for fungal community analyses. When determining species-area relationships, samples should be taken from the same contiguous area. If subsamples are amassed from scattered samples across a habitat, the species-area curve will bow out and the line will have a steeper slope due to sampling artifact (Rosenzweig 1995).

Even in the light of the information provided by the species-area relationship, investigators have been constrained to proceed with sampling regimes that are inadequate. When sampling soil fungi from three locations within a sagebrush-grassland ecosystem in Wyoming, USA (within clumps of sagebrush, under-grass, and between plants), Christensen (1989) observed that new fungal species continued to be added with each incremental group of 100 isolates even when a total of 1100 isolates had been taken. These results suggest that the degree of effort required to determine species richness of soil fungal communities can best be provided for within the framework of long-term ecological research efforts.

An alternative explanation to the pattern observed by Christensen (1989), however, is that more than one community was being sampled within the various plant associations. That is, the fungal communities in this system were organized at a scale different from that of the plant communities sampled. Frankland (1989) also cautioned about equating a specific fungal community with a specific vegetational unit and not realizing that within any given plant community there can be numerous substrate successions occurring that reflect the various types of organic matter inputs into the system.

B. Diversity and Abundance

An important aspect of communities, namely the commonness and rarity of species, is completely ignored when a community is described in terms of the numbers and the composition of species. The simplest measure of the character of a community that takes into account both species abundance patterns and species richness is a diversity index (Magurran 1988). Use of diversity indices should be tempered by the realization that important ecological information is lost when taxonomic and community data are incorporated into a single measure. Additionally, differences in diversity indices among communities cannot be analyzed using parametic statistics. Therefore, differences between two indices are left to subjective evaluation.

Zak (1992) has argued that species abundance distributions, plotting of the relative importance of each species against rank, will provide the most complete analysis of fungal communities that can be interpreted ecologically. Moreover, the four main species abundance distributions provide models against which data can be statistically tested (Magurran 1988). Differences in species abundance distributions can then be used to assess possible abiotic and biotic factors affecting the spatial and temporal patterns of fungal community development. When collecting species abundance information, taxa should not be grouped into genera. Furthermore, while species identifications are unnecessary for community analyses, it is crucial that morphological species groups be established to separate one isolate from another (Zak 1992). The designation of morphological species groups becomes crucial when fungal communities contain taxa that do not readily sporulate.

C. Ordination and Gradient Analyses

While there is a certain simplistic appeal to conceptualizing communities as discrete entities with well-defined boundaries, where groups of species exist adjacent to each other and do not intergrade, such a state of affairs almost never exists. The spatially defined community is inadequate for understanding and elucidating the mechanisms that determine fungal community structure and organization. Species within a fungal community vary in their response to various environmental factors, resulting in the observed spatial and temporal distribution of abundance of species along a gradient of specific environmental factors. Gradient analysis, which originated as a technique for analyzing plant community patterns (Gosz 1992), can be used to examine spatial and temporal patterns in fungal communities. Multivariate methods, such as ordination (principle components analysis), have been used by ecologists to examine the spatial relationships among species along an environmental gradient (Kenkel and Booth 1992). These techniques provide the means to examine qualitative and quantitative variation in a given community in response to the multidimensional variation in abiotic variables.

Several fungal community studies have used this approach to examine the degree to which various abiotic factors contributed to observed fungal community composition. Christensen (1969) found that litter calcium level, not vegetation, was positively correlated to the species composition of fungal communities associated with conifer-hardwood forests located along a moisture gradient in Wisconsin. In an extensive study of abiotic factors determining the distribution of soil fungi in alpine environments, Bissett and Parkinson (1979) reported that temperature, moisture, available potassium, and pH were the major determinants of community structure from three alpine habitats. The distributions of the dominant species in the alpine sites were most affected by abiotic constraints – perhaps not surprisingly so, for these extreme environments – while distributions of the subdominant taxa were regulated by the activity of the dominant species. The importance of using gradient analyses and multivariate approaches is not to provide for the description of another special case. Rather, the impact of these approaches is in allowing one to ask questions that can generate predictions concerning the roles of abiotic and biotic factors in regulating community composition and structure (Allen and Hoekstra 1992).

D. The Problem of Destructive Sampling

The data required for fungal community analyses are usually obtained from the destructive subsampling of organic substrates that may be heterogeneously distributed in space and time. Community assessments based on epigeous fruitbody production are one of the few exceptions. Fungal ecologists are therefore left with the dilemma of whether fungal community dynamics can be accurately inferred by assessing the dynamics within or on subsamples of the resource (e.g., individual leaves)? The hypothesis has been advanced that when substrates are destructively sampled to estimate community composition and dynamics, separate communities on the same substrate develop in a similar manner over time Kinkel et al. (1992). Added corollaries, according to Kinkel and coleagues, of this hypothesis are that (1) separate communities are described by the same species abundance distribution, and (2) differences in species number among samples correspond to

similar differences in species richness among communities.

A simulation model describing microbial communities on leaves as fitting either a negative binomial or Poisson lognormal was used by Kinkel et al. (1992) to evaluate the relationship between the number of species in a community (individual leaves), and the species abundance distribution that described the community. They found that the mean number of species detected among separate communities that "developed" on their simulated leaf surfaces significantly differed for many combinations of sample size and species abundance distributions (Fig. 1). Furthermore, the study suggested that standardizing the size of subsamples does not increase the probability that samples will contain the same proportion of the total numbers of species from each community.

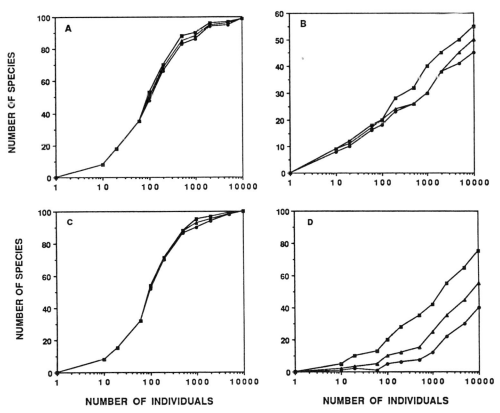

Fig. 1A–D. A simulation of the ability to estimate species richness of microbial communities based on various sample sizes for communities that are described by different species abundance distributions. Each graph depicts three different communities that are replicates of the same algorithm. Community types are as follows: **A** negative bionomal with k = 1; **B** negative bionomal with k = 0.1; **C** Poisson lognormal with S^2 = 1; **D** Poisson lognormal with S^2 = 3. A large k-value describes a community in which numbers of species are relatively evenly distributed among the abundance levels, while a small k-value describes a community in which most of the isolates are from a few species and the remainder of the species contain a small number of isolates. A small S^2 describes a community in which most species are moderately abundant. For large values of S^2, a community would be comprised of a larger number of taxa with either high or low population densities and few moderately abundant species. (Kinkel et al. 1992)

Clearly, all sampling methods are limited. The simulation by Kinkel and colleagues (1992) raises the question again as to what constitutes appropriate space/time boundaries when sampling fungal communities. The questions that remains to be addressed for each community study is at what scale can discrete resource units be considered a functional community sensu Swift (1984), and what part does this community play in mediating the process of nutrient and energy flow in ecosystems?

Indeed, in ecology generally, identifying the appropriate scale and establishing the boundaries for data collection are critical challenges. Rather than consider ecological systems as composed of subunits of increasing size and complexity, biological systems may be more productively envisioned as a cake with many layers (Allen and Hoekstra 1992; Fig. 2). That is, ecological systems (populations, communities, ecosystems) occur at various scales of observation and these eco-logical units are defined by such strictly scale-defined criteria as grain and extent. Grain represents the resolving power of the data; that is, the ability to detect the smallest and most transient unit. Those aspects of the data that are related to scope (i.e., the highest level of ecological organization that can be accessed) are termed extent. By conceptualizing ecological systems in this layer-cake metaphor one can compare the properties of fungal communities that develop on individual leaves, for example, to the dynamics of the fungal community for an entire tree since the layers of the "cake" are strictly scale-dependent. Each type of ecological system (population, community) can occur on every level within the larger-scaled entity, such as leaves on a tree. The investigator is not limited to making comparisons horizontally across a defined level, but can move, up and down the cake, allowing for comparisons to be made between differently scaled entities of the same system. For example, we can examine how individual

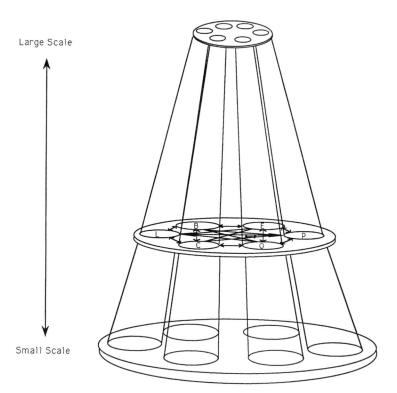

Large Scale

Small Scale

Fig. 2. The layer-cake metaphor for ecological criteria and ecological scale. The *wide base* indicates a large number of small entities; the *narrow top* indicates a small number of large entities. Although only one is shown, any number of cross sections could have been drawn, each at its own scaled level. Each *letter* on the cross section indicates a different ecological criterion: *O* organism; *P* population; *C* community; *E* ecosystem; *L* landscape; *B* biome. Individually, the columns represent a criterion for looking at the material system, e.g., the abstract notion of community. The disk labeled *C* is an actual community with a particular scale. In the *C* column, large-scaled concepetual communities occur above that level while smaller-scaled community subsystems occur below. (Allen and Hoekstra 1992)

leaves (smaller entities) contribute to the fungal community associated with a tree, and how an individual tree contributes to the larger community associated with the forest in which the tree resides. Furthermore, by investigating the lower levels, we may elucidate the underlying mechanisms that contribute to the spatial/temporal patterns observed in the larger system.

V. Community Dynamics and Biodiversity Patterns

Despite our knowledge of the roles of fungi in decomposition and nutrient cycling (Parkinson and Coleman 1991), there are still many gaps in our knowledge of the effects of fungal community dynamics on ecosystem processes. In attempting to understand the mechanisms contributing to patterns of fungal community development and biodiversity, the researcher is confronted with trying to answer such questions as (1) what controls fungal community development? (2) are there relationships between fungal species richness and decomposition rates? (3) does fungal biodiversity matter? and (4) what role does fungal biodiversity and community dynamics play in ecosystem functioning and stability? These questions are certainly not unique to fungal community ecology, but represent critical areas for research in much of community and ecosystem level ecology (Freckman 1994).

Although current thinking about global biodiversity patterns with regards to fungi is focused on taxonomic richness (Hawksworth 1991), emphasis should be placed on understanding the relationship between taxonomic richness and functional processes. Simply collecting inventories of fungal species without integrating this information into an ecosystem context does not address the important questions stated above. Solbrig (1991) emphasized that biodiversity is composed of three interrelated elements: genetic, functional, and taxonomic. Little is understand about the degree to which taxonomic diversity is controlled by genetic diversity, and even less about the manner in which genetic and taxonomic diversity determines fungal functional diversity and ecosystem processes (Zak et al. 1994). The relationship between genetic and taxonomic diversity has been examined for a limited number of economically important plant pathogenic fungi (Brown et al.

1991). In a recent review on the roles of fungi in woodland ecosystems, Rayner (1995) emphasized that the functioning and biodiversity of woodland ecosystems are intimately related to the functional and taxonomic diversity of the fungal component of these systems. In turn, fungal biodiversity and community dynamics may be regulated by trophic interactions and food web dynamics (Moore and de Ruiter 1991). The implications of trophic interactions, at least as they influence fungal biodiversity and community dynamics, have not been evaluated effectively, with the exception of certain highly focused studies (Newell 1984).

VI. Conclusions

In all likelihood, a better understanding of the dynamics of fungal communities and the importance of fungal biodiversity, as they relate to decomposition, will come from research designed to address these phenomena at the ecosystem and landscape level. As this chapter has indicated, mycologists involved in communities attempt to understand fungal community dynamics by using scaleup approaches. These methodologies include the search for causal factors by analyzing and documenting community structure, and by comparing the structure of communities in similiar habitats. Perturbation experiments, which consist of disturbing the species composition and observing the consequent effects on fungal community dynamics, have also been productive. These experimental designs, as well as the approaches and methodologies we documented in this chapter, have provided a sound mycological and ecological foundation. However, we suggest that the information needed to link our knowledge of communities to processes may come from a better understanding of the functional aspects of fungal species at the ecosystem and landscape level.

The problem of ecological scale may present a more difficult problem to overcome for fungal community ecology, primarily because of our unwillingness to let go of old outdated ecological paradigms of community, hierarchy, and ecological complexity. While mainstream ecology has fully recognized the implications of lack of attention to scale, the recognition of this paradigm shift has been slow to appear in the mycological literature. Not only does our lack of attention to scale hinder our development of the discipline, but it also interferes with the integration of mycological research with mainstream ecology.

References

Allen TFH, Hoekstra TH (1992) Toward a unified ecology. Columbia University Press, New York

Ammirati S, Ammirati JF, Bledsoe C (1987) Spatial and temporal distributions of ectomycorrhizal fungi in a Douglas-fir plantation. In: Sylvia DM, Hung LL, Graham JH (eds) Mycorrhizae in the next decade, 7th NACOM. IFAS, University of Florida, Gainesville, pp 81

Andrews JH (1984) Life history strategies and plant parasites. In: Ingram DS, Williams PH (eds) Advances in plant pathology, vol 2, Academic Press, New York, pp 105–130

Andrews JH (1992) Fungal life history strategies. In: Carroll GC, Wicklow DT (eds) The fungal community: its organization and role in the ecosystem, 2nd edn. Dekker, New York, pp 119–146

Andrews JH, Harris RF (1986) Plant pathogens and the theory of r- and K-selection. Am Nat 120:283–296

Andrews JH, Kinkel LL, Barbee FM, Nordheim EV (1987) Fungi, leaves, and the theory of island biogeography. Microb Ecol 14:277–290

Angel K, Wicklow DT (1983) Coprophilous fungal communities in semiarid to mesic grasslands. Can J Bot 61:594–602

Arnolds E (1988) The changing macromycete flora in The Netherlands. Trans Br Mycol Soc 90:391–406

Bååth E (1988) A critical examination of the soil washing technique with special reference to the effect of the size of the soil particle. Can J Bot 66:1566–1569

Bills GF, Holtzman GI, Miller OK (1986) Comparison of ectomycorrhizal-basidiomycete communities in red spruce versus northern hardwood forests of West Virginia. Can J Bot 64:760–768

Bissett J, Parkinson D (1979) Functional relationships between soil fungi and environment in alpine tundra. Can J Bot 57:1642–1659

Bissett J, Widden P (1972) An automatic, multi-chamber soil-washing apparatus for removing fungal spores from soil. Can J Micro 18:1399–1409

Boddy L (1992) Development and function of fungal communities in decomposing wood. In: Carroll GC, Wicklow DT (eds) The fungal community: its organization and role in the ecosystem, 2nd edn. Dekker, New York, pp 749–782

Boddy L, Griffith GS (1989) Role of endophytes and latent invasion in the development of decay communities in sapwood of angiospermous trees. Sydowia 41:41–73

Boddy L, Gibbon OM, Grundy MA (1985) Ecology of Daldina concentrica: effect of abiotic variables on mycelial extension and interspecific interactions. Trans Br Mycol Soc 85:201–211

Brown JKM, Jessop AC, Resanoor HN (1991) Genetic uniformity in barley and its powdery mildew pathogen. Proc R Soc Lond [B] 246:83–90

Carroll GC, Wicklow DT (eds) (1992) The fungal community: its organization and role in the ecosystem, 2nd edn. Dekker, New York

Cooke RC, Rayner ADM (1984) Ecology of saprotrophic fungi. Longman, London

Christensen M (1969) Soil microfungi of dry to mesic conifer-hardwood forests in northern Wisconsin. Ecology 50:9–27

Christensen M (1989) A view of fungal ecology. Mycologia 81:1–19

Dickinson CH, Preece TF (1976) Microbiology of aerial plant surfaces. Academic Press, London

Durall DM, Parkinson D (1990) Initial fungal community development on decomposing timothy (Phleum pratense) litter from a reclaimed coal-mine spoil in Alberta, Canada. Mycol Res 95:14–18

Fogel R (1976) Ecological studies of hypogeous fungi. II. Sporocarp phenology in a western Oregon Douglas fir stand. Can J Bot 54:1152–1162

Fogel R (1981) Quantification of sporocarps produced by hypogeous fungi. In: Wicklow DT, Carroll GC (eds) The fungal community, its organization and role in the ecosystem. Dekker, New York, pp 553–567

Fogel R, Trappe JM (1978) Fungus consumption (mycophagy) by small animals. Northwest Sci 53:1–31

Fokkema NJ (1981) Fungal leaf saprophytes, beneficial or detrimental? In: Blakeman JP (ed) Microbial ecology of the phylloplane. Academic Press, New York, pp 433–454

Frankland JC (1966) Succession of fungi on decaying bracken petioles. J Ecol 57:25–36

Frankland JC (1981) Mechanisms in fungal succession. In: Wicklow DT, Carroll GC (eds) The fungal community, its organization and role in the ecosystem. Dekker, New York, pp 403–426

Frankland JC (1989) Fungal succession: myth or reality? In: Hattori T (ed) Recent advances in microbial ecology. Proc 5th Int Symp Micro Ecol, Kyoto. Japan Scientific Societies Press, Tokyo, pp 255–259

Frankland JC (1992) Mechanisms in fungal succession. In: Carroll GC, Wicklow DT (eds) The fungal community: its organization and role in the ecosystem, 2nd edn. Dekker, New York, pp 383–402

Frankland JC, Dighton J, Boddy L (1990) Methods for studying fungi in soil and forest litter. In: Grigorova R, Norris JR (eds) Methods in microbiology, vol 22. Academic Press, London, pp 343–404

Freckman D (ed) (1994) Soil biodiversity: its importance to ecosystem processes. Workshop report, Natural History Museum, London

Garrett SD (1952) The soil fungi as a microcosm for ecologists. Sci Prog (Lond) 40:436–450

Gochenaur SE (1984) Fungi of a Long Island oak-birch forest II. Population dynamics and hydrolase patterns for the soil penicillia. Mycologia 76:218–231

Gosz JR (1992) Gradient analysis of ecological change in time and space: implications for forest management. Ecol Appl 2:575–587

Grime JP (1977) Evidence for the existence of three primary strategies in plants and its relevance to ecological and evolutionary theory. Am Nat 111:1169–1194

Harley JL, Waid JS (1955) A method for studying active mycelia on living roots and other surfaces in the soil. Trans Br Mycol Soc 38:104–118

Hawksworth DL (1991) The fungal dimension of biodiversity: magnitude, significance, and conservation. Mycol Res 95:641–655

Hoekstra TW, Allen TFH, Flather CH (1991) Implicit scaling in ecological research: on when to make studies of mice and men. Bioscience 41:148–154

Hudson HJ (1968) The ecology of fungi on plant remains above the soil. New Phytol 67:837–874

Hunt GA, Trappe JM (1987) Seasonal hypogeous sporocarp production in western Oregon Douglas-fir stand. Can J Bot 65:438–445

Iqbal SH, Webster J (1973) Aquatic hyphomycete spora of the River Exe and its tributaries. Trans Br Mycol Soc 61:33–346

Kenkel N, Booth T (1992) Multivariate analysis in fungal ecology. In: Carroll GC, Wicklow DT (eds) The fungal community: its organization and role in the ecosystem, 2nd edn. Dekker, New York, pp 209–227

Kinkel LL (1991) Fungal community dynamics. In: Andrews JH, Hirano SS (eds) Microbial ecology of leaves. Springer, Berlin Heidelberg New York, pp 253–270

Kinkel LL, Andrews JH, Nordheim EV (1989a) Fungal immigration dynamics and community development on apple leaves. Micro Ecol 18:45–58

Kinkel LL, Andrew JH, Nordheim EV (1989b) Microbial introductions to apple leaves: influences of alterd immigration on fungal community dynamics. Micro Ecol 18:161–173

Kinkel LL, Nordheim EV, Andrews JH (1992) Microbial community analysis in incompletely or destructively sampled systems. Micro Ecol 24:227–242

Kjøller A, Struwe S (1980) Microfungi of decomposing red alder leaves and their substrate utilization. Soil Biol Biochem 12:425–431

Kuthubutheen AJ, Webster J (1986) Water availability and the coprophilous fungal succession. Trans Br Mycol Soc 86:63–76

Luoma DL, Frenkel RE, Trappe JM (1991) Fruiting of hypogeous fungi in Oregon Douglas-fir forests: seasonal and habitat variation. Mycologia 83:335–353

Lussenhop J (1992) Mechanisms of microarthropod-microbial mediated interactions in soils. In: Begon M, Fitter AH (eds) Advances in ecological research, vol 23. Academic Press, London, pp 1–133

Magurran AE (1988) Ecological diversity and its measurement. Princeton University Press, Princeton

Molina R, Palmer JG (1982) Isolation, mainteneance, and pure culture manipulation of ectomycorrhizal fungi. In: Schenck NC (ed) Methods and principles of mycorrhizal research. The American Phytopathological Society, St Paul, pp 115–130

Moore JC, de Ruiter PC (1991) Temporal and spatial heterogeneity of trophic interactions with in below-ground food webs. Agri Ecosys Environ 34:371–397

Nakasone KK (1990) Cultural studies and identifications of wood-inhabiting Corticiaceae and selected hymenomycetes from North America. Cramer, Berlin (Mycologia memoir no 15)

Newell K (1984) Interactions between two decomposer basidiomycetes and a collembolan under Sitka spruce: grazing and its potential effects on fungal distribution and litter decomposition. Soil Biol Biochem 16:235–240

Ohenoja E (1984) Fruit body production of larger fungi in Findland. I. Introduction to the study in 1976–1978. Ann Bot Fenn 21:349–355

Ohtonen R (1986) The effect of forest fertilization on the nitrogen content of the fruit-bodies of two mycorrhizal fungi, *Lactarius rufus* and *Suillus variegatus*. Ann Bot Fenn 23:189–203

Parkinson D (1981) Ecology of soil fungi. In: Cole GT, Kendrick B (eds) Biology of conidial fungi, vol 1. Academic Press, New York, pp 277–294

Parkinson D, Coleman DC (1991) Microbial communities, activity, and biomass. Agri Ecosys Environ 34:3–33

Preece TF, Dickinson CH (1971) Ecology of leaf surface microorganisms. Academic Press, London

Pugh GJF, Boddy L (1988) A view of disturbance and life strategies in fungi. Proc R Soc (Edinb) 4B:3–11

Rayner ADM (1995) Fungi, a vital component of ecosystem function in woodland. In: Allsopp D, Hawksworth DL, Colwell RR (eds) Microbial diversity and ecosystem function. CAB International, Wallingford, pp 231–251

Rayner ADM, Boddy L (1988) Fungal decomposition of wood: its biology and ecology. Wiley, New York

Rayner ADM, Todd NK (1979) Population and community structure and dynamics of fungi in decaying wood. Adv Bot Res 7:333–420

Robinson CH, Dighton J, Frankland JC, Roberts JD (1994) Fungal communities on decaying wheat straw of different resource quality. Soil Biol Biochem 26:1053–1058

Rosenzweig ML (1995) Species diversity in space and time. Cambridge University Press, Cambridge

Ruhling A, Bååth E, Nordgren A, Soderstrom B (1984) Fungi in metal contaminated soils. Ambio 13:34–36

Sagara N (1975) Ammonia fungi – a chemoecological grouping of terrestrial fungi. Contr Biol Lab Kyto Univ 24:205–276

Sagara N (1981) Occurrence of *Laccaria proxima* in the gravesite of a cat. Trans Mycol Soc Japan 22:271–275

Sagara N (1992) Experimental disturbances and epigeous fungi. In: Carroll GC, Wicklow DT (eds) The fungal community: its organization and role in the ecosystem, 2nd edn. Dekker, New York, pp 427–454

Shearer CA, Lane LC (1983) Comparison of three techniques for the study of aquatic hyphomycete communities. Mycologia 75:498–508

Shearer CA, Webster J (1985) Aquatic hyphomycete communities in the River Teign III. Comparison of sampling techniques. Trans Br Mycol Soc 84:509–518

Solbrig OT (1991) From genes to ecosystems: a research agenda for biodiversity. Report of a IUBS-SCOPE-UNESCO Worksh, The International Union of Biological Sciences, Paris

Sousa WP (1984) The role of disturbance in natural communities. Ann Rev Ecol Syst 15:353–391

Staplers JA (1978) Identification of wood-inhabiting Aphyllophorales in pure culture. Stud Mycol (Baarn) 16:1–48

Stevens RB (1974) Mycological guidebook. University Washington Press, Seattle

Suberkropp K (1984) Effect of temperature on seasonal occurrences of aquatic hyphomycetes. Trans Br Mycol Soc 82:53–62

Suberkropp K (1992) Aquatic hyphomycete communities. In: Carroll GC, Wicklow DT (eds) The fungal community: its organization and role in the ecosystem, 2nd edn. Dekker, New York, pp 729–747

Swift MJ (1976) Species diversity and the structure of microbial communities. In: Anderson JM, MacFadyen A (eds) The role of aquatic and terrestrial organisms in decomposition processes, Blackwell Scientific, London, pp 185–222

Swift MJ (1984) Microbial diversity and decomposer niches. In: Klug MJ, Reddy CA (eds) Current prespectives in microbial ecology. American Society for Microbiology, Washington DC, pp 8–16

Thompson W, Boddy L (1983) Decomposition of suppressed oak tress in even-aged plantations. II. Colonization of tree roots by cord- and rhizomorph-producing basidiomycetes. New Phytol 93:277–291

Villeneuve N, Grandtner MN, Fortin JA (1989) Frequency and diversity of ecotmycorrhizal and saprophytic macrofungi in the Laurentide Mountains of Quebec. Can J Bot 67:2616–2629

Vogt KA, Edmonds RL, Grier CC (1981) Biomass and nutrient concentrations of sporocarps produced by mycorrhizal and decomposer fungi in *Abies amabilis* stands. Oecologia (Berl) 50:170–175

Vogt KA, Bloomfield J, Ammirati JF, Ammirati SR (1992) Sporocarp production by basidiomycetes, with emphasis on forest ecosystems. In: Carroll GC, Wicklow DT (eds) The fungal community: its organization and role in the ecosystem, 2nd edn. Dekker, New York, pp 563–581

Warcup JH (1955) Isolation of fungi from hyphae present in soil. Nature 175:953–954

Webster J (1988) The coprophilous fungal succession: a model system. Proc R Soc (Edinb) 94B:45–46

Wicklow DT (1981) The coprophilous fungal community: a mycological system for examining ecological ideas. In: Wicklow DT, Carroll GC (eds) The fungal community: its organization and role in the ecosystem. Dekker, New York, pp 47–76

Wicklow DT (1992) The coprophilous fungal community: an experimental system. In: Carroll GC, Wicklow DT (eds) The fungal community: its organization and role in the ecosystem, 2nd edn. Dekker, New York, pp 715–728

Wicklow DT, Carroll GC (eds) (1981) The fungal community: its organization and role in the ecosysystem. Dekker, New York

Wildman HG (1992) Fungal colonization of resource islands: an experimental approach. In: Carroll GC, Wicklow DT (eds) The fungal community: its organization and role in the ecosystem, 2nd edn. Dekker, New York, pp 885–900

Wildman HG, Parkinson D (1979) Microfungal succession on living leaves of *Populas tremuloides*. Can J Bot 57:2800–2811

Williams ST, Parkinson D (1964) Studies of fungi in a podzal I. Nature and fluctuations of the fungus flora of the mineral horizons. J Soil Sci 15:331–341

Wilson DS (1980) Natural selection of populations and communities. Benjamin/Cummings, Menlow Park

Worrall JJ (1991) Media for selective isolation of hymenomycetes. Mycologia 83:296–302

Yocom DH, Wicklow DT (1980) Community differentiation along a dune succession: an experimental approach with coprophilous fungi. Ecology 61:868–880

Zak JC (1992) Response of soil fungal communities to disturbance. In: Carroll GC, Wicklow DT (eds) The fungal community: its organization and role in the ecosystem, 2nd edn. Dekker, New York, pp 403–426

Zak JC, Willig MR, Moorhead DL, Wildman HG (1994) Functional diversity of microbial communities: a quantative approach. Soil Biol Biochem 26:1101–1108

Zak JC, Sinsabaugh R, MacKay WP (1995) Windows of opportunity in desert ecosystems: their implications to fungal community development. Can J Bot 73(Suppl I): S1407–S1414

4 Disturbance in Natural Ecosystems: Scaling from Fungal Diversity to Ecosystem Functioning

C.F. Friese[1], S.J. Morris[2], and M.F. Allen[3]

CONTENTS

I. Introduction

Disturbance may be the single most important process regulating the structure and functioning of fungal communities because of the unique physiology, morphology and reproductive biology of fungi. Disturbance, as defined by Pickett and White (1985), includes elements of both scale and process. Specifically, "a *disturbance* is any relatively discrete event in time that disrupts ecosystem, community or population structure and changes resources, substrate availability, or the physical environment." The structure and functioning of fungal communities is influenced by disturbances through changes in vegetation, substrate, and a number of other components characteristic of the ecosystem in which the fungal community is located. Many of the descriptive studies of fungal communities come from the development of associations with particular vegetation types, which is dependent upon the successional sere, or with specific decomposing substrates. These, in turn, are composed of continuously and gradually changing substrates that have demonstrable trajectories. This led to relatively deterministic descriptions of fungal communities and their changing compositions (sensu Clements 1916, 1936; Christensen 1981, 1989).

Fungi are structurally unique organisms. This makes attempts to describe their dynamics difficult in the rather conventional terms used for higher plants and animals. Fungi, because of the size of individual hyphae or yeast cells, and because they are predominantly studied in the laboratory under a microscope, are viewed as microorganisms that primarily respond to minute quantities of substrates in their environment. In mass and spatial extent, many fungi are macroorganisms, often extending across large patches. These patches may be rather static in space, existing over long time periods, such as the mat-forming ectomycorrhizal fungi or the *Armillaria* mycelial networks (Allen et al. 1995a). However, because these large organisms are still made up of microscopic hyphae with the potential to function independently, detailing their life-history characteristics means understanding both the macroscopic and microscopic aspects of their existence.

As a result of this dichotomy in the structure of fungal "individuals," any disturbance, from an

[1] Microbial and Environmental Ecology Research Group, Department of Biology, University of Dayton, 300 College Park, Dayton, Ohio 45469-2320, USA
[2] Department of Plant Biology, The Ohio State University, 1735 Neil Ave., Columbus, Ohio 43210, USA
[3] Soil Ecology and Restoration Group, Department of Biology, San Diego State University, San Diego, California 92182-4614, USA

The Mycota IV
Environmental and Microbial Relationships
Wicklow/Söderström (Eds.)
© Springer-Verlag Berlin Heidelberg 1997

individual gopher mound to a volcano, has important ramifications to the structure and functioning of the fungal community. A volcano will destroy the entire community, requiring subsequent reinvasion and establishment subject to classical models of succession (Allen et al. 1992a). A gopher mound will disrupt the mycelial networks within a patch and within an "individual." This allows some reinvasion subject to the rebuilding of a mycelium from the hyphal fragments. Each of these disturbance extremes directly (e.g., mycorrhizal hyphae) and indirectly (nutrient immobilization) regulates the composition of the plant community and, in turn, the animal community. It is this range in the activity that we will attempt to describe. We will then attempt to develop a set of conceptual models that will allow us to begin to link our understanding of fungal biodiversity and ecosystem functioning.

II. Ecosystem Disturbance: a Conceptual Framework

A. Conceptual Model Overview

Understanding the factors influencing the diversity, distribution, and abundance of organisms in a community has long been a goal of ecologists and mycologists. Disturbance has been proposed to influence community diversity and structure through a variety of mechanisms. A single disturbance type, e.g., gopher mounds, can enhance (Allen et al. 1984b, 1992a) or retard (Koide and Mooney 1987) the rate of succession. Disturbances may also affect communities by creating new sites for colonization (Platt 1975; Collins 1987), or creating new substrates (Cooke and Rayner 1984; Gams 1992). Ultimately, the ability of a disturbance to modify the substrate on which a fungal community exists will determine the extent to which the community will change. This modification can be the substrate chemistry (e.g., differing litter types), or can be the result of disturbances which affect the environmental factors under which those substrates can exist, such as changes in soil moisture content with removal of the litter layer. If various disturbances differentially affect both substrate and the abiotic and biotic environment that influence species' colonization ability, then different disturbances caused by a variety of agents will alter fungal species diversity within a community in response to changes in spatial environmental heterogeneity.

We have developed a conceptual model of natural disturbance scale and patch recovery that focuses on patch and ecosystem level processes (Fig. 1). Our overview will include elements of these disturbance scales and the factors that affect the various states of the disturbance cycle. This conceptual model also indicates our goal of integrating processes operating at the level of the patch disturbance, such as soil enrichment, with those at the ecosystem level, such as spatial variation in natural disturbance density (e.g., fire, animal digging), soils, microbial and plant community structure (Fig. 1). We believe that our emphasis on the feedback between patch dynamics and local ecosystem processes is an important key to increasing our understanding of the role of natural disturbance in fungal community structure and ecosystem function.

B. Disturbance and Species Diversity

Disturbances can influence the composition and species richness of communities by a variety of mechanisms. Some disturbances affect the entire community simultaneously, such as volcanic eruptions (Andersen and MacMahon 1985; Allen 1988; Allen et al. 1984b) and catastrophic winds (e.g., Dunn et al. 1983), while other disturbances influence only a portion of the community at a time and thus result in a patchy environment, such as animal diggings (Allen et al. 1992a; Koide and Mooney 1987; Friese and Allen 1993). The scale and intensity of the disturbance can, in turn, significantly affect the response of organisms and the resulting successional patterns at a site (e.g., Bazzaz 1983; Mooney and Godron 1983; Sousa 1984; Pickett and White 1985; Pickett et al. 1989; McClendon and Redente 1990).

Large-scale disturbances affect fungal communities in different ways. Volcanoes, the most extreme of the large-scale disturbances, can completely destroy fungal communities, leaving topsoil buried under sterile tephra (Allen et al. 1984b; Hendrix and Smith 1986) or lava (Gemma and Koske 1990) or by the creation of new lands (Henrikssen and Henrikssen 1988). Fungal communities in the most severely damaged areas are destroyed completely. The recovery of these areas is driven by wind or small animal vectors capable of bringing new propagules from surviving areas,

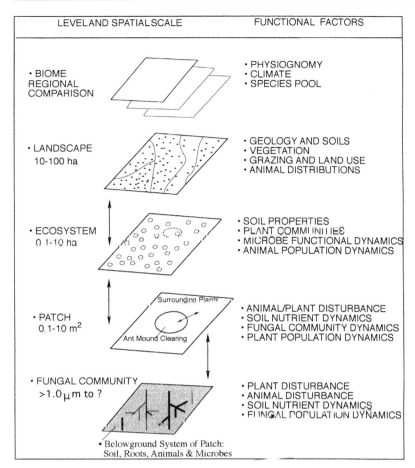

LEVEL AND SPATIAL SCALE	FUNCTIONAL FACTORS

- BIOME
  REGIONAL
  COMPARISON

- PHYSIOGNOMY
- CLIMATE
- SPECIES POOL

- LANDSCAPE
  10-100 ha

- GEOLOGY AND SOILS
- VEGETATION
- GRAZING AND LAND USE
- ANIMAL DISTRIBUTIONS

- ECOSYSTEM
  0 1-10 ha

- SOIL PROPERTIES
- PLANT COMMUNITIES
- MICROBE FUNCTIONAL DYNAMICS
- ANIMAL POPULATION DYNAMICS

- PATCH
  0.1-10 m²

Surrounding Plants

Ant Mound Clearing

- ANIMAL/PLANT DISTURBANCE
- SOIL NUTRIENT DYNAMICS
- FUNGAL COMMUNITY DYNAMICS
- PLANT POPULATION DYNAMICS

- FUNGAL COMMUNITY
  >1.0 μm to ?

- PLANT DISTURBANCE
- ANIMAL DISTURBANCE
- SOIL NUTRIENT DYNAMICS
- FUNGAL POPULATION DYNAMICS

- Belowground System of Patch:
  Soil, Roots, Animals & Microbes

Fig. 1. Conceptual framework illustrating how the ecological concepts of heirarchy and scale can be used to integrate and model how small-scale phenomenon, such as the impact of disturbance on microbial functional groups, can be related to larger-scale spatial patterns and processes (e.g., ecosystem dynamics). (After C.F. Friese and T.O. Crist, unpubl.)

patches, or from source areas at differing distances. Tephra often has a different pH, which leads to the creation of new communities. On the most devastated area of Mt. St. Helens following the blast of 1980, Ascomycotina were found to colonize the tephra within the first year followed by Basidiomycotina (Carpenter et al. 1987). After 10 years, AM and ectomycorrhizal fungi reestablished on the site mediated by gophers and wind dispersal (Allen et al. 1992a), although the ectomycorrhizal fungi were initially poorly developed.

The rate at which the reinvasion progresses on a volcano or most severely disturbed sites depends on the availability of sources of fungal inoculum. With the existence of nearby or internal source patches of inoculum, fungal communities can begin reinvading and establishing. The rates of reinvasion of mycorrhizal fungi onto the pumice plain of Mt. St. Helens to Krakatau following the 1883 eruption (Allen 1991) and from Hawaii (Gemma and Koske 1990) demonstrate that reinvasion of these fungi is dependent on location and type of inocula, and upon the reinvasion of vegetation. Mt. St. Helens, by having mycorrhizae in the most damaged areas 1 year after the eruption, recovered more quickly than Krakatau, where facultatively mycorrhizal species were reported 3 years after the eruption. This is because the most severely damaged areas of Mt. St. Helens were surrounded by vegetation rather than water as in the island of Krakatau. On the Hawaiian islands, there was a rapid invasion of mycorrhizal species from adjacent kipukas, or isolated patches of vegetation that remain untouched by disturbance (Gemma and Koske 1990). Volcanic eruptions such as the ones above occur relatively frequently and are considered predictable and therefore subject to selective evolutionary pressure. Presumably, dispersal strategies and life histories of the plants on these islands are adapted to these disturbances, and it is probable that the same is true for the fungi.

Fire presents another example of large-scale disturbance which has the ability to affect fungal communities in different ways. Following a severe fire, there is an initial drop in the number of propagules (Wright and Bollen 1961) and diversity of fungi present in an area (Wicklow 1973). Changes in soil pH and mineralization rates caused by fire regulate the fungi that can initially establish the area (Gochenaur 1981). Following the initial decrease in fungal propagules there occurs a rapid increase in fungal biomass, often to more than ten times the prefire number (Ahlgren and Ahlgren 1965; Wicklow 1973). These species, often referred to as pyrophilous fungi, are capable of taking advantage of the new resources. Mycorrhizal species may decrease in number or diversity following fire (Vilarino and Arines 1991) or remain unaffected within the plant root (Molina et al. 1992).

Fire can have a number of effects on mycorrhizal spores, depending on the maximum ground temperature reached while burning. Vilarino and Arines (1991) found that, following fire, the number and viability of spores decreased. They also determined that on at least one site the dominant species of mycorrhizal fungus changed from *Acaulospora laevis* to *Acaulospora scrobiculata*. Percent infection of root by VA mycorrhizae was found to increase over the year following the burn but did not reach the levels found before the fire. The depression in level of soil infectivity following fire in this site was detected for longer than in other similar research, such as that of Dhillion et al. (1988) on prairie soils. The authors suggested that the temperatures reached on these soils with a shrub and tree vegetation were greater than the temperatures reached with a herbaceous vegetation producing the longer-lasting affects. Therefore, the frequency and intensity of a fire, which is determined largely by the structure of the plant community (e.g., forest vs. grassland), can determine both the spatial and temporal patterns of fungal community development.

Small-scale disturbance can also affect fungal communities in a variety of ways. These distur-

bances are often poorly characterized because they create a mosaic of heterogenous patches within the landscape that are often functionally and structurally different from the landscape that surrounds them. Patches are defined ecologically as discrete spatial patterns with easily identifiable boundaries (Pickett and White 1985). Disturbances such as mound-building by animals create or alter the patchiness of a landscape (Allen 1988; Friese and Allen 1993; C.F. Friese et al., in prep.). These patch disturbances disrupt existing external soil mycelial networks such as those described by Finlay and Read (1986) for ectomycorrhizal fungi and Friese and Allen (1991) for arbuscular mycorrhizal fungi (e.g., absorptive hyphal networks and hyphal bridges). Disruptions of any type of soil hyphal network will create openings for the colonization and spread of new fungi, thus increasing fungal biodiversity, just as will occur for higher plants. Gophers and ants are examples of animals that are capable of overturning soil and moving mycorrhizal propagules within that soil to new patches in the soil matrix. Additionally, gophers trap spores within their fur and can transport these fungi to new areas. Ingestion and excretion of viable propagules at new locations by large mammals are also considered small-scale disturbances which have the capacity to change the community composition of a patch. The deposition of dung containing viable mycorrhizal spores from areas adjacent to the blast zone on the tephra at Mt. St. Helens by elk, days following the blast, allowed the return of fungal propagules to a biotically sterile area (Allen 1987).

Chronic small-scale disturbances can also be caused by large ungulates. Serengeti ecosystems are heavily grazed, resulting in increased nitrogen and dung applications to the soil (Seagle et al. 1992). This increases nitrogen levels and mineralizable carbon sources for the microorganisms. Studies of mycorrhizal distribution in this ecosystem demonstrated an inverse relationship between soil fertility and the presence of mycorrhizal fungi (McNaughton and Oesterheld 1990). This relationship is also associated with a smaller gradient of nutritional status associated with the vegetation. The mycorrhizae allow the plants to maintain a high nutritional status across a broad range of soil nutritional ranges, which ultimately results in better forage for the animals and a return of nutrients to the soil in the form of urine and dung.

III. Fungal Community Dynamics: a Natural Disturbance Model

The change in fungal species composition following disturbance is dependent on the disturbance event itself, the substrates that exist following the disturbance, the subsequent microclimate and its effects on substrate, and the surrounding organisms. The disturbance event directly alters the fungal community by destroying the hyphae and propagules of exposed species, but that is modified by the intensity of the disturbance and the seasonality of the event (Gochenaur 1981). Clearly, a hot fire during a drought will devastate many litter fungi, whereas a cool one initiated during a wet season may affect only smoke-sensitive species or those that exclusively exist on highly flammable materials. As the soil environment changes, such as decreased soil moisture in response to a lack of cover from direct radiation, the survival and growth of many hyphae is compromised (Boddy 1984). The growth changes of a single important individual ramify through the entire community by changing the competitive balance among species and altering the ecosystem dynamics due to altered enzymatic activities.

An additional dilemma in determining the effects of disturbance on communities is determining the extent of a fungal community (Cooke and Rayner 1984). Using the interpretation that communities are made of individuals with definable sets of species interactions (MacMahone et al. 1978), these effects can be studied. Large-scale disturbances can disrupt entire communities. Small-scale disturbances may affect only a patch within the community. For this reason, the model predicts the effects of disturbance within the patch, yet in some circumstances the patch may encompass an entire community. Following disturbance, the changes within the patch may affect the community of which it was a part, or may become a community of its own.

A. Disturbance Types and Characteristics

There are many types of disturbance that will affect a fungal community. In the fungal literature, disturbances are often broken into two main groups, enrichment and destructive disturbances (Cooke and Rayner 1984). For understanding fungal community dynamics, these classifications are too broad because disturbances at the scale that relates to the fungal community can never be entirely enriching or entirely destructive. For example, the effect of a forest fire on a fungal community is dependent on the type and size of fungal community being described and the state that community is in when the fire occurs. The fire is destructive to the phylloplane fungal community, whereas soil fungi 5 to 10 cm deep may find the fire an enrichment disturbance.

B. Biotic and Abiotic Characteristics of the Disturbance Model

The greatest effect of disturbance in the community will be on the key parameters that impact the growth of the fungal hyphae and germination of spores. These parameters, denoted as biotic and abiotic characteristics in Figs. 2 and 3, include physical characteristics, resources, soil flora and fauna, and fungal propagules or hyphal fragments (Boddy 1984; Gams 1992). The physical components of the fungal community affected by disturbance includes light, temperature, moisture, and pH (Gentry and Stiritz 1972; Rogers and Lavigne 1974; Mandel and Sorensen 1982). The resources include mostly mineral nutrients for growth and sporulation, for which some may be required in higher quantities than for vegetative growth (Moore-Landecker 1990), water, and oxygen. Another key resource considered here is the litter layer which can be a fungal substrate or it can modify the soil quality below it. Plants and animals also regulate the composition of fungi in soils but in a nonlinear manner. Plants provide primary production for a carbon source but also provide many inhibitory substances. Animals can remove (graze), disperse, or provide substrate (defecate or die) for different members of the fungal community. Additionally, changes in more than one of the parameters can act synergistically or antagonistically, to further change the emergent fungal community. With the loss of the litter layer by environmental factors such as fire, not only is there loss of substrate, but temperature increases and water-holding capacity of the soil declines. As this happens, mites and collembolans will migrate deeply into the soil to escape drought, reducing grazing but also reducing dispersal near the surface (Klironomos and Kendrick 1995). Thus,

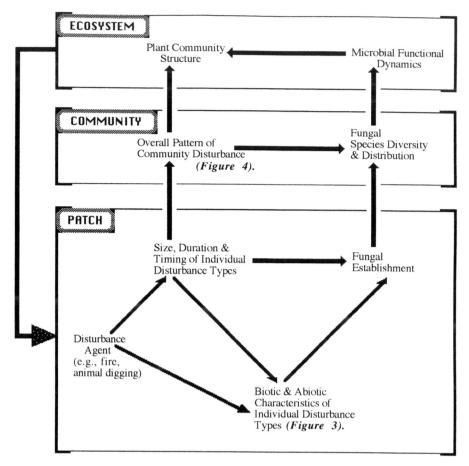

Fig. 2. Theoretical conceptual model depicting how patch-level disturbance processes can affect microbial functional dynamics, which, in turn, can shape the larger-scale patterns and processes of community- and ecosystem-level dynamics. (After C.F. Friese and S.S. Dhillion, in prep.)

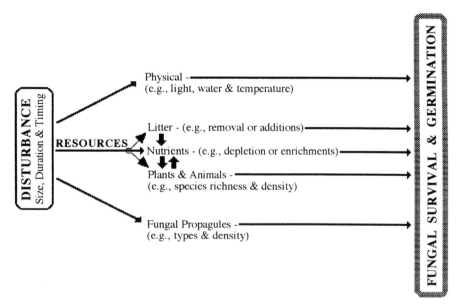

Fig. 3. Detailed model of the biotic and abiotic characteristics of individual disturbance types. This model is a subset of the complete conceptual model depicted in Fig. 2. The disturbance agent (e.g., harvester ants) determines such aspects as the size, duration, and timing of an individual disturbance event. All of these variables create a wide diversity of disturbance patches with unique biotic and abiotic characteristics. If each disturbance type creates a unique set of biotic and abiotic characteristics, then it is hypothesized that the fungal community will also be differentially affected within each of these disturbance types. (After C.F. Friese and S.S. Dhillion, in prep.)

changes in each of the above parameters will be dependent on the type and intensity of the disturbance and the interaction among the response variables.

C. Fungal Establishment Following Disturbance

The ability of the vegetative hyphae to survive the disturbance and germination of new individuals is not the only process that ultimately determines the community composition. The composition is also dependent on the competitive interactions among residuals and immigrants. A good deal of work has been undertaken studying the interactions of competitive (C), stress-tolerant (S), or ruderal (R) strategies in saprophytes following disturbance (Cooke and Rayner 1984). Immediately following a disturbance, the R-strategy is likely to predominate although, depending on the disturbance, many S-strategists may remain as residuals. Presumably the C-, and more S-strategies will predominate later. Fungi such as *Mucor* and *Rhizopus* are presumed to be ruderals because they exploit simple carbohydrates rapidly (R-strategy). Alternatively, *Phanerochyte* grows slowly but can degrade almost any type of substrate (S-strategy). *Cephalosporium* is an outstanding competitor because it expends a large amount of resources to produce antibiotics that restrict access to its own resource base. However, these separations are highly artificial and organisms exist along gradients of these extremes. For example, a common fungus of burned pine forests is *Morchella*. Is that because it tolerates the fires and harsh conditions following fires (S), competes well with other residuals and immigrants by growing hyphae fast and utilizing resources in the early spring before other saprophytes are getting started (C), produces a massive sporulating fungus that disperses spores by wind and animals (R), or (most probably) has some effective combination of all strategies?

Mycorrhizal fungi may be more problematic. In these communities, the frequency of disturbance may well be as important as the intensity. Puppi and Tartaflini (1991) evaluated the effects of fire on mediterranean communities. They found that although the communities were under similar environmental constraints, the vegetative and mycorrhizal community structure differed. These differences were attributed to the recurrence time of fire. In the more disturbed commu-

nity, arbuscular mycorrhizae were more common, whereas in the less disturbed community, EM were prevalent. EM were hypothesized to be more stress-tolerant than AM. Alternatively, in many grasslands, the plants forming AM were tolerant of fire, whereas those forming EM tended not to be. The fungi may simply be locked to the host strategy.

D. Community Level Effects

Change in the species composition within the patch may cause changes within the larger fungal community. The establishment of fungi within the patch can affect the diversity and distribution of fungi within the community or it can cause the establishment of a new community that exists only within the patch. The exact outcome will be determined by the characteristics of the disturbance, especially the size of the disturbance, but also by the heterogeneity of the landscape prior to disturbance. Figure 4 is the schematic representation of the interwoven matrix that is established by a series of disturbances of increasing scale. From an area of no disturbance, the smaller-scale disturbances are overlaid by the increasing intensity of the larger-scale disturbance. The impact of the smaller-scale disturbances may alleviate the effects of the larger-scale disturbances by acting as islands of inoculum or dispersal agents. This was the case for the arbuscular mycorrhizal (AM) fungi at Mt. St. Helens. Disturbances by gopher digging and elk droppings returned the AM fungal inoculum to the site that was completely decimated by the volcano blast (Allen et al. 1984b, 1992a). Initially, EM fungi were predominantly dispersed back onto the site by wind (Allen 1987). Thus, across the pyroclastic flow zone, small AM fungal patches reformed along animal pathways, whereas the EM plants initially established at random locations.

E. Ecosystem Processes: Feedback Loop Effects

Changes in the fungal community structure will be most important to the ecosystem if the functioning of the community changes along with the composition, or if the fungi affect the types of vegetation present (e.g., Reeves et al. 1979). Changes in the rates of decomposition and nutrient pool conversion within a community can affect the stability,

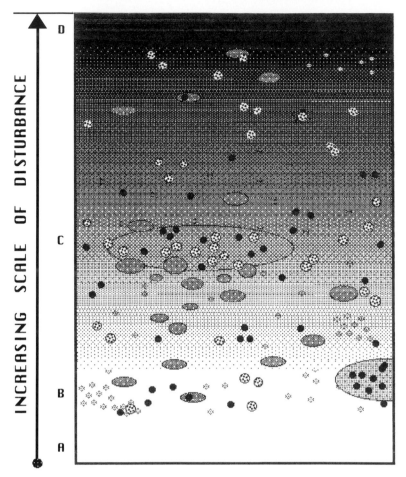

Fig. 4. Diagram illustrating the complexity of disturbance within the mosaic of the landscape, with the scale and intensity of disturbance increasing *from the bottom to the top* of the figure. This figure is a depiction of the overall pattern of community disturbance conceptualized in Fig. 2. The drawing demonstrates the complex pattern of disturbances that can be encountered in natural ecosystems; this disturbance mosaic can easily include the scenario of several disturbances occurring at a given location over time (spatial and temporal scale). Disturbances can range from *A* (none at all – hypothetical) to *B* (a mosaic of different animal disturbances) to *C* (disturbances over disturbances; e.g., fire combined with animal disturbances) to *D* (total disruption to the ecosystem; e.g., volcanic eruption or human impact such as strip-mining)

productivity and, ultimately, the functioning of ecosystems (Remacle 1981). Consequently, disturbances that alter these processes in soil will alter the corresponding ecosystem dynamics.

Ecosystem level feedback loops (Figs. 1 and 2) can influence patch structure through effects on disturbance types, characteristics, and the biotic and abiotic characteristics of the patch. Ecosystem dynamics will affect small-scale disturbance by influencing such things as animal types and densities, and large-scale disturbances by fuel loads and litter layer thickness. All scales of ecosystem disturbance ranging from landslides to animal burrowing to hyphal grazing by microarthropods can disrupt critical points in the hyphal network that exist in the soil. As illustrated for arbuscular mycorrhizal fungi in Fig. 5, there are numerous ways in which natural or anthropogenic disturbances can disrupt key functional areas of hyphal networks that grow and function within the soil matrix.

The end result of disturbance on a community will be a change in the community structure following a disturbance. It is predicted that a disturbed site tends to return to a community structure that does not entirely resemble the predisturbance state (Gochenaur 1981). The existence of a new assemblage of species, even of the same species of different age or density, may restrict the ability of the community to return to a predisturbance state. This was observed in the

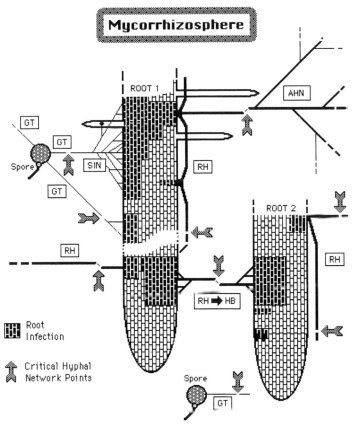

Mycorrhizosphere

Root Infection

Critical Hyphal Network Points

Fig. 5. Illustration of some critical hyphal network points (critical hyphal network points, CHNP, identified by *arrows* and *broken hyphae*) for arbuscular mycorrhizal fungi (modified from Friese and Allen 1991). CHNP each have their own key functional role (from infection to nutrient absorption and/or translocation), and each can be disrupted by any scale of ecosystem disturbance. The CHNP illustrated above are the runner hyphae (*RH*), hyphal bridges (*HB*), absorptive hyphal networks (*AHN*), germ tubes (*GT*), and spore infection networks (*SIN*) of AM fungi

simple experiment on cultured *Penicillium* and *Aspergillus* by Armstrong (1976), which demonstrated that, although *Aspergillus* would exclude *Penicillium* in plated cultures of the same age, in *Aspergillus* spores plated with *Penicillium* spores of a younger culture, both would be maintained. It is also likely that the changes induced by disturbance have differential effects on the species present, affecting the composition of the postdisturbance community.

IV. Harvester Ants: a Case Study in Natural Disturbance

A. Ant Disturbances

In arid and semiarid ecosystems of North America, seed-harvesting ants (*Pogonomyrmex*

spp., *Pheidole* spp., *Novomessor* spp., *Messor* spp.) can be quite abundant. The combined density of three *Pogonomyrmex* spp. in the Chihuahuan desert, for example, exceeds 180 colonies ha^{-1} (Whitford and Ettershank 1975). Throughout the semiarid rangelands of the West, disturbances created by nests of *Pogonomyrmex occidentalis* and the closely related northern species, *P. owyheei*, have been considered rangeland pests where they attained high densities (Costello 1944; Sharp and Barr 1960; Lavigne 1966; Kirkham and Fisser 1966; Wight and Nichols 1966). The western harvester ant, *P. occidentalis*, is particularly widespread in its distribution, including the shortgrass steppe, intermountain shrub-steppe, and Great Basin (Cole 1968; Rogers 1987). Several research efforts have documented rates of plant clipping (Clark and Comanor 1975), vegetation patterns near nest clearings (Wight and

Nichols 1966; Rogers and Lavigne 1974; Coffin and Lauenroth 1990; Nowak et al. 1990), soil nutrient enrichment within nest mounds (Rogers and Lavigne 1974; Mandel and Sorensen 1982), or rates of seed harvest by ants (Rogers and Lavigne 1974; Fewell 1990; Crist and MacMahon 1992). Relatively little is known, however, about how disturbance by *P. occidentalis* varies among soils, land use, and vegetation, and how these disturbances influence patch dynamics in semiarid landscapes (cf. Rogers 1987; Coffin and Lauenroth 1988; Friese and Allen 1993).

In the shortgrass steppe of Colorado, Coffin and Lauenroth (1988) used a simulation approach to estimate rates of turnover by ants based on average colony densities that were obtained by Rogers and Lavigne (1974). They found that disturbances by *P. occidentalis* were considerable ($>2\,m^2$/ha/year) and generally exceeded those of other animal disturbances (e.g., burrows) except fecal deposition by cattle. These turnover rates suggest that a dynamic view should be considered in small-scale disturbance processes (e.g., Sneva 1979), despite the fact that colonies are long-lived and persistent (Porter and Jorgensen 1988). Thus, whether the portion of the area occupied by harvester ants is >10% (Sharp and Barr 1960; Willard and Crowell 1965) or <5% (Kirkham and Fisser 1966; Rogers and Lavigne 1974), most studies measure only active ant colonies and do not consider the spatial extent of abandoned nest clearings. In the shrub-steppe of southwestern Wyoming, for example, there may be >30 active ant colonies ha^{-1} and another 25 abandoned sites ha^{-1} undergoing various stages of succession (Crist 1990).

B. Plant Establishment and Soil Enrichment (Abiotic)

The spatial distributions of disturbance by ants will affect the mosaic pattern of plant succession following colony abandonment or mortality. Disturbance opens colonization sites (Platt 1975; Collins 1987) and may alter the biotic and abiotic components of microsites for seedling establishment (McGinley et al. 1996). As with other disturbances, establishment on or near ant mounds is more favorable for some plants than for others (Culver and Beattie 1983; Woodell and King 1991; Friese and Allen 1993). Changes in the abiotic components of the disturbance patch include soil

physical properties (bulk density, texture), soil nutrient levels (N, P, K, Mn) and organic matter (Gentry and Stiritz 1972; Rogers and Lavigne 1974; Mandel and Sorensen 1982). Forms of N that plants can utilize, such as nitrate, are considerably higher in ant mounds (see preliminary findings; cf. Rogers and Lavigne 1974; McGinley et al. 1996). Ant colony clearings are higher in water availability (Rogers and Lavigne 1974; Laundré 1990) and undergo substantial daily fluctuations in soil surface temperature (Crist and MacMahon 1991).

C. Plant Establishment and Soil Enrichment (Biotic)

The biotic components of soil enrichment in ant disturbances are also important to subsequent plant establishment. These are primarily bacteria (e.g., Petal 1978; McGinley et al. 1996; C.F. Friese et al., in prep.) and fungi (Allen and MacMahon 1985; Friese and Allen 1993; C.F. Friese et al., in prep.), but may involve soil invertebrates such as termites (Crist and Friese 1994). Soils with higher levels of saprophytic bacteria and fungi may lead to increased nutrient mineralization rates (Haynes 1986; Paul and Clark 1989). Availability of the appropriate mycorrhizal fungi will differentially affect plant establishment and subsequent growth on the disturbance patch (Allen et al. 1989a, 1992b). Friese and Allen (1993) expanded upon the concept of nest enrichment to the field of microbial ecology when their studies revealed that the root mat region of these harvester ant mounds also exhibited significantly increased population densities of mutualistic VA mycorrhizal fungi. Friese and Allen (1993) proposed that harvester ant mounds also exhibit nest microbial enrichment which increases nutrient availability to establishing seedings following mound abandonment.

D. Patch Enrichment and Root Proliferation

Nutrient-rich microsites also enhance root proliferation (Jackson and Caldwell 1989) that can lead to the formation of dense root mats in ant mounds, adding to the organic matter accumulated by ants (Friese and Allen 1993). Additional organic matter from surrounding roots potentially increases the retention of water and nutrients, creating a positive feedback loop between the biotic and abiotic components of nest enrichment. Root growth into ant clearings would explain why surrounding plants often show increased biomass production over nonmound areas (Wight and Nichols 1966; Rogers and Lavigne 1974; Coffin and Lauenroth 1990).

E. Case Study Preliminary Findings

The objective of this particular case study was to further test the nest microbial enrichment hypothesis (Friese and Allen 1993) by investigating the saprophytic microbial dynamics of harvester ant mounds in two distinct ecosystems located in Colorado and Wyoming, USA (C.F. Friese et al., in prep.). The Central Plains Experimental Range and the Central Plains LTER, 40 km northeast of Fort Collins, was the study site in Colorado. The shortgrass prairie vegetation is dominated by perennial grasses including blue grama (*Bouteloua gracilis*), buffalo grass (*Buchloe dactyloides*), and three-awn (*Aristida longiseta*). Scattered shrubs include *Atriplex canescens*, *Artemisia frigida*, *Xanthocephellun sarothae*, and *Chrysothamnus nauscosus*. The study site in Wyoming is located in a semiarid shrub-steppe near Elkol, Wyoming. The dominant shrubs on the site are big sagebrush (*Artemisia tridentata* Nutt.) and rabbitbrush [*Chrysothamnus viscidiflorus* (Hook.) Nutt.]. Grasses common to the site include Indian rice-grass [*Oryzopsis hymenoides* (R. & S.) Ricker], cheatgrass (*Bromus tectorum* L.), and foxtail barley (*Hordeum jubatum* L.). Sets of samples were collected from harvester ant mounds at both the Colorado and Wyoming study sites in September 1991. At each site, samples were collected from mound-associated root mat, separated into soil and root components, seed cache, soil from between shrubs, and soil from under a shrub closest to the mound. Mounds had a disk diameter of 3.95–4.3 m.

Preliminary nutrient analyses of soil from ant mound and nonmound areas at the Wyoming site (Table 1) show some similarities from previous work at the shortgrass-steppe Colorado site (Rogers and Lavigne 1974). For example, soil P had comparable levels of enrichment in nests at the two sites. However, there appear to be much higher levels of NO_3-N in nests at the Wyoming site than in Colorado (4.0 mg/kg; Rogers and Lavigne 1974). Soil pH in nests is reduced considerably compared to that reported for a vertical profile by Mandel and Sorensen (1982) at a different Colorado site.

Biotic enrichment of microorganisms in ant nests is pronounced at the two sites (C.F. Friese et al., in prep.). Higher total numbers of bacteria and fungi were found in soil from ant nests than in soil from blue grama grass (*Bouteloua gracilis*) at the Colorado site or under shrubs (*Artemisia tridentata*) at the Wyoming site (Table 2). Pseudomonad counts were similar in ant nests and nonmound soil. Bacterial functional groups (decomposers of chitin, cellulose, and protein) were all greater within nests than in surrounding soil from both sites (Table 2). The assemblages and dominant species of fungi also differed in the two types of samples, with more fungi (characteristic of mesic sites) occurring in ant nests (C.F. Friese et al., in prep.). These results suggest that the biotic components of enrichment are potentially significant soil characteristics of ant colony disturbances.

Preliminary organic matter and nutrient analyses of nest soil from this site by T.O. Crist (1990, unpubl. data) found organic matter levels near the 10% level (versus <7% in undershrub

Table 1. Preliminary soil analysis from harvester ants nests at the Elkol, Wyoming, site. Soil samples (n = 3) were taken from the center of the nest at ground surface, and from random locations under shrubs of *Artemisia tridentata* and bare ground 3 m away from ant colonies. (C.F. Friese, S.S. Dhillion, and T.O. Crist, unpubl. data)

Soil characteristic	Mound	Under shrubs	Bare ground
N, %	0.177 ± 0.017	0.193 ± 0.014	0.107 ± 0.030
P, %	0.120 ± 0.012	0.060 ± 0.012	0.050 ± 0.060
NO_3-N (mg kg^{-1})	211 ± 123	3.1 ± 2.4	4.0 ± 1.2
Organic matter, %	3.85 ± 0.22	6.98 ± 0.14	3.18 ± 0.17
pH	5.2 ± 0.1	6.3 ± 0.1	6.7 ± 0.1

Table 2. Comparisons of colony-forming units of functional groups of microorganisms in soil from ant nests and surrounding areas at two sites. The Colorado site is shortgrass-steppe dominated by blue grama grass, and the Wyoming site is shrub-steppe dominated by big sagebrush. Off-mound samples were taken from under the dominant plant species at each site. All comparisons within a site are significantly different ($P < 0.05$) except for pseudomonads. (C.F. Friese, S.S. Dhillion, and T.O. Crist, unpubl. data)

Microorganisms	Colorado site		Wyoming site	
	Mound	Off-mound	Mound	Off-mound
Total bacteria	20×10^7	13×10^7	24×10^7	12×10^7
Total fungi	18×10^3	10×10^3	18×10^3	13×10^3
Functional groups				
Pseudomonads	7×10^2	7×10^2	3×10^2	4×10^2
Chitin decomposers	27×10^5	21×10^5	25×10^5	16×10^5
Cellulose decomposers	11×10^2	3×10^2	11×10^2	3×10^2
Protein decomposers	16×10^6	9×10^6	14×10^6	7×10^6

soils), and nitrate levels as high as 100 times that found within undershrub and interspace areas. Fungal species richness was higher from mound-associated material than from either soil adjacent to mounds, or soil collected under shrubs at the Wyoming site, but not from soil adjacent to the mounds at the Central Plains Experimental Range location (C.F. Friese et al., in prep.). For both locations, each microenvironment selected a distinct assemblage of dominant fungi with *Fusarium* spp. dominating the root material, and *Aspergillus* and *Penicillium* species predominating in seed cache soil (C.F. Friese et al., in prep.). However, *Aspergillus fumigatus* had high densities in off-mound soil from the Colorado site. Mucoraceous taxa, (i.e., *Cunninghamella*, *Rhizopus*, and *Syncephalastrum*) were isolated primarily from mound material, suggesting that the ant mounds may represent refugia for these more mesic adapted fungi. All microbial numbers were significantly higher in material obtained from the ant mounds than in adjacent soil, except for pseudomonad colony-forming units, for which there were no significant differences (Tables 1 and 2). Bacterial and fungal colony-forming units were generally highest in the root-mat-(root) material than in root-mat-(soil) or seed-cache-(soil) (Table 2).

These data further support and expand upon the hypothesis of nest microbial enrichment proposed by Friese and Allen (1993). It was found that mounds of the western harvester ant can exhibit nest microbial enrichment for both saprophytic bacteria and fungi, as well as mutualistic mycorrhizal fungi. The greater species richness of fungi from off-mound soil at the Central Plains location may be in response to the plant type (grass versus shrub) and the importance of a rhizosphere effect at this location. The greater bacterial numbers and functional groups in mound material than from adjacent soils could account for the higher nutrient levels through increased mineralization rates. Certainly carbon, necessary for nitrogen mineralization, would not be limiting in the mound habitats. In summary, the mounds of the western harvester ant, *Pogonomyrmex occidentalis*, contain a varied and diverse assemblage of microfauna and microflora whose interactions have the potential to significantly affect such factors as nutrient availability for seedlings establishing in disturbed soils.

It is clear that animal disturbances such as the mounds of the western harvester ant can create a mosaic of fungal community patches. Across a landscape, seed-harvesting ants can be quite abundant. The combined density of three *Pogonomyrmex* spp. in the Chihuahuan desert, for example, exceeds 180 colonies ha^{-1} (Whitford and Ettershank 1975). However, most studies measure only active ant colonies and do not consider the spatial or temporal extent of abandoned nest clearings. In the semiarid shrub-steppe of southwestern Wyoming there may be greater than 30 active ant colonies ha^{-1} and another 25 abandoned sites ha^{-1} undergoing various stages of succession (Crist 1990). While little is known regarding the turnover rates of harvester ant colonies, the average life span of a harvester ant colony has been estimated to range anywhere from 10–20 years (T.O. Crist, pers. comm.). The disks (areas cleared of vegetation by the ants) at the Wyoming site have an average range in diameter from 2–4 m, which allows us to estimate a total soil community turnover rate of somewhere between 100 and 1000 years. Projections of this nature allow us to speculate on the long-term implications of disturbance to the nutrient dynamics, fungal community structure, and soil heterogeneity of various ecosystems. Such soil turnover rates suggest that a dynamic, long-term view should be considered in small-scale disturbance processes. Fungal community dynamics are not static processes that can simply be studied in a single time frame. Such experiments provide us with only a snapshot of what was taking place at the time of the study, when in reality the environment is constantly changing over time. Disturbance is one of the dynamic processes that drives environmental change. Understanding how disturbances such as harvester ant activity affect the fungal community over longer time periods will provide us with a more dynamic and realistic view of the role of fungi in ecosystem structure and function.

V. Conclusions

A. The Impact of Disturbance on the Functional Role of Fungi

This chapter was designed to demonstrate that disturbance may be the single most important process regulating the structure and functioning of fungal communities. This is due to the unique physiology, morphology, and reproductive biology of fungi. While microbial ecologists are beginning to describe the importance of individual disturbance events, we know far less about the interactions of small- and large-scale perturbations set in the larger landscape of single or multiple plant communities. This distinction becomes of even greater importance in the light of the global dimensions of anthropogenic influences on ecosystems such as N fertilization, exotic species migrations, habitat fragmentation, and global climate change. We suggest that understanding the roles of disturbance in fungal communities, and the feedbacks from the fungal communities to

ecosystem functioning are crucial to understanding the results from these larger global concerns.

Developing models for linking the range of scales that comprise disturbance dynamics depends on linking two distinct types of studies. First, we must begin to build an array of case studies from particular ecosystems in which we know the natural history of the fungi and how these natural histories contribute to the existing community composition and functioning. Second, we must develop a conceptual framework that integrates all of these various case studies into a comprehensive view of how communities work and what factors regulate them. Finally, we must continuously reevaluate those conceptual models to develop a quantitative model of the complex roles of fungi within landscapes that are undergoing anthropogenic and natural change.

We have tried to present examples of the first two aspects of this process. We set the stage by defining the wide array of scales and interacting factors whereby disturbance affects fungal communities. We then defined a single case study, the harvester ant mounds in western North America. These ants disturb distinct patches of soil to build their nests. In that process, they alter the substrate composition, thereby enriching a soil patch both in terms of the chemical/physical structure and by altering the fungal composition and activity. After these mounds are abandoned, because of the altered fungal activity, they become enriched sites for establishing and renewing the plant communities. These mounds are continuously created and abandoned, leading to an estimated turnover rate of 100 to 1000 years. We believe that this renewal process is critical to the productive maintainence of these ecosystems at the larger landscape scale.

B. Future Research Directions

As a result of the widespread interest in anthropogenic changes to the earth's atmosphere and the effects that these changes may have on the biosphere, it is important to study and "tease out" a better understanding of how small-scale microbial processes (such as nutrient mobilization and immobilization) fit into the larger global picture. The diversity and biomass of microbial communities is a direct indicator of the extent of the functional role that these organisms play in the dynamics of different ecosystems. If anthropogenic change alters the structure and biodiversity of microbial communities, then it is also likely that their critcal functional roles in ecosystem and global level nutrient cycling are also impacted.

As is the case with other groups of organisms, it is just as important to understand how the functional role of fungi is affected by various forms of human impact on the environment (e.g., Allen et al. 1993, 1995b; Meyer 1994; Read 1994). Future research needs to focus on linking the issues of fungal biodiversity and functionality with both natural disturbance and anthropogenic change. This research direction is critical for us to completely understand and explain the importance of fungi in ecosystem dynamics. Attempts to explore and integrate all of the above factors are crucial if we are ever to gain a comprehensive understanding of the functional role of fungi in diverse ecosystems and the biosphere as a whole.

Acknowledgments. The authors wish to thank Shivcharn Dhillion and Tom Crist for technical assistance in the areas of field/laboratory work and input on the development of the conceptual models. Partial grant funding for this research was provided to C. Friese through a Strengthening Award (Grant #95-37101-1901) from the United States Department of Agriculture (USDA) Rangeland Ecosystems Program and a seed grant from the University of Dayton Research Council.

References

Agnew WD, Uresk W, Hansen RM (1986) Flora and fauna associated with prairie dog colonies and adjacent ungrazed mixed-grass prairie in western South Dakota. J Range Manag 39:135–139

Ahlgren IF, Ahlgren CE (1965) Effects of prescribed burning on soil microorganisms in a Minnesota jack pine forest. Ecology 46:304–310

Allen MF (1987) Re-establishment of mycorrhizas on Mount St. Helens: migration vectors. Trans Br Mycol Soc 88:413–417

Allen MF (1988) Re-establishment of VA mycorrhizae following severe disturbance: comparative patch dynamics of a shrub desert and a subalpine volcano. Proc R Soc Edinb 94:63–71

Allen MF (1991) The ecology of Mycorrhizae. Cambridge University Press, Cambridge

Allen MF, MacMahon JA (1985) Impact of disturbance on cold desert fungi: comparative microscale dispersion patterns. Pedobiologia 28:215–224

Allen MF, Allen EB, Stahl PD (1984a) Differential niche response of *Bouteloua gracilis* and *Pascopyrum smithii* to VA mycorrhizae. Bull Torrey Bot Club 111:361–325

Allen MF, MacMahon JA, Andersen DC (1984b) Reestablishment of Endogonaceae on Mount St. Helens: survival of residuals. Mycologia 76:1031–1038

Allen MF, Allen EB, Friese CF (1989a) Responses of the non-mycotrophic plant *Salsola kali* to invasion by vesicular-arbuscular mycorrhizal fungi. New Phytol 111:45–69

Allen MF, Hipps LE, Wooldridge GL (1989b) Wind dispersal and subsequent establishment of VA mycorrhizal fungi across a successional arid landscape. Landscape Ecol 2:165–171

Allen MF, Crisafulli C, Friese CF, Jeakins SJ (1992a) Reformation of mycorrhizal symbioses on Mount St Helens, 1980–1990: interactions of rodents and mycorrhizal fungi. Mycol Res 96:447–453

Allen MF, Klouse SD, Weinbaum BS, Jeakins SJ, Friese CF, Allen EB (1992b) Mycorrhizae and the integration of scales: from molecules to ecosystems. In: Allen MF (ed) Mycorrhizal functioning. Chapman and Hall, New York, pp 488–515

Allen MF, Allen EB, Dahm CN, Edwards FS (1993) Preservation of biological diversity in mycorrhizal fungi: importance and human impacts. In: Sundnes G (ed) International Symposium on human impacts on self-recruiting populations. The Royal Norwegian Academy of Sciences, Trondheim, Norway, pp 81–108

Allen MF, Allen EB, Helm DJ, Trappe JM, Molina R, Rincon E (1995a) Patterns and regulation of arbuscular and ectomycorrhizal plant and fungal diversity. Plant Soil 170:47–62

Allen, MF, Morris SJ, Edwards F, Allen EB (1995b) Microbe-plant interactions in mediterranean-type habitats: shifts in fungal symbiotic and saprophytic functioning in response to global change. In: Moreno JM, Oechel WC (eds) Global change and mediterranean-type ecosystems. Ecological Studies 117: Springer, Berlin Heidelberg New York, pp 287–305

Andersen AN (1991) Parallels between ants and plants: implications for community ecology. In: Huxley CA, Cutler DA (eds) Ant-plant interactions. Oxford Science, Oxford, pp 539–558

Andersen DC (1987) Below-ground herbiovry in natural communities: a review emphasizing fossorial animals. Q Rev Biol 62:261–286

Andersen DC, MacMahon JA (1985) Plant succession following the Mount St. Helens volcanic eruption: facilitation by a burrowing rodent, *Thomomys talpoides*. Am Mid Nat 114:52–69

Andrews JH (1992) Fungal life-history strategies. In: Carrol GC, Wicklow DT (eds) The fungal community: its organization and role in the ecosystem. Dekker, New York, pp 119–146

Armstrong RA (1976) Fugitive species: experiments with fungi and some theoretical considerations. Ecology 57:953–963

Bazzaz FA (1983) Characteristics of populations in relation to disturbance in natural and man-modified environments. In: Mooney HA, Godron M (eds) Disturbance and ecosystems. Ecological Studies 44. Springer, Berlin Heidelberg New York, pp 259–275

Beattie AJ (1985) The evolutionary ecology of ant-plant mutualisms. Cambridge University Press, Cambridge

Beattie AJ, Culver DC (1983) The nest chemistry of two seed-dispersing ant species. Oecologia 56:99–103

Boddy L (1984) The micro-environment of basidiomycete mycelia in temperate deciduous woodlands. In: Jennings DH, Rayner ADM (eds) The ecology and physiology of the fungal mycelium. Cambridge University Press, Cambridge, pp 261–269

Buckley RC (1982) Ant-plant interactions: a world review. In: Buckley RC (ed) Ant-plant interactions in Australia. Junk, The Hague, pp 111–162

Carpenter SE, Trappe JM, Ammirati J (1987) Observations of fungal succession in the Mount St. Helens devastation zone, 1980–1983. Can J Bot 65:716–722

Chambers JC, MacMahon JA, Brown RW (1990) Alpine seedling establishment: the influence of disturbance type. Ecology 71:1323–1341

Chambers JC, MacMahon JA, Haefner JH (1991) Seed entrapment in alpine ecosystems: effects of soil particle size and diaspore morphology. Ecology 72:1668–1677

Chapin FS (1983) Patterns of nutrient absorption and use by plants from natural and man-modified environments. In: Mooney HA, Godron M (eds) Disturbance and ecosystems. Ecological Studies 44. Springer, Berlin Heidelberg New York, pp 175–187

Christensen M (1981) Species diversity and dominance in fungal communities. In: Wicklow DT, Carroll GC (eds) The fungal community. Dekker, New York, pp 201–232

Christensen M (1989) A view of fungal ecology. Mycologia 81:1–19

Clark WH, Comanor PL (1975) Removal of annual plants from the desert ecosystem by western harvester ants, Pogonomyrmex occidentalis. Env Ent 4:52–56

Clements FE (1916) Plant succession. Carnegie Institution of Washington Press, Washington

Clements FE (1936) Nature and structure of the climax. J. Ecol 24:252–284

Coffin DP, Lauenroth WK (1988) The effects of disturbance size and frequency on a shortgrass plant community. Ecology 69:1609–1617

Coffin DP, Lauenroth WK (1990) Vegetation associated with nest sites of western harvester ants (Pogonomyrmex occidentalis Cresson) in a semiarid grassland. Am Mid Nat 123:226–235

Cole AC (1968) Pogonomyrmex harvester ants: a study of the genus in North America. University of Tennessee Press, Knoxville

Collins SL (1987) Interactions of disturbances in tallgrass prairie: a field experiment. Ecology 68:1243–1250

Connell JH (1978) Diversity in tropical rain forests and coral reefs. Science 199:1302–1310

Connell JH (1980) Diversity and the coevolution of competitors, or the ghost of competition past. Oikos 35:131–138

Cooke RC, Rayner ADM (1984) Ecology of saprotrophic fungi. Longman, New York

Costello DF (1944) Natural revegetation of abandoned plowed land in the mixed prairie association of northwestern Colorado. Ecology 25:312–326

Crist TO (1990) Granivory in a shrub-steppe ecosystem: interactions of harvester ant foraging and native seeds. PhD dissertation, Utah State University, Logan

Crist TO, Friese CF (1993) The impact of fungi on soil seeds: implications for plants and granivores in a semi-arid shrub-steppe. Ecology 74(8):2231–2239

Crist TO, Friese CF (1994) The use of ant nests by subterranean termites in two semi-arid ecosystems. Am Mid Nat 131:370–373

Crist TO, MacMahon JA (1991) Foraging patterns of Pogonomyrmex occidentalis (Hymenoptera: Formicidae) in a shrub-steppe ecosystem: the roles of temperature, trunk trails, and seed resources. Env Ent 20:265–275

Crist TO, MacMahon JA (1992) Harvester ant foraging and shrub-steppe seeds: interactions of seed resources and seed use. Ecology 73:1768–1779

Crist TO, Wiens JA (1996) Scale effects of vegetation on forager movement and seed harvesting by ants. Oikos (in press)

Culver DC, Beattie AJ (1983) Effects of ant mounds on soil chemistry and vegetation patterns in a Colorado montaine meadow. Ecology 64:485–492

Dayton PK (1971) Competition, disturbance and comunity organization: the provision and subsequent utilization of space in a rocky intertidal community. Ecol Monogr 41:351–389

Denslow JS (1985) Disturbance-mediated coexistence of species. In: Pickett STA White PS (eds) The ecology of natural disturbance and patch dynamics. Academic Press, New York, pp 307–324

Dhillion SS, Anderson RC (1993) Growth dynamics and associated mycorrhizal fungi of little bluestem grass (Schizachyrium scoparium) on burned and unburned sand prairies. New Phytol 123:77–91

Dhillion SS, Anderson RC, Liberta AE (1988) Effect of fire on the mycorrhizal ecology of little bluestem (Schizachyrium scoparium). Can J Bot 66:706–713

Dunn CP, Guntenspergen GR, Dorney JR (1983) Catastrophic wind disturbance in and old-growth hemlock-hardwood forest, Wisconsin. Can J Bot 61:211–217

Fewell JH (1990) Directional fidelity as a foraging constraint in the western harvester ant, Pogonomyrmex occidentalis. Oecologia 82:45–51

Finlay RD, Read DJ (1986) The structure and function of the vegetative mycelium of ectomycorrhizal plants. I. Translocation of ^{14}C-labelled carbon between plants interconnected by a common mycelium. New Phytol 103:143–156

Friese CF (1991) The interaction of harvester ants and vesicular-arbuscular mycorrhizal fungi in a patchy environment: the effects of mound structure on fungal dispersion and establishment. PhD dissertation. Utah State University, Logan

Friese CF, Allen MF (1991) The spread of VA mycorrhizal fungal hyphae in the soil: inoculum types and external hyphal architecture. Mycologia 83(4):409–418

Friese CF, Allen MF (1993) The interaction of harvester ants and vesicular-arbuscular mycorrhizal fungi in a patchy semi-arid environment: the effects of mound structure on fungal dispersion and establishment. Funct Ecol 7:13–20

Friese CF, Koske RE (1991) The spatial dispersion of spores of vesicular-arbuscular mycorrhizal fungi in a sand dune: microscale patterns associated with the root architecture of American beachgrass plants. Mycol Res 95(8):952–957

Friese CF, Dhillion SS, Crist TO, Zhang Q, Zak JC (in prep) Microbial nest enrichment within the mounds of the western harvester ant (Pogonomyrmex occidentalis): comparative patch dynamics between two sites in Colorado and Wyoming

Gams W (1992) The analysis of communities of saprophytic microfungi with special reference to soil fungi. In: Winterhoff W (ed) Fungi in vegetation science. Kluwer Academics, Amsterdam, pp 183–223

Gemma JN, Koske RE (1990) Mycorrhizae in recent volcanic substrates in Hawaii. Am J Bot 77:1193–1200

Gentry JB, Stiritz KL (1972) The role of the Florida harvester ant, Pogonomyrmex badius, in old field nutrient relationships. Env Ent 1:39–41

Gochenaur SE (1981) Response of soil fungal communities to disturbance. In: Wicklow DT, Carroll GC (eds) The fungal community: its organization and role in the ecosystem. Dekker, New York, pp 459–479

Grubb PJ (1977) The maintenance of species richness in plant communities: the importance of the regeneration niche. Biol Rev 52:107–145

Haynes RJ (1986) The decomposition process: mineralization, immobilization, humus formation, and degradation. In: Haynes RJ (ed) Mineral nitrogen in the plant-soil system. Academic Press, Orlando, pp 52–126

Hendrix LB, Smith SD (1986) Post-eruption revegetation of Isla Fernandina, Galapagos: II. Natl Geogr Res 2:6–16

Henriksson E, Henriksson LE (1988) Fungi in Surtsey soils. Proc R Soc Edinb 94B:61

Hobbs RJ, Hobbs VJ (1987) Gophers and grasslands: a model of vegetation response to patchy soil disturbance. Vegetatio 69:141–146

Hobbs RJ, Mooney HA (1985) Community and population dynamics of serpentine grassland annuals in relation to gopher disturbance. Oecologia 67:342–351

Huntly N, Inouye R (1987) Pocket gophers in ecosystems: patterns and mechanisms. Bioscience 38:786–793

Jackson RB, Caldwell MM (1989) The timing and degree of root proliferation in fertile-soil microsites for three cold-desert perennials. Oecologia 81:149–153

Janos DP (1980) Mycorrhizae influence tropical succession. Biotropica 12:56–64

Janzen DH (1970) Herbivores and the number of tree species in tropical forests. Am Nat 104:501–528

Kirkham DR, Fisser HG (1966) Rangeland relations and harvester ants in north-central Wyoming. J Range Manage 25:55–60

Klironomos JN, Kendrick B (1995) Relationships among microarthropods, fungi, and their environment. Plant Soil 170:183–197

Koide RT, Mooney HA (1987) Spatial variation in inoculum potential of vesicular-arbuscular mycorrhizal fungi caused by formation of gopher mounds. New Phytol 107:173–182

Laundré JW (1990) Soil moisture patterns below mounds of harvester ants. J Range Manage 43:10–12

Lavigne RJ (1966) Individual mound treatments for control of the western harvester ant, *Pogonomyrmex occidentalis*, in Wyoming. J Econ Entomol 59:525–539

MacKay WP (1991) The role of ants and termites in desert communities. In: Polis GA (ed) The ecology of desert communities. University of Arizona Press, Tucson, pp 113–150

MacMahon JA, Phillips DL, Robinson JV, Schimpf DJ (1978) Levels of biological organization: an organism-centered approach. Bioscience 28:700–704

Mandel RD, Sorenson CJ (1982) The role of the western harvester ant (*Pogonomyrmex occidentalis*) in soil formation. Soil Sci Soc Am J 46:785–788

McGinley MA, Dhillion SS, Neumann J (1996) Environmental heterogeneity and seedling establishment: ant-plant-microbe interactions. Functional Ecology (in press)

McLendon T, Redente EF (1990) Succession patterns following soil disturbance in a sagebrush steppe community. Oecologia 85:293–300

McNaughton SJ, Oesterheld M (1990) Extramatrical mycorrhizal abundance and grass nutrition in a tropical grazing ecosystem, the Serengeti National Park, Tanzania. Oikos 59:92–96

Meyer O (1994) Functional groups of microorganisms. In: Schulze ED, Mooney HA (eds) Biodiversity and ecosystem function. Springer, Berlin Heidelberg New York, pp 67–96

Molina R, Massicotte H, Trappe J (1992) Specificity phenomena in mycorrhizal symbioses: community-ecological consequences and practical implications. In: Allen MF (ed) Mycorrhizal functioning. Chapman and Hall, New York, pp 357–423

Mooney HA, Godron M (1983) Disturbance and ecosystems: components of response. In: Mooney HA, Godron M (eds) Disturbance and ecosystems. Springer, Berlin Heidelberg New York (Ecological studies, vol 44)

Moore-Landecker E (1990) Fundamentals of the fungi. Prentice Hall, New Jersey

Naiman RJ (1988) Animal influences on ecosystem dynamics. Bioscience 38:750–752

Nowak RS, Nowak CL, DeRocher T, Cole N, Jones MA (1990) Prevalence of *Oryzopsis hymenoides* near harvester ant mounds: indirect facilitation by ants. Oikos 58:190–198

Orians GH (1982) The influence of tree falls in tropical forests on tree species richness. Trop Ecol 23:255–279

Paine RT (1966) Food web complexity and species diversity. Am Nat 100:65–75

Paul EA, Clark FE (1989) Soil microbiology and biochemistry. Academic Press, San Diego

Petal J (1978) The role of ants in ecosystems. In: Brian MV (ed) Production ecology of ants and termites. Cambridge University Press, Cambridge, pp 293–325

Petraitas PS, Latham RE, Niesenbaum RA (1989) The maintenance of species diversity by disturbance. Q Rev Biol 64:393–418

Pickett STA, White PS (1985) Patch Dynamics: a synthesis. In: Pickett STA, White PS (eds) The ecology of natural disturbance and patch dynamics. Academic Press, New York, pp 371–384

Pickett STA, Kolasa J, Armesto JJ, Collins SL (1989) The ecological concept of disturbance and its expression at various hierarchical levels. Oikos 54:129–136

Pisarski B (1978) Comparison of various biomes. In: Brian MV (ed) Production ecology of ants and termites. Cambridge University Press, Cambridge, pp 326–332

Platt WJ (1975) The colonization and formation of equilibrium plant species associations on badger disturbances in a tall-grass prairie. Ecol Monogr 45:285–305

Porter SD, Jorgensen CD (1988) Longevity of harvester ant colonies in southern Idaho. J Range Manage 41:104–107

Puppi G, Tartaglini N (1991) Mycorrhizal types in three mediterranean communities affected by fire to different extents. Acta Oecol 12:295–304

Read DJ (1994) Plant-microbe mutualisms and community structure. In: Schulze ED, Mooney HA (eds) Biodiversity and ecosystem function. Springer, Berlin Heidelberg New York, pp 181–209

Reeves FB, Wagner DW, Moorman T, Kiel J (1979) The role of Endomycorrhizae in revegetation practices in the semi-arid west. I. A comparison of incidence of mycorrhizae in severly disturbed vs. natural environments. Am J Bot 66:1–13

Remacle J (1981) The impact of fungi on environmental biogeochemistry. In: Wicklow DT, Carroll GC (eds) The fungal community: its organization and role in the ecosystem. Dekker, New York, pp 595–606

Rogers LE (1987) Ecology and management of harvester ants in the shortgrass plains. In: Capinera JL (ed) Integrated pest management on rangelands. Westview Press, Boulder, pp 261–270

Rogers LE, Lavigne RJ (1974) Environmental effects of western harvester ants on the shortgrass plains ecosystem. Env Ent 3:994–997

Seagle SW, McNaughton SJ, Ruess RW (1992) Simulated effects of grazing on soil nitrogen and mineralization in contrasting Serengeti grasslands. Ecology 73:1105–1123

Sharp LA, Barr WF (1960) Preliminary investigation of harvester ants on southern Idaho rangeland. J Range Manage 12:131–134

Smith SE, Smith FA, Nicholas DJD (1981) Effects of endomycorrhizal infection on phosphate and cation uptake by *Trifolium subterraneum*. Plant Soil 63:57–64

Sneva FA (1979) The western harvester ants: their density and hill size in relation to herbaceous productivity and big sagebrush cover. J Range Manage 32:46–47

Sousa WP (1984) The role of disturbance in natural communities. Annu Rev Ecol Syst 15:353–391

Vilarino A, Arines J (1991) Numbers and viability of vesicular-arbuscular fungal propagules in field soil samples after wildfire. Soil Biol Biochem 23:1083–1087

Vitousek PM (1985) Community turnover and ecosystem nutrient dynamics. In: Pickett STA, White PS (eds) The ecology of natural disturbance and patch dynamics. Academic Press, New York, pp 325–334

Vitousek PM, Gosz JR, Grier CC, Melillo JM, Reiners WA, Todd RL (1979) Nitrate losses from disturbed ecosystems. Science 204:469–474

Whicker AD, Detling JK (1988) Ecological consequences of prairie dog disturbances. Bioscience 38:778–785

Whitford WG (1986) Decomposition and nutrient cycling in deserts. In: Whitford WG (ed) Pattern and process in desert ecosystems. University of New Mexico Press, Albuquerque, pp 93–118

Whitford WG, Ettershank G (1975) Factors affecting foraging activity in Chihuahuan desert harvester ants. Environ Entomol 4:689–696

Wicklow DT (1973) Microfungal populations in surface soils of manipulated prarie stands. Ecology 54:1302–1310

Wight JR, Nichols JT (1966) Effects of harvester ants on production of a saltbush community. J Range Manage 19:68–71

Willard JR, Crowell HH (1965) Biological activities of the harvester ant, *Pogonomyrmex owyheei*, in central Oregon. J Econ Entomol 58:484–489

Wilson EO (1987) Causes of ecological success: the case of the ants. J Anim Ecol 56:1–9

Wood TG, Sands WA (1978) The role of termites in ecosystems. In: Brian MV (ed) Production ecology of ants and termites. Cambridge University Press, Cambridge, pp 245–292

Woodell SRJ, King TJ (1991) The influence of mound-building ants on British lowland vegetation. In: Huxley CA, Cutler DA (eds) Ant-plant interactions. Oxford Science, Oxford, pp 521–535

Wright EW, Bollen WB (1961) Microflora of Douglas-fir forest soil. Ecology 42:825–828

Zak JC (1992) Response of soil fungal communities to disturbance. In: Carroll GC, Wicklow DT (eds) The fungal community: its organization and role in the ecosystem. Dekker, New York, pp 403–425

5 Fungal Responses to Disturbance: Agriculture and Forestry

R.M. MILLER[1] and D.J. LODGE[2]

CONTENTS

I. Introduction

In a comprehensive review of the effects of disturbance on fungal communities, Zak (1992) identified fungal responses to management practices as a neglected area of research. He pointed out that much of the past research on fungal responses to management practices has been descriptive, being concerned with the composition and richness of fungal species. Furthermore, Zak indicated that this approach contributes little to our understanding of the role of fungi in nutrient cycling and accumulation of organic matter. More recent studies using trophic structure and food-web approaches to understanding the effects of tillage and crop rotation on fungi and associated soil biota appear to be more informative (e.g., Beare et al. 1992; Wardle 1995); for example, by integrating fungal responses with tillage practices, particularly informative linkages of fungi to processes associated with the dynamics of soil organic matter (SOM) have been identified (e.g., Beare et al. 1992; Hendrix et al. 1986). Also, investigations have corroborated the importance of fungi in the hierarchical model of soil structure (Tisdall and Oades 1982; Oades 1984) by demonstrating the structural role of hyphae and the annealing properties of the polysaccharides that they exude to form and maintain a stable aggregate structure (e.g., Gupta and Germida 1988; Chenu 1989; Miller and Jastrow 1990; Cambardella and Elliott 1994).

Although there are many studies concerned with management practices in agriculture and forestry, a neglected area of research is the integration of fungal responses with these practices, especially as the responses relate to nutrient cycling and organic matter accumulation. One reason for this neglect is that plant and fungal responses and disturbance responses associated with land management practices are often studied at different spatial and temporal scales. Nevertheless, from a conceptual viewpoint, fungi do contribute to system processes and functions at various hierarchical organizational levels, indicating linkages and feedbacks between fungi and system responses (O'Neill et al. 1991; Miller and Jastrow 1994; Beare et al. 1995). The difficulty lies in our ability to focus questions and to measure responses or processes that function as control points.

[1] Terrestrial Ecology Group, Environmental Reseach Division, Argonne National Laboratory, 9700 S. Cass Avenue, Argonne, Illinois 60439, USA
[2] Center for Forest Mycology, USA – Forest Service, Forest Products Laboratory*, Box 1377, Luquillo, Puerto Rico 00773-1377, USA
* The Forest Products Laboratory is maintained in cooperation with the University of Wisconsin. This paper was written and and prepared in part by a US Government employee on official time, and is therefore in the public domain and not subject to copyright.

The Mycota IV
Environmental and Microbial Relationships
Wicklow/Söderström (Eds.)
© Springer-Verlag Berlin Heidelberg 1997

In this chapter, we will elaborate on the impacts of the various management practices associated with agriculture and forestry as they influence fungal structure and function. Contributions of fungi to nutrient cycling, organic matter accumulation, and the formation of soil structure will be discussed. Furthermore, we will discuss the hierarchical nature of soils and how this nature influences a systems response to disturbance. We will use examples from agroecosystems, soil restorations, and forest systems.

II. Disturbance as a General Phenomenon

Ubiquitous as the idea of "disturbance" is in ecology, it is not easily defined (DeAngelis et al. 1985). Disturbance is a common feature of many systems, occurring at all levels of ecological organization and at numerous temporal and spatial scales (Zak 1992). For our discussion, we mean by disturbance the physical or chemical phenomena that disrupt communities and ecosystems. Disturbances may be either anthropogenic or natural, but it is the biota and the variation in terms of disturbance severity, frequency, and scale that result in different pathways of ecosystem response (rather than being the source of the disturbance) (Waide and Lugo 1992). Some disturbances disrupt the physical structure of communities, such as soil tillage, clearcut harvesting, and storm damage; while other disturbances involve chemical additions, such as acid rain, fertilization, salinization, or addition of biocides. Fungal communities may be directly affected by these physical or chemical disturbances. Fungi also respond to the indirect effects of disturbance, such as the mortality of litter decomposers that results from the increased drying frequency of the forest floor following disturbance to forest canopies.

Fungi that are favored by disturbance generally have ruderal characteristics, including effective dispersal, rapid uptake of nutrients, and rapid extension for resource capture; they may also be stress-tolerant (Pugh and Boddy 1988). In contrast, fungi that are favored by relatively nonstressed undisturbed conditions are often combative or highly competent in competition with other species (Pugh and Boddy 1988). These patterns in fungi are similar to those identified in higher plants (e.g., Grime 1979); however, major differences exist between plants and fungi (Pugh 1980).

Substrates for decomposition can be viewed as discrete or continuous. Fungi of continuous substrates inhabit a niche which continually receives new resources, whereas fungi of discrete substrates inhabit the substrate. In his classical study on ecological groupings of soil fungi, Garrett (1951) states that the succession of fungi on a substrate causes a progressive deterioration in the capacity of the substrate to support further growth. He further states that the substrate comes directly to the soil microorganism, i.e., roots grow through soil and die in it; dead leaves fall upon the soil. The first example views a substrate as discrete, whereas the last example views a substrate as being continuous. Accordingly, disturbance can be a perturbation to both substrate and fungus. For a substrate, disturbance disrupts its delivery rate, its quality and its accessibility, whereas for the fungus, disturbance represents the physical or chemical disruption of the mycelial network.

III. Fungi as Control Points in Management Practices

Other than bacteria, fungi are the most numerically abundant organisms in the terrestrial ecosystem and are the primary decomposers of organic residues in soil. Although fungi may be numerically less abundant than bacteria, fungi can account for as much as 70 to 80% by weight of the soil microbial biomass (Shields et al. 1973; Lynch 1983). These fungi may be free-living or in mycorrhizal associations with plant roots. Studies of agricultural soils indicate that fungal biomass typically outweighs bacterial biomass (Anderson and Domsch 1978; Jenkinson and Ladd 1981; Beare et al. 1992), although studies to the contrary exist (Holland and Coleman 1987; Hunt et al. 1987; Hassink et al. 1993). It has been suggested that the size and composition of microbial communities upon and in soils are primarily controlled by the quality, quantity, and distribution of substrates, all of which are influenced by land management practices (Anderson and Domsch 1985; Schnürer et al. 1985). Differences in fungal: bacterial ratios most likely reflect changes in the composition in response to management effects on the retention of litter and its quality, e.g., fungi typically dominate under no-till conditions (e.g.,

Hendrix et al. 1986; Holland and Coleman 1987; Beare et al. 1992). Furthermore, many of the discrepancies reported in the literature may be explained by where bacteria and fungi tend to live in soil; e.g., a primary reason for fungal biomass being larger than bacterial biomass in surface litter is related to the ability of fungal hyphae to traverse the gap between surface litter and soil more readily than bacteria (Holland and Coleman 1987; Beare et al. 1992).

A. The Habitats of the Fungus

In an attempt to better conceptualize soil systems into biologically relevant regions on the basis of their spatial and temporal heterogeneity, the concept of "sphere of influence" has been proposed (Coleman et al. 1994; Beare et al. 1995). Figure 1 presents the five areas of concentrated activity in soils; the areas include: (1) the detritusphere, com-

posed of the litter, fermentation, and humification layers above the soil surface, which have considerable root, mycorrhizal, and saprophytic fungal activity, and grazing by the soil fauna; (2) the drilosphere, or that portion of the soil that is influenced by the activities of earthworms and their casts; (3) the porosphere, which is a region of water films occupied by bacteria, protozoa, and nematodes and of channels between aggregates occupied by microarthropod and the aerial hyphae of fungi; (4) the aggregatusphere, the region where the activity of microbes and fauna is concentrated in the voids between microaggregates and even macroaggregates; and (5) the rhizosphere, or the zone of soil influenced by roots, associated mycorrhizal hyphae, and their products. The spheres are formed and maintained by biological influences that operate at different spatial and temporal scales. Moreover, each sphere has distinct properties that regulate interactions among organisms and the biogeochemical proper-

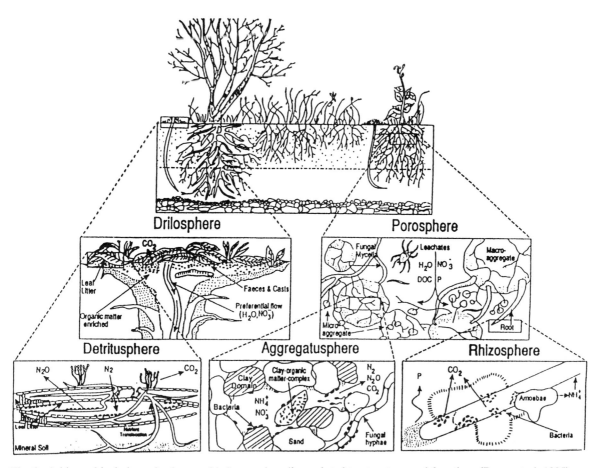

Fig. 1. A hierarchical view of spheres of influence in soil as related to structure and function. (Beare et al. 1995)

ties that they mediate (Coleman et al. 1994; Beare et al. 1995).

B. Effects of Disturbance on Fungi

Soils are spatially heterogeneous because of their biologically mediated properties (Beare et al. 1995). A driving force for creating a spatially heterogeneous environment is the process of bioturbation (Hole 1982). These biologically mediated disturbances result in creation of a spatially heterogeneous environment, an outcome to system processes very different from the disturbances associated with land management practices, such as tillage or clearcut harvesting, that usually result in a loss of spatial heterogeneity.

For example, in the drilosphere, the grazing activities of microarthropods and millipedes change the size and distribution of litter on the soil surface, thereby increasing the surface area for fungal colonization and mixing the fragments with other debris (Beare et al. 1995). Macrofauna redistribute the litter in and upon the soil, creating a patchwork of both substrates and refugia for soil fungi, bacteria, and fauna (Lee and Pankhurst 1992). In the porosphere, the physical rearrangement of soil particles by growing roots and earthworm burrowing creates macropores that influence the preferential flow of water and nutrients. In the aggregatusphere and detritusphere, the relationship between soil organisms and bioturbation is considerable. A large amount of faunal feeding occurs on surface litter and associated fungal hyphae, resulting in the accumulation of particulate and fecal aggregates in surface soils. Earthworm casts also accumulate in the surface soils; however, the primary agent of aggregate stabilization is through the deposition of bacterial and fungal polysaccharides and by hyphal entanglement of particles. The continued inputs of microbial gums and glues, as well as fungal hyphae, are required for the long-term maintenance of the aggregate structure.

Conventional practices of land management usually result in a loss of the spatial heterogeneity of soil. With more sustainable practices, the goal of management is to create a more spatially heterogeneous habitat; e.g., tillage results in the physical disruption of the more transient fungal hyphal network. Furthermore, tillage mixes surface residues vertically within the soil profile while

usually doing little direct damage to the aggregate structure (Angers et al. 1992); however, without the physical continuity of the hyphal network in place, the rewetting of dried aggregates will cause their disruption or slaking. Another phenomenon associated with plowing is a temporary flush in mineralization, which appears to be directly related to the exposure of organic residues to soil biota as a result of slaking (e.g., Elliott and Coleman 1988; Van Veen and Kuikman 1990). Some of the nutrient flush is attributed to disruption of the hyphal network (D.J. Lodge, unpubl. data). The amount of nutrient flush depends on (1) the overall amount of organic matter in the soil, (2) the quality of organic residues sequestered within the aggregated portion of the soil, and (3) the amount of microbial biomass and its activity.

C. Contributions of Fungi to Nutrient Cycling

In agricultural and forest ecosystems the primary role of saprophytic fungi is as contributors to nutrient cycling and soil organic matter (SOM) dynamics: SOM dynamics are influenced by saprophytic fungi through their regulation of the decomposition of plant residue (Swift et al. 1979; Beare et al. 1992), the production of polysaccharides (Chenu 1989), and the stabilization of soil aggregates (Van Veen and Kuikman 1990). Fungi are especially good decomposers of nutrient-poor plant polymers. Furthermore, saprophytic fungi possess several growth habits that enable them to grow in nitrogen-deficient environments (Paustian and Schnürer 1987). These habits include (1) lysis and reassimilation of nitrogen from degenerated hyphae (Levi et al. 1968), (2) directed growth to locally enriched nutrient sites (Levi and Cowling 1969; St. John et al. 1983; Boddy 1993), and (3) the translocation of cytoplasm to hyphal apices from mycelium in nitrogen-depleted regions (Cooke and Rayner 1984). Fungal hyphae may also translocate mineral nitrogen to nitrogen-poor substrates. Evidence for such translocation is indirect. The most convincing evidence comes from the observation that the absolute amount of nitrogen in a decomposing substrate increases during the early stages of decomposition (Berg and Söderström 1979; Aber and Melillo 1982; Holland and Coleman 1987). Also, the lateral and upward movement of ^{15}N-labeled inorganic nitrogen from mineral soil to decomposing litter has been demonstrated (Schimel and Firestone 1989). Furthermore, the net immobilization of nitrogen in surface litter can be relieved by the application of fungicide (Beare et al. 1992).

Fungal biomass represents a significant pool of available nutrients in soils. The turnover of this component has important consequences for SOM and nutrient cycling. The availability of hyphal components and cell products in the form of cell walls, cytoplasm, and extracellular polysaccharides represents a relatively labile organic pool in soils. Anderson and Domsch (1980) have estimated from a survey of arable soils that the biomass of the microflora in agricultural soils has about 108 and 83 kg ha^{-1} of nitrogen and phosphorus, respectively. Hence, along with bacteria, fungal hyphae represent a major sink for plant nutrients.

The mycorrhizal fungus represents a considerable portion of the fungal biomass in most terrestrial ecosystems. The importance of mycorrhizal fungi in nutrient cycling is that, as symbionts, they provide a direct physical link between primary producers and decomposers. Processes influenced by mycorrhizal fungi include such host functions as phytosynthesis, nutrient uptake, and water usage. Mycorrhizal fungal influences on host functions can affect nutrient accumulation and alter nutrient ratios in plant tissues. In addition, mycorrhizae can affect plant nutrient uptake by affecting a host's growth rate and by influencing mineral ion uptake (Barea 1991; Marschner and Dell 1994).

It has been suggested that a fundamental role for arbuscular mycorrhizal (AM) fungal hyphae is to bridge the annular space within soil, producing a physical connection between the root surface and surrounding soil particles (Miller 1987). In creating such bridges, the hyphae increase the effective surface area of the root and decrease resistance to water flow to the root surface by allowing closer contact with the soil. This physical relationship between the root surface, hyphae, and soil matrix could be especially important to plants growing in soils of high conductive resistance or where drought is commonplace, and would also allow continued uptake of nutrients from the soil solution during a drought cycle.

The contributions of externally produced AM hyphae to nutrient cycling are considerable (Miller and Jastrow 1994). Although studies of below ground carbon allocation are few, they indicate that external AM hyphae represent approxi-

mately 26% of the labeled extraradical organic carbon pool (Jakobsen and Rosendahl 1990). Annual production of external hyphae in prairie soils is estimated to be 28 m cm^{-3} of soil, with a calculated annual hyphal turnover of 26% (Miller et al. 1995). Turnover rates appear to be faster for extraradical hyphae than for intraradical hyphae of roots (Hamel et al. 1990). This may be true for the relatively thin-walled, small-diameter hyphae; but a substantial portion of the AM hyphal network is composed of thick-walled runner or arterial hyphae (Friese and Allen 1991; Read 1992). These runner hyphae are likely to be longer lived and more recalcitrant than the thinner walled hyphae that are probably directly involved in nutrient acquisition. Also, a considerable proportion of the hyphae extracted from soil is either nonviable or highly vacuolated (Schubert et al. 1987; Sylvia 1988; Hamel et al. 1990), suggesting considerable persistence for these hyphae. Because hyphal cell walls also contain chitin, they may be a relatively passive source of nitrogen.

Although the amount of external mycorrhizal hyphae is probably an important factor in nutrient ion uptake, the positioning of the hyphae around the root and into the soil may be equally important; e.g., just as root architecture influences nutrient scavenging (Fitter 1985), the architecture of the external hyphae may also affect nutrient uptake (Read 1992). Furthermore, the reported differences in phosphorus uptake among isolates of AM fungi were independent of the amount of external hyphae produced, with the differences apparently related to the placement of hyphae in the surrounding soil (Abbott and Robson 1985; Jakobsen et al. 1992).

The external hyphae of AM fungi can also influence nutrient cycling within the rhizosphere by influencing soil pH and by producing organic acids. The usual explanation for the elevated phosphorus levels found in shoots of mycorrhizal plants is the increased surface area provided by the external hyphae associated with their root systems; however, it now appears that, like roots, hyphae apparently can form zones of phosphorus depletion and altered pH in the surrounding soil (Li et al. 1991). Acidification of the surrounding soil is believed to be a mechanism for enhancing the mobility of calcium-bound phosphates and possibly trace nutrients. In addition to proton extrusion, the production of low-molecular-weight organic acids by external hyphae may be a mechanism behind soil acidification and increased phosphorus uptake (Li et al. 1991; Paris et al. 1995).

D. Contributions of Fungi to Soil Aggregation

An important component of a successful soil management strategy is the creation and maintenance of the soil aggregate structure (Miller and Jastrow 1992b). The degree to which a soil has been degraded will determine the extent of formation versus maintenance of aggregates. The importance of aggregated soils in soil management comes not only from the role of aggregates in controlling soil erosion, but also because aggregates facilitate the maintenance of nutrient cycles. In arable systems, the nutrient reserve of a soil typically is maintained by inputs from crop residues and fertilizers. In grassland and forest systems, the nutrient reserve is typically maintained by inputs from litter. In temperate forest systems, litter accumulates mainly from leaf fall, while root turnover is more important in grasslands and the few tropical forests that have been studied. Unless organic inputs are protected within the soil aggregate structure, the accumulation of organic matter and the concomitant buildup of soil nutrients is usually minimal (Elliott and Coleman 1988). Without the physical protection afforded within stable aggregates, organic matter and associated nutrients may be rapidly lost via both mineralization and erosion (Elliott 1986; Beare et al. 1994a).

The binding substances that hold soil particles together have both mineral and organic origins. In soils where organic matter is the major binding agent, several types of substances contribute to creating stable soil aggregates. Inorganic and relatively persistent organic binding agents are important for the stabilization of microaggregates (<250 μm diameter), but microaggregates subsequently are bound together into macroaggregates (>250 μm diameter) by a variety of primary organic mechanisms (Tisdall and Oades 1982; Oades 1984). The hyphae of both mycorrhizal and saprophytic fungi, along with fibrous roots, bind soil particles and microaggregates into larger aggregated units (Gupta and Germida 1988; Miller and Jastrow 1990, 1992b; Tisdall 1991, 1994). Polysaccharides produced by bacteria, fungi, and roots can act as the gums and glues that bind and stabilize aggregates (Tisdall and Oades 1982; Oades 1984; Foster 1994). The breakdown products of plant residues resulting from the actions of

bacteria and fungi are also important in contributing to the formation of soil aggregates (Elliott and Papendick 1986; Beare et al. 1994b; Cambardella and Elliott 1994).

The contribution of AM hyphae to the formation of soil aggregates is believed to be primarily due to a mechanism of physical entanglement. In this mechanism, hyphae enmesh microaggregates to create larger aggregated units. Furthermore, this mechanism can occur at higher levels of organization, whereby small macroaggregate units are packaged together to form larger macroaggregates (Miller and Jastrow 1992b). The relative importance of the various kinds of intermicroaggregate binding agents associated with the stabilization of macroaggregates may differ depending upon whether the soil region is rhizosphere, aggregatusphere, or detritusphere (Fig. 1). Several studies have demonstrated that polysaccharides and other organic compounds serve as the most important intermicroaggregate binding agents (e.g., Sparling and Cheshire 1985; Elliott 1986; Cambardella and Elliott 1994). Although other studies have suggested that polysaccharides and other organic materials function as persistent binding agents for microaggregates (Tisdall and Oades 1982; Oades 1984), the contribution of organic materials in stabilizing macroaggregates appears to be transient. The contribution of polysaccharides as binding agents may be more important in both detritusphere and aggregatusphere soils, where the saprophytic fungi are most active; while mycorrhizal hyphae may be more important in both the aggregatusphere and porosphere soils.

Visual evidence suggests that fungal hyphae are intimately involved in both the physical and chemical binding of soil particles into stable aggregates (Tisdall and Oades 1980; Gupta and Germida 1988; Perry et al. 1989; Oades and Waters 1991; Foster 1994). Additional evidence supports the role of AM hyphae in physical and chemical binding of soil aggregates (Tisdall and Oades 1980; Gupta and Germida 1988; Miller and Jastrow 1990). It appears that a simple hyphal entanglement mechanism apparently contributes to the formation of soil aggregates during an initial aggregative phase. Once they are formed, cementation of fungal hyphae to soil particles by organic or amorphous materials is a mechanism apparently involved in a later stabilization phase (e.g., Gupta and Germida 1988). This formation allows for intense biological activity to occur within these macroaggregates, and substantial amounts of organic materials are deposited, further stabilizing both the macroaggregates and the micro-aggregates that comprise them.

Electron microscopy studies further suggest that the decomposition process leads to the development of an aggregate hierarchy (Tiessen and Stewart 1988; Oades and Waters 1991). Particulate organic matter (POM) is often found at the core of microaggregates, where it is protected from rapid decomposition by encrustation with inorganic material. POM is now being recognized as an important pool of SOM and in some systems may represent up to 36% of the whole soil C regardless of tillage (Beare et al. 1994a). Originating as debris from roots, hyphae, and fecal matter, POM becomes incorporated into macroaggregates by physical and biological mechanisms (Coleman et al. 1994). This POM is eventually decomposed, leaving cavities surrounded by smaller microaggregates believed to be stabilized by the metabolic products and bodies of microbes that used the POM as a substrate (Oades and Waters 1991). Indirect evidence for the physical protection of SOM in both macro- and micro-aggregates comes from observations of increased mineralization when intact aggregates are disrupted (Elliott 1986; Gregorich et al. 1989; Beare et al. 1994a).

Very little information exists on the contributions of ectomycorrhizal fungi to soil aggregation. It has been suggested that ectomycorrhizal hyphae may create a very stable soil structure by producing stronger bonds with clays than do other hyphal types (Emerson et al. 1986). What is known is that ectomycorrhizal hyphae have the ability to extend considerable distances into the soil; exude polysaccharides and organic acids; and enmesh soil particles between them (Skinner and Bowen 1974; Foster 1981). Furthermore, the release of organic acids by hyphae may be a factor in the dissolution of clays (Leyval and Berthelin 1991).

IV. Fungi and Agriculture

Because of the various management practices used in agroecosystems, fungi experience vastly different disturbance regimes. The impacts of these disturbances are expressed directly and by their effects on food-web structure and function (Hendrix et al. 1986; Paustian et al. 1990; Beare et al. 1992). Under conventional management practices, the activities of the soil biota, including saprophytic and mycorrhizal fungi, have been largely marginalized by the use of agrochemicals, such as fungicides, herbicides, pesticides, and fertilizers, which suppress pests or bypass nutrient cycles; however, with increased societal pressures to reduce the use of agrochemicals and fertilizers, a greater reliance on processes influenced by soil biota, and especially fungi, will increase. Several areas of research with fungi offer promise for reducing the use of agrochemicals and intensive tillage regimes. These areas include the contributions of mycorrhizal fungi in plant production and soil aggregation, the use of fungi as biological agents for control of plant pathogens and insect and nematode pests (not discussed in this chapter), and the management of nutrient cycling through better use of saprophytic processes (Elliott and Coleman 1988; Bethlenfalvay and Linderman 1992; de Leij et al. 1995).

A. Tillage and Crop Rotation Effects on Fungi

Agricultural practices such as tillage, crop rotation, crop residue retention, and fertilizer use all affect the ecological niches available for occupancy by the soil biota. Simply put, an agricultural field is basically an experiment in natural selection where those organisms best adapted to those habitats and niches gradually replace those individuals not so well adapted (Rovira 1994). Of these practices, tillage most disrupts the soil fungal community (Table 1), resulting in a reduction of the soil's macroaggregate structure (e.g., Gupta and Germida 1988). The loss or reduction of this basic structural component of soil results in the destruction of many of the ecological niches suitable for soil fungi and bacteria. The practice of tillage serves many purposes, including preparation of seed bed, mechanical weed control, accelerated mineralization of nutrients from organic matter, and improved water capture and storage in the soil profile (Cook 1992). Unfortunately, tillage also sets the stage for soil erosion and loss of organic matter.

A major control point in conservation tillage is the management of crop residues. In conventional tillage practices, plowing results in the mixing of soil profiles and the burial of crop residues, whereas in no-tillage systems the soil is not plowed and residues are placed on the soil surface as

Table 1. Fungal and microbial properties of soil aggregate size classes from a native grassland soil and adjacent soil subjected to cultivation for 69 years. (Data from Gupta and Germida 1988)

Size class (mm)	Fungi ($10^4 \times$ cfu g^{-1})	Fungal biomass		Microbial biomass	
		Length (m g^{-1})	Biovolume (mm^3 g^{-1})	C (μg g^{-1})	N (μg g^{-1})
Native soil					
>1.00	11.7	874	3.74	1538	139
0.50–1.00	9.9	1276	12.89	1862	155
0.25–0.50	19.5	2163	10.85	1463	133
0.10–0.25	9.3	508	2.06	1161	124
<0.10	1.2	511	3.59	1106	124
Cultivated soil					
>1.00	2.8	180	0.98	886	86
0.50–1.00	2.5	166	11.12	946	92
0.25–0.50	6.5	543	3.63	859	90
0.10–0.25	0.5	144	1.70	655	82
<0.10	0.4	66	0.75	645	80

mulch. These differences in soil disturbance and residue placement can influence the composition and activity of the fungal community (Beare et al. 1993). Hendrix et al. (1986) and others (e.g., Beare et al. 1992), using a detritus food-web approach, demonstrated that no-tillage systems favored the fungal component of the soil microflora, resulting in the buildup of fungivorous microarthropods, nematodes, and earthworms. Alternatively, conventional tillage practices favored the bacterial component, resulting in the buildup of a completely different group of organisms by favoring protozoa, bacterivorous nematodes, and enchytraeids. The direct effects of the various tillage practices on fungi are related to physical disruption of the hyphal network and to the mixing of surface residues within the soil profile. Crop residues placed upon the soil surface are colonized predominately by saprophytic and facultatively pathogenic fungi, whereas residues mixed within the soil also have a significant bacterial component.

Tillage and crop rotation practices affect the growth of facultatively pathogenic fungi, e.g., the adoption of conservation tillage practice increases the incidence of foliar disease. This increase would be expected because the survival of foliar pathogens is often dependent on their association with host residues, and the rate of residue decomposition is slower if left on the soil surface (Rothrock 1992). Conservation tillage and cropping practices also influence disease incidence by indirect means; e.g., many grass species can be the primary host of

the take-all fungus *Gaeumannomyces graminis* var. *tritici*, a major pathogen of wheat. Hence, when soils are placed into pasture to improve nutrient storage and soil structure, an unanticipated effect caused by the density of the pasture's grass content occurs with resumption of cropping, where a relationship exists between prior grass content and the amount of the take-all pathogen *G. graminis* var. *tritici* in the subsequent wheat crop (Fig. 2).

An association has also been demonstrated between the fungal pathogen *Rhizoctonia solani* and conservation tillage. In this case, tillage practices that conserve crop residues on or near the soil's surface create conditions ideal for the growth of *R. solani*. The relationship is caused by the strong competitive growth capability of *Rhizoctonia* on particulate organic matter (Rovira 1986, 1990). The practice of direct drilling allows for optimal conditions for the proliferation of the fungus. A primary mechanism for control of *Rhizoctonia* in conventionally tilled soils is that the plowing breaks up the hyphal network in the soil and either kills portions of the hyphae or breaks the hyphae into less infective units (Neate 1994).

Tillage and crop rotation practices can also influence the AM fungus (Sieverding 1991; Johnson and Pfleger 1992; Abbott and Robson 1994; Kurle and Pfleger 1994). Many studies have demonstrated that past cropping and tillage practices influence the mycorrhizal fungi (e.g., Kruckelmann 1975; McGraw and Hendrix 1984;

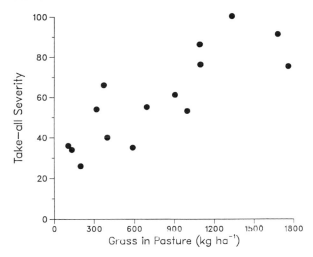

Fig. 2. A significant positive association exists between grass content of pastures and the density of the take-all fungus *Gaeumannomyces graminis* var. *tritici* ($r^2 = 0.71$, $P < 0.001$). (After MacNish and Nicholas 1987)

Dodd et al. 1990; Johnson et al. 1991; Thompson 1991). Tillage results in maximum disturbance to the fungal propagules and hyphal network, reducing the effectiveness of the symbiosis (Read and Birch 1988). Studies in Ontario demonstrate that a reduction in phosphorus uptake by maize following the plowing of no-tillage soils is caused by the disruption of the AM hyphal network (O'Halloran et al. 1986). This disruption caused both a delay and a reduction in mycorrhizal colonization and subsequently in phosphorus uptake (Evans and Miller 1988). Furthermore, the effect of disturbance was not dependent on the length of time the soil had been under a no-till system (Fairchild and Miller 1990). The use of fallow rotations can also result in the expression of phosphorus deficiency, even though there may be no decrease in available phosphorus (Thompson 1987). The mechanism for the response appears to be caused by a reduction in AM propagules associated with the fallow practice.

Although most crops are dependent upon mycorrhizal fungi, roots of crops belonging to the chenopod and crucifer families usually do not possess mycorrhizal fungi. These two families include such crops as spinach, sugar beet, canola, rapeseed, and mustards. When such crops are used in rotations, they tend to lead to a reduction in mycorrhizal propagules, and appear to act in rotations like a fallow year (Thompson 1991). Hence, the yield of mycorrhiza-dependent crops after a

nondependent crop may decline, if the needs of that crop for symbiont-supplied nutrients cannot be fulfilled (Fig. 3).

Cropping and tillage practices can also select for less efficient or even pathogenic mycorrhizal fungi. Rotations with sod crops are necessary to eliminate decline symptoms associated with the proliferation of a certain glomalean fungus when grown with tobacco as the host crop (Modjo and Hendrix 1986; Hendrix et al. 1995). Furthermore, those AM fungi that proliferate under tillage conditions may be less beneficial or even detrimental to those crops in which they proliferate (Johnson et al. 1992).

B. Role of Fungi in Soil Restorations

Depending on the degree of degradation of a soil, an important component of a successful restoration is the reestablishment of a nutrient reserve (Bradshaw et al. 1982). This reserve is initiated by the combination of atmospheric deposition, weathering, and detrital inputs from vegetation or by additions of organic amendments and fertilizers. Unless these organic inputs are stabilized, accrual of organic matter and microbial biomass and the concomitant buildup of nutrient reserves in soils are usually minimal. As described previously, organic residues are generally protected or stabilized within soils through the formation of soil aggregates. Hence, a major goal of any soil restoration should be to establish conditions that favor formation of stable soil macroaggregates, thereby facilitating an important step in the creation of a nutrient reserve (Miller and Jastrow 1992a,b).

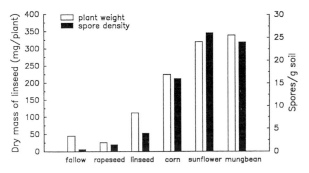

Fig. 3. The effects of a preceding crop on linseed production are related to mycorrhizal spore density. (Data from Thompson 1991)

Mycorrhizal fungal dynamics appear to be a good indicator for determining the consequences of different crop rotations on soil stability. Studies in Australia indicated that 50 years of crop rotation decreased the amount of stable macroaggregates and simultaneously decreased the lengths of roots and AM hyphae in the soil, as compared with soils from long-term pasture and natural sites (Tisdall and Oades 1980). The study found that the positive association between the amount of macroaggregates and the length of external mycorrhizal hyphae was related to the type of crop rotation and the frequency of fallow rotations (Fig. 4). A similar relationship also was found between macroaggregates and root length. This study indicated that frequent use of fallow in crop rotations can significantly decrease the amount of soil held as macroaggregates. Conversely, the longer a soil has a cover crop, the greater the amount of soil held as macroaggregates. The study also suggested that the loss of stable soil macroaggregates associated with fallow rotations may be caused, at least in part, by a reduction in the mycorrhizal fungus population that is brought on by both tillage and fallow disturbances.

Using a series of prairie reconstructions in the central Unitied States as a means of investigating the aggradative phase of a soil indicates that a stable soil aggregate structure can develop rapidly under prairie and pasture vegetation (Jastrow 1987; Miller and Jastrow 1990, 1992b). The soil studied had been under cultivation for over 100 years; however, within 8 years after being planted to prairie species, the proportions of stable macroaggregates had approached that of a nearby prairie remnant (Jastrow 1987). The rapid recovery of soil aggregates was most likely due to a well-developed microaggregate structure that remained relatively intact during cropping in combination with a rapid reestablishment of a relatively dense root and hyphal network after the cessation of tillage (Miller and Jastrow 1992b). Both total root length and the length of roots colonized by AM fungi increased with time from disturbance (Cook et al. 1988). Furthermore, root and soil hyphal lengths were associated with increases in the proportion of soil held as water-stable macroaggregates (Miller and Jastrow 1990, 1992b).

The observed associations between plant species composition and aggregate formation appear to be related to the diameter size class of roots produced by the various species. The study found that a large portion of the effects of root size class on soil aggregation was due to indirect effects of root associations with mycorrhizal fungi (Miller and Jastrow 1990). Because of the effects of plant life-form on root morphology, some life-forms appear to be more effective than others in promoting aggregate formation. Hence, one of the mechanisms behind plant community composition acting as a control point for organic matter accumulation, and hence nutrient cycling, is based on the relationship between the morphology of the root and mycorrhizal fungi (Miller and Jastrow 1994).

In addition to the changes in hyphal lengths, Miller and Jastrow (1992b) found that the composition of the AM fungal community changed with the cessation of tillage and reconstruction of the prairie. Using spore biovolume as a measure of AM species contributions revealed that soils under conventional tillage practices were dominated by *Glomus constrictum* and *Glomus etunicatum*. However, with the cessation of cultivation, *Gigaspora gigantea* replaced *Glomus* as the dominant fungal group by the fifth growing season (Fig. 5). The spore biovolume of *Glomus* species was negatively associated with recovery time since disturbance, external hyphal length, and percent of soil held as macroaggregates; in contrast, *Gigaspora* was positively associated with extra-

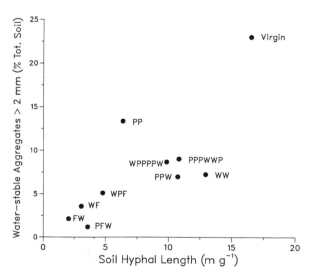

Fig. 4. The relationship between extraradical hyphae and percentage of water-stable aggregates for a soil under different crop rotations: *PP* old pasture; *PPW* pasture-pasture-wheat; *WW* wheat every year; *WPF* wheat-pasture-fallow; *PFW* pasture-fallow-wheat; *WF* wheat-fallow; *FW* fallow-wheat; *PPPWWP* and *WPPPPW* 2 years wheat and 4 years pasture. (After Tisdall and Oades 1980)

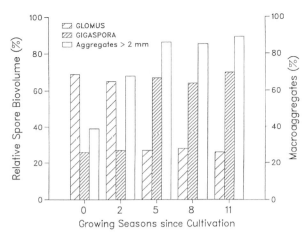

Fig. 5. Relative spore biovolumes of AM fungi and water-stable aggregates (macroaggregates) >2 mm (% of total soil) for row crop and reconstructed prairie soils (n = 10). (After Miller and Jastrow 1992b)

radical hyphal length and macroaggregation. Other investigators have also found *Gigaspora* to be more effective than *Glomus* in producing aggregates (Schreiner and Bethlenfalvay 1995). These trends suggest that AM fungi may differ in their sensitivities to disturbance and in their abilities to produce extraradical hyphae, both of which can influence the development of soil aggregates.

V. Fungi and Forestry

Forests that are managed for recreation, timber, pulpwood, or other forest products differ widely in tree diversity, their dominant mycorrhizal fungal associates, the seral stage that is being managed, climate, seasonality, and decomposability of litterfall. For example, tree plantations and many native temperate and boreal forests are dominated by one or a few tree species, whereas many tropical forests are speciose; a single hectare of lowland Amazonian forest in Ecuador was reported to have over 470 tree species greater than 10 cm in diameter at breast height (Valencia et al. 1994). Such underlying differences among forest ecosystems will greatly influence how fungi respond to disturbances related to forestry practices.

A. Nutrient Additions in Forest Systems

Fertilization and atmospheric inputs of nitrogen can influence the rate at which fungal decompos-

ers recycle nutrients from organic matter, but such effects may differ among forest types. Rates of leaf decomposition differ among tree species and forest ecosystems, primarily because of limitations on the growth and activity of fungal decomposers imposed by differences in concentrations of mineral nutrients in the fallen leaves, climate, and abundance of lignin and other recalcitrant or toxic secondary plant compounds (e.g., Meentemeyer 1978; Swift et al. 1979; Aber and Melillo 1982; Vogt et al. 1986). Low nitrogen concentrations in fallen litter are often more limiting to nutrient cycling in temperate forests, whereas the availability of phosphorus is often more important in lowland tropical forests (Vitousek and Sanford 1986), although phosphorus concentrations can also be important in temperate coniferous systems (Dyer et al. 1990). Many successional forest trees produce highly decomposable leaf litterfall containing higher concentrations of nutrients and labile carbon and lower concentrations of secondary chemicals than species from later successional stages (Bazzaz 1979; Grime 1979; Marks 1974). Therefore, the effects of episodic fertilization or chronic additions of nitrogen via air pollutants on litter decomposer fungi and the rate of decomposition may depend on the forest species composition and may be correlated with climate.

In a late successional subtropical wet forest in Puerto Rico, complete fertilization accelerated decomposition of fine roots that had very low nutrient concentrations, thus lowering the dead root standing stocks after 9 months (Parrotta and Lodge 1991). Fertilization apparently also accelerated decomposition of leaf litter in the same experiment, as indicated by a higher turnover rate. Production of leaf litter was significantly increased by fertilization, while standing stocks of litter did not differ among treatments at low elevation; litterfall was greater and litter standing stocks lower in fertilized than in control plots at high elevation (Zimmerman et al. 1995).

Despite higher rates of leaf decomposition, certain decomposer fungi were apparently negatively affected by fertilization in the previous study. Superficial and interstitial mycelia of decomposer basidiomycetes completely disappeared from the litter layer in fertilized plots, and the diversity and abundance of their fruitifications were also reduced compared with control plots (DJ Lodge, S Cantrell, and O Oscar, unpublished data), suggesting that the fungal decomposer community may have changed in response to repeated fertilization. Similarly, Heinrich and Wojewoda (1976) found that fertilization of forest plots in Poland significantly decreased the number of fruiting bodies of basidiomycetes that decompose SOM and wood. Shifts in the composition of the fungal decomposer community in response to fertilization may reflect differential sensitivity among species to salt stress (Castillo Cabello et al. 1994) or reduced competitive advantage of fungi that use hyphal cords and rhizomorphs to translocate nutrients into nutrient-depauperate food bases (Boddy 1993; Lodge 1993). The disappearance of cord-forming basidiomycetes may have a negative impact on the ecosystem in terms of loss of SOM and nutrients. Absence of cords formed by basidiomycetes in the litter layer had previously been found to significantly increase the rate of litter export from steep slopes during storms and subsequent soil erosion from the exposed surfaces (Lodge and Asbury 1988). Thus, there may be contrasting responses to fertilization in different ecosystem-level processes that are mediated or influenced by fungi. In standard forestry practices, fertilization generally occurs only at planting or post-thinning at 10 to 15 years (Barrett 1962; Oliver 1986), so negative long-term impacts on litter fungi are probably less than in the experiments cited previously.

Addition of mineral nutrients that are limiting to the growth of fungal decomposers might reasonably be expected to consistently increase the rate of decomposition, but this is not always the case. Although higher nitrogen and phosphorus concentrations were correlated with higher initial rates of decomposition of coniferous litter in Sweden, later decomposition and N-mineralization were inhibited (Berg et al. 1982, 1987). As noted in a review by Berg (1986), later decomposition may be slowed by ammonium and amino acid repression of fungal ligninolytic enzymes and by the complexing of nitrogen with phenolic compounds to form particularly recalcitrant products (Nommik and Vahtras 1982; Stevenson 1982). Palm and Sanchez (1990) found that leguminous leaves that naturally contained high concentrations of both nitrogen and polyphenolic compounds decayed more slowly in tropical agroecosystems than predicted from their nitrogen concentrations alone, indicating that the negative interaction between nitrogen and polyphenolics is a widespread phenomenon. Berg and Ekbohm (1991) hypothesized that nitrogen and phosphorus initially stimulate microbial decompo-

sition and nitrogen mineralization, thereby increasing the rate of release of nitrogen compounds that react with aromatic compounds to form resistant residues. Söderström et al. (1983) found that annual fertilization of Scots pine plantations caused slowing of the long-term decomposition rate and a 50% reduction in microbial respiration and abundance in humus. The addition of nitrogen through fertilization or air pollutants may therefore be expected to increase the rate of total decomposition in forests where litter contains low concentrations of both nitrogen and polyphenolic compounds, but may slow complete decomposition in forests where litter contains high concentrations of polyphenolics and low concentrations of nitrogen (Berg and Ekbohm 1991).

B. Effects of Air Pollution and Fertilization on Fungi

The decline in fruiting bodies of ectomycorrhizal fungi in Europe (Derbsch and Schmitt 1987; Termorshuizen and Schaffers 1987; Jakucs 1988; Arnolds 1989; Fellner 1989, 1993; Nauta and Vellinga 1993) has been attributed to direct and indirect effects of air pollution (Arnolds 1991). While ectomycorrhizal fungi have decreased, fruiting by wood decay fungi has increased in central Europe (Fellner 1993). Application of nitrogen fertilizer was found to cause similar declines in the diversity or abundance (or both) of ectomycorrhizal fungal species (Fiedler and Hunger 1963, Heinrich and Wojewoda 1976, Schlechte 1991). Ectomycorrhizal and ericoid mycorrhizal fungi are thought to confer special advantages to plants in the uptake of nitrogen from organic residues in environments in which the availability or uptake of nitrogen is limited (Read 1991), which may help explain why chronic additions of nitrogen from air pollutants are associated with declines in ectomycorrhizal fungi and their associated host trees.

Some fungal responses to air pollution may be related to soil pH. Bååth et al. (1979) found that active hyphae of soil fungi were decreased by application of artificial acidified rain. Not all ectomycorrhizal fungi have been negatively affected by air pollution. For instance, Fellner (1993) found that certain *Cortinarius* species in the subgenus *Dermocybe* that are favored by acidic conditions were increasing in highly acidified forests. Schlechte (1991) found that moderate amelioration of soil pH through application of dolomitic limestone had a stabilizing effect on ectotrophic fungal communities in chronically acidified acid forests. However, addition of some forms of lime have been found to adversely affect the abundance or diversity of ectomycorrhizal fungal fruiting bodies (Fiedler and Hunger 1963; Heinrich and Wojewoda 1976; Hora 1959).

C. Forestry Practices for Pulpwood and Lumber Production

In sustainable pulpwood forestry, monospecific stands are often planted and harvested on short rotations, using tree species from an early seral stage such as pines or other conifers, certain eucalyptus, white birch, aspen and poplars. Such even-aged management practices have been predominant in the United States for pulpwood and lumber since 1950 (Oliver 1986). In such cases, clearcut harvesting, sometimes followed by burning of slash, is often used to regenerate the species. Though such forestry practices are often viewed as severe by environmentalists, these management practices often mimic natural disturbances to which those early seral stage tree species and their associated fungi are adapted (Oliver 1986). For example, some forest types, especially certain pines, are dependent on fire for their establishment and the maintenance of their integrity (Oliver 1986).

D. Effect of Site Preparation on Fungi

The effects of different intensities of site preparation and silvicultural practices on microbial biomass and microbial nutrient stores before replanting of pine were studied in the southeastern United States by Vitousek and Matson (1984). They compared (1) whole-tree versus stem-only harvest; (2) chopping the harvest debris versus shearing and piling of all debris and forest floor into wind rows followed by disking of the mineral soil; and (3) application of herbicide. Vitousek and Matson (1984, 1985) found that microbial nutrient immobilization was effective in preventing leaching losses of nitrogen when the forest floor was left intact (chopping) but not when the forest floor was removed (shearing/piling/disking). Losses of nitrate to streamwater and groundwater via leaching (Borman and Likens 1979; Vitousek and Melillo

1979; Robertson and Tiedje 1984) and atmospheric losses through denitrification occur in forests that have been cleared, but microorganisms in the forest floor and soil immobilize and conserve most of the nitrogen that is mineralized in response to the disturbance (Marks and Borman 1972; Vitousck and Matson 1984, 1985). Large woody debris (slash) and whole-tree harvesting had little effect on microbial immobilization of nitrogen, at least during the short-term study (2 years), but herbicides may have been toxic to microorganisms, thus increasing leaching losses of nitrogen (Vitousek and Matson 1985).

Although burning of coarse woody debris (slash) is considered an appropriate management practice for regeneration of Douglas fir in the Pacific Northwest (Oliver 1986), broadcast burning following clearcuts (Harvey et al. 1980a) and partial cuts (Harvey et al. 1980b) has been found to greatly decrease the abundance of active ectomycorrhizae in the forest floor, and levels of ectomycorrhizal fungal inoculum became insufficient for regeneration if replanting was delayed until the next dry season (July). Broadcast burning has been misapplied to old-growth true fir forests, resulting in mortality of young trees, predisposition of the firs to fungi that decay stems (Oliver 1986), and destruction of the litter fungal community. Highly decayed coarse woody debris is important for maintenance of ectomycorrhizal fungi in the northern Rocky Mountains during the hot dry summers (Harvey et al. 1980a,b). In addition to affording a moist refugium to mycorrhizae during drought, ectomycorrhizal fungi apparently obtain part of their nitrogen from organic sources including rotting wood (Read et al. 1989). Thus removal of coarse woody debris from forests by whole-tree harvesting or burning of woody debris following harvest may have negative effects on ectomycorrhizal fungi and tree nutrition, especially in areas with seasonal drought (Harvey et al. 1980a,b).

Others have also reported that the severity of a particular management activity can affect the persistance of ectomycorrhizal fungi in the northwestern United States (e.g., Perry et al. 1982; Schoenberger and Perry 1982; Amaranthus and Perry 1987). Moreover, studies of younger clearcuts stands indicate that prompt regeneration may be important to secure adequate formation of indigenous fungi; suggesting that mycorrhizal nursery stock may be more useful on older and more severely burned sites (Pilz and Perry 1984).

Studies of site preparation effects on mycorrhizal fungi from tropical forest systems are practically nonexistent. The few studies that have been conducted demonstrate disturbances due to site preparation reduce AM fungal propagule levels and species composition compared to undisturbed forest levels (Alexander et al. 1992; Mason et al. 1992; Wilson et al. 1992). In a Cameroon study, mycorrhizal spore loss appeared to be related to the severity of disturbance, with decreases being greatest in stands with vegetation completely removed. Reduced spore numbers from manually cleared stands were not as severe as those cleared by mechanical methods, and planting of the stands with *Terminalia* seedlings allowed for their recovery, although mycorrhizal species composition remained altered (Mason et al. 1992). Similar findings were made in a *Terminalia* plantation in the Côte d'Ivoire, where, even though site preparation effects on spore numbers were less easily differentiated, manual clearing of vegetation resulted in less spore reduction than did mechanical means (Wilson et al. 1992). Also, spore numbers recovered more rapidly in the manually cleared stand; AM fungal richness increased greatly under both methods of clearcutting, although species balance under manual clearcutting was closer to the undisturbed forest. These studies suggest that selective harvest practices do minimal damage to the mycorrhizal community in that they support extensive mycelial networks, which for natural systems are the primary source of infection for tree seedlings (Alexander et al. 1992). The reduced levels for mycorrhizal roots, hyphae, and spores in heavily logged systems are due to the mortality of root fragments containing mycorrhizal hyphae. Hence, clearing practices that allow for the continued existence of a mycelial network may be preferable ecologically and silviculturally (Table 2).

Table 2. The most probable number (MPN propagules per 100 g fresh soil ± SE) of AM fungi in the top 15 cm of soil for Malaysian forest sites under different logging intensities (n = 5). (Data from Alexander et al. 1992)

	MPN
Undisturbed	307.5 ± 13.8a
Selectively logged	291.2 ± 7.8a
Heavily logged	75.3 ± 2.6b
Heavily logged (eroded)	6.0 ± 1.5c

Values followed by the same letter are not significantly different ($p < 0.05$, Duncan's Multiple Range Test).

Stumps of felled trees, roots in the soil, and other woody residues left from forestry operations often provide essential resources for pathogenic fungi and those that cause root or stem decay (Boyce 1961), e.g., stumps and woody debris have been found to increase mortality from root rots by certain *Armillaria* species in tropical and temperate plantations (Rishbeth 1985; Wargo and Harrington 1991), although raised water tables and shock following thinning and partial cuts may be predisposing factors (Wargo and Harrington 1991). Increased availability of nutrients, especially nitrogen, after logging and fire is known to improve the growth and survival of many root pathogens and to increase the disease incidence (Matson and Boone 1984). In contrast, increased nitrogen mineralization following logging apparently increases the resistance of *Tsuga mertensiana* to laminated root rot caused by *Phellinus weirii* (Matson and Boone 1984). Feeder root rots caused by *Fusarium* and *Phytophthora* species are favored by soil conditions following fire, as well as by increased soil moisture following harvesting.

E. Effects of Woody Debris on Ecosystem Processes

The removal of woody storm debris dramatically affected forest recovery and productivity in Puerto Rico following damage from hurricane Hugo by altering fungal competition for nutrients (Zimmerman et al. 1995). The results of this study are relevant to management of woody debris (slash) following a partial or selective harvest. Plots from which woody debris was removed and those that received complete fertilization recovered their canopies and productivity more rapidly than the control plots (Zimmerman et al. 1995). The massive deposition of organic matter carbon on the forest floor apparently stimulated microbial production and nutrient immobilization, reducing the availability of nutrients to trees in the control plots, whereas fertilization released the trees from nutrient competition with decomposers, predominantly fungi (Lodge et al. 1994; Zimmerman et al. 1995). However, bole increment growth was greater only in the fertilization treatment, and removal of woody debris is expected to have a longer-term negative impact on forest productivity through its effects on reduced SOM inputs from decomposing wood and consequent lower phosphorus availability (Sanford et al. 1991). The removal of woody debris following a forest disturbance may therefore have opposing short- and long-term consequence for forest recovery and productivity, so it is the balance of the short- and long-term goals that will determine the appropriate management strategy.

F. Effect of Opening the Canopy and Moisture Fluctuations on Fungi

Opening of a forest canopy by storms or from logging causes dramatic environmental changes on the forest floor. Canopy removal reduces transpiration and is often accompanied by higher soil moistture or raised water tables; however, the litter and humus layers may experience more rapid drying as a result of greater exposure to solar irradiation and wind. Moisture is often cited as a primary factor controlling fungal biomass (Parkinson et al. 1968; Söderström 1979; Lodge 1993; Lodge et al. 1994) but temperature can also be important (Flanagan and Van Cleve 1977). Fluctuations in fungal and microbial biomass in response to normal wetting and drying cycles are thought to be important in determining the fate of limiting nutrients and maintaining forest productivity (Lodge et al. 1994). Following a major disturbance, stress tolerance in some fungal species allows them to replace other species that are less tolerant, such as the replacement of *Collybia johnstonii* by marasmioid species of basidiomycete litter decomposers following hurricane Hugo in Puerto Rico (Lodge and Cantrell 1995). Thus, the overall response of fungal biomass to factors such as moisture fluctuations may be more obscure than the responses of individual species and may depend on the severity and scale of the disturbance.

VI. Conclusions

Unfortunately, few examples exist for fungi and their response to management practices used in agriculture and forestry. Nevertheless, this chapter represents the gleaning from a wide variety of sources ranging from decomposition studies, to detrital food-web investigations, to research on soil aggregation, all of which suggest an important role for fungi in managed systems. Moreover, obstacles to identifying the contribution of fungi in managed systems have been just as much methodological as conceptual.

Although our ability to quantify the various fungal components (especially their activities) is still limited, major conceptual breakthroughs have occurred on how we view soils. Two important conceptual advances are the hierarchical theory of soil structure (Tisdall and Oades 1982) and the "spheres of influence" view of soil systems (Coleman et al. 1994; Beare et al. 1995). In both of these conceptual views of soils, the contributions of fungi are considerable. Furthermore, the pertubations imposed by management practices have played an important role in the development of these views. Fungal responses to disturbance offer an opportunity to test these two compelling theories of soil structure and function.

A necessary step in developing sustainable management practices in agriculture and forestry will require identifying practices that allow for controlled manipulations of the fungal community. Nevertheless, our ability to manipulate fungi is rather limited, but such manipulations are not impossible. Research is needed to better understand the response of both the saprophytic hyphae and the mycorrhizal hyphal network to disturbances associated with different management practices. Of crucial importance is the development of management practices that maximize nutrient uptake but are not detrimental to the litter layer and the soil aggregation process. Specifically, research is needed on how disruption of the mycorrhizal hyphal network takes place via tillage and rotation effects in agriculture (e.g., fallow or nonmycorrhizal host crops); and harvesting and site preparation effects in forestry (e.g., clearcut or slash removal) to the soil aggregate structure. Research also needs to be directed at developing management practices that take advantage of the nutrient pools associated with the fungal hyphae. These kinds of studies would represent important steps in controlling the soil's labile nutrient pools.

Acknowledgments. We thank C. Loehle for his comments on the manuscript and P. Weaver for assistance in locating references. The support of RMM for the preparation of this chapter was by the US Department of Energy, Office of Energy Research, Office of Health and Environmental Research, Environmental Sciences Division, under contract No. W-31-109-ENG-38.

References

Abbott LK, Robson AD (1985) Formation of external hyphae in soil by four species of vesicular-arbuscular mycorrhizal fungi. New Phytol 99:245–255

Abbott LK, Robson AD (1994) The impact of agricultural practices on mycorrhizal fungi. In Pankhurst CE, Doube BM, Gupta VVSR, Grace PR (eds) Soil biota management in sustainable farming systems. CSIRO Melbourne, Australia, pp 88–95

Aber JD, Melillo JM (1982) Nitrogen immobilization in decaying hardwood leaf litter as a function of initial nitrogen and lignin content. Can J Bot 60:2263–2269

Alexander I, Ahmad N, See LS (1992) The role of mycorrhizas in the regeneration of some Malaysian forest trees. Philos Trans R Soc Land (B) 335:379–388

Amaranthus MP, Perry DA (1987) Effect of soil transfer on ectomycorrhiza formation and survival and growth of conifer seedlings in disturbed forest sites. Can J For Res 17:944–950

Anderson JPE, Domsch KH (1978) Mineralization of bacteria and fungi in chloroform-fumigated soils. Soil Biol Biochem 10:207–213

Anderson JPE, Domsch KH (1980) Quantities of plant nutrients in the microbial biomass of selected soils. Soil Sci 130:211–216

Anderson T-H, Domsch KH (1985) Determination of ecophysiological maintenance carbon requirements of soil microorganisms in a dormant state. Biol Fertil Soils 1:81–89

Angers DA, Pesant A, Vigneux J (1992) Early cropping-induced changes in soil aggregation, organic matter, and microbial biomass. Soil Sci Soc Am J 56:115–119

Arnolds E (1989) The changing macromycete flora in The Netherlands. Trans Brit Mycol Soc 90:391–406

Arnolds E (1991) Decline of ectomycorrhizal fungi in Europe. Agric Ecosyst Environ 35:209–244

Bååth E, Lundgren B, Söderström B (1979) Effects of artificial rain on microbial activity and biomass. Bull Environ Contamination Toxicol 23:737–740

Barea JM (1991) Vesicular-arbuscular mycorrhizae as modifiers of soil fertility. Adv Soil Sci 15:1–40

Barrett JW (1962) Regional silviculture of the United States. Ronald Press, New York

Bazzaz FA (1979) The physiological ecology of plant succession. Annu Rev Ecol Syst 10:351–372

Beare MH, Parmelee RW, Hendrix PF, Cheng W, Coleman DC, Crossley DA (1992) Microbial and faunal interactions and effects on litter nitrogen and decomposition in agroecosystems. Ecol Monogr 62:569–591

Beare MH, Pohlad BR, Wright DH, Coleman DC (1993) Residue placement and fungicide effects on fungal communities in conventional- and no-tillage soils. Soil Sci Soc Am J 57:392–399

Beare MH, Cabrera ML, Hendrix PF, Coleman DC (1994a) Aggregate-protected and unprotected organic matter pools in conventional- and no-tillage soils. Soil Sci Soc Am J 58:787–795

Beare MH, Hendrix PF, Coleman DC (1994b) Water-stable aggregates and organic matter fractions in conventional- and no-tillage soils. Soil Sci Soc Am J 58:777–786

Beare MH, Coleman DC, Crossley DA Jr, Hendrix PF, Odum EP (1995) A hierarchical approach to evaluating the significance of soil biodiversity to biogeochemical cycling. In: Collins HP, Robertson GP, Klug MJ (eds) The significance and regulation of soil biodiversity, Kluwer, Amsterdam, pp 5–22

Berg B (1986) Nutrient release from litter and humus in coniferous forest soils – a mini-review. Scand J For Res 1:359–369

Berg B, Ekbohm G (1991) Litter mass-loss rates and decomposition patterns in some needle and leaf litter types. Long-term decomposition in a Scots pine forest VII. Can J Bot 69:1449–1456

Berg B, Söderström B (1979) Fungal biomass and nitrogen in decomposing Scots pine needle litter. Soil Biol Biochem 11:339–341

Berg B, Wessen B, Ekbohm G (1982) Nitrogen level and lignin decomposition in Scots pine needle litter. Oikos 38:291–296

Berg B, Staaf H, Wessen B (1987) Decomposition and nutrient release in needle litter from nitrogen-fertilized Scots pine, (Pinus sylvestris) stands. Scand J For Res 2:399–415

Bethlenfalvay GL, Linderman RG (eds) (1992) Mycorrhizae in sustainable agriculture. ASA spcial publication 54. Agronomy Society of America, Crop Science Society of America, and Soil Science Society of America, Madison, Wisconsin, pp 71–99

Boddy L (1993) Cord-forming fungi: warfare strategies and other ecological aspects. Mycol Res 97:641–655

Borman FH, Likens GE (1979) Pattern and process in a forested ecosystem. Springer, Berlin, Heidelberg, New York

Boyce JS (1961) Forest pathology, 3rd edn. McGraw Hill, New York

Bradshaw AD, Marrs RH, Roberts RD, Skeffington RA (1982) The creation of nitrogen cycles in derelict land. Philos Trans R Soc Lond (B) 296:557–561

Cambardella CA, Elliott ET (1994) Carbon and nitrogen dynamics of soil organic matter fractions from cultivated grassland soils. Soil Sci Soc Am J 58:123–130

Castillo Cabello G, Georis P, Demoulin V (1994) Salinity and temperature effects on growth of three fungi from Laing Island (Papua New Guinea). Abstracts 5th Int Mycological Congr, Vancouver, BC, Canada, 14–21 Aug 1994, p 31

Chenu C (1989) Influence of a fungal polysaccharde, scleroglucan on clay microstructure. Soil Biol Biochem 21:299–305

Coleman DC, Hendrix PF, Beare MH, Crossley DA, Hu S, van Vliet PCJ (1994) The impacts of management and biota on nutrient dynamics and soil structure in subtropical agroecosystems: impacts on detritus food webs, In: Pankhurst CE, Doube BM, Gupta VVSR, Grace PR (eds) Soil biota management in sustainable farming systems. CSIRO, Melbourne, Australia, pp 133–143

Cook BD, Jastrow JD, Miller RM (1988) Root and mycorrhizal endophyte development in a chronosequence of restored tallgrass prairie. New Phytol 110:355–362

Cook RJ (1992) Wheat root health management and environmental concern. Can J Plant Pathol 14:76–85

Cooke RC, Rayner ADM (1984) Ecology of saprophytic fungi. Longman, London

DeAngelis DL, Waterhouse JC, Post WM, O'Neill RV (1985) Ecological modelling and disturbance evaluation. Ecol Mod 29:399–419

de Leij FAAM, Whipps JM, Lynch JM (1995) Traditional methods of detecting and selecting functionally important microorganisms from soil and the rhizosphere, In: Allsopp D, Colwell RR, Hawksworth DL (eds) Microbial diversity and ecosystem function. CAB International, Wallinford, pp 321–336

Derbsch H, Schmitt JA (1987) Atlas der Pilze des Saarlandes, Teil 2. Nachweise, Okologie, Vorkommen und Beschreibungen. Aus Natur und Landschaft im Saarland, vol 3. Minister für Umwelt des Saarlandes, Saarbrücken

Dodd JC, Arias I, Kooman I, Hayman DS (1990) The management of populations of vesicular-arbuscular mycorrhizal fungi in acid-infertile soils of a savanna ecosystem. II. The effects of pre-crops on the spore populations of native and introduced VAM-fungi. Plant Soil 122:241–247

Dyer ML, Meentemeyer V, Berg B (1990) Apparent controls of mass loss rate of leaf litter on a regional scale. Litter quality vs. climate. Scand J For Res 5:311–323

Elliott ET (1986) Aggregate structure and carbon, nitrogen, and phosphorus in native and cultivated soils. Soil Sci Soc Am J 50:627–633

Elliott ET, Coleman DC (1988) Let the soil work for us. Ecol Bull 39:23–32

Elliott LF, Papendick RI (1986) Crop residue management for improved soil productivity. Biol Agric Hort 3:131–143

Emerson WW, Foster RC, Oades JM (1986) Organomineral complexes in relation to soil aggregation and structure. In: Huang PM, Schnitzer M (eds) Interactions of soil minerals with natural organics and microbes. Soil Science Society of America, Madison, Wisconsin, pp 521–548

Evans DG, Miller MH (1988) Vesicular-arbuscular mycorrhizas and the soil-disturbance-induced reduction of nutrient absorption in maize. I. Causal relations. New Phytol 110:67–74

Fairchild GL, Miller MH (1990) Vesicular-arbuscular mycorrhizas and the soil-disturbance-induced reduction of nutrient absorption in maize. III. Influence of P amendments to soil. New Phytol 114:641–650

Fellner R (1989) Mycorrhizae-forming fungi as bioindicators of air pollution. Agric Ecosyst Environ 28:115–120

Fellner R (1993) Air pollution and mycorrhizal fungi in Central Europe. In: Pegler DN, Boddy L, Ing B, Kirk PM (eds) Fungi of Europe: investigation, recording and conservation, Royal Botanic Gardens, Kew, pp 239–250

Fiedler HJ, Hunger W (1963) Über den Einfluss einer Kalkdüngun auf Vorkommen, Wachstum und Nährelamentgehalt höherer Pilze im Fichtenbestand. Arch Forstw 12:936–962

Fitter AH (1985) Functioning of vesicular-arbuscular mycorrhizas under field conditions. New Phytol 99:257–265

Flanagan PW, Van Cleve K (1977) Microbial biomass, respiration and nutrient cycling in a black spruce taiga ecosystem. In: Lohm U, Persson T (eds) Soil organisms as components of ecosystems. Ecol Bull 25:261–273

Foster RC (1981) Polysaccharides in soil fabrics. Science 214:665–667

Foster RC (1994) Microorganisms and soil aggregates. In: Pankhurst CE, Doube BM, Gupta VVSR, Grace PR (eds) Soil biota management in sustainable farming systems. CSIRO Melbourne, Australia, pp 144–155

Friese CF, Allen MF (1991) The spread of VA mycorrhizal fungal hyphae in the soil: inoculum types and external hyphal architecture. Mycologia 83:409–418

Garrett SD (1951) Ecological groups of soil fungi: a survey of substrate relationships. New Phytol 50:149–166

Grime JP (1979) Plant strategies and vegetation processes. Wiley, New York

Gregorich EG, Kachanoski RG, Voroney RP (1989) Carbon mineralization in soil size fractions after various amounts of aggregate disruption. J Soil Sci 40:649–659

Gupta VVSR, Germida JJ (1988) Distribution of microbial biomass and its activity in different soil aggregate size classes as affected by cultivation. Soil Biol Biochem 20:777–786

Hamel C, Fyles H, Smith DL (1990) Measurement of development of endomycorrhizal mycelium using three different vital stains. New Phytol 115:297–302

Harvey AE, Jurgensen MF, Larsen MJ (1980a) Clearcut harvesting and ectomycorrhizae: survival of activity on residual roots and influence on bordering forest stance in western Montana. Can J For Res 10:300–303

Harvey AE, Larsen MJ, Jurgensen MF (1980b) Early effects of partial cutting on the number and distribution of active ectomycorrhzae in steep, rocky soils of Douglas-fir/larch forests in western Montana. Can J For Res 10:436–440

Hassink J, Bouwman LA, Zwart KB, Brussaard L (1993) Relationships between habitable pore space, soil biota and mineralization rates in grassland soils. Soil Biol Biochem 25:47–55

Heinrich Z, Wojewoda W (1976) The effect of fertilization on a pine forest ecosystem in an industrial region. IV. Macromycetes. Ecol Polska 24:319–330

Hendrix JW, Guo BZ, An Z-Q (1995) Divergence of mycorrhizal fungal communities in crop production systems. In: Collins HP, Robertson GP, Klug MJ (eds) The significance and regulation of soil biodiversity, Kluwer, Amsterdam, pp 131–140

Hendrix PF, Parmelee RV, Crossley DA Jr, Coleman DC, Odum EP, Groffman P (1986) Detritus food webs in conventional and no-tillage agroecosystems. Bioscience 36:374–380

Hole F (1982) Effects of animals in soil. Geoderma 25:75–112

Holland EA, Coleman DC (1987) Litter placement effects on microbial and organic matter dynamics in an agroecosystem. Ecology 68:425–433

Hora FB (1959) Presidential address: quantitative experiments on toadstool production in woods. Trans Br Myco Soc 42:1–14

Hunt HW, Coleman DC, Ingham ER, Ingham RE, Elliott ET, Moore JC, Rose SL, Reid CPP, Morley CR (1987) The detrital food web in a short grass prairie. Biol Fert Soils 3:57–68

Jakobsen I, Rosendahl L (1990) Carbon flow into soil and external hyphae from roots of mycorrhizal cucumber plants. New Phytol 115:77–83

Jakobsen I, Abbott LK, Robson AD (1992) External hyphae of vesicular-arbuscular mycorrhizal fungi associated with Trifolium subterraneum L. Spread of hyphae and phosphorus inflow into roots. New Phytol 120:371–380

Jakucs P (1988) Ecological approach to forest decay in Hungary. Ambio 17:267–274

Jastrow JD (1987) Changes in soil aggregation associated with tallgrass prairies restoration. Am J Bot 74:1656–1664

Jenkinson DS, Ladd JN (1981) Microbial biomass in soil, measurement and turnover. In: Paul EA, Ladd JN (eds) Soil biochemistry, vol 5. Marcel Dekker, New York, pp 415–472

Johnson NC, Pfleger FL (1992) Vesicular-arbuscular mycorrhizae and cultural stresses. In: Bethlenfalvay GL, Linderman RG (eds) Mycorrhizae in sustainable agriculture. ASA special publication 54. Agronomy Society of America, Crop Science Society of America, and Soil Science Society of America, Madison, Wisconsin, pp 71–99

Johnson NC, Pfleger FL, Crookston RK, Simmons SR, Copeland PJ (1991) Vesicular-arbuscular mycorrhizas respond to corn and soybean cropping history. New Phytol 117:657–663

Johnson NC, Copeland PJ, Crookston RK, Pfleger FL (1992) Mycorrhizae: possible explanation for yield decline with continuous corn and soybean. Agron J 84:387–390

Jurinak JJ, Dudley LM, Allen MF, Knight WG (1986) The role of calcium oxalate in the availability of phosphorus in soils of semiarid regions: a thermodynamic study. Soil Sci 142:255–261

Kruckelmann HW (1975) Effects of fertiliser, soils, soil tillage and plant species on the frequency of Endogone chlamydospores and mycorrhizal infection in arable soils. In: Sanders FE, Mosse B, Tinker PB (eds) Endomycorrhizas. Academic Press, London, pp 511–525

Kurle JE, Pfleger FL (1994) The effects of cultural practices and pesticides on VAM fungi. In: Pfleger FL, Linderman RG (eds) Mycorrhizae and plant health. APS Press, American Phytopathological Society, St. Paul, Minnesota, pp 101–131

Lee KE, Pankhurst CE (1992) Soil organisms and sustainable productivity. Aust J Soil Res 30:855–892

Levi MP, Cowling EB (1969) Role of nitrogen in wood deterioration. VII. Physiological adaptation of wood-destroying and other fungi to substrates deficient in nitrogen. Phytopathology 59:460–468

Levi MP, Merrill W, Cowling EB (1968) Role of nitrogen in wood deterioration. VI. Mycelial fractions and model nitrogen compounds as substrates for growth of *Polyporus versicolor* and other wood-destroying and wood-inhabiting fungi. Phytopathology 58:626–634

Leyval C, Berthelin J (1991) Weathering of mica by roots and rhizospheric microorganisms of pine. Soil Sci Soc Am J 55:1009–1016

Li XL, George E, Marschner H (1991) Phosphorus depletion and pH decrease at root-soil and hyphae-soil interfaces of VA mycorrhizal white clover fertilized with ammonium. New Phytol 119:397–404

Lodge DJ (1993) Nutrient cycling by fungi in wet tropical forests. In: Isaac S, Frankland JC, Watling R, Whalley AJS (eds) Aspects of tropical mycology. British Mycological Society symposium series 19. Cambridge University Press, Cambridge, pp 37–57

Lodge DJ, Asbury CE (1988) Basidiomycetes reduce export of organic matter from forest slopes. Mycologia 80:888–890

Lodge DJ, Cantrell S (1995) Fungal communities in wet tropical forests: variation in time and space. Can J Bot 73 (Suppl):S1391–S1398

Lodge DJ, McDowell WH, McSwiney CP (1994) The importance of nutrient pulses in tropical forests. Trends Ecol Evol 9:384–387

Lynch JM (1983) Soil biotechnology. Blackwell, London

MacNish GC, Nicholas DA (1987) Some effects of field history on the relationship between grass production in subterranean clover pasture, grain yield and take-all (*Gaeumannomyces graminis* var. *tritici*) in a subsequent crop of wheat at Bannister, Western Australia. Aust J Agric Res 38:1011–1018

Marks PL (1974) The role of in cherry (*Prunus pensylvanica* L.) in the maintenace of stability in northern hardwood ecosystems. Ecol Monogr 44:73–88

Marks PL, Borman FH (1972) Revegetation following forest cutting: mechanisms for return to steady state nutrient cycling. Science 176:914–915

Marschner H, Dell B (1994) Nutrient uptake in mycorrhizal symbiosis. Plant Soil 159:89–102

Mason PA, Musoko MO, Last FT (1992) Short-term changes in vesicular-arbuscular mycorrhizal spore populations in *Terminalia* plantations in Cameroon. In: Read DJ, Lewis DH, Fitter AH, Alexander IJ (eds) Mycorrhizas in ecosystems. CAB International, Oxon, pp 261–267

Matson PA, Boone RD (1984) Natural disturbance and nitrogen mineralization: wave-form dieback of mountain hemlock in the Oregon Cascades. Ecology 65:1511–1516

McGraw A-C, Hendrix JW (1984) Host and soil fumigation effects on spore population densities of species of endogonaceous mycorrhizal fungi. Mycologia 76:122–131

Meentemeyer V (1978) Macroclimate and lignin control of litter decomposition rates. Ecology 59:465–472

Miller MH, McGonigle TP, Addy HD (1995) Functional ecology of vesicular arbuscular mycorrhizas as influenced by phosphate fertilization and tillage in an agricultural ecosystem. Crit Rev Biotech 15:241–255

Miller RM (1987) The ecology of vesicular-arbuscular mycorrhizae in grass- and shrublands. In: Safir GR (ed) Ecophysiology of VA mycorrhizal plants, CRC Press, Boca Raton, pp 135–177

Miller RM, Jastrow JD (1990) Hierarchy of root and mycorrhizal fungal interactions with soil aggregation. Soil Biol Biochem 22:579–584

Miller RM, Jastrow JD (1992a) The application of VA mycorrhizae to ecosystem restoration and reclamation. In: Allen MF (ed) Mycorrhizal functioning. Chapman and Hall, New York, pp 438–467

Miller RM, Jastrow JD (1992b) The role of mycorrhizal fungi in soil conservation. In: Bethlenfalvay GL, Linderman RG (eds) Mycorrhizae in sustainable agriculture. ASA special publication 54. Agronomy Society of America, Crop Science Society of America, and Soil Science Society of America, Madison, Wisconsin, pp 29–44

Miller RM, Jastrow JD (1994) VA mycorrhizae and biogeochemical cycling. In: Pfleger FL, Linderman RG (eds) Mycorrhizae and plant health. American Phytopathology Society, St. Paul, Minnesota, pp 189–212

Miller RM, Reinhardt DR, Jastrow JD (1985) External hyphal production of vesicular-arbuscular mycorrhizal fungi in pasture and tallgrass prairie communities. Oecologia 103:17–23

Modjo HS, Hendrix JW (1986) The mycorrhizal fungus *Glomus macrocarpum* as a cause of tobacco stunt disease. Phytopathology 76:688–691

Nauta M, Vellinga EC (1993) Distribution and decline of macrofungi in The Netherlands. In: Pegler DN, Boddy L, Ing B, Kirk PM (eds) Fungi of Europe: investigation, recording and conservation. Royal Botanic Gardens, Kew, pp 21–46

Neate SM (1994) Soil and crop management practices that affect root diseases of crop plants. In: Pankhurst CE, Doube BM, Gupta VVSR, Grace PR (eds) Soil biota management in sustainable farming systems. CSIRO, Melbourne, Australia, pp 96–106

Nommik H, Vahtras K (1982) Retention and fixation of ammonium and ammonia in soils. In: Stevenson FJ (ed)

Nitrogen in agricultural soils. Agronomy monographs 22. Agronomy Society of America, Madison, Wisconsin, pp 123–171

Oades JM (1984) Soil organic matter and structural stability: mechanisms and implications for management. Plant Soil 76:319–337

Oades JM, Waters AG (1991) Aggregate hierarchy in soils. Aust J Soil Res 29:815–828

O'Halloran IP, Miller MH, Arnold G (1986) Absorption of P by corn (Zea mays L.) as influenced by soil disturbance. Can J Soil Sci 66:287–302

Oliver CD (1986) Silviculture, the next 30 years, the past 30 years. Part I. Overview. J For 84:32–42

O'Neill EG, O'Neill RV, Norby RJ (1991) Hierarchy theory as a guide to mycorrhizal research on large-scale problems. Environ Pollution 73:271–284

Palm CA, Sanchez PA (1990) Decomposition and nutrient release patterns of leaves of three tropical legumes. Biotropica 22:330–338

Paris F, Bonnaud P, Ranger J, Robert M, Lapeyrie F (1995) Weathering of ammonium- or calcium-saturated phyllosilicates by ectomycorrhizal fungi in vitro. Soil Biol Biochem 27:1237–1244

Parkinson D, Balasooriya I, Winterhalder K (1968) Studies on fungi in a pinewood soil. III. Application of the soil sectioning technique to the study of amounts of fungal mycelium in the soil. J Soil Sci 16:258–269

Parrotta JA, Lodge DJ (1991) Fine root dynamics in a subtropical wet forest following hurricane disturbance in Puerto Rico. Biotropica 23:343–347

Paustian K, Schnürer J (1987) Fungal growth response to carbon and nitrogen limitation; application of a model to laboratory and field data. Soil Biol Biochem 19:621–629

Paustian K, Andrén O, Clarholm M, Hansson AC, Johansson G, Lagerlöf J, Lindberg T, Pettersson R, Sohlenius B (1990) Carbon and nitrogen budgets of four agro-ecosystems with annual and perennial crops, with and without N fertilization. J Appl Ecol 27:60–84

Perry DA, Meyer MM, Egeland D, Rose SL, Pilz D (1982) Seedling growth and mycorrhizal formation in clearcut and adjacent, undisturbed soils in Montana: a greenhouse bioassay. For Ecol Manage 4:261–273

Perry DA, Amaranthus MP, Borchers JG, Borchers SL, Brainerd RE (1989) Bootstrapping in ecosystems. Bioscience 39:230–237

Pilz DP, Perry DA (1984) Impact of clearcutting and slash burning on ectomycorrhizal associations of Douglas-fir seedlings. Can J For Res 14:94–100

Pugh GJF (1980) Strategies in fungal ecology. Trans Br Mycol Soc 75:1–14

Pugh GJF, Boddy L (1988) A view of disturbance and life strategies in fungi. Proc R Soc Edinb 94B:3–11

Read DJ (1991) Mycorrhizas in ecosystems – nature's response to the "law of the minimum". In: Hawksworth DL (ed) Frontiers in mycology. 4th Int Mycological Congr, Regensburg 1990, CAB International, Kew, pp 101–130

Read DJ (1992) The mycorrhizal mycelium. In: Allen MF (ed) Mycorrhizal functioning. Chapman and Hall, New York, pp 102–133

Read DJ, Birch P (1988) The effects and implications of disturbance of mycorrhizal mycelial systems. Proc R Soc Edinb 94B:13–24

Read DJ, Leake JR, Langdale AR (1989) The nitrogen nutrition of mycorrhizal fungi and their host plants. In: Boddy L, Marchant R, Read DJ (eds) Nitrogen, phosphorus and sulphur utilization by fungi. Symp British Mycological Society, Birmingham 1988. Cambirdge University Press, Cambridge, pp 181–204

Rishbeth J (1985) Armillaria: resources and hosts. In: Moore D, Casselton LA, Wood DA et al. (eds) Developmental biology of higher fungi. Cambridge University Press, Cambridge, pp 87–101

Robertson GP, Tiedje JM (1984) Denitrification and nitrous oxide production in successional and old-growth Michigan forests. Soil Sci Soc Am J 48:383–389

Rothrock CS (1992) Tillage systems and plant disease. Soil Sci 154:308–315

Rovira AD (1986) Influence of crop rotation and tillage on Rhizoctonia bare patch of wheat. Phytopathology 76:669–673

Rovira AD (1990) The impact of soil and crop management practices on soil-borne root diseases and wheat yields. Soil Use Manage 6:195–200

Rovira AD (1994) The effect of farming practices on the soil biota. In: Pankhurst CE, Doube BM, Gupta VVSR, Grace PR (eds) Soil biota management in sustainable farming systems. CSIRO Melbourne, Australia, pp 81–87

Sanford RL, Parton WJ, Ojima DS, Lodge DJ (1991) Hurricane effects on soil organic matter dynamics and forest production in the Luquillo Experimental Forest, Puerto Rico: results of simulation modelling. Biotropica 23:364–372

Schimel JP, Firestone MK (1989) Nitrogen incorporation and flow through a coniferous forest profile. Soil Sci Soc Am J 53:779–785

Schlechte G (1991) Zur Struktur der basidiomyzetenflora von unterschiedlich immissionsbelasteten Waldstandorten in Südniedersachsen unter besonderer Berücksichtigung der mycorrhizabildung, Jahn und Ernst, Hamburg

Schnürer J, Clarholm M, Rosswall T (1985) Microbial activity in an agricultural soil with different organic matter contents. Soil Biol Biochem 17:612–618

Schoenberger MM, Perry DA (1982) The effect of soil disturbance on growth and ectomycorrhizae of Douglas-fir and western hemlock seedlings: a greenhouse bioassay. Can J For Res 12:343–353

Schreiner RP, Bethlenfalvay GJ (1995) Mycorrhizal interactions in sustainable agriculture. Crit Rev Biotech 15:271–285

Schubert A, Marzachi C, Mazzitelli M, Cravero MC, Bonfante-Fasolo P (1987) Development of total and viable extraradical mycelium in the vesicular-arbuscular mycorrhizal fungus Glomus clarum Nicol. and Schenck. New Phytol 107:183–190

Shields JA, Paul EA, Lowe WE, Parkinson D (1973) Turnover of microbial tissue in soil under field conditions. Soil Biol Biochem 5:753–764

Sieverding E (1991) Vesicular-arbuscular mycorrhiza management in tropical agosystems. Deutsche Gesellschaft für Technische Zusammenarbeit (GTZ) GmbH, Eschborn, Germany

Skinner MF, Bowen GD (1974) The uptake and translocation of phosphate by mycelial strands of pine mycorrhizas. Soil Biol Biochem 6:53–56

Söderström B (1979) Some problems in assessing the fluorescein diacetate-active fungal biomass in soil. Soil Biol Biochem 11:147–148

Söderström B, Bååth E, Lundgren B (1983) Decrease in soil microbial activity and biomasses owing to nitrogen amendments. Can J Microbiol 29:1500–1506

Sparling GP, Cheshire MV (1985) Effect of periodate oxidation on the polysaccharide content and microaggregate stability of rhizosphere and non-rhizosphere soils. Plant Soil 88:113–122

Stevenson FJ (1982) Humus chemistry. Genesis, composition, reactions. Wiley, New York

St. John TV, Coleman DC, Reid CPP (1983) Growth and spatial distribution of nutrient-absorbing organs: selective exploitation of soil heterogeneity. Plant Soil 71:487–493

Swift MJ, Heal OW, Anderson JM (1979) Decomposition in terrestrial ecosystems. Studies in ecology, vol 5. Blackwell, Oxford

Sylvia DM (1988) Activity of external hyphae of vesicular-arbuscular mycorrhizal fungi. Soil Biol Biochem 20:39–43

Termorshuizen AJ, Schaffers AP (1987) Occurrence of carpophores of ectomycorrhizal fungi in selected stands of Pinus sylvestris in The Netherlands in relation to stand vitality and air pollution. Plant Soil 104:209–217

Thompson JP (1987) Decline of vesicular-arbuscular mycorrhizae in long fallow disorder of field crops and its expression in phosphorus deficiency of sunflower. Aust J Agric Res 38:847–867

Thompson JP (1991) Improving the mycorrhizal condition of the soil through cultural practices and effects on growth and phosphorus uptake in plants. In: Johansen C, Lee KK, Sahrawat KL (eds) Phosphorus nutrition of grain legumes in the semi-arid tropics. International Crops Research Institute for the Semi-arid Tropics (ICRISAT), Patancheru, India, pp 117–137

Tiessen H, Stewart JWB (1988) Light and electron microscopy of stained microaggregates: the role of organic matter and microbes in soil aggregation. Biogeochemistry 5:312–322

Tisdall JM (1991) Fungal hyphae and structural stability of soil. Aust J Soil Res 29:729–743

Tisdall JM (1994) Possible role of soil microorganisms in aggregation in soils. Plant Soil 159:115–121

Tisdall JM, Oades JM (1980) The effect of crop rotation on aggregation in a red-brown earth. Aust J Soil Res 18:423–433

Tisdall JM, Oades JM (1982) Organic matter and water-stable aggregates in soils. J Soil Sci 33:141–163

Valencia R, Balsev H, Pax Y, Miño G (1994) High tree alpha-diversity in Amazonian Ecuador. Biodiv Conserv 3:21–28

Van Veen JA, Kuikman PJ (1990) Soil structural aspects of decomposition of organic matter by micro-organisms. Biogeochemistry 11:213–233

Vitousek PM, Matson PA (1984) Mechanisms of nitrogen retention in foret ecosystems: a field experiment. Science 225:51–52

Vitousek PM, Matson PA (1985) Disturbance, nitrogen availability, and nitrogen losses in an intensively managed loblolly pine plantation. Ecology 66:1360–1376

Vitousek PM, Melillo JM (1979) Nitrate losses from disturbed ecosystems: patterns and mechanisms. For Sci 25:605–619

Vitousek PM, Sanford R (1986) Nutrient cycling in moist tropical forest. Annu Rev Ecol Syst 17:137–167

Vogt KA, Grier CC, Vogt DJ (1986) Production, turnover, and nutrient dynamics of above- and below-ground detritus of world forests. Adv Ecol Res 15:303–377

Waide RB, Lugo AE (1992) A research perspective on disturbance and recovery of a tropical montane forest. In: Goldammer JG (ed) Tropical forests in transition: ecology of natural and anthropogenic disturbance processes. Berkhouser, Basel, pp 173–190

Wardle DA (1995) Impacts of disturbance on detritus food webs in ago-ecosystems of contrasting tillage and weed management practices. Adv Ecol Res 26:105–185

Wargo PM, Harrington TC (1991) Host stress and susceptibility. In: Shaw GS II, Kile GA (eds) Armillaria root disease. USDA-Forest Service Agriculture Handbook no 691, Washington DC, pp 88–101

Wilson J, Ingleby K, Mason PA, Ibrahim K, Lawson GJ (1992) Long-term changes in vesicular-arbuscular mycorrhizal spore populations in Terminalia plantations in Côte d'Ivoire. In: Read DJ, Lewis DH, Fitter AH, Alexander IJ (eds) Mycorrhizas in ecosystems. CAB International, Oxon, pp 268–275

Zak JC (1992) Response of soil fungal communities to disturbance. In: Carroll G, Wicklow DT (eds) The fungal community, 2nd edn. Dekker, New York, pp 403–425

Zimmerman JK, Pulliam WM, Lodge DJ, Quiñones-Orfila V, Fetcher N, Guzman-Grajales S, Parrotta JA, Asbury CE, Walker LR, Waide RB (1995) Nitrogen immobilization by decomposing woody debris and the recovery of tropical wet forest from hurricane damage. Oikos 72:314–322

6 Fungi and Industrial Pollutants

M. WAINWRIGHT[1] and G.M. GADD[2]

CONTENTS

I. Introduction

Fungi and other microorganisms are exposed to a wide variety of pollutants when growing in the environment. Principal amongst these are air pollutants, which include acid rain, and toxic metals. The aim of this chapter is to highlight the effects of such pollutants on fungal populations. Emphasis will be placed on the effects of air pollutants and metals on soil fungi, but reference will also be given to the effects of pollutants on phylloplane fungi and on the occurrence and distribution of higher fungi.

Studies on the effects of pollutants on fungi are fraught with difficulty, largely because of the inadequacy of the techniques which are currently available to study fungi in the environment. However, an appreciation of the effect which pollutants can have on fungi can be obtained by a combination of the following measurements: (1) pollutant concentration, composition and distribution; (2) the concentration of the pollutant which is available to the fungal population; (3) the concentration of the pollutant which causes a toxic or physiological response in vitro; (4) the effects of the pollutant loads on fungal population/community size and composition; and finally (5) secondary changes resulting from pollution effects on fungal populations, e.g. impact on leaf litter decomposition. While pollutant concentration and composition are relatively easily determined using standard analytical techniques, many of the chemical extractants in common use do not accurately indicate the degree of pollutant availability.

The effect of pollutants on fungal population/community size and composition is particularly difficult to assess. Many workers have turned to the dilution plate count to obtain a so-called fungal count, which is then used in the hope of assessing any changes in fungal community composition. The shortcomings of this technique have been criticised at length (see, for example Cooke and Rayner 1984). Some authors, however, are less critical of the dilution plate count method; States (1981) and Fritze and Baath (1993), for example, recently defended the use of what they call the sporulation capacity of soils since it provided a picture of the effects of alkaline deposition on the fungal population similar to that obtained by the soil washing technique, which is supposed to favour the isolation of fungal hyphae. However, the dilution technique is perhaps best avoided, especially since methods are available for determining the fungal biomass of soil (Baath and Soderstrom 1980), while recent developments in the use of ergosterol measurements promise to provide a means for accurately determining the response of the total population to the impact of pollutants (West et al. 1987). In one attempt to overcome problems relating to the use of plate counts, Frostegard et al. (1993) analysed the phospholipid fatty acid (PLFA) composition of soil, in

[1] Department of Molecular Biology and Biotechnology, University of Sheffield, Sheffield, S10 6UH, UK
[2] Department of Biological Sciences, University of Dundee, Dundee, DD1 4HN, UK

The Mycota IV
Environmental and Microbial Relationships
Wicklow/Söderström (Eds.)
© Springer-Verlag Berlin Heidelberg 1997

order to detect changes in the overall composition of the microbial community and provide more reliable information on fungal populations than can be produced using plate counts. Two soils were amended with Cd, Cu, Ni, Pb and Zn and analysed after 6 months. PLFA 18:2ω6 is regarded as an indicator of fungal biomass, and this increased with increasing metal contamination for all metals except Cu, possibly reflecting the well-known mycotoxicity of Cu. However, in forest soils, such an increase in PLFA 18:2ω6 was not observed because of masking by identical PLFAs derived from plant material (Frostegard et al. 1993).

Total microbial activity in soils as affected by pollutants can also be determined by respirometric methods, while the proportion of the total respiration contributed by fungi and bacteria can be assessed with the aid of selective inhibitors (Anderson and Domsch 1975). The effects of pollutants on litter decomposition can be readily achieved by determining the rate of breakdown of cellulose or leaf litter in the field. The transfer of tubes containing soil from polluted to unpolluted sites, and vice versa, allows the impact of a pollutant on unpolluted soils to be determined, as well as providing a means of measuring the rate at which microbial communities in a polluted soil return to the prepollution condition (Killham and Wainwright 1981).

There exists a vast literature on the effect of pollutants such as toxic metals, SO_2 and fungicides on fungi when grown in vitro. Unfortunately, many of these data have little, if any, relevance to the potential effects of pollutants on microfungi growing in the environment. It is unlikely that a meaningful picture of how fungi respond to pollutants in the environment can be gained from determining the response of fungi to pollutants added to solid or liquid growth media. The effects of toxic metals on soil fungi growing in vitro, for example, is markedly influenced by the composition of the medium used; metals are likely to be more toxic to fungi in low carbon media than in carbon-rich media where the production of large amounts of extracellular polysaccharides and chemical interactions with the medium will tend to reduce metal availability. The medium components may also complex metals out of solution, making them unavailable (Hughes and Poole 1991). Culture studies also omit the complexities of natural ecosystems; the addition of clay minerals to medium for example, can markedly alter the view that is gained of the toxicity of pollutants such as toxic metals from culture studies in which no clay is added. Finally, synergism between pollutants and organisms is obviously excluded from in vitro studies, while interactions between different pollutants may have a major influence on the toxicity of a pollutant in the natural environment.

Fungi growing in culture may also show atypical responses which rarely occur in the environment. For example, a linear relationship between the toxicity of a pollutant and increasing pollutant concentration is generally assumed. In some cases, however, small amounts of a pollutant may stimulate growth (the so-called oligodynamic effect), while in other cases non-linear responses at various concentrations may occur (so-called paradoxical effect; Schatz et al. 1964).

II. Predicted Effects of Pollutants on Fungal Populations

Environmental pollution might be expected to lead to both toxic (destructive) and enrichment disturbance on fungal populations (Wainwright 1988a). Although toxic disturbance is likely to predominate, instances will occur where both types of disturbance are found together. Toxic disturbance of fungal populations is likely to be particularly damaging to ecosystem functioning, while the more rare enrichment disturbance may occasionally produce beneficial effects on soil processes. Toxic disturbance is likely to lead to a reduction in fungal numbers and species diversity, as well as biomass and activity changes which may detrimentally influence fundamentally important processes such as litter decomposition. The resultant degree of toxic disturbance will depend upon both toxicant concentration and its availability to the fungal population, as well as to the susceptibility of the individuals involved. Toxicants may be selective and affect only a few species, or they may have a more generalised effect. Selective inhibition may have less of an impact on overall soil fungal activity than might be imagined, since susceptible species can be replaced by more resistant fungi, some of which may be more active in a given physiological process than the original population. This helps to explain why the detrimental impact of pollutants on fungal processes is often less severe than might be expected from information gained from the use of pure culture studies.

While concentration effects are generally emphasised, it is surprising how often the question of toxicant availability is avoided in studies on the effects of pollutants on microorganisms. In soils, for example, availability of a pollutant will generally depend upon factors such as (1) adsorption to organic and inorganic matter; (2) the form in which the pollutant occurs (i.e. as a soluble or insoluble form); (3) microbial breakdown; and finally (4) leaching.

Fungal populations are unlikely to remain genetically static when confronted with a toxic agent, and resistant populations are likely to develop which will be a major factor in determining population responses to the pollutant. On the other hand, a number of studies have shown that fungi isolated from metal-contaminated soils show less adaptation to toxic metals, such as copper, than might be expected (Yamamoto et al. 1985; Arnebrant et al. 1987). Mowll and Gadd (1985) also found no differences in the sensitivity of *Aureobasidium pullulans* to lead when isolates from either contaminated and uncontaminated phylloplanes were compared.

Another factor of paramount importance in relation to the effects of toxicants on soil fungi concerns nutrient availability. Fungi are generally thought to be already stressed by the low levels of available carbon present in most soils and other environments (Wainwright 1992). They will grow slowly, if at all, under these conditions, and may be more susceptible to pollutants than when growing in high nutrient media.

The response of the soil fungal population/community is unlikely to remain static because both the concentration and composition of any pollution load impacting on the environment will vary. As a result, the community in a non-polluted soil may reach a new dynamic equilibrium in response to a pollutant load, which might be again disturbed by any further variation in the composition or concentration of a pollutant. Fungal populations/communities in polluted environments will therefore alter continuously in response to changes in the composition of the pollutant load.

Enrichment disturbances may also be either selective or non-selective. Non-selective enrichment disturbance might theoretically result from the input into the ecosystem of a pollutant which can be universally used as a nutrient source. Since such enrichment is rare, most examples of this form of disturbance will be selective. Reduced forms of sulphur from sour gas plants are, for example, likely to enrich the soil for S-oxidizing fungi, while phenolic and hydrocarbon pollutants will favour species capable of utilising these compounds as substrates.

A. Effects of Acid Rain and Airborne Pollutants on Fungal Populations

Considerable research effort has been devoted to determining the effects of atmospheric pollutants, notably acid rain, on the chemistry and microbiology of soils (Tabatabai 1985). However, most of these studies have focussed on the effects of air pollutants on bacteria and relatively few have considered the effects of pollution impact on soil fungi. This neglect largely reflects the problems involved in determining the identity, biomass and activity of these organisms in soil. Fungi play a major role in the carbon and other nutrient cycles in the environment (Wainwright 1988b), the impairment of which could have important consequences for functioning of ecosystems. It is obviously desirable that more is known about the impact of pollutants on these organisms. Unfortunately, while it is easy to speculate on the likely effects of pollutants on fungi, it is far more difficult to demonstrate such effects.

Ruhling et al. (1984) employed various methods to study the effects on soil fungi of toxic metals emitted from a Swedish brass mill. They found that the soil respiration rate, fluorescein diacetate active mycelium (FDA) and mycelial standing crop were all reduced with increasing copper concentration. Colony counts, on the other hand, were not affected, a result which once again illustrates the unsuitability of this technique for determining the effects of pollutants on fungi. A major criticism of this study is that, because total soil copper concentrations were measured, it is unclear how much of the high measured concentrations (up to $10000\,\mu g\,g^{-1}$) of copper were actually available to inhibit the fungi. However, since there is likely to be a correlation between the total amount of metal present in a soil and the biologically available concentration, this study clearly confirms that high concentrations of airborne metals can markedly affect the soil fungal community.

Nordgren et al. (1983, 1985) also showed that fungal biomass and soil respiration decreased by some 75% along an increasing concentration gradient of metal pollution. The dilution plate method again detected no measurable effect in terms of total fungal counts, but *Penicillium* and *Oidodendron* species decreased in isolation frequency while an increase in the frequency of *Geomyces* species occurred; it remains unclear, however, what such frequency changes tell us about the impact of metal pollution on soil fungal populations and communities.

Although acid rain is generally regarded as a long-range pollution phenomenon, it is inevitable that high concentrations of mineral acids will pollute ecosystems close to point source emissions. The effects of acid rain on soil fungi have recently been studied in some detail. Baath et al. (1984), for example, showed that soil biological activity, as determined by respiration rate, was significantly reduced following treatment with simulated acid rain. Mycelial lengths (FDA active) were also reduced by the treatment, while plate counts again showed no response. Studies by Fritze (1987), on the other hand, showed that urban air pollution had no effect on the total length of fungal hyphae in the surface horizons of soils supporting Norway spruce (*Picea abies*). Bewley and Parkinson (1985) showed that the contribution which fungi make to the total respiration of a soil was reduced by acid rain, while, in contrast, Roberts et al. (1980) concluded that the addition of acid rain to forest soils did not affect the normal 9:1 balance of fungal to bacterial respiration; such studies clearly illustrate how difficult it is to generalise about the effects of atmospheric pollutants on soil microorganisms. Among higher fungi, simulated acid rain has been shown to increase the dominance of some ectomychorrizal fungi, while decreasing species diversity among saprophytic species (Sastad and Jenssen 1993). Shaw et al. (1992) also showed that fumigation with sulphur dioxide or ozone had no effect on mycorrhizal populations.

The effect of acid rain on leaf litter decomposition was reviewed by Francis (1986). Acid treatments have been shown to impair the decomposition of both deciduous leaves and conifer needles (Baath et al. 1984; Prescott and Parkinson 1985). Small-scale inhibitory effects were common, although stimulatory effects were also observed. Pollution in the form of alkaline dust from iron and steel works was shown by Fritze (1991) to lead to a doubling of the total length of fungal hyphae. The addition of lime has been shown to variously decrease soil fungal populations (Nodar et al. 1992) or to have no measurable effect (Persson et al. 1989).

The measurement of leaf litter and cellulose decomposition provides a particularly useful way of assessing the impact of atmospheric pollutants on soils. However, in the absence of a means of partitioning the relative impact of the toxicants on fungi, bacteria and soil animals, these methods provide only a measure of the effects of the pollutants on the total soil community. On the other hand, it could be argued that such partitioning is unnecessary, since what is required is a measure of the overall impact of pollutants on important ecosystem processes, such as leaf litter decomposition. Since soil microarthropods play a crucial role in the initial stages of litter degradation, the effects of air pollutants on their populations should never be ignored. Atmospheric pollutants from coking works can, for example, reduce populations of soil microarthropods, a response which retards the rate of litter decomposition in deciduous woodland soils (Killham and Wainwright 1981).

A number of reports have also appeared on the effects of air pollutants from point sources on phylloplane microorganisms, including fungi (Helander et al. 1993). Bewley (1979) for example, showed that microbial populations on leaves of *Lolium perenne* were generally unaffected by high levels of metals, although fungal "numbers" were found to be negatively correlated with lead concentration. The many pitfalls which can arise during the study of the complex interactions between air pollutants and microorganisms were unwittingly demonstrated by Bewley (1980). In these studies, oak (*Quercus*) leaves at an unpolluted site were sprayed with oxides of cadmium, lead and zinc in order to assess the effects of these toxicants on the phylloplane fungi "without the complicating effects of other pollutants found at the polluted site". Unfortunately, such other pollutants would include mineral acids which have an important effect on the availability of the metal ions present in these oxides (Wainwright et al. 1982). Thus, while Bewley (1980) reported that metal oxides had no effects on phylloplane fungi, one wonders what the result would have been had metal availability been increased by spraying the leaves with a solution of dilute sulphuric acid, thereby simulating the effects of acid rain.

Few examples of the effects of enrichment disturbance by air pollutants on fungal populations can be found in the literature. However, fungi have been reported to utilise atmospheric pollution deposits from coking works as a nutrient source, as well as being able to oxidise the reduced sulphur which these particles contain (Killham and Wainwright 1982, 1984).

III. Effects of Toxic Metals on Fungi

The effects of heavy metal pollutants on fungi have long been of scientific interest largely because metal toxicity is the basis of many fungicidal preparations (Ross 1975). In an ecological context, an increase in the rate of pollution of natural environments by toxic metals, metalloids, radionuclides and organometal(loid)s has led to increased interest in their effects on fungi; such interest is based on (1) the ubiquitous and sometimes dominant presence of fungi in metal-polluted habitats; (2) the uptake and translocation of toxic metals and radionuclides to fruiting bodies of edible higher fungi, and (3) the important role which mycorrhizal fungi play in polluted habitats (Fig. 1) (Gadd and White 1989; Brown and Hall 1990; Gadd 1993a). Fungi can be used in

environmental biotechnology of polluted environments; for example the amelioration of metal phytotoxicity by mycorrhizal fungi can be useful in land reclamation (Colpaert and Van Assche 1987), while fungal biomass can be used to detoxify metal/radionuclide-containing industrial wastes (Gadd 1990, 1992a, 1993a).

The ability of fungi to survive in the presence of potentially toxic metals depends on a number of biochemical and structural properties, including physiological and/ or genetical adaptation, morphological changes, and finally environmental modification of the metal in relation to speciation, availability and toxicity (Gadd and Griffiths 1978; Gadd 1990, 1992b). Terms such as resistance and tolerance are often used interchangeably in the literature, and may be arbitrarily based on the ability to grow on a certain metal concentration in laboratory media (Baath 1991; Gadd 1992b). "Resistance" is probably more appropriately defined as the ability of an organism to survive metal toxicity by means of a mechanism produced in direct response to the metal species concerned; the syn-

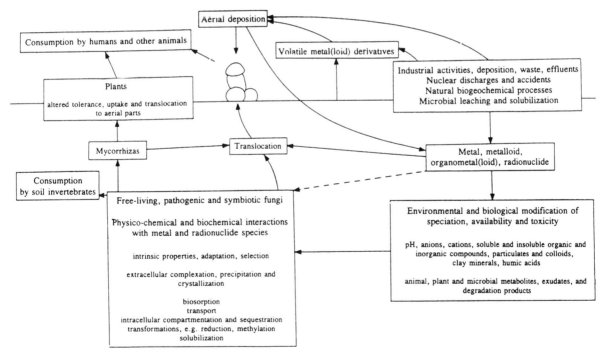

Fig. 1. Diagrammatic representation of the interactions of toxic metals and radionuclides with fungi in the terrestrial environment. In almost every case, there will be environmental and biological modification of metal/radionuclide behaviour. The *dotted line* shows direct effects of metal species on fungi; this may sometimes occur and is more likely for metal species, such as Cs^+, which are highly mobile. Release of metal/radionuclide species from dead and

decomposing animal, plant and microbial biomass is not shown but will be an important part of metal cycling. Fungal roles in metal solubilization from naturally occurring substrates and/or industrial materials is indicated (see Burgstaller and Schinner 1993). For more detailed information regarding physiological and cellular interactions, see Mehra and Winge (1991) and Gadd (1993a); for organometal (loid) transformations, see Gadd (1993b)

thesis of metallothionine and γ-glutamyl peptides in response to the presence of metals providing perhaps the best examples (Mehra and Winge 1991). Metal tolerance may be defined as the ability of an organism to survive metal toxicity by means of intrinsic properties and/or environmental modification of toxicity (Gadd 1992b). Intrinsic properties that can determine survival include possession of impermeable pigmented cell walls, extracellular polysaccharide and metabolite excretion, especially where this leads to detoxification of the metal species by, e.g. binding or precipitation (Gadd 1990, 1993a). However, such distinctions are often difficult to recognise because of the involvement in fungal survival in response to metal toxicity of several direct and indirect physicochemical and biological mechanisms. Biological mechanisms implicated in fungal survival (as distinct from environmental modification of toxicity) include extracellular precipitation; complexation and crystallisation; the transformation of metal species by, e.g. oxidation, reduction, methylation and dealkylation; biosorption to cell walls, pigments and extracellular polysaccharide; decreased transport or impermeability, efflux; intracellular compartmentation; and finally precipitation and/or sequestration (Ross 1975; Gadd and Griffiths 1978; Brown and Hall 1990; Gadd 1990, 1992a,b; Mehra and Winge 1991). It needs to be emphasised that most of the information which is available on cellular responses to toxic metals has been obtained from in vitro studies using pure cultures, and it is often unclear whether the mechanisms listed above operate in natural environments.

A. Effects of Metals on Fungal Populations

A range of fungi from all the major groups may be found in metal-polluted habitats (Ross 1975; Gadd 1990, 1993a). In general terms, toxic metals are believed to affect fungal populations by reducing abundance and species diversity and selecting for a resistant/tolerant population (Jordan and Lechevalier 1975; Babich and Stotzky 1985; Arnebrant et al. 1987). In a paddy field contaminated with Cd, Cu and Zn, the relative degree of metal tolerance exhibited by the microflora was in the order fungi > bacteria > actinomycetes (Hiroki 1992) and, while there may be considerable differences in responses between individual species and strains, it is generally accepted that fungi are less

sensitive to metal pollution than bacteria (Doelman 1985). However, the effect of toxic metals on microbial abundance in natural habitats varies with the metal species and organisms present and also depends on a variety of environmental factors (Gadd and Griffiths 1978).

General reductions in fungal "numbers" (as assessed by the dilution plate count) have often been noted in soils polluted with Cu, Cd, Pb, As and Zn (Bewley and Stotzky 1983; Babich and Stotzky 1985). However, numerical estimates alone may provide little meaningful information unless possible changes in fungal groups and species are considered. There is some evidence that such changes can occur in response to metal exposure. In Cu- and Zn-polluted soil, for example, *Geomyces* and *Paecilomyces* spp. and some sterile forms have been found to increase with increasing pollution, whereas plate counts of *Penicillium* and *Oidodendron* spp. declined at polluted sites (Nordgren et al. 1983). *Trichocladium asperum*, *Trichoderma hamatum*, *Zygorrhynchus moelleri* and *Chrysosporium pannorum* were isolated more frequently from an organomercurial-treated golf green than from untreated location, whereas species of *Chaetomium*, *Fusarium*, *Penicillium* and *Paecilomyces* were greatly reduced (Williams and Pugh 1975). On the other hand, some of the best examples of microbial metal tolerance are found within members of the genus *Penicillium*, which underlines the fact that metal responses may be strain-specific. *Penicillium ochro-chloron* can grow in a saturated solution of $CuSO_4$ (Gadd et al. 1984) and *Penicillium lilacinum* dominated fungal isolates in soil polluted by acid mine drainage (Tatsyama et al. 1975). In soil polluted with cadmium dust, *Strobilurus tenacellus*, *Mycene ammoniaca*, *Auriscalpium vulgare* and *Armillaria lutea* were the most common basidiomycetes (Turnau 1991).

Metal pollution of plant surfaces is widespread and while there may be significant decreases in total microbial numbers (including bacteria) on phylloplanes, numbers of filamentous fungi and non-pigmented yeasts appear to be little affected (Bewley 1979, 1980; Bewley and Campbell 1980; Mowll and Gadd 1985). On polluted oak leaves, *Aureobasidium pullulans* and *Cladosporium* species were the most numerous organisms and a greater proportion were metal-tolerant when compared with control isolates (Bewley 1980). In fact, numbers of *A. pullulans* showed a good positive correlation with lead concentrations, whether derived from industrial or vehicular sources, and this fungus was frequently the dominant microorganism present, in some cases comprising up to 97% of the phylloplane microflora (Bewley and Campbell 1980; Mowll and Gadd 1985).

In contrast, the ballistospore-producing yeast *Sporobolomyces roseus* and heterotrophic bacteria were infrequently isolated from polluted leaves and numbers of *S.roseus* showed a significant negative correlation with increasing lead concentrations (Mowll and Gadd 1985). Smith (1977) also studied heavy metal effects on phylloplane fungi and found that *A. pullulans*, *Epicoccum* spp. and *Phialophora verrucosa* were tolerant; *Gnomia platani*, *Cladosporium* spp. and *Pleurophomella* showed intermediate tolerance, while species of *Pestalotiopsis* and *Chaetomium* were sensitive. Mercury-tolerant fungi have been isolated from the surfaces of seeds treated with mercury compounds; including *Pyrenophora avenae*, *Penicillium crustosum*, *Cladosporium cladosporoides*, *Syncephalastrum racemosum* and *Ulocladium atrum* (Ross 1975). From the above observations, it appears that elevated concentrations of toxic metals can affect both the qualitative and quantitative composition of fungal populations although it is often extremely difficult to separate their effects from those of other environmental pollutants.

It is apparent that certain fungi can exhibit considerable tolerance towards toxic metals and can become dominant microorganisms in some polluted habitats. However, while species diversity may be reduced in certain cases, resistance/tolerance can be exhibited by fungi from both polluted and non-polluted habitats. Fungal numbers were reduced and there were alterations in species composition in polluted soil near a zinc smelter (Jordan and Lechevalier 1975). However, there was little difference in the zinc tolerance of fungi isolated from either site and most could exhibit 50% growth at $700\,\mu M$ Zn. At control sites, zinc-tolerated genera included *Bdellospora*, *Verticillium* and *Paecilomyces*, with *Penicillium*, *Torula* and *Aureobasidium* dominating at polluted sites. In another study, fungal populations in soil contaminated with nicked, copper, iron and cobalt were not significantly different from those at control sites (Freedman and Hutchinson 1980). Copper- and nickel-tolerant fungi (defined as being capable of growth at approximately 1.6 mM Cu and/or Ni) were isolated from both control and contaminated sites, the predominant general being *Penicillium* (60%) followed by species of *Trichoderma*, *Rhodotorula*, *Oidodendron*, *Mortierella* and *Mucor*. Examination of microfungi isolated from unpolluted and copper-polluted forest soil confirmed that species from the polluted site were usually copper-tolerant, although there was little evidence for adaptation in isolates from sites with short or long histories of pollution (Arnebrant et al. 1987). Such studies indicate that, in many cases, survival must be dependent on intrinsic properties of the organisms, rather than adaptive changes (which are generally studied under laboratory conditions). Physicochemical properties of the environment, including changes associated with the metal pollution, may also influence metal toxicity and thereby affect species composition (Gadd 1984, 1990, 1992b, 1993a; Baath 1989).

B. Mycorrhizal Responses Towards Toxic Metals

The responses of mycorrhizal fungi to toxic metals are of importance because of the importance of this symbiosis to plant productivity and to both the natural and artificial revegetation of polluted sites. The amelioration of metal phytotoxicity by mycorrhizal fungi has been widely demonstrated. In ericaceous plants, for example, little if any growth occurs in the presence of copper and zinc when the plants lack mycorrhizas; hyphal complexes of the endophyte apparently being effective in preventing both metal translocation to the shoots and resulting toxic symptoms (Bradley et al. 1982). Similar explanations have been proposed for metal tolerance in ectomycorrhizal fungi (Denny and Wilkins (1987). The importance of vesicular-arbuscular mycorrhizas in conferring metal tolerance varies between different hosts and fungal species. In some cases, protection from toxicity is apparent, whereas in others, increased metal uptake may reduce growth (Killham and Firestone 1983). The ectomycorrhizal species *Amanita muscaria* and *Paxillus involutus* increased zinc tolerance of *Betula* sp.; the latter fungus also ameliorates aluminium toxicity towards Norway spruce when growing in a free-living state in the rhizosphere (Brown and Wilkins 1985). Other ectomycorrhizal fungi capable of enhancing survival of *Betula papyrifera* include *Lactarius hibbardae*, *L. rufus*, *Laccaria proxima* and

Scleroderma flavidum, although overall responses cannot be generalised with respect to either metal or fungus (Jones and Hutchinson (1986)).

The isolation of metal-tolerant mycorrhizal fungi from polluted sites has been widely documented (Brown and Wilkins 1985a; Jones and Hutchinson 1986; Colpaert and Van Assche 1987). However, no correlation has been reported between the apparent metal tolerance of mycorrhizal fungi and the toxicity of the soil of origin (Brown and Wilkins 1985b; Denny and Wilkins 1987). In addition, other studies have shown that a relationship between laboratory-determined metal tolerance of ectomycorrhizal fungi and their ability to increase the metal tolerance of *Betula papyrifera* does not always exist (Jones and Hutchinson 1988b).

C. Accumulation of Metals and Radionuclides by Macrofungi

It is well known that elevated concentrations of toxic metals and radionuclides can occur in fruiting bodies of higher fungi sampled from polluted environments. This phenomenon is of significance in relation to the possible use of macrofungi as bioindicators of metal pollution (discussed later) and because of human toxicity resulting from the consumption of edible wild fungi. The physiology of the process of metal accumulation by sporophores is poorly understood, but represents an important and characteristic fungal attribute. The ability to transport and translocate both essential and inessential solutes along the hyphae, possibly over considerable distances, must be a prime determinant of the ecology of these organisms.

In general, levels of Pb, Cd, Zn and Hg found in macrofungi from urban or industrial areas are higher than from corresponding rural areas, although there are wide differences in uptake abilities between different species; the type of metal also influences the degree to which it is accumulated (Tyler 1980; Bressa et al. 1988; Lepsova and Mejstrik 1989; Brown and Hall 1990). Cadmium is accumulated to quite high levels in macrofungi, averaging around 5 mg $(kg\,dry\,wt.)^{-1}$ although levels of up to 40 mg $(kg\,dry\,wt.)^{-1}$ have also been recorded (Byrne et al. 1976). Average concentrations of As, Cu, Hg, Mn, Se and Zn present in a range of up to 27 species of higher fungi were 1, 50, 2, 3, 30, 1.5 and 114 mg $(kg\,dry\,wt.)^{-1}$, respectively (Byrne et al. 1976). However, a wide range of accumulation values can occur, depending on the species and location of sporophore growth; *Laccaria amethystina* caps exhibited total As concentrations of 100–200 mg $(kg\,dry\,wt.)^{-1}$ (Stijve and Porette 1990; Byrne et al. 1991). Accumulation of ^{110}Ag and ^{203}Hg was studied in *Agaricus bisporus* and concentration factors (metal concentration in mushroom: metal concentration in substrate) were found to be up to 40 and 3.7 respectively, with the highest Ag and Hg contents recorded being 167 and 75 mg $(kg\,dry\,wt.)^{-1}$ respectively (Byrne and Tusek-Znidaric 1990).

As well as fruiting bodies, rhizomorphs (e.g. of *Armillaria* species) can concentrate metals up to 100 times the level found in soil. Concentrations of Al, Zn, Cu and Pb in rhizomorphs were 3440, 1930, 15, 680 mg $(kg\,dry\,wt.)^{-1}$ respectively, with the metals primarily located in extracellular portions. It has been suggested that the formation of a metal-rich coat may increase the longevity and survival of the rhizomorphs (Rizzo et al. 1992).

D. Accumulation of Radiocaesium by Macrofungi

In recent years, particularly following the Chernobyl accident in 1986, there have been several studies on radiocaesium (mainly ^{137}Cs) accumulation by macrofungi. Free-living and mycorrhizal basidiomycetes can accumulate radiocaesium (Haselwandter 1978; Elstner et al. 1987; Byrne 1988; Dighton and Horrill 1988; Haselwandter et al. 1988; Clint et al. 1991; Dighton et al. 1991); these organisms appear to have a slow turnover rate for Cs, and comprise a major pool of radiocaesium in soil (Clint et al. 1991). Mean activities of 25 Ukrainian, 6 Swedish and 10 North American collections were 4660, 9750 and 205 Bq $(kg\,dry\,wt.)^{-1}$, respectively (Smith et al. 1993). Deviations in the $^{137}Cs:^{134}Cs$ ratio attributable to Chernobyl have revealed considerable accumulation of pre-Chernobyl Cs in macrofungi, probably as the result of weapons testing (Byrne 1988; Dighton and Horrill 1988). It appears that about 20% of the ^{137}Cs in Eastern Europe (Moscow area, Belarus, Ukraine) is of non-Chernobyl origin (Smith et al. 1993). Radiocaesium accumulation in basidiomycetes appears to be species-dependent, with influences exerted by soil properties. Significantly higher activites may be found in mycorrhizal species compared to saprotrophic and parasitic fungi (Smith et al. 1993). Some species, e.g. *Suillus granulatus* and *Lactarius hatsudake*, preferentially accumulate radiocaesium rather than potassium (Muramatsu et al. 1991). Levels of ^{137}Cs in about 25 basidiomycete species from Japan varied widely, and ranged from <3–1520 Bq (kg dry wt.)[-1] (Muramatsu et al. 1991). It has been reported that some edible mushrooms (e.g. *Agaricus campestris* and *Agaricus bisporus*) have little or no affinity for [137]Cs (Stijve and Porette 1990). Concentrations of ^{137}Cs in Japanese edible mushrooms (e.g. *Lentinus edodes, Flammulina velutipes, Pleurotus ostreatus* and *Pholiota nameko*, were low at around <50 Bq (kg dry wt.)[-1]) (Muramatsu et al. 1991). However, Smith et al. (1993) found that many prized edible mycorrhizal fungi may contain unacceptably high levels of ^{137}Cs, that is at levels of greater than 1000 Bq (kg dry wt.)[-1]. The influx rates of ^{137}Cs into the hyphae of several basidiomycetes showed considerable variation, with saprotrophic species exhibiting highest rates and mycorrhizal species the lowest. Such observations may partly explain the wide differences in radiocaesium levels encountered in fungal fruiting bodies following the Chernobyl disaster (Clint et al. 1991; Heinrich 1992). It has also been demonstrated that the fungal component of soil can immobilize the total Chernobyl radiocaesium fallout received in upland grasslands (Dighton et al. 1991) although grazing of fruiting bodies by animals may lead to radiocaesium transfer along the food chain (Bakken and Olsen 1990).

E. Fungi as Bioindicators of Metal and Radionuclide Contamination

As has been mentioned above, higher fungi growing on contaminated sites show significantly elevated concentrations of metals in their fruiting bodies, and several experiments have demonstrated a correlation between the quantities of metals added to a growth substrate and the amounts subsequently found in the fruiting bodies (Wondratschek and Roder 1993). The concept of bioindicators is usually discussed in terms of reaction indicators and accumulation indicators. Reaction indicators may comprise individual organisms and/or communities which reveal the extent of metal pollution. Thus, there may be a decline or disappearance of sensitive species and increases in tolerant species, resulting in changes in the overall community composition. In the case of accumulation indicators, the pollutant is measured by analysing the indicator organism. Some organisms therefore can serve as both reaction and accumulation indicators.

The use of fungi as reaction indicators has received little attention although alteration of macrofungal communities by metal pollution has frequently been recorded. Ruhling et al. (1984) noted a decline from about 40 species per 1000 m[2] to about 15 species near the source of metal contamination (smelter emissions), with only *Laccaria laccata* increasing in frequency at more polluted locations. Other higher fungi which are apparently tolerant of high metal pollution include *Amanita muscaria* and several species of *Boletus*; some *Russula* species, on the other hand, appear sensitive to metal toxicity (Wondratschek and Roder 1993).

Fungi possess several advantages over plants as metal accumulation indicators. The fruiting bodies may accumulate greater amounts of metals than plants, while the large area of mycelium ensures contact with and translocation from a large area of soil. Furthermore, fruiting bodies project above the ground for only a short period, thereby minimising contamination from aerial or wet deposition of metal pollutants. Sporophores are also compact, easily harvested, and amenable to rapid chemical analysis. Many fungal species are also widely distributed and numerous, all these factors being desirable attributes for a useful bioindicator (Mejstrik and Lepsova 1993).

However, despite these apparent advantages, it is debatable whether a sufficiently clear relationship exists between the indicator species and the metal pollution under examination. In the case of mercury, for example, a wide variation in metal content in fruiting bodies occurs in different species sampled at the same site, ranging over as much as three orders of magnitude, with higher concentrations occurring in fungi from polluted sites. Some species show extremely high Hg accumulation values. Mercury concentrations in fungi generally occur in the range 0.03–21.6 mg (kg dry wt.)[-1] although concentrations greater than 100 mg (kg dry wt.)[-1] have been recorded from polluted sites. A wide variation also occurs within species at the same site, possibly due to a variable proportion of the metal being derived from aerial contamination. Despite these problems, several macrofungi have been suggested as being suitable bioindicators of mercury pollution (Table 1).

Table 1. Higher fungi proposed as suitable bioindicators for metal pollution based on metal analyses of fruiting bodies. (See Mejstrik and Lepsova 1993; Wondratschek and Roder 1993)

Species	Metal(s)
Agaricus arvensis	Hg, Cd
Agaricus campestris	Hg, Cd
Agaricus edulis	Hg, Cd
Agaricus haemorrhoidarius	Hg
Agaricus xanthodermus	Hg
Agaricus sp.	Pb, Zn, Cu
Amanita rubescens	Hg
Amanita strobiliformis	Hg
Coprinus comatus	Hg
Lycoperdon perlatum	Hg
Lycoperdon sp.	Pb, Zn, Cu
Marasmius oreadus	Hg
Mycena pura	Hg, Cd

A wide variation in Cd content has also been recorded in macrofungi with ranges of reported values from <0.1–229 mg (kg dry wt.)$^{-1}$ (Tyler 1980); not surprisingly, fungi sampled from polluted sites generally contain higher concentrations of metals than sporophores sampled from unpolluted sites. However, there is frequently a lack of correlation between the fungal Cd content and the Cd content of the soil, thereby questioning the suitability of macrofungi as indicators of Cd (Wondratschek and Roder 1993). Compared to other common metal pollutants, lower concentrations of Pb tend to be found in macrofungi, with much of the Pb content being derived from aerial sources, such as industrial and vehicular emissions. Levels of Pb in of around 0.4–36 mg (kg dry wt.)$^{-1}$ have been reported in sporophores, with higher levels occurring in urban areas (Tyler 1980). Difficulties in distinguishing between internal and aerially deposited Pb probably preclude higher fungi as bioindicators for the presence of this metal, other than as an indicator of generalized Pb pollution. Zinc, an essential metal for fungal growth and metabolism, occurs at high concentrations within fungi, 50–300 mg (kg dry wt.)$^{-1}$ (Tyler 1980), with a few genera apparently showing high affinities for the metal (Table 1). Copper may also be found at high levels (20–450 mg (kg dry wt.)$^{-1}$) in higher fungi (Tyler 1980), with some genera again showing particularly high accumulation values. However, with both Cu and Zn, there is a tendency for metal concentrations in fruiting bodies to be independent of soil concentrations which reduces their value as bioindicators (Gast et al. 1988).

Many factors contribute to the wide variations in recorded metal contents of macrofungal fruiting bodies, even in the same species sampled at the same site. Such factors may be organism-related, including the age, and therefore the size and developmental stage of the fruiting body as well as the dimensions and physiological activity of the fungal mycelium. Despite numerous studies, almost all investigations tend to be contradictory and provide little useful information (Wondratschek and Roder 1993); this is perhaps not surprising in view of the complexity of fruiting body formation, and frequent difficulties experienced in standardising sampling and analysis. Additional complicating problems include the existence of different races of fungi having differing affinities for heavy metals as well as showing differential partitioning of any accumulated metal within the fruiting bodies. So far, such phenomena have been little investigated.

Apart from organism-related factors, environmental factors are of paramount importance in relation to metal accumulation by higher fungi, and include physicochemical soil properties like moisture and temperature, all of which influence metal availability as well as the physiological activity of the fungus. Furthermore, metal accumulation may be influenced by additive, competitive or even synergistic interactions between different essential and inessential metal pollutants (Wondratschek and Roder 1993). However, results obtained in this area to date are often inconsistent, again probably reflecting the enormous complexity of the physicochemical and biological processes involved. It can be concluded, therefore, that a perfect macrofungal bioindicator does not exist, although macrofungi may be useful in determining the extent of a polluted or unpolluted area, and perhaps also indicate the sources of pollution.

IV. Conclusions

It is clear from the above that while the theoretical response of fungi to pollutants can readily be speculated upon, these effects are difficult to demonstrate and quantify largely because, until recently, suitable techniques have not been available to determine accurately changes in fungal populations/communities and their activity in the environment, notably in soils. However, newly developed methods, such as techniques for determining active fungal biomass, should allow a better appreciation to be gained of the effects of pollutants on soil fungi. Growth media containing low, and therefore more realistic, concentrations of available carbon should also be used if in vitro techniques are employed to help determine the effects of pollutants on fungal growth. By using such approaches, a more accurate assessment should be possible of the impact of pollutants on the growth and activity of fungi in the environment.

References

Anderson JPE, Domsch KH (1975) Measurement of bacterial and fungal contribution to respiration of selected agricultural and forest soils. Can J Microbiol 21:314–321

Arnebrant K, Baath E, Nordgren A (1987) Copper tolerance of microfungi isolated from polluted and unpolluted forest soil. Mycologia 79:890–895

Baath E (1989) Effects of heavy metals in soil on microbial processes and populations (a review). Water Air Soil Pollut 47:335–379

Baath E (1991) Tolerance of copper by entomogenous fungi and the use of copper-amended media for isolation of entomogenous fungi from soil. Mycol Res 95:1140–1152

Baath E, Soderstrom B (1980) Comparisons of the agar-film and membrane filter methods for the estimation of hyphal lengths in soil with particular reference the effect of magnification. Soil Biol Biochem 12:385–387

Baath E, Lundgren B, Soderstrom B (1984) Fungal populations in podzolic soil experimentally acidified to simulate acid rain. Microbial Ecol 10:197–203

Babich H, Stotzky G (1985) Heavy metal toxicity to microbe-mediated ecologic processes: a review and potential application to regulatory policies. Environ Res 36:11–137

Bakken LR, Olsen RA (1990) Accumulation of radiocaesium in fungi. Can J Microbiol 36:704–710

Bewley RJF (1979) The effects of zinc, lead and cadmium pollution on the leaf surface microflora of Lolium perenne L. J Gen Microbiol 110:247–254

Bewley RJF (1980) Effects of heavy metal pollution of oak leaf microorganisms. Appl Eviron Microbiol 40:1053–1059

Bewley RJF, Campbell R (1980) Influence of zinc, lead and cadmium pollutants on microflora of hawthorn leaves. Microbial Ecol 6:227–240

Bewley RJF, Parkinson D (1985) Bacterial and fungal activity in sulphur dioxide polluted soils. Can J Microbiol 31:13–15

Bewley RJF, Stotzky G (1983) Effects of cadmium and zinc on microbial activity in soils: influence of clay minerals, part 1: metals added individually. Sci Total Environ 31:41–45

Bradley R, Burt AJ, Read DJ (1982) The biology of mycorrhiza in the Ericaceae. VIII. The role of mycorrhizal infection in heavy metal resistance. New Phytol 91:197–209

Bressa G, Cima L, Costa P (1988) Bioaccumulation of Hg in the mushroom Pleurotus ostreatus. Ecotoxicol Environ Safety 16:85–89

Brown MT, Hall IR (1990) Metal tolerance in fungi. In: Shaw J (ed) Heavy metal tolerance in plants: evolutionary aspects. CRC Press, Boca Raton, pp 94–104

Brown MT, Wilkins DA (1985a) Zinc tolerance of mycorrhizal Betula. New Phytol 99:101–106

Brown MT, Wilkins DA (1985b) Zinc tolerance of Amanita and Paxillus. Trans Br Mycol Soc 84:367–369

Burgstaller W, Schinner F (1993) Leaching of metals with fungi. J Biotechnol 27:91–116

Byrne AR (1988) Radioactivity in fungi in Slovenia, Yugoslavia, following the Chernobyl accident. J Environ Radioact 6:177–183

Byrne AR, Tusek-Znidaric M (1990) Studies of the uptake and binding of trace metals in fungi, part 1: accumulation and characterisation of mercury and silver in the cultivated mushroom, Agaricus bisporus. Appl Organometal Chem 4:43–48

Byrne AR, Ravnik V, Kosta L (1976) Trace element concentrations in higher fungi. Sci Total Environ 6:65–78

Byrne AR, Tusek-Znidaric M, Puri BK, Irgolic KJ (1991) Studies of the uptake and binding of trace metals in fungi, part II: arsenic compounds in Laccaria amethystina. Appl Organometal Chem 5:25–32

Clint GM, Dighton J, Rees S (1991) Influx of 137_{Cs} into hyphae of basidiomycete fungi. Mycol Res 95:1047–1051

Colpaert JV, Van Assche JA (1987) Heavy metal tolerance in some ectomycorrhizal fungi. Funct Ecol 1:415–421

Cook RC, Rayner ADM (1984) Ecology of saptotrophic fungi. Longman, London

Denny HJ, Wilkins DA (1987) Zinc tolerance in Betula spp. IV. The mechanism of ectomycorrhizal amelioration of zinc toxicity. New Phytol 106:545–553

Dighton J, Horrill AD (1988) Radiocaesium accumulation in the mycorrhizal fungi Lactarius rufus and Inocybe longicystis in upland Britain following the Chernobyl accident. Trans Br Mycol Soc 91:335–337

Dighton J, Clint GM, Poskitt J (1991) Uptake and accumulation of ^{137}Cs by upland grassland soil fungi: a potential pool of Cs immobilization. Mycol Res 95:1052–1056

Doelman P (1985) Resistance of soil microbial communities to heavy metals. In: Jensen V, Kjoller A, Sorensen LH (eds) Microbial communities in soil. Elsevier, London, pp 369–384

Elstner EF, Fink R, Holl W, Lengfelder E, Ziegler H (1987) Natural and Chernobyl-caused radioactivity in mushrooms, mosses and soil samples of defined biotopes in S.W. Bavaria. Oecologia 73:553–558

Francis AJ (1986) Acid rain effects on soil and aquatic processes. Experientia 42:455–465

Fritze H (1987) The influence of urban air pollution on soil respiration and fungal hyphal length. Ann Bot Finn 24:251–256

Freedman B, Hutchinson TC (1980) Effects of smelter pollutants on forest leaf litter decomposition near a nickel-copper smelter at Sudbury, Ontario. Can J Bot 58:1722–1736

Fritze H (1991) Forest soil microbial responses to emissions from an iron and steel smelter. Soil Biol Biochem 23:151–155

Fritze H, Baath E (1993) Microfungal species composition and fungal biomass in a coniferous forest soil polluted by alkaline deposition. Microbial Ecol 25:83–92

Frostegard A, Tunlid A, Baath E (1993) Phosopholipid fatty acid composition, biomass, and activity of microbial communities from two soil types experimentally exposed to different heavy metals. Appl Environ Microbiol 59:3605–3617

Gadd GM (1984) Effect of copper on *Aureobasidium pullulans* in solid medium: adaptation not necessary for tolerant behaviour. Trans Br Mycol Soc 82:546–549

Gadd GM (1990) Metal tolerance. In: Edwards C (ed) Microbiology of extreme environments. Open University Press, Milton Keynes, pp 178–210

Gadd GM (1992a) Molecular biology and biotechnology of microbial interactions with organic and inorganic heavy metal compounds. In: Herbert RA, Sharp RJ (eds) Molecular biology and biotechnology of extremophiles. Blackie, Glasgow, pp 225–257

Gadd GM (1992b) Metals and microorganisms: a problem of definition. FEMS Microbiol Lett 100:197–204

Gadd GM (1993a) Interactions of fungi with toxic metals. New Phytol 124:25–60

Gadd G (1993b) Microbial formation and transformation of organometallic and organometalloid compounds. FEMS Microbiol Rev 11:297–316

Gadd GM, Griffiths AJ (1978) Microorganisms and heavy metal toxicity. Microbial Ecol 4:303–317

Gadd GM, White C (1989) Heavy metal and radionuclide accumulation and toxicity in fungi and yeasts. In: Poole RK, Gadd GM (eds) Metal-microbe interactions. IRL Press, Oxford, pp 19–38

Gadd GM, Chudek JA, Foster R, Reed RH (1984) The osmotic responses of *Penicillium ochro-chloron*: changes in internal solute levels in response to copper and salt stress. J Gen Microbiol 130:1969–1975

Gast CH, Jansen E, Bierling J, Haanstra L (1988) Heavy metals in mushrooms and their relationship with soil characteristics. Chemosphere 17:789–799

Haselwandter K (1978) Accumulation of the radioactive nuclide ^{137}Cs in fruitbodies of basidiomycetes. Health Phys 34:713–715

Haselwandter K, Berreck M, Brunner P (1988) Fungi as bioindicators of radiocaesium contamination: pre- and post-Chernobyl activities. Trans Br Mycol Soc 90:171–174

Heinrich G (1992) Uptake and transfer factors of ^{137}Cs by mushrooms. Radiat Environ Phys 31:39–49

Helander ML, Ranta H and Neuvonen S (1993) Responses of phyllosphere microfungi to simulated sulphuric and nitric acid deposition. Mycol Res 97:533–537

Hiroki M (1992) Effects of heavy metal contamination on soil microbial population. Soil Sci Plant Nutr 38:141–147

Hughes MN, Poole RK (1991) Metal speciation and microbial growth – the hard (and soft) facts. J Gen Microbiol 137:725–734

Jones MD, Hutchinson TC (1986) The effects of mycorrhizal infection on the response of *Betula papyrifera* to nickel and copper. New Phytol 102:429–442

Jones MD, Hutchinson TC (1988a) Nickel toxicity in mycorrhizal birch seedlings infected with *Lactarius rufus* or *Scleroderma flavidum*. I. Effects on growth, photosynthesis, respiration and transpiration. New Phytol 108:451–459

Jones MD, Hutchinson TC (1988b) Nickel toxicity in mycorrhizal birch seedlings infected with *Lactarius rufus* or *Scheroderma flavidum*. II. Uptake of nickel, calcium, magesium, phosphorus and iron. New Phytol 108:461–470

Jordan MJ, Lechevalier MP (1975) Effects of zinc-smelter emissions on forest soil microflora. Can J Microbiol 21:1855–1865

Killham K, Firestone MK (1983) Vesicular-arbuscular mycorrhizal mediation of grass responses to acidic and heavy metal depositions. Plant Soil 72:39–48

Killham K, Wainwright M (1981) Deciduous leaf litter and cellulose decomposition in soil exposed to heavy atmospheric pollution. Environ Pollut [A] 26:69–78

Killham K, Wainwright M (1982) Microbial release of sulphur ions from atmospheric pollution deposits. J Appl Ecol 18:889–896

Killham K, Wainwright M (1984) Chemical and microbiological changes in soil following exposure to heavy atmospheric pollution. Environ Pollut [A] 33:122–131

Lepsova A, Mejstrik V (1989) Trace elements in fruit bodies of fungi under different pollution stress. Agric Ecosyst Environ 28:305–312

Mehra RK, Winge DR (1991) Metal ion resistance in fungi: molecular mechanisms and their related expression. J Cell Biochem 45:30–40

Mejstrik V, Lepsova A (1993) Applicability of fungi to the monitoring of environmental pollution by heavy metals. In: Market B (ed) Plants as biomonitors. VCH Verlagsgesellschaft, Weinheim, pp 365–378

Mowll JL, Gadd GM (1985) The effect of vehicular lead pollution on phylloplane mycoflora. Trans Br Mycol Soc 84:685–689

Muramatsu Y, Yoshida S, Sumiya M (1991) Concentrations of radiocesium and potassium in basidiomycetes collected in Japan. Sci Total Environ 105:29–39

Nodar R, Acea MJ, Carballas T (1992) Microbiological response to Ca(OH)$_2$ treatments in a forest soil. FEMS Microbiol Ecol 86:213–219

Nordgren A, Baath E, Soderstrom B (1983) Microfungi and microbial activity along a heavy metal gradient. Appl Environ Microbiol 45:1839–1837

Nordgren A, Baath E, Soderstrom B (1985) Soil microfungi in an area polluted by heavy metals. Can J Bot 63:448–455

Persson T, Lundkvist H, Wiren A, Hyvonen R, Wessen B (1989) Effect of acidification and liming on carbon and nitrogen mineralization and soil organisms in mor humus. Water Air Soil Pollut 45:77–79

Prescott CE, Parkinson D (1985) Effects of sulphur pollution on rates of litter decomposition in a pine forest. Can J Bot 63:1436–1443

Rizzo DM, Blanchette RA, Palmer MA (1992) Biosorption of metal ions by *Armillaria* rhizomorphs. Can J Bot 70:1515–1520

Roberts TM, Clarke TA, Ineson P, Gray TRG (1980) Effects of sulphur deposition on litter decomposition and nutrient leaching in coniferous forest soils. In: Hutchinson TC, Hava M (eds) Effects of acid precipitation on terrestrial ecosystems. Dekker, New York, pp 381–393

Ross IS (1975) Some effects of heavy metals on fungal cells. Trans Br Mycol Soc 64:175–193

Ruhling A, Baath E, Nordgren A, Soderstrom B (1984) Fungi in metal contaminated soil near the Gusum brass mill, Sweden. Ambio 13:34–36

Sastad SM, Jensenn HB (1993) Interpretation of regional differences in the fungal biota as effects of atmospheric pollution. Mycol Res 12:1451–1458

Schatz A, Schalscha EB and Schatz V (1964) The occurrence and importance of paradoxical concentration effects in biological systems. Compost Sci 5:26–30

Shaw PJA, Dighton J, Poskitt J, McCleod AR (1992) The effects of sulphur dioxide and ozone on the mycorrhizas of Scots pine and Norway spruce in a field fumigation system. Mycol Res 96:785–791

Smith ML, Taylor HW and Sharma HD (1993) Comparison of the post-Chernobyl ^{137}Cs contamination of mushrooms from Eastern Europe, Sweden, and North America. Appl Environ Microbiol 59:134–139

Smith WH (1977) Influence of heavy metal leaf contaminants on the in vitro growth of urban-tree phylloplane fungi. Microbial Ecol 3:231–239

States JS (1981) Useful criteria in the description of fungal communities. In: Wicklow DT, Carroll GC (eds) The fungal community. Dekker, New York, pp 185–199

Stijve T, Porette M (1990) Radiocaesium levels in wild-growing mushrooms from various locations. Mushroom J (Summer 1990):5–9

Tabatabai M (1985) Effect of acid rain on soils. CRC Crit Rev Environ Contr 15:65–109

Tatsuyama K, Egawa H, Senmaru H, Yamamoto H, Ishioka S, Tamatsukuri T, Saito K (1975) *Penicillium lilacinum*: its tolerance to cadmium. Experientia 31:1037–1038

Turnau K (1991) The influence of cadmium dust on fungi on a *Pino-Quercetum* forest. Ekol Polska 39:39–57

Tyler G (1980) Metals in sporophores of basidiomycetes. Trans Br Mycol Soc 74:41–49

Wainwright M (1988a) Effect of point source atmospheric pollution on fungal communities. Proc R Soc Edinb 94B:97–104

Wainwright M (1988b) Metabolic diversity of fungi in relation to growth and mineral cycling in soil–a review. Trans Br Mycol Soc 23:85–90

Wainwright M (1992) Oligotrophic growth of fungi-stress or natural state? In: Jennings DH (ed) Stress tolerance of fungi. Dekker, New York, pp 127–144

Wainwright M, Supharungsun S, Killham K (1982) Effects of acid rain on the solubility of heavy metal oxides and fluorspar added to soil. Sci Total Environ 23:85–90

West AW, Grant WD, Sparling GP (1987) Use of ergosterol, diampimelic acid and glucosamine content of soils to monitor changes in microbial populations. Soil Biol Biochem 19:607–612

Williams JI, Pugh GJF (1975) Resistance of *Chrysoporium pannorum* to an organomercury fungicide. Trans Br Mycol Soc 64:255–263

Wondratschek I, Roder U (1993) Monitoring of heavy metals in soils by higher fungi. In: Markert B (ed) Plants as biomonitors. VCH Verlagsgesellschaft, Weinheim, pp 345–363

Yamamoto H, Tatsuyama K, Uchiwa T (1985) Fungal flora of soil polluted with copper. Soil Biol Biochem 17:785–790

7 Fungi in Extreme Environments

N. MAGAN[1]

CONTENTS

I. General Introduction

In nature, fungi are ubiquitous, having evolved over time to occupy a wide range of ecological niches. These niches are determined by competition between species, and may overlap. Thus certain species may occupy similar general niches because of their ecologically similar behaviour within a community. The occupation of specific niches may be determined by the availability of nutrients, temperature, water availability, pH or salt concentration. Such stress factors may be a long-lived or permanent feature of a habitat while other disturbance factors may be only transient, having a short-lived impact. Based on work in plant communities, it has been suggested that fungi employ different primary strategies to survive and prosper in different environments. These are combative (C-selected) strategies which maximize occupation and exploitation of resources in relatively non-stressed and undisturbed conditions; stress (S-selected) strategies which have involved the development of adaptations which allow survival and endurance of continuous environmental stress; and ruderal (R-selected) strategies characterized by a short life span with a high reproductive potential, which often enables success in severely disturbed but nutrient-rich conditions. These three stategies can merge to give secondary strategies (C-R, S-R, C-S, C-S-R) which form part of a continuum with transition zones between them. The main attributes of these three primary groups are summarized in Fig. 1. In this chapter we are specifically concerned with S-selected strategies for growth and survival in a range of so-called extreme environments. It should, however, be remembered that in nature, fungal activity will be influenced by a range of biotic and abiotic factors, both spatially and temporally.

II. Thermotolerance

Fungi have the ability to grow over a wide range of temperature conditions. Fungi, and microorganisms generally, have been classified as psychrophiles, mesophiles, thermotolerant or true thermophiles. A thermophilic fungus is defined as one which has minimum growth at 20 °C or above and a maximum growth at 50 °C or above. Optima for thermophilic fungi thus occur in the range 40–50 °C. Figure 2 shows the optimal, maximum and minimum temperature range for some thermotolerant and thermophilic fungi. While there are well-known examples of bacteria which are able to grow in a variety of natural environments including hot springs and geysers where temperature can reach 100 °C, eukaryotes are much more sensitive because at temperatures

[1] Applied Mycology Group, Cranfield Biotechnology Centre, Cranfield University, Cranfield, Bedford MK43 0AL, UK

The Mycota IV
Environmental and Microbial Relationships
Wicklow/Söderström (Eds.)
© Springer-Verlag Berlin Heidelberg 1997

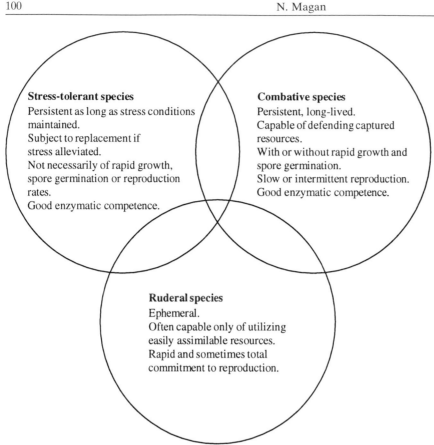

Fig. 1. Summary of attributes of fungi in relation to the three major ecological strategies. (Cooke and Rayner 1984)

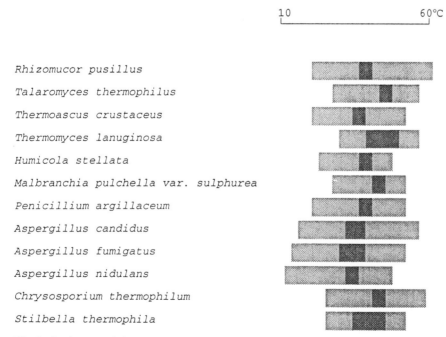

Fig. 2. Optimum, minimum and maximum temperatures and range of conditions for growth of some thermotolerant and thermophilic fungi

above 65 °C, their membranes become irreparably damaged. However, many mesophilic thermotolerant fungi do exist, with, for example, some deuteromycetes isolated from thermal springs having maximum growth temperature of 61.5 °C (Tansey and Brock 1973). One must, however, distinguish between the ability to actively grow as a thermophile at such high temperatures and survival. Often, thermotolerant species are found as components of communities of fungi colonizing a range of damp organic substrates, particularly hay, straw-based composts and moist temperature and tropical cereals, birds' nests and tropical soils. They thus form important components of the succession of fungi colonizing a wide variety of substrates.

Most vegetative yeast cells, fungal mycelium and asexual conidia are killed by exposure to 80 °C for only 1 min. By contrast, sexually produced ascospores of some food spoilage fungi are able to survive readily even for 1 h exposure at this temperature. Such fungi include *Talaromyces flavus* and *Neosartorya fischeri* var. *glaber* (Beuchat 1988; Baggerman and Samson 1988). Heat resistance has usually been described by the determination of two types of values: D and z. A D-value is defined as the decimal reduction time, and indicates the periods of time required to reduce a certain number of living organisms by a factor of 10 under standard temperature and other environmental conditions. If the D-values at different temperature are plotted on a logarithmic scale, then a straight line graph should be obtained, the slope of which is known as the z-value. This defines the increase in temperature (°C) necessary to decrease or increase the D-value by a factor of 10. Figure 3 shows the heat resistance of ascospores, conidia and some yeast cells at different temperatures. This shows clearly the gradation of sensitivity of yeasts and spores of filamentous fungi to increasing temperature. It should, however, be borne in mind that the D- and z-values will also be influenced by pH, water availability and the actual nature of the substratum.

A. Mechanisms of Thermotolerance

There have been a number of hypothesis proposed for explaining the basis of thermophilly. Crisan (1973) suggested four main possibilities; (1) lipid solubilization; (2) rapid resynthesis of essential metabolites; (3) molecular thermostability and (4) ultrastructural thermostability. The latter may be of particular importance because there is the possibility that solubilization of cellular lipids can occur at high temperature to an extent where cells lose their integrity. An increase in temperature may result in cellular lipids containing more saturated fatty acids which have a higher melting point than those present in mesophiles. They would thus be able to maintain cellular integrity at

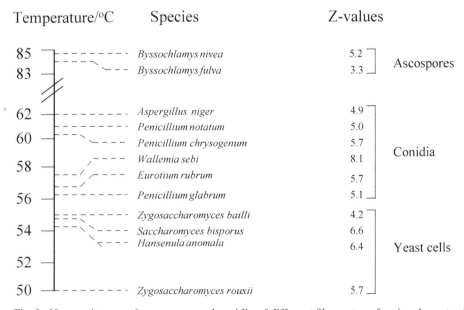

Fig. 3. Heat resistance of ascospores and conidia of different filamentous fungi and yeast cells. The D- and z-values were determined in phosphate buffer (0.2 mll⁻¹; pH 5.5) containing sucrose (400 g l⁻¹). (Baggerman and Samson 1988)

higher temperatures than mesophiles, which contain markedly less saturated lipids. It has also been suggested that the increased fluidity of saturated lipids at high temperatures may enable metabolic activity and cell functioning to enable active growth at >40 °C.

Recently, attention has been focused on the ability of organisms generally to produce a specific range of proteins, so-called heat shock proteins, when exposed to extremes of environmental factors, particularly of temperature. However, very little work has been carried out on heat shock proteins in fungi. These have to a large extent been carried out on *Saccharomyces cerevisiae*, *Neurospora crassa*, *Aspergillus nidulas*, *Achlya ambisexualis* and *Schizophillum commune*. In almost all cases, studies have involved exposure of strains of these fungi to elevated temperatures of, for example, 45–55 °C for a period of 1–3 h. Such conditions have been found to reduce growth, but not aerial hyphal development, and often resulted in the appearance of a number of proteins resolvable by SDS-PAGE that were newly synthesized or whose synthesis increased. In the basidiomycete *S. commune* it was also found that proteolytic processes were significantly affected by such exposure (Higgins and Lilly 1993). In the last 5 years, intensive investigations into the activities of heat shock proteins have produced significant advances in their cellular role. It has been discovered that many of these (heat shock) proteins are, in fact, essential proteins that are synthesized normally by cells at temperatures optimal for growth. Sometimes this has been accompanied by concomitant production of the low molecular sugar alcohol, glycerol, which will be considered in more detail in Section IV. Detailed information of the type of heat shock proteins and on cellular effects of heat shock have been extensively reviewed recently (see Plesofsky-Vig and Brambl 1993).

III. Psychrophiles

There are a wide range of natural habitats where low temperatures occur continuously or fluctuate due to seasonal effects. These regions include oceans, the tundra and subarctic regions. For example, although the oceans have a stable temperature of <5 °C (Morita 1974), there are distinct difficulties in actually studying such ecological systems because of the effects of pressure. Also, often clear distinctions have not been made between true psychrophiles and cold-tolerant species of fungi. Generally, members of all classes of fungi exist in areas such as the Arctic Tundra region. However, most information has been gathered on the survival and growth of fungi in subarctic and dry cold desert regions of the world. Fungi may be present because they are true psychrophiles or are psychrotolerant, with the ability to survive but not actively grow at temperatures <5 °C. Schmidt-Nielsen (1902) proposed the term psychrophilic, which has been defined as an organism which has optimum growth at not greater than 16 °C, and a maximum of about 20 °C (Morris and Clarke 1987). For example, detailed studies of the yeast community of cold dry deserts showed that species such as *Cryptococcus visnniacii* were common (Vishniac and Hempfling 1979). They also, interestingly, showed that many of these psychrophilic yeasts in fact had temperature ranges up to 20 °C or more. However, in vitro studies do not often give an accurate indication of the conditions over which active growth might occur in a natural habitat. This does, however, differ considerably from work in marine environments with bacteria where the temperature ranges were significantly narrower (Fuhrman and Anzam 1980).

However, species of *Aureobasidium* (dimorphic yeasts) have been isolated from rocks in Antarctica which can grow and reproduce at 0–5 °C and are able to tolerate temperatures as low as −80 °C. Other stable man-made environments such as domestic fridges provide temperature in the range 4–10 °C and allow spoilage fungi, particularly *Penicillium* spp. to grow on foodstuffs. There is also evidence that spores of common deuteromycetes such as *Cladosporium* spp. and of *Sporotrichum* sp. are able to germinate and grow on meat in cold storage at temperatures below 0 °C (Cochrane 1958). Some fungi, the aptly named snow moulds such as *Fusarium* spp. *Sclerotinia borealis*, *Phacidium infestans* and *Typhula* spp. are common in northern British Columbia and Sweden and able to actively colonise foliage, particularly of grasses, to cause significant disease. For example, *S. borealis* in culture has optimum temperature for growth of 0 °C and a minimum of <−7 °C and does not grow at >15 °C. It is thus a true psychrophile. However, *Typhula idahoensis* and *T. trifolii* grow at −5 °C and have an optimum of 5 °C. They do grow at near 20 °C, but no growth occurs at 25 °C. Some of these fungi are considered to be psychrophilic, while others which actively grow and infect at approx. 3 °C (e.g. *Monographella nivalis*) are considered to be a cold-tolerant mesophiles because they are able to grow at 20 °C. There may be some niche overlap between such fungi, with the latter being a very effective organism at 3 °C by being at a competitive advantage in such a very specialized ecological environment.

As mentioned earlier, there are many climatic regions where very cold winter temperatures are followed by quite warm temperature in the summer months, e.g. subarctic zones. Here, overwintering and survival are of critical importance for effective competition in the subsequent season. A large number of phyllosphere, endophytic and soil fungi are able to survive very severe winter temperatures of <–50 °C. They are able to remain viable by surviving as resting structures, often thick-walled chlamydospores and sclerotia, or thin-walled conidia. Even thin-walled asexually produced conidia are resistant to freezing temperatures (Mazur 1966). For example, conidia of *Aspergillus flavus* were found to be resistant to freezing in water at –73 °C. This survival may partially be due to a very low water content so that little or no ice formation occurs, which can affect the integrity of the spore.

A. Survival at Low Temperatures

Cellular processes obviously slow down significantly as temperature is reduced. This is manifested by cessation of growth, and the slowing down of enzyme activity, denaturing of proteins and their synthesis, and perhaps transport or membrane integrity. When freezing occurs, the effect depends on whether rapid or slow freezing occurs. Rapid freezing results in the cell contents becoming more concentrated and coming into equilibrium with the extracellular solution, resulting in shrinkage in the cell. Where slow cooling occurs, extensive supercooling can occur. If this occurs at –39 °C or below, homogeneous nucleation of ice occurs. Particles can act as a nucleus, resulting in heterogeneous nucleation before this temperature is reached. Cooling and freezing will thus affect the structure and metabolism of cells significantly. Mazur (1966) proposed a two-factor hypothesis; that at relatively fast cooling rates injury by intracellular ice formation occurs, while at slow cooling rates injury is caused by prolonged

exposure to solution effects due to a concentration of extracellular solution or cell dehydration. By contrast, Steponkus (1984) suggests that there is no evidence that intracellular ice causes injury. Other hypotheses have been reviewed at length by Smith (1993).

It is possible to summarize the type of injury which can occur based on work on plant protoplasts (Steponkus 1984). These include expansion-induced lysis during warming which occurs as dilution of the suspending medium, allowing expansion of the protoplasts; loss of osmotic responsiveness during slow cooling and warming due to changes in electrolyte concentration, and effects on the lipid-protein membrane affecting osmotic responsiveness; leakage of intracellular solutes through plasma membrane; and intracellular ice formation during fast cooling (>–3 °C min^{-1}). Thus, cryoinjury is dependent on the rate of cooling, cell type and the internal and external composition of the cell and its substrate.

Compounds are, however, often produced within the cell to enable survival and growth at low temperature and, indeed, other stresses. Of particular importance for survival of fungi are the production of osmoregulators or compatible solutes which are often synthesized when the fungus is placed under general environmental stress, but particularly of water availability or low temperature stress. These will be referred to in more detail in a later section and include arabitol, erythritol, glycerol, proline and trehalose. Trehalose has, for example, been demonstrated to effectively improve cryotolerance of *Saccharomyces cerevisiae*, baker's yeast, to withstand temperatures of –20 to –70 °C. They are able to reduce the amount of ice formation at low temperature, thus reducing water loss from cells. However, during freezing, significant shrinkage of the cell membrane can occur which can influence subsequent survival. For example, it has been shown that diameter of the hyphae of *Phytophthora nicotianae* was reduced by up to 60% when compared to untreated controls. This can cause direct physical injury by dissolution of the cell membrane, as ice formation results in an increase in volume of about 10%. Thus, cryoprotectants have the effect of preventing such significant shrinkages, and reduce ice-crystal formation. This can be achieved by exogenous addition of cryoprotectants to fungi prior to freezing for culture collection purposes (Smith 1993). The effects of cooling rates and storage temperatures on recovery of strains have been studied in detail to determine the best methods of preservation (see Smith 1993).

In the 1980s the use of cryogenic light microscopy to examine the effect of freezing and thawing on fungal mycelium and propagules proved particularly useful. Studies of *Penicillium expansum* showed that shrinkage of hyphae at slow cooling rates occurred, and intracellular ice formation at fast rates of cooling. Interestingly, hyphal septa were found to form no barriers to such ice nucleation. Comparison of *P. expansum* with *P. nicotianae* showed that the former was extremely resistant to such freezing and thawing, while the latter was sensitive and failed to recover from such treatment (Smith et al. 1986). Extensive and rapid shrinkage of mycelium occurred at all cooling rates up to $-120\,°C\,min^{-1}$. Culture age, growth phase and nutrient status of medium have all been found to influence the effect on fungi. There appear to be two distinct groups of fungi, those which show shrinkage at slow rates of cooling but less at high rates accompanied by intracellular ice formation, while other fungi shrink at all cooling rates with no intracellular ice formation occuring. These groups cut across taxonomic groupings. For example, *P. nicotianae* (oomycete), *Aschersomia alleyrodis* (hyphomycete) and *Lentinus edodes* and *Volvariella volvacea* (basidiomycetes) all respond without any ice nucleation occurring.

Some contrasting effects have been observed with these groups of fungi. For example, fungi such as *P. nictianae* and *P. expansum* react by shrinkage of the mycelium and deletion of plasma and nuclear membranes and cytoplasm between cell wall and membrane. On subsequent thawing, the hyphae often reexpand to their original shape and size and are able to grow again. For other fungi, e.g. *L. edodes*, the hyphae do not return to their original size or shape. However, such species still retain viability. Smith (1993) has suggested that the membrane is a critical structure in tolerance to freezing and thawing cycles. Because it is not very elastic and does not therefore fold easily, material must be lost from the structure during shrinkage due to freezing. Thus, rapid shrinkage can cause damage to the hyphal cells although quite a number of fungi can survive during rapid cooling, even if intracellular ice formation occurs (Morris et al. 1988). Generally, information is to a large extent available on the impact of freezing and thawing in relation to cryopreservation. However, more information is required on the ecology of psychrophilic fungi in natural habitats, what

type of niche overlap may exist between different species, their competence to survive such conditions and actively grow where the cycle may involve both slow and fast cooling or thawing.

IV. Water Relations of Fungi

Microorganisms all require a source of water to enable cellular functioning to occur effectively. They all have a semipermeable cell membrane, which allows water molecules to enter the cell through osmosis to come to equilibrium with its environment. However, often conditions prevail where water may be scarce either due to the presence of a high concentration of salts, e.g. a marine environment or a dry arid desert region, or in intermediate moisture agricultural products. Because all cellular processes occur in water solutions, cells must physiologically be able to adjust to such osmotic alterations to be able to grow and reproduce in these environments. Certain groups of yeasts and filamentous fungi have over time evolved the capability of adaptation to such environments and exploiting niches occupied by few other microorganisms, or dominanting them by preemptive exclusion of others.

A. Concept of Water Availability

Scott (1957) was the first to identify the importance of water availability and try to relate this to the total water content of substrates. For example, in solid substrates such as agricultural commodities, water content consists of bound water (water of constitution), which is held in chemical union with the absorbing substrate by very strong forces, and free water, which is weakly bound. Free water is more readily available for microbial growth and metabolism than bound water, but the ease with which it can be removed depends on the water content of the substrate. The degree of binding also varies with type of substrate, thus total water content is not a good indicator of water availability for microbial growth. Scott (1957) suggested that water acitivity (a_w) would best describe the water availability for microbial activity. Thus a_w is the ratio between the vapour pressure of water in a substrate (P) and the vapour pressure of pure water (Po) at the same temperature and pressure; thus $a_w = P/Po$. The a_w of pure water is 1. A substrate containing no free water has a smaller vapour pressure than pure water and its a_w is consequently less. An alternative measure to a_w is that of water potential (Ψ), which is often used in soil microbiology and is measured in pascals (Pa). This is the sum of the osmotic, matric and turgor potentials and is related directly to a_w by the following formula:

$$\text{water potential} \left(\Psi\right) = RT/V \log_n a_w \left(+P\right),$$

where R is the ideal gas constant, T the absolute temperature, P the atmospheric pressure and V is the volume of 1 mol of water. The advantage of Ψ is that it is possible to pantition osmotic and matric components and their influence on growth and physiological functioning of microbes. The relationship between a_w and Ψ is shown in Table 1. In cells, the Ψ of the environment normally almost always equals that of the cell: $\Psi env = \Psi cell$. In most cases, the Ψ is a function of the osmotic component. In addition, because fungi have a very rigid cell wall, this prevents swelling of the cytoplasm and the total Ψ is the combination of the osmotic and the turgor pressure of the cell wall.

Microorganisms which are able to tolerate and actively grow under conditions of water stress have been described by various terms. The most common have included halophilic, osmophilic, osmotolerant, xerotolerant or xerophilic. The two most appropriate terms for fungi are probably osmophilic, which describes specialized groups of yeasts which are able to grow in high salt environments, and xerophilic (from the Greek; dry-loving). Pitt (1975) defined a xerophile as a fungus which is able to grow in some phase of its life cycle at -22.4 MPa ($0.85 a_w$), and this has now become generally accepted.

Table 1. Water activity, equilibrium relative humidity and water potentials at 25 °C

Water activity	E.R.H. (%)	Water potential (−MPa)
1.00	100	0
0.99	99	1.38
0.98	98	2.78
0.97	97	4.19
0.96	96	5.62
0.95	95	7.06
0.90	90	14.50
0.85	85	22.40
0.80	80	30.70
0.75	75	39.60
0.70	70	40.10
0.65	65	59.30
0.60	60	70.30

B. Fungal Growth and Water Potential

The effect of Ψ experimentally has been usually determined by measurement of growth responses to different steady-state osmotic and matric potentials in vitro by modifications of media using ionic or non-ionic solutes (e.g. NaCl, KCl, glycerol, polyethylene glycol: 200–6000 mw). This has enabled data to be obtained on optimum, maximum and minimum conditions of Ψ for growth and survival. Figure 4 shows a general example of the pattern of relative growth rate in relation to Ψ of a growth medium. Osmotolerant yeasts such as *Zygosaccharomyces rugosus*, *Z. rouxii*, *Torulopsis*

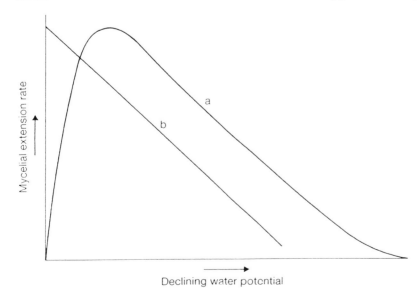

Mycelial extension rate

Declining water potential

Fig. 4. Idealised relationship between growth and water potential; *a* solute potential; *b* matric potential. (Griffin 1981)

halonitatophila and *Saccharomyces mellis* or *S. cerevisiae* all grew in this manner with media modified with PEG 200 (Anand and Brown 1968). For these yeasts the Ψ_{opt} is quite close to the Ψ_{max}. However, for some xerophilic fungi the Ψ_{opt} may be very different from the Ψ_{max}. For example *Eurotium* spp., *Xeromyces bisporous* and *Chrysosporium fastidium* have Ψ_{opt} values at 25 °C of −7.0, −5.6 and −2.8 MPa, respectively (Pitt and Hocking 1977; Magan and Lacey 1984a). However, it should be noted that optimum Ψ for growth will also be influenced by interaction with other environmental factors, particularly temperature, pH and gas composition (Magan and Lacey 1984a,b). Interaction between temperature and Ψ have been studied in detail, and Fig. 5 shows examples of the effect Ψ x temperature interactions on growth rates of a range of fungi. The isopleths represent points of similar rates of minimum, maximum and optimum growth temperature and Ψ for growth (mm day^{-1}). This shows a two-dimensional relationship between two very important environmental factors which can influ-

ence the ecological competence and competitive ability of different fungi (Magan and Lacey 1984c). It has been found that germination usually occurs at a slightly lower Ψmin than that for growth (Magan and Lacey 1984a). Indeed, Ayerst (1969) and Smith and Hill (1982) found the reciprocal of germination time to be significantly correlated with linear growth rates. However, others have surprisingly found that for some *Fusarium* spp. growth occurred over a wider Ψ range than for germination.

It is particularly important to consider not just single, but two- and three-way interactions between environmental factors, as they might affect the community structure within an ecosystem. This has seldom been examined in detail. However, using the stored grain ecosystem as a model, attempts have been made to predict dominance of individual fungi, niche overlap, and temporal changes in community structure (Magan and Lacey 1984c, 1985). These studies showed that a_w, temperature, substrate nutrient status all have a profound influence on antagonism, competitive-

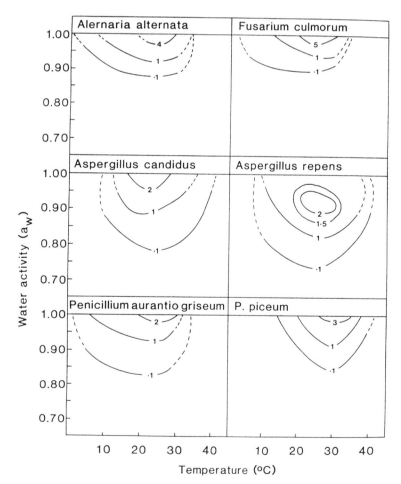

Fig. 5. Diagrammatic representation of the effect of water activity and temperature on growth of different filamentous fungi. The *figures on the isopleths* represent growth rates in mm/day

ness, and dominance of individual fungi. The production of secondary metabolites, mycotoxins, are also similarly affected and can contribute to the success of individual fungi by preemptive exclusion from a common resource.

Scott (1957) suggested that the optimum Ψ for growth of osmotolerant fungi was independent of the predominant solute used in the medium, although a range of filamentous spoilage fungi have been found by Pitt and Hocking (1977) and Andrews and Pitt (1987) to grow similarly with different solutes. For some soil fungi, distinct differences in germination and growth optima have been found for germination and growth in relation to ionic/non-ionic osmotic solutes and matric potential (PEG 6000) alterations (Magan 1988). Scott's (1957) suggestion that absolute growth rate was, however, related to solute type has been borne out in a number of studies. The radial growth rates of fungi over a range of Ψ have been found to vary considerably for species of decomposer fungi with lower Ψ_{min} for growth with glycerol as solute than with KCl to modify the medium, perhaps because of the role of glycerol as a compatible solute at lowered Ψ (Luard 1983; Magan and Lynch 1986).

C. Adaptation to Water Potential

Growth under water stress due to osmotic or matric potential effects requires the maintenance of cell turgor for cell functioning and growth and reproduction to occur. A shift to a high osmotic/water potential affects nutrient uptake, protein biosynthesis and a number of enzyme activities. To enable internal cell functioning, particularly of essential enzymes, fungi produce so-called compatible solutes, often polyhydric alcohols or organic acids. The polyols include glycerol, arabitol, erythritol and mannitol. The low-molecular-weight polyol, glycerol, is particularly important as it is able to protect hydrated biopolymers and allow structural integrity under low water potential conditions. Glycerol is more important relative to the other polyols because it produces a lower a_w than the other polyols at the same molar concentration, followed by arabitol, erythritol, and then the higher molecular weight polyol, mannitol. The physiology of osmophilic yeasts and of filamentous xerophilic fungi will be considered here in more detail, as the mechanisms for adaptation to water stress can be different.

D. Yeast Physiology and Osmotic Stress

In yeasts, polyols are the main compatible solutes, glycerol being the predominant polyol and arabitol the minor one (Edgeley and Brown 1978). However, there are differences in the way in which yeast species use glycerol as a compatible solute. Sensitive isolates of *S. cerevisiae* have been found to both synthesize and secrete glycerol, thus maintaining approximately the same ratio between intra- and extracellular glycerol concentrations (Edgeley and Brown 1980). Experiments in media containing 10% salt (approx. −7.0 MPa water potential) showed an almost 40-fold increase in the enzyme glycerol-3-phosphate dehydrogenase. By contrast, the osmotolerant yeast *Zygosaccharomyces rouxii* retains a higher proportion of the same synthesized glycerol within the cell, indicative of a lower permeability of the plasma membrane. Van Zyl and Prior (1990) demonstrated that in *Z. rouxii*, glycerol was actively transported into the cell, via a carrier-mediated system with a high specificity for glycerol. An active glycerol transport system has also been demonstrated for another osmotolerant yeast *Debaryomyces hansenii*, which shows an increase in production and accumulation of glycerol in cells at lowered water potentials (Andre et al. 1988; Larsen et al. 1990) although the importance of the carrier system has not yet been determined.

Studies with *S. cerevisae* and *Z. rouxii* have shown that the trigger for glycerol synthesis may be K^+ depletion, because of the transient rapid decrease when these yeasts are transferred from high to low osmotic potential media. By contrast, in the osmotolerant yeast *D. hansenii* Na^+ is excluded and K^+ accumulated, so that the internal $K^+:Na^+$ ratio is much higher than that of the medium. However, glycerol accumulation is still probably more important in overcoming such stress. Recently, work has been concentrated on the way in which yeast cells may sense and respond to osmotic stress. Mager and Varela (1993) have produced a hypothesis that change in external osmolarity is probably sensed at the plasma membrane as a result of disturbance of ion gradients (e.g. Na^+, K^+, H^+; see Fig. 6). This results in a loss in turgor pressure and a complex series of molecular events, including protein kinase cascade, that leads to modification of enzyme activities and changes in gene expression. Part of this process may include synthesis of trehalose and certain heat shock proteins to help recovery pro-

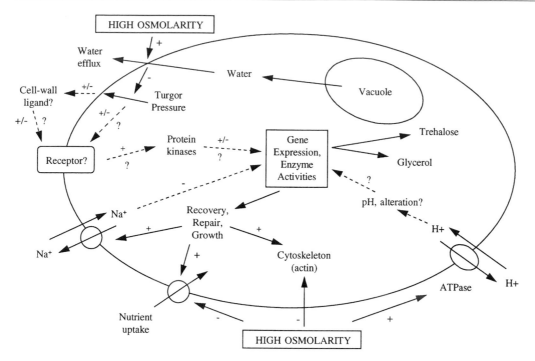

Fig. 6. Schematic relationship of how yeasts may respond to osmotic stress. *Bold arrows* represent flux of molecules, biosynthesis of compatible solutes and physical forces.

Dashes and plain arrows represent putative pathways. (Mager and Varela 1993)

cesses of the cell. At the same time, polyols such as glycerol are produced to restore the turgor pressure in the yeast cell.

E. Xerophilic Filamentous Fungi

It is interesting to note that, compared with the wealth of information on physiological adaptation of yeasts to water stress, much less work has been carried out on filamentous fungi. Lower filamentous fungi have been found to lack polyols under water stress but instead accumulate organic compounds such as proline in response to lowered water potentials. Luard (1982b) demonstrated that *Mucor hiemalis*, *Phytophthora cinnamomi* and *Pythium debaryanum* all synthesized proline when grown on media containing either ionic or non-ionic solutes to modify water potential. Although these fungi are not xerophilic and are quite sensitive to water stress, proline may act in a way similar to glvcerol in yeasts, enabling enzymes to function efficiently.

For other filamentous fungi, work by Luard (1982a,b) has been particularly useful in understanding the relative importance of different polyols in enabling growth at low water potentials. This showed that the type of solute present in a

medium may influence the major polyol accumulated in the mycelium of fungi. For example, the marine fungus *Dendryphiella salina* accumulates glycerol, mannitol and inositol when grown on media containing NaCl, MgCl and inositol to modify water potential. Luard showed that for xerophilic fungi such as *Chrysosporium fastidium* and xerotolerant *Penicillium chrysogenum*, glycerol was the major polyol accumulated in mycelium, with lower concentrations of arabitol and erythritol. As water potential was reduced mannitol levels decreased, suggesting that it may function as a carbohydrate source or energy reserve for glycerol production. Studies by Hocking (1986) have also shown that xerophilic fungi accumulate glycerol during active growth phases, but that when sporulation occurs glycerol depletion can occur rapidly, suggesting that it could be acting as an energy reserve for the production of conidia. Recent work on non-xerophilic entomopathogenic fungi has demonstrated that culture age, temperature, pH, C:N ratio and solute, used for modifying water potential, significantly influence accumulations of glycerol, mannitol and erythritol in conidia of *B. bassiana*, *M. anisopliae* and *P. farinosus* (Hallsworth and Magan 1994a,b, 1996) and can give an ecological advantage in colonizing insects in the environment, particularly at lowered a_w (Hallsworth and Magan 1995). However, such information on xerophilic fungi is lacking. Generally, it is possible to say that for both yeasts and filamentous fungi, glycerol may be the most effective compatible solute and that those fungi which are able to produce and retain glycerol may be at a competitive advantage under water stress conditions and colonise substrates preferentially. At the present time, little information is available on whether active transport mechanisms exist in filamentous fungi, as has been found for some osmophilic yeasts. This may be one area where more detailed study is needed to elucidate the mechanisms involved in the rapid accumulation of polyols in xerotolerant and xerophilic filamentous fungi in relation to water stress.

V. Anaerobic Fungi

In natural substrates, fungi are often exposed to atmospheric conditions which are aerobically not ideal. Thus many fungi are microaerophilic and able to actively colonize substrates under low oxygen conditions. Such environments include aquatic habitats, wood, and man-made grain ecosystems where elevated CO_2 is sometimes used as part of controlled atmosphere storage systems. Dematiaceus hyphomycetes, and some *Aspergillus* and *Penicillium* spp. are good examples of such fungi. Many other fungi are considered to be facultative anaerobes. Such fungi include *Saccharomyces cerevisiae*, and members of the Chytridiomycetes and Oomycetes including *Blastocladiella ramosa* and *Aqualinderella fermentans* (Held 1970). For such fungi to actively grow anaerobically, specific exogenous supplies of nutrients such as fatty acids, sterols and vitamins are necessary (Bull and Bushell 1976; Gibb and Walsh 1980).

In 1975, true anaerobic fungi, morphologically similar to Chytridiomycete fungi, were found in the rumen of sheep (Orpin 1975). Since then, several genera of anaerobic fungi have been isolated and identified from the rumen of herbivores. These fungi appeared to generally have a vegetative stage and sporulate by the production of zoospores, which can be mono- or polyflagellate. Because of their unusual morphological characters, including the presence of hydrogenosomes and fine structure, they have been placed in a separate family, aptly called the Anaeromycetales.

They are not parasites, but symbiotic, as they utilize nutrients and also provide nutrients to the herbivore by production of volatile fatty acids to the animal (Trinci et al. 1994).

Fungi such as *Neocallimastix frontallis*, *Piromonas communis* and *Spaeromonas communis* are confined to the herbivore rumen and thus need to compete effectively with a range of other microflora and fauna, including anaerobic and facultative anerobic bacteria, anaerobic protozoa and each other. The rumen is a very complex ecosystem and it is worthwhile comparing the populations of different organisms in the rumen to obtain a perspective on the possible interactions and the competitive nature of the different components in this type of environment. The rumen contents commonly contain 10^6 protozoa, 10^{10} bacteria and 10^0 anaerobic fungi g^{-1}. Thus, in this ecosystem, active competition occurs for the readily utilisable sugars that may be present from the structural components of plant material (Orpin and Ho 1992; Trinci et al. 1994). Effective competition by anaerobic fungi is also partially influenced by the pH of rumen and its contents. The pH is usually in the range 5.7–7.0 and optimum growth of anaerobic fungi occurs at pH 6.0–7.0 (Orpin 1975), while the temperature range is relatively narrow and fluctuates only between 37–41 °C, depending on herbivore species.

One of the factors which has made these fungi such effective inhabitants of this special habitat is their ability to rapidly catabolize plant material, predominantly cellulose and hemicellulose, to actively grow and multiply in the rumen. The life cycle thus consists of two main stages: a vegetative stage when attached to fragments of digested plant material, and a sporulation stage when motile zoospores are released. These zoospores are very important in enabling anaerobic fungi to have a competitive advantage over other organisms. The zoospores, via a chemotactic response, are able to rapidly utilize the nutrients, particularly glucose, sucrose and fructose, in freshly ingested plant material. They are able to attach themselves to the material, enabling them to encyst rapidly. The hemicellulose and cellulose are utilized efficiently while the more recalcitrant lignified tissue is often less effectively metabolized.

Metabolism in anaerobic fungi has been studied predominantly in *Neocallimastix patriciarum* and operates predominantly via a mixed acid fermentation, resulting in the production of volatile fatty acids and lactate. Other compounds produced include ethanol and succinate and, of course, CO_2 and hydrogen. There appears to be a close interrelationship between some bacteria, particularly methanogens, which can result in cross-feeding of fermentation products and more effective breakdown of the plant material.

The effectiveness of anaerobic fungi is linked to their capability for producing a range of enzymes including cellulases, cellodextrinase, xylanases, glycosidases and aryl esterases, which are all critical to efficient utilization of plant material in the herbivore rumen in a competitive habitat. Orpin (1993) and Trinci et al. (1994) have detailed the importance of the production of these enzymes by anaerobic fungi which place them at a competitive advantage in this very specialized ecological niche.

VI. Acidophiles and Alkalophiles

There are distinct ranges of hydrogen ion concentration over which biochemical and chemical processes in cells will effectively and efficiently be completed. Because the hydrogen ion concentration affects the ionic state and therefore the possible availability of inorganic ions and metabolites to the organism, it is critical in determining the metabolic activity of cells. Very high concentrations of hydrogen ions (acidic) or very low concentrations (alkaline) will have a profound effect on the activity and thus ability of organisms to effectively live in an environment. They function best at pHs close to neutral. For example, extreme pH can result in the primary and secondary structure of proteins to be irreversibly damaged. Thus even if the external pH is extreme, the internal cellular pH must be maintained at close to neutral for efficient cellular functioning. However, the possession of a osmotic barrier to the external environment can maintain the cytoplasmic components at a different pH to the surrounding substrate. It is this ability which has enabled fungi to become established in both extremely low (close to 1) and high pH (11) environments (Longworthy 1978).

A. Acidophiles

Acidophiles have been defined as oganisms which are able to grow down to pH 1 and be able to grow actively at <pH 4. Although acidophilic microorganisms are predominantly bacteria, particularly thiobacilli (Ingledew 1990), a range of yeasts and filamentous fungi have been found to grow in very low pH substrates. Good examples of low pH environments include geothermal regions with high hydrogen sulphide emissions, coal refuse tips and acidic copper mine wastes. Most yeasts grow optimally at pH 5.5–6, some, including *Candida krusei*, *Rhodotorula mucilaginosa* and *Saccharomyces exigua*, can grow at pH 1.5–2 (Recca and Mrak 1952; Battley and Bartlett 1966). Species of *Saccharomyces*, *S. ellipsoideus*, *S. guttulata* and *S. cerevisiae* were demonstrated to grow actively at, respectively, pH 2.5, 2 and 1.9 (Shirfine and Phaff 1958).

The most acidophilic filamentous fungi reported are *Acontium velutium* and a *Cephalosporium* sp. which were isolated from laboratory media containing 2.5n H2S04 (Starkey and Waksman 1943). Many other filamentous fungi have been shown to be able to grow at very low pH values. Many *Aspergillus*, *Eurotium*, *Fusarium* and *Penicillium* spp. can grow down to pH values of 2 but also have pH optima of up to 10. *Acontium pullulans* was isolated at pH 2.5 from acidic coal waste and acid streams by Belly and Brock (1974). It is, however, very difficult to correlate the activity of fungi at low pH values in laboratory media with their activity in natural habitats, especially where pH is a key ecological factor impacting on the microbial community structure. Furthermore, while acidic soils and streams may contain specific groups of fungi, this does not always relate directly to their activity in laboratory media. Thus, some care is needed in interpreting the capabilities of fungi to grow in extreme acidic environments. More detailed in situ studies are necessary to determine the importance of such abiotic factors.

B. Alkalophiles

Akaline environments include soda lakes, desert soils and alkaline springs, where the pH can often be consistently at about pH 10. In many cases, the presence of ammonium carbonate, potassium carbonate, sodium borate or sodium orthophosphate are responsible for the alkaline nature of these environments. Alkalotolerant organisms have been defined as those which grow optimally at approx. pH 7 but are able to grow actively at pH values of up to 9–9.5. Alkalophilic organisms are defined as those which do not grow at <pH 8.5 or have optimum growth at two pH units above neutral (Kroll 1990).

Many fungi are able to grow over a very wide range of pH values and often between pH 2 and pH 11. However, many of these fungi, which come from 15 different genera including species of *Botrytis, Colletotrichum, Cladosporium, Fusarium, Penicillium* and *Paecillomyces*, are most probably alkalotolerant. For example, *Paecilomyces lilacimus* was described as being alkalophilic and able to grow very well between pH 7.5 and 9. True alkaliphilic *Chrysosporium* spp. were isolated and described from birds' nests having a pH maximum for growth of pH 11. These fungi are specialized keratinolytic organisms living in a very specialized environment. Among the yeasts, *Exophila alcaliphila, Candida pseudotropicalis* and *Saccharomyces fragilis* have been described as being alkalotolerant. However, few examples of truly alkalophilic yeasts exists.

C. Mechanisms of Survival in Extreme pH Environments

The hydrogen ion (H⁺) is a very special cation because it is a proton with no electrons. In solution it becomes hydrated to form the hydronium ion (H$_3$O⁺). At acid pH this predominates, while at alkaline pH the hydroxyl ion (OH⁻) is dominant. The protons and the membrane transport and bioenergetic processes are critical to the ability of acidophile and alkalophiles to colonise such specialised environments. The ability to occupy these niches is largely determined by the ability of microorganisms, including fungi, to have pH-controlling systems. This involves efficient transmembrane transport systems so that solutes needed to achieve intracellular modifications can be effectively utilized to maintain the membrane potential with respect to the outside environment. This involves the efficient control of proton movement into and out of the cells, and the supply of the necessary energy requirements. It is also linked with the control of osmotic pressure because of the involvement of cations and anions. However, practically no work has been carried out on the mechanisms involved in yeasts or filamentous fungi. Work has predominantly been carried out with acidophilic and alkalophilic bacteria. This work has been extensively reviewed by

Longworthy (1978), Ingledew (1990) and Kroll (1990). Mechanisms of internal pH control have been ascribed for bacteria such as *Streptococcus faecalis*, where the control has been found to be completely due to the operations of the ATPase to increase proton pump efficiency. For example, in extreme acidic conditions, the internal pH falls quickly, ATP is used to rapidly pump protons out of the bacterial cells via the ATPase to increase the internal pH of the cell. In alkalophilic bacteria such as *Bacillus alcalophilus*, sodium (Na⁺) is utilised to reverse the pH gradient in extreme alkaline conditions. Adaptations of bacteria to enable growth in these environments have included the possession of flagella, modifications to cell walls and membranes, and in biochemical activity including respiration and oxidative phosphorylation. The extensive studies on bacteria need to be extended to yeasts and filamentous fungi to enable a more clear understanding of their occupation of such ecological niches. However, the possible impact of pH alone must not be seen in isolation, and their activity will also be influenced by interactions between pH, water availability and osmotic potential, and temperature.

VII. Irradiation and Fungi

The two types of radiation which fungi are exposed to are firstly, non-ionizing radiation due to solar radiation and ultraviolet (UV) light, and secondly, ionizing radiation from natural and man-made sources. The most important component of non-ionizing radiation in the environment is the UV-B light. Because of depletion of the ozone layer, exposure to UV-B light (290–320 nm) has increased and its impacts on plants and microorganisms are receiving increasing attention. Man-made irradiation sources include gamma-irradiation, which has been used for a long time as a method of preserving food, particularly that intended for human consumption (IAEA-FAO 1978). Microwave radiation has also been used increasingly to enable more rapid drying of substrates and the destruction of pests and microorganisms, including fungi. Irradiation splits water molecules into free radicals which damage DNA, and are highly toxic to a wide range of organisms. However, specific groups of fungi which, for example, occupy environments where they are exposed particularly to natural radiation from UV-light, such as plant surfaces (phyllosphere), often have dark pigmentation of both mycelium and spores to increase survival potential.

Work on the effects of UV-B radiation on saprophytes and pathogens of plants has recently received particular attention. For example, Ayres et al. (1996) showed that yeasts such as *Sporobolomyces roseus* and *Crytococcus* spp. had different senstivities to UV-B light. *S. roseus* was less sensitive than the *Cryptococcus* sp. Spore germination in cereal plant pathogens such as *Septoria nodorum*, but not *S. tritici*, was inhibited by low doses of UV-B light. Interestingly, isolates from warmer climatic regions in North Africa were found to be less sensitive to UV-B fluxes than UK isolates (Ayres et al. 1996). Thus, different strains of the same fungus may have evolved in very different ways in temperate and subtropical environments where exposure to UV-B occur for longer periods of time. While radiation damages DNA of microorganisms to some extent, some are able to repair radiation-induced damage relatively quickly to enable essential metabolic functions to occur. Thus, some fungi may protect themselves from natural radiation by protection, e.g. pigmentation; others may have developed rapid mechanisms for repair of damaged DNA.

Man-made radiation from sources such as gamma rays or microwaves splits water into free radicals which damage DNA and are highly toxic to a wide range of fungi. Therefore the higher the water content, the more free radicals produced and the higher the level of DNA damage (Kiss and Farkas 1977). As mentioned earlier, quite a lot of information has been obtained on the effect on fungi alone or those present on agricultural substrates exposed to various irradiation doses. Saleh et al. (1988) demonstrated that ten species of fungi from the genera *Alternaria*, *Aspergillus*, *Cladosporium*, *Curvularia*, *Fusarium* and *Penicillium* were inactivated by doses of gamma-irradiation between 0.6 and 1.7 kGy. However, dematiaceous fungi with a high level of pigmentation were more resistant than moniliaceous spp. when tested in aqueous suspensions or on enriched agar media (Saleh et al. 1988). Other workers have found that the concentration required for destruction of fungi present on substrates, e.g. temperate and tropical grains, varied with both grain type and water content (Cuero et al. 1986; Ramakrishna et al. 1991). On maize, barley and wheat, *Penicillium* and *Aspergillus* spp. were killed by doses of 0.3 and 1.2 kGy, respectively, while 12 kGy was necessary to eliminate all filamentous fungi and yeasts (Cuero et al. 1986; Ramakrishna et al. 1991; Hamer 1993).

Microwave energy has also been used particularly in the agricultural and food industries for disinfection processes. Exposure of fungi directly or fungi contaminating different agricultural substrates has been carried out (More et al. 1992; Cavalcante and Muchovej 1993). The microwave energy used is dependent on the frequency (often 1250 MHz) and the time periods of exposure. For instance, a 30-s exposure at different power levels can represent energy inputs of between 2.25 and 18 kJ. As for gamma irradiation, moisture content of the medium can significantly influence resistance or sensitivity to the chosen energy level. In naturally contaminated sorghum grain, between 9–18 kJ of energy completely destroyed *Eurotium* spp., *Aspergillus candidus*, *A. niger*, *Penicillium* spp. and species of *Cladosporium* and *Alternaria* (More et al. 1992). Interestingly, germination of spores of *Aspergillus flavus*, *A. niger*, *Colletotrichum* spp., *Fusarium oxysporum* and *Bipolaris sorokiniana* on slides directly at 6, 9 and 18 kJ for periods of 0–7 min significantly reduced viability. Single-celled spores were more sensitive to microwave exposure than multicelled spores, and dark pigmented spores (*A. niger*, *B. sorkiniana*) were less affected by exposure to such concentrations of microwaves than hyaline spores (Cavalcane and Muchovej 1993).

It may be that dark pigmentation formed sites which absorbed ionising or non-ionising radiation, thus preventing more direct damage to DNA. However, bacterial mutants devoid of the pigments were not found to be more sensitive than wild types having the pigments (Nazim and James 1978). It is now believed that DNA repair mechanisms are predominantly responsible for radiation resistance. The mechanisms of DNA repair involve both photo and dark repair, depending on whether visible light is involved. The reader is referred to a recent review by Strike and Osman (1993) for more detailed information on DNA repair mechanisms.

VIII. Conclusions

This chapter has covered a wide range of highly complex stressed ecosystems which contained specialized communities of fungi. For some extreme environments, e.g. water and temperature stress, a wealth of information is available. However, in other environments, very little of the ecophysiology of the fungi has been unravelled. While it is often difficult to carry out experiments in situ in some extreme environments, for the future it is critical that experiments in vitro be carried out which better simulate or mimic naturally fluctuating parameters so that their impact on germination, growth, survival, competeveness and dominance in such highly complex ecosystems can be effectively interpreted.

References

Anand JC, Brown AD (1968) Growth patterns of so-called osmophilic and non-osmophilic yeasts in solutions of polyethylene glycol. J Gen Microbiol 52:205–212

Andre L, Nilssson A, Adler L (1988) The role of glycerol in osmotolerance of the yeast *Debaryomyces hansenii*. Arch Microbiol 110:110–116

Andrews S, Pitt JI (1987) Further studies on the water relations of xerophilic fungi, including the characterisation of some halophiles. J Gen Microbiol 133:233–238

Aycrst G (1969) The effects of moisture and temperature on growth and spore germination in some fungi. J Stored Prod Res 5:127–141

Ayres PG, Rasanayagam MS, Paul ND, Gunesekera, T (1996) UV B effects on leaf surface saprophytes and pathogens. In: Frankland JC, Magan N, Gadd GM (eds) Fungi and environmental change. Cambridge University Press, Cambridge, pp 32 50

Baggerman WI, Samson RA (1988) Heat resistance of fungal spores. In: Samson RA, van Reenen-Hoekstra ES (eds) Introduction to food-borne fungi. Centraalureau voor Schimmelcultures, Baarn, The Netherlands, pp 262–267

Battley EM, Bartlett EJ (1966) A convenient pH-gradient method for the determination of the maximum and minimum pH for microbial growth. Antonie van Leeuwenhoek 32:245–255

Belly RT, Brock TD (1974) Widespread occurrence of acidophilic strains of *Bacillus coagulans* in hot springs. J Appl Bacteriol 37:175–177

Beuchat LR (1988) Thermal tolerance of *Talaromyces flavus* ascospores as affected by growth medium and temperature and age and sugar content in the activation medium. Trans Br Mycol Soc 90:359–364

Bull AT, Bushell ME (1976) Environmental control of fungal growth. In: Smith JE, Berry DB (eds) The filamentous fungi, vol II: biosynthesis and metabolism. Arnold, London, pp 1–31

Cavalcante MJB, Muchovej JJ (1993) Microwave irradiation of seeds and selected fungal spores. Seed Sci Technol 21:247–253

Crisan EV (1973) Current concepts of thermophilism and the thermophilic fungi. Mycologia 65:1171–1198

Cochrane VW (1958) Physiology of fungi. Wiley, New York

Cooke RC, Rayner ADM (1984) Ecology of saprophytic fungi. Longman, London

Cuero R, Smith JE, Lacey J (1986) The influence of gamma irradiation and sodium hypochlorite sterilisation on maize microflora and germination. Food Microbiol 3:107–113

Edgley M, Brown AD (1978) Response of xerotolerant and non-tolerant yeasts to water stress. J Gen Microbiol 104:343–345

Fuhrman JA, Anzam F (1980) Bacterioplankton secodary production estimates in coastal waters of British Columbia, Antarctica, and California. Appl Environ Microbiol 39:1085–1095

Gibb E, Walsh JH (1980) Effect of nutritional factors and carbon dioxide on growth of *Fusarium moniliforme* and other fungi in reduced oxygen concentrations. Trans Br Mycol Soc 74:111–118

Griffin DM (1981) Water and microbial stress. In: Alexander M (ed) Advances in microbial ecology, vol 9. Plenum, New York, pp 91–136

Hallsworth JE, Magan N (1994a) Effect of KCl concentration on accumulation of acyclic sugar alcohols and trehalose in conidia of three entomopathogenic fungi. Lett Appl Microbiol 18:8–11

Hallsworth JE, Magan N (1994b) Effect of carbohydrate type and concentration on polyhydroxy alcohol and trehalose content of conidia of three entomopathogenic fungi. Microbiology 140:2705–2713

Hallsworth JE, Magan N (1995) Manipulation of intracellular glycerol and erythritol enhances germination of conidia at low water availability. Microbiology 141:1109–1115

Hallsworth JE, Magan N (1996) Culture age, temperature and pH affect the polyol and trehalose contents of fungal propagules. Appl Environ Microbiol 62:2435–2447

Hamer A (1993) Dynamics of mould growth in stored grain. PhD Thesis, Cranfield Univeristy, Cranfield, Bedford, UK

Held AA (1970) Nutrition and fermentative energy metabolism of the water mould *Aqualinderella fermentans*. Mycologia 62:339–358

Higgins SM, Lilly WW (1993) Multiple responses to heat stress by the basidiomycete *Schizophyllum commune*. Curr Microbiol 26:123–127

Hocking AD (1986) Effect of water activity and culture age on the glycerol accumulation patterns in five fungi. J Gen Microbiol 132:269–275

IAEA-FAO (1978) Food preservation by irradiation, vol 1. Symp Proc, Wageningen, The Netherlands

Ingledew WJ (1990) Acidophiles. In: Edwards C (ed) Microbiology of extreme environments. Open University Press, Milton Keynes, pp 33–54

Kiss I, Farkas J (1977) The storage of wheat and corn of high moisture content as affected by ionising radiation. Acta Aliment 6:193–214

Kroll, RG (1990) Alkalophiles. In: Edwards C (ed) Microbiology of extreme environments. Open University Press, Milton Keynes, pp 55–92

Larsen C, Morales C, Gustafsson L, Adler L (1990) Osmoregulation of the salt-tolerant yeast *Debaryomyces hansenii* grown in a chemostat at different salinities. J Bacteriol 172:1769–1774

Longworthy TA (1978) Microbial life in extreme pH values. In: Kushner DJ (ed) Microbial life in extreme environments. Academic Press, London, pp 279–315

Luard EJ (1982a) Accumulation of intracellular solutes by two filamentous fungi in response to growth at low steady state osmotic potential. J Gen Microbiol 128:2563–2574

Luard EJ (1982b) Growth and accumulation of solutes by *Phytophthora cinnamomi* and other lower fungi in response to changes in external solute potential. J Gen Microbiol 128:2583–2590

Luard EJ (1983) Activity of isocitrate dehydrogenase from three filamentous fungi in relation to osmotic and solute effects. Arch Microbiol 134:233–237

Magan N (1988) Effect of water potential and temperature on spore germination and germ-tube growth in vitro and on straw leaf sheaths. Trans Br Mycol Soc 90:97–107

Magan N, Lacey J (1984a) Effect of temperature and pH on water relations of field and storage fungi. Trans Br Mycol Soc 82:71–81

Magan N, Lacey J (1984b) Effect of gas compositionand water activity on growth of field and storage fungi and their interactions. Trans Br Mycol Soc 82:305–314

Magan N, Lacey J (1984c) Effect of water activity, temperature and substrate on interactions between field and storage fungi. Trans Br Mycol Soc 82:83–93

Magan N, Lacey J (1985) Interactions between field and storage fungi on wheat grain. Trans Br Mycol Soc 85:29–37

Magan N, Lynch JM (1986) Water potential, growth and cellulolysis of soil fungi involved in decomposition of crop residues. J Gen Microbiol 132:1181–1187

Mager WH, Varela JCS (1993) Osmostress response of the yeast *Saccharomyces*. Mol Microbiol 10:253–258

Mazur P (1966) Studies of the effects of subzero temperatures on the viability of spores of *Aspergillus flavus*. I. The effect of rate of warming. J Gen Physiol 39:869–874

More HG, Magan N, Stenning BC (1992) Effect of microwave heating on quality and mycoflora of sorghum. J Stored Prod Res 28:251–256

Morita RY (1974) Hydrostatic pressure efects on microorgansism. In: Colwell RR, Morita RY (eds) Effect of the ocean environment on microbial activities. University Park Press, Baltimore, pp 133–138

Morris GJ, Clarke A (1987) Cells at low temperature. In:Grout BWW, Morris GJ (eds) The effects of low temperatures on biological systems. Arnold, London, pp 72–84

Morris GJ, Smith D, Coulsen GE (1988) A comparative study of the changes in the morphology of hyphae during freezing and viability upon thawing for twenty species of fungi. J Gen Microbiol 134:2897–2906

Nazim A, James AP (1978) Life under conditions of high irradiation. In: Kushner DJ (ed) Microbial life in extreme environments. Academic Press, London, pp 409–439

Orpin CG (1975) Studies on the rumen flagellate *Neocallimastix frontalis*. J Gen Microbiol 91:249–262

Orpin CG (1993) Anaerobic fungi. In: Jennings DH (ed) Stress tolerance of fungi. Dekker, New York, pp 257–273

Orpin CG, Ho YW (1992) Ecology and function of the anaerobic rumen fungi. In:Ho YW, Wong HK, Addullah N, Tajuddin ZA (eds) Recent advances on the nutrition of herbivores. Malaysian Society of Animal Production, Kuala Lumpar, pp 163–170

Pitt JI (1975) Xerophilic fungi and the spoilage of food of plant origin. In: Duckworth RB (ed) Water relations of food. Academic Press, London, pp 273–307

Pitt JI, Hocking AD (1977) Influence of solutes and hydrogen ion concentration on the water relations of some xerophilic fungi. J Gen Microbiol 101:35–40

Plesofsky-Vig N, Brambl R (1993) Heat shock proteins in fungi. In: Jennings DH (ed) Stress tolerance of fungi. Dekker, New York, pp 45–68

Ramakrishna N, Lacey J, Smith JE (1991) Effect of surface sterilisation, fumigation and gamma irradiation on the microflora and germination of barley seeds. Int J Food Microbiol 13:47–54

Recca J, Mrak EM (1952) Yeast occurring in citrus products. Food Technol 6:450–454

Saleh YG, Mayo MS, Ahearn DG (1988). Resistance of some common fungi to gamma irradiation. Appl Environ Microbiol 54:2134–2135

Schmidt-Nielsen S (1902) Über einige psychrophile Mikroorganismen und ihr Vorkommen. Z Bakteriol Parasitenk D Infectionskr Hyg Abt II 19:145

Scott WJ (1957) Water relations of food spoilage microorganisms. Adv Food Res 7:83–127

Shirfine M, Phaff HJ (1958) On the isolation, ecology and taxonomy of *Saccharomyces guttulatta*. Antonie van Leeuwenhoek 24:193–209

Smith D (1993) Tolerance to freezing and thawing. In: Jenning DH (ed) Stress tolerance of fungi. Dekker, New York, pp 145–172

Smith D, Coulson GE, Morris GJ (1986) A comparative study of the morphology and viability of hyphae of *Penicillium expansum* and *Phytophthora nicotianae* during freezing and thawing. J Gen Microbiol 132:2013–2021

Smith SL, Hill ST (1982) Influence of temperature and water activity on germination and growth of *Aspergillus restrictus* and *A. versicolor*. Trans Br Mycol Soc 49:558–560

Starkey RL, Waksman SA (1943) Fungi tolerant to extreme acidity and high concentration of copper sulphate. J Bacteriol 45:509–519

Steponkus PL (1984) Role of the plasma membrane in freezing injury and cold acclimatization. Annu Rev Plant Physiol 35:543–569

Strike P, Osman F (1993) Fungal response to DNA damage. In: Jennings DH (ed) Stress tolerance of fungi. Dekker, New York, pp 297–338

Tansey MR, Brock TD (1973) Dactylaria gallopava, a cause of avian encephalitis, in hot spring effluents, thermal soils and self-heated coal waste piles. Nature 242:202–203

Trinci APJ, Davies DR, Gull K, Lawrence MI, Bonde Nielsen B, Rickers A, Theodorou MK (1994) Anaerobic fungi in herbivorous animals. Mycol Res 98:129–152

van Zyl PJ, Prior BA (1990) Water relations of polyol accumulation by *Zygosaccharomyces rouxii* in continuous culture. Appl Microbiol Biotechnol 33:231–235

Vishniac HS, Hempfling WP (1979) Evidence of an indigenous microbiota (yeast) in the dry valleys of Antarctica. J Gen Microbiol 112:301–314

8 Biogeography and Conservation[1]

E.J.M. ARNOLDS[2]

CONTENTS

I. Introduction

Biogeography is the study of distribution patterns of organisms in space and time, including the study of factors determining these patterns. Factors involved are actual conditions such as climate, soil and the availability of hosts and substrates, as well as historical conditions, including geological and evolutionary processes. Biogeographical studies strongly depend on knowledge of taxonomy, floristics, ecology, geology and genetics.

Biogeographical studies on fungi are relatively scarce, mainly due to methodological problems. A bibliography on distribution maps of fungi has been published by Kreisel in Feddes Repertorium from 1970 onwards. Information on distribution patterns is becoming increasingly important for our understanding of evolutionary processes and patterns of biodiversity, but also in such practical disciplines as control of crop pests and nature conservation. In view of the increasing disturbance of ecosystems by human activities, the conservation of fungi is becoming a topic of increasing interest.

II. Mapping of Fungi

A. Methods

Mapping of fungi is in principle a simple procedure: the collecting of records of a certain taxon in a certain area from different sources and plotting them on a topographical map. However, in practice, this work, and in particular the interpretation of maps, is hampered by a number of complications proper to fungi (Kreisel 1985; Redhead 1989; Arnolds 1992):

1. Problems in detection. Most saprotrophic microfungi live hidden in the substrate and cannot be observed directly in the field. They can only be isolated and identified by special techniques. The same applies to mycelia of most macrofungi, but they produce macroscopic sporocarps which are detectable in the field. However, sporocarps in most groups of macrofungi are ephemeral and subject to strong periodicity. In addition, the formation of sporocarps fluctuates from year to year. Some groups of ascomycetes and basidiomycetes produce hypogeous sporocarps which are difficult to detect for that reason. Molecular-biological techniques provide promising tools for the identification of mycelia in future.

2. Defective taxonomic knowledge. Many groups of fungi are in need of critical revision. In addition, species concepts are still changing. For instance, among the 50 species mapped in Europe by Lange (1974), selected because they were "well defined and easy to identify", at

---

[1] Communication no. 528 of the Biological Station Wijster
[2] Biological Station, Centre for Soil Ecology, Agricultural University Wageningen, Kampsweg 27, 9418 PD Wijster, The Netherlands

least four species are now regarded as species complexes, e.g. *Armillaria mellea* (Vahl.: Fr.) Kumm. This fungus proved to be a complex, comprising in Europe five biological species with different distribution patterns (Kile et al. 1991). Problems on conspecificity are becoming more important when distant areas are compared, such as North America, Europe and East Asia (Redhead 1989).

3. Availability of data. Distributional data come from herbarium collections, records in literature and unpublished observations. Reliable data are scanty in most parts of the world. Even in relatively well-investigated areas, the geographical range of data is often inappropriate to provide realistic distribution patterns.

4. Accessibility of data. Even if data are available, it may be problematic to trace all relevant information. This problem can be solved by establishment of central data bases for records of fungi. Indeed, the gaps in our knowledge are still enormous, but the situation at present is not so hopeless that "writing an essay on the geographical distribution of fungi is to attempt to accomplish an impossible task" (Pirozynski 1968).

In practice, two different, although not clearly separated, approaches exist to mapping of fungi. In the **monographic** approach, specialized taxonomists collect records on a particular group of fungi, usually only herbarium collections and in combination with a taxonomic revision (Kreisel 1967; Demoulin 1971; Lawrynowicz 1989, 1990). The conclusions on distribution patterns are therefore reliable, but at the same time relatively inaccurate because of the limited availability of data. The second approach concerns **mapping programmes**, which are usually carried out on a national or regional (occasionally continental) scale. Such programmes are coordinated from a certain central point and attempt to collect as many records as possible on the occurrence of selected (but not necessarily taxonomically related) species (Lange 1974) or all species in a certain area (Krieglsteiner 1991; Nauta and Vellinga 1993). The data come from different sources, including literature records and unpublished observations, and are collected by many mycologists, both professional and amateur. As a result, the maps are less reliable from a taxonomic point of view, but more accurate because of the larger number of observations.

B. Maps and Scales

Distributional studies are carried out and presented on different geographical scales: (1) global patterns may reveal relations to macroclimate, host distribution and historical factors (Fig. 1); (2) continental patterns may, in addition, be related to mesoclimate (e.g. mountain ranges) and large-scale edaphic patterns, e.g. zonal soils (Figs. 2, 3, 4); (3) regional patterns, often studied within political borders of a country, may express more subtle differences on the scale of the landscape, for instance patterns of alluvial deposits in river valleys or human influence by agriculture or forestry (Figs. 5, 6); (4) local patterns are studies within a single landscape or stand and may be related to microclimate, distribution patterns of plant communities and individual plants (Fig. 7).

Results of biogeographical studies can be presented on different kinds of maps (Kreisel 1985): (1) **Outline maps**, where borders of the known distribution areas are indicated with a line (Fig. 1). A strong disadvantage is that the original records are not indicated. Consequently, it is impossible to detect possible differences in density inside the area and to give an alternative interpretation of the distribution. (2) **Dot maps**, where each record or locality is indicated by a separate sign (Fig. 2). This method is more accurate and objective than an outline map, the accuracy being slightly influenced by the size of the dot (Lange 1974). However, when research efforts are not evenly divided over the mapped area – which is often the case – concentrations of dots may be artefacts, marking rather the activities of mycologists than the abundance of the fungus (Kreisel 1985). Dot maps can be easily combined with outline maps. (3) **Grid maps**, where the records are plotted in a topographical grid with units of a certain constant size (Figs. 3, 5, 6). The number of records per grid unit is not relevant, which to a certain extent prevents the artificial concentrations in dot maps. The accuracy of grid maps is intermediate between dot maps and outline maps and is strongly dependent on the mesh size of grid. It varies with the size of the investigated area: usual grid units are, for instance, 50×50 km for Europe (Fig. 4), 12×13 km in Germany (Fig. 5) and 5×5 km in Belgium (Fig. 6). The applied grid may be a national grid indicated on official topographical maps of a country (Fig. 5) or an international grid, for instance in Europe the UTM grid (Figs. 4, 6), also used for mapping of vascular plants and invertebrates. The lines of the grids can be afterwards deleted on published maps (Fig. 4).

C. Statistics

Numerical analysis of distribution patterns is gaining increasing importance, but has hardly been applied to fungi at present. A notable exception is the analysis of patterns of 43 polypore species in Finland by Väisanen et al. (1992), using several multivariate techniques. They distinguished eight different species assemblages in quadrats of 100 × 100 km. The distribution of these assemblages reflected in the main the phytogeographical division in the hemiboreal and several boreal zones, based on vascular plants.

Cladistic approaches are becoming increasingly important in the study of biogeographical and historical relationships between areas and of coevolutionary patterns between taxa. Methods used were critically reviewed by Wiley (1987). Such studies must be based on detailed knowledge of distribution patterns of a number of taxonomically related taxa. Fungi were so far not involved.

III. Distribution Patterns

A. Global Distribution

Maps on the world distribution of plant pathogenic fungi are published by CMI and revised with irregular intervals, for example the map of *Phaeosphaeria nodorum* (E. Müller) Hedjar, an ascomycete occurring on various grasses, including cereals (Fig. 1). Nowadays it has a subcosmopolitan distribution, following cultivated host plants almost everywhere. The original range is difficult to trace, however, as in many other pathogens on cultivated crops.

Data on the global distribution of nonpathogenic fungi depend largely on the availability of world monographs. Even then, the studied herbarium material is usually not adequately divided over different continents, so that conclusions must necessarily be preliminary. It is, for instance, interesting to notice the strongly changing opinions on distribution patterns in the ectomycorrhizal agaric genus *Rozites* P. Karst. in a period of 30 years (Redhead 1989).

General ideas on distribution patterns of fungi have evolved in the course of years

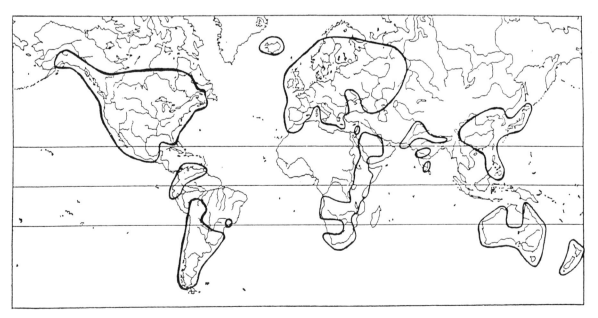

Fig. 1. World distribution of *Phaeosphaeria nodorum* (E. Müller) Hedjar. (CAB International Mycological Institute 1992)

(Demoulin 1971). In the 19th century, it was thought that most fungal species inhabit small areas, comparable to those of phanerogams. In contrast, in the first decades of this century, it was generally assumed that most fungi are (sub)cosmopolitan. Since 1930, these opposite viewpoints have merged to the notion that few species are truly cosmopolitan, but that nevertheless distribution areas of fungi are larger than the ranges of most vascular plants (Diehl 1937). In addition, distribution areas of soil-inhabiting and marine microfungi seem to be generally larger than those of larger fungi (Pirozynski 1968; Volkmann-Kohlmeyer and Kohlmeyer 1993).

Data on global patterns in some recently monographed groups of macrofungi are summarized in Table 1. The division into distribution types is necessarily strongly simplified. Ryvarden (1991) stressed the large distribution ares of genera of polypores, being mostly either (sub)cosmopolitan (23%), circumpolar (26%) or pantropical (17%). Relatively few genera are endemic to one continent where they are usually widely distributed, in contrast to many genera of phanerogams with very restricted distribution areas. Redhead (1989) stated that transatlantic disjunction patterns are, in general, exhibited by vascular plants at the generic or family level, but by agarics at the species level, just as they are for lichenized ascomycetes and bryophytes. Hallenberg (1991) drew similar conclusions on the basis of distribution patterns of Corticiaceae.

In *Lentinus* Fr., most species (78%) are tropical, but only very few species (3%) are pantropical or (sub)cosmopolitan (5%) (Pegler 1983a). Pantropical species seem to be lacking in *Thelephora* Ehrh.: Fr., whereas 39% of the species are endemic to Southeast Asia (Corner 1968). This proportion may be overestimated by intensive collecting by the author in this area compared to other tropical regions. The proportions of supracontinental distribution areas are also low in *Bovista* Pers.: Pers. and *Scleroderma* Pers.: Pers. although these puffballs produce enormous amounts of airborne spores (Kreisel 1967; Guzman 1970). The number of (sub)cosmopolitan species among *Ascobolus* Pers. and *Saccobolus* Boud. is very high (Van Brummelen 1967). This may be due to their specialized ecology: most species grow on dung, and ascospores must pass though the intestines of grazing animals. On the other hand, dispersal of the heavy spores over large distances seems unlikely. Many species may,

in fact, have spread by introduction of cattle in other continents. In contrast, the proportion of European species is also very high; this may be due to undercollecting in other parts of the world.

Distribution patterns of fungi are, in general, so similar to the ranges of phanerogams, albeit on different taxonomic levels, that they are likely to be determined by the same environmental and historical factors. Some data supporting the importance of geological phenomena are:

1. Some species and genera are common to southern South America and Australia or New Zealand. They may be considered as remnants of populations fragmented by the breakup of Gondwanaland approximately 100 million years ago (Horak 1983; Ryvarden 1991).
2. A number of species show a disjunct distribution in Europe and (eastern) North America, or two closely related vicariant species exist in these areas. These patterns may be explained by the opening of the Atlantic in the Eocene (Demoulin 1973).
3. The species diversity of most taxonomic groups is considerably larger in North America than in Europe, which may be caused by different possibilities for reaching refugia during Pleistocene glacial periods (Redhead 1989).

The ideas on historical events are not conflicting with palaeontological evidence. Fossil basidiomycetes are known from the Middle Pennsylvanian, approximately 300 million years B.P. (Dennis 1970).

An alternative hypothesis for the explanation of supracontinental ranges of fungi is the existence of efficient long-distance spore dispersal (Redhead 1989; Hallenberg 1991; Ryvarden 1991). Arguments against this hypothesis are (1) the large majority of spores are deposited in the near surroundings of the source; (2) most spores are not viable after a stay in higher atmospheric strata; (3) the chance of reaching an appropriate substrate and microhabitat is small in many cases; (4) many basidiomycetes must establish a dikaryon; the chance that two compatible colonies are formed close to each other is extremely small; (5) wind dispersal by spores does not necessarily imply genetic exchange between allopatric populations.

Consequently, also for most fungi, dispersal is likely to be restricted by geographic barriers, such as oceans and mountain ranges. Circumstantial

Table 1. Percentages of species or genera of some groups of macrofungi in different parts of the world

Reference	Ryvarden (1991)	Pegler (1983a)	Correr (1968)	Kreisel (1967)	Guzman (1970)	Van Brummelen (1967)
Taxonomic group	Polypores	Lentinus	Thelephora	Bovista	Scleroderma	Ascobolus
	Basidiom.	Basidiom.	Basidiom.	Basidiom.	Basidiom.	Ascom.
Principal substrate	Wood	Wood	Soil	Soil	Soil	Dung
Principal way of life	Saprotroph Necrotroph	Saprotroph	Mycorrhizal Saprotroph	Saprotroph	Mycorrhizal	Saprotroph
Taxonomic units	Genera	Species	Species	Species	Species	Species
Number of taxa	132	63	49	46	20	61
Known from one locality	?	12	13	13	10	17
1. (Sub)cosmopolitan	23	5	4	2	5	28
2. Temperate/mediterranean						
a. Bipolar	–	–	–	4	15	–
b. Boreocircumpolar	26	5	6	2	5	–
c. Southern temperate	1	2	–	11	–	–
d. Eurasia	1	–	2	11	–	–
e. Europe	2	3	10	2	–	23
f. Asia	2	3	6	13	–	1
g. Europe and N. America	–	–	4	4	20	20
h. E. Asia and N. America	3	–	4	–	–	–
i. N. America	3	4	10	22	25	10
3. Tropical						
a. Pantropical	17	3	–	4	–	5
b. Paleotropical	4	11	–	–	–	–
c. Africa	5	14	2	7	15	–
d. S.E. Asia	4	13	39	2	–	1
e. Tropical America	7	24	8	9	15	7
f. Australia/Pacific	1	13	2	7	–	5

evidence is given by some pathogenic fungi which were introduced to other continents and caused catastrophic epidemics in newly available host plants. Apparently, they were unable to spread so far without human assistance.

B. Local Endemics

Many species and even genera of vascular plants are restricted to small areas in the order of 1 to 1000 km². Concentrations of such local endemics are found in isolated areas surrounded by effective barriers for dispersal, for instance remote islands and isolated mountains. The percentage of local endemic plant species may be over 50%. There is no evidence that this is also true for fungi. The mycoflora of some islands was adequately described, e.g. of the Lesser Antilles by Pegler (1983b), but comparable information on surrounding areas is not available.

It is generally thought that local endemics are non-existent or scarce among fungi, except for species restricted to endemic host plants. However, the situation is rather paradoxical, since many species of fungi are known only from their type locality, and consequently are potential local endemics. In the monographs listed in Table 1, the percentage of species known from one locality only ranges from 10% in *Scleroderma* to 28% in *Ascobolus* and *Bovista*. Even the monograph of the striking genus *Hygrophorus* sensu lato (Agaricales) in relatively well-investigated North America includes 96 taxa out of 244 (39%), which are known only from the type locality. Among them, 21 species were collected exclusively before 1920 (Hesler and Smith 1963).

The questions arise: are these species truly endemic, do they fruit sporadically, or have they simply escaped the attention of mycologists? Only more intensive research on distribution patterns can provide the answer. It is striking that some fungi are recorded a subsequent time quite far away from the original location. For instance, *Bovista verrucosa* (G.H. Cunn.) G.H. Cunn. was described by Kreisel (1967) on the basis of two collections, one from Piemont, Italy (1857), and one from South Australia (1922). *Squamanita odorata* (Cool) Bas is a very remarkable agaric, discovered in The Netherlands in 1915 and observed there since then in approximately 20 localities. It was long thought to be endemic to The Netherlands, but it seems to be spreading now in

West Europe. It was reported from Denmark in 1948, Germany (1963), Norway (1968), Switzerland (1989), but, remarkably enough, also from the northwestern United States in 1951. Such patterns may be explained by introductions, but evidence is lacking. Other examples of odd disjunct distribution patterns were given by Pirozynsky (1968).

C. Continental Patterns

Continental patterns of fungi were mainly studied in Europe and North America. The second Congress of European Mycologists initiated a mapping programme for 100 selected species of macrofungi. Maps of 50 species were published by Lange (1974). The interpretation of distribution patterns was hampered by the absence of data from large parts of southern and eastern Europe, and concentrations of records in some areas. Lange distinguished seven climatologically determined distribution patterns. In addition, the occurrence of some species appeared to be determined by edaphic factors. Only one species showed a more or less disjunct distribution, viz. *Phallus hadriani* Vent.: Pers., with centres in the coastal dunes of West Europe and continental steppe areas in Central Europe. Some species were restricted to subarctic and alpine regions in North Europe, others to the Central European mountains, although there are many similarities in climate and soil. These differences may be caused by isolation of these areas since the Pleistocene glaciations, but also actual environmental factors (e.g. differences in day length and soil temperatures in summer) may play a role.

Redhead (1989), in his outstanding overview of the distribution of Canadian fungi, provided dot maps of 78 species in North America. He distinguished 13 main types of distribution patterns, largely coinciding with those of phanerogams. Four types are shown in Fig. 2. Many fungi appear to be restricted to parts of the continent east and west of the Rocky Mountains. This mountain range and the prairie area of the midwest may be an effective barrier for some species. Some species occupying the entire boreal zone show subtle differences in morphology and/or ecology. Redhead (1989) suggested that these differences may be the results of survival of eastern and western populations in different refugia during Pleistocene glaciations.

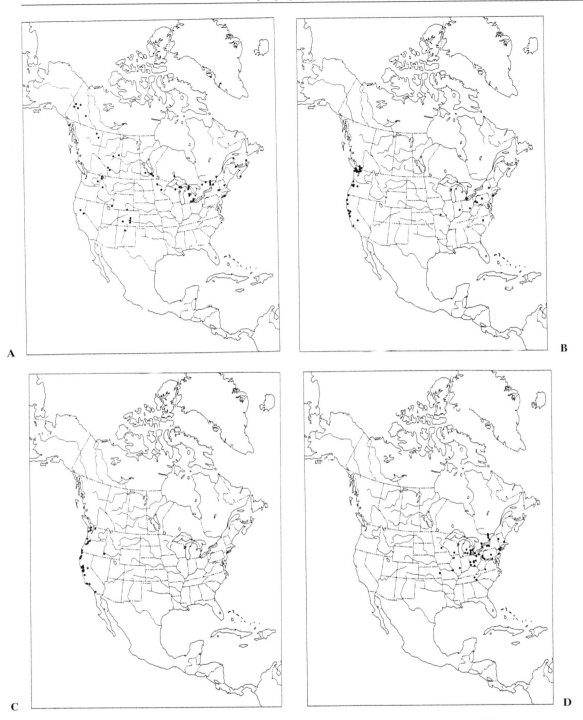

Fig. 2A–D. Examples of distribution patterns of agarics in North America. **A** *Marasmius epiphyllus* (Pers.: Fr.) Fr. with boreotemperate distribution. **B** *Marasmiellus candidus* (Bolt.: Fr.) Sing. with bicoastal distribution. **C** *Marasmius plicatulus* Peck with western temperate distribution. **D** *Marasmius pyrrhocephalus* Berk. with eastern temperate distribution. (Redhead 1989)

D. Regional Patterns

In principle, studies on regional distribution patterns are less complicated than those on large-scale distribution: data are easier to collect and subject to less taxonomic and nomenclatural problems. However, useful interpretation of the patterns is often dependent on large numbers of records, since patterns are more subtle and the relations with environmental factors are frequently less obvious. Regional studies have mainly been carried out in Europe and published in a wide variety of journals and books. In some countries, systematic mapping programmes have been initiated, e.g. in Germany, Great Britain and The Netherlands. The first distribution atlas on macrofungi has been published in Germany (Krieglsteiner 1991, 1993; Fig. 5). It contains grid maps of 5500 taxa, based on about 3 000 000 records. Unfortunately, texts providing an ecological interpretation of the maps and an evaluation of the methods are lacking. An extensively annotated atlas of 370 selected macrofungi in The Netherlands was published recently (Nauta and Vellinga 1995).

As an example, distribution maps on different scales are presented of the saprotrophic, well-characterized puffball *Lycoperdon echinatum* sensu lato (Figs. 3–6). They may illustrate some problems encountered with mapping and the interpretation of the results. The world distribution of *L. echinatum* sensu lato covers temperate areas in Europe and eastern North America (Fig. 3). Demoulin discovered that the North American populations are morphologically different from the European ones, and described the new species, *Lycoperdon americanum*. Consequently,

what seemed to be a species with a disjunct area proved to be two vicariant species (although the name *L. echinatum* is still often used in American literature).

The distribution of *L. echinatum* in Europe is almost restricted to the nemoral zone, dominated by deciduous, broad-leaved trees, such as *Quercus robur* L., of which the area is indicated on the map (Fig. 4). The species is almost lacking in the boreal zone of coniferous forests and in the mediterranean area. It is tempting to interpret this pattern in terms of ecological dependence on deciduous forests, but within the nemoral zone it can also be found in conifer plantations (e.g. Gross et al. 1980). Consequently, it is more likely to be limited by climatological factors, in particular temperature. The species seems to be much more common in northwestern Europe than in the southwest, southeast and east, but this might very well be an artefact due to undercollecting in these areas (Demoulin 1987).

The distribution in Belgium on the basis of a 5 × 5-km grid (Fig. 6) provides again more detailed information. The species is lacking in the north and west of the country, as well as in the Ardennes mountains above 400 m. Demoulin (1987) suggested that climatological factors caused its ab-

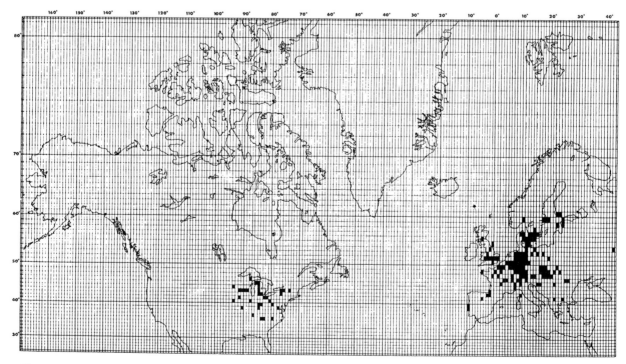

Fig. 3. World distribution of *Lycoperdon echinatum* Pers.: Pers. (in Europe) and its vicariant *L. americanum* Demoulin (in America). Each *rectangle* represents one degree latitude × one degree longitude. (Demoulin 1987)

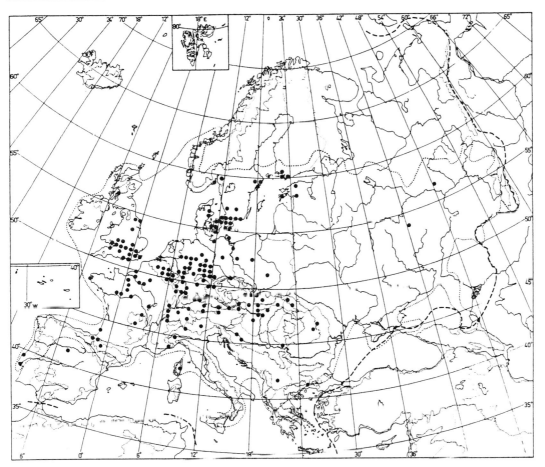

Fig. 4. Distribution of *Lycoperdon echinatum* Pers.: Pers. in Europe according to UTM grid. Each *dot* represents a square of 50 × 50 km. The *dashed line* indicates the distribution area of *Quercus robur* L. (Demoulin 1987)

sence in the Ardennes. However, *L. echinatum* is found in adjacent western Germany in many areas above 500 m (Fig. 5). It seems that its regional distribution is primarily determined by soil conditions, since *L. echinatum* is absent or scarce in all areas with acid, sandy or peaty soils with humus of the mor type: western Belgium, northwestern Germany and also The Netherlands. Instead, it is widespread in areas with forests on calcareous, loamy soils producing mull humus. On the other hand, the occurrence of *L. echinatum* in mountains above 1000 m on the mediterranean island of Corse, also noticed by Demoulin (1987), seems to be determined by the climate and not by soil conditions.

E. Local Patterns

Distribution patterns of fungal species in a stand are usually not regarded as the subject of biogeographic research, but rather as part of mycocoenological studies. For instance, Jansen (1984) studied the distribution of macrofungi in oak forests in The Netherlands using a grid of 40 subplots of 5 × 5 m within 1000-m^2 plots. She was able to demonstrate correlations between the occurrence of certain plants and fungi. Murakami (1987) did the same for *Russula* species in 32 subplots of 25 m^2 in a Japanese *Castanopsis-Pasania* forest (Fig. 7). He demonstrated that the three dominant species (*R. densifolia* Gill., *R. laurocerasi* Melz., *R. lepida* Fr.) showed little spatial overlap, whereas two other species (*R. castanopsidis* Hongo and *R. lilacea* Quél.) are fruiting in much lower numbers scattered over the plot.

The analysis of the distribution of individual mycelia or genets in stands is considered to belong to the domain of population biology (Chap. 2, this Vol.).

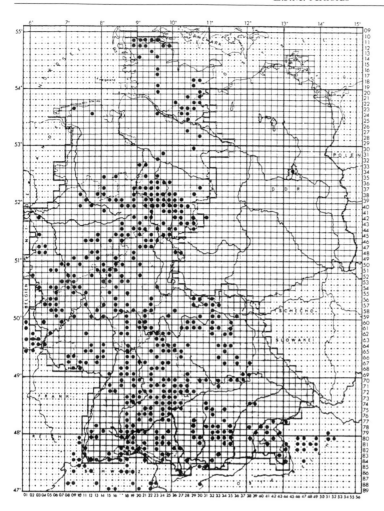

Fig. 5. Distribution of *Lycoperdon echinatum* Pers.: Pers. in Germany. Each *square* represents 12 × 13 km. (Krieglsteiner 1991)

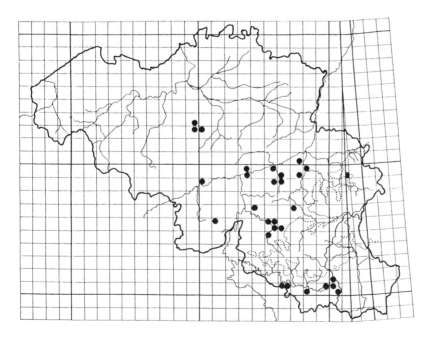

Fig. 6. Distribution of *Lycoperdon echinatum* Pers.: Pers. in Belgium according to UTM grid. Each *square* represents 5 × 5 km (except in the correction area). The *dotted line* represents the altitude of 400 m. (Demoulin 1987)

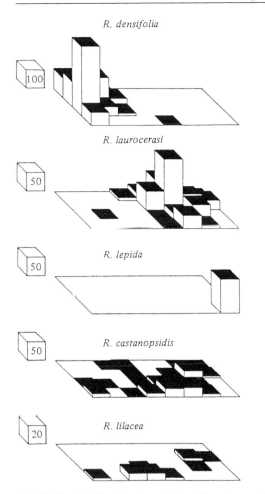

R. densifolia

100

R. laurocerasi

50

R. lepida

50

R. castanopsidis

50

R. lilacea

20

Fig. 7. Spatial distribution of five *Russula* species in 32 subquadrats (5 × 5 m) in a forest of *Castanopsis cuspidata* and *Pasania edulis* near Fukuoka, Japan. The *numbers to the left* indicate the abundance of basidiocarps, over a period of 2 years. (Murakami 1987)

F. Recent Changes in Distribution Patterns

The study of changes in distribution patterns depends on the availability of accurate data on geographical ranges from different periods. Such data are known on a global scale only from some pathogenic fungi on plants. Changes in regional distribution patterns or frequencies of macrofungi have been investigated by (1) repeated mapping, (2) comparison of representative samples of the local mycoflora over the years, for instance foray reports, (3) comparison of sporocarp counts in selected stands or plots in different periods and (4) in some cases data on the supply of edible fungi to local markets (Arnolds 1988, 1992). Direct comparison between numbers of records or occupied

grid units of a certain species in different periods (methods 1 and 2) is usually inadequate to draw conclusions on increase or decrease. The use of correction factors was described by Arnolds and Jansen (1992) and Nauta and Vellinga (1993).

IV. Expansion of Distribution Areas

Expansion of the range of fungi may be caused by natural colonization of new areas by propagules, introduction of host plants outside their natural range and introduction of fungi outside their natural range. Introductions may be intentional or accidental.

Few well-documented examples exist of spontaneously expanding fungi. A striking phenomenon is the recent establishment and increase of some polypores in the Northwest European lowland, for instance *Fomes fomentarius* (L.: Fr.) Fr., *Pycnoporus cinnabarinus* (Jacq.: Fr.) Donk (Kreisel 1985) and *Schizopora carneolutea* (Rodw. & Clel.) Kotl. & Pouz. The latter species seems to be superseding its close and widespread relative *S. paradoxa* (Schrad.: Fr.) Donk in places (Keizer 1990). The causes of these extensions of range are unknown. There have been suggestions attributing them to climatological change, changes in forestry practices and decline of tree vitality. Introduction in Europe cannot be excluded in advance for *S. carneolutea*.

Another group of increasing macrofungi in Europe comprises agarics such as *Stropharia rugosoannulata* Farlow, *S. aurantiaca* (Cooke) P.D. Orton and *Psilocybe cyanescens* Wakef. These species are characteristic of disturbed environments such as parks and roadsides, where they prefer mixtures of fertile soil and wood chips. Such habitats were very rare in the past, but are now spreading in urban areas. However, it is also quite possible that these fungi were introduced by accident (Kreisel 1985).

A. Introductions of Plants

The artificial extension of ranges of plants by introduction creates a new potential environment for associated parasites, symbionts and saprotrophs. Fungi may colonize the new area spontaneously by propagules or may be introduced together with plant material.

Many trees are planted on a large scale all over the world far outside their original range, in northwest Europe, for instance, *Picea abies* (L.) Karsten and *Larix decidua* Miller, native to North Europe and the Central European mountains, and *Pseudotsuga menziesii* (Mirb.) Franco and *Picea sitchensis* (Bong.) Carrière, introduced from North America. Only part of the host-specific ectomycorrhizal symbionts follow their tree outside its natural range, the number of species decreasing with increasing differences from the original habitat. For instance, in *Picea* plantations in The Netherlands, only 12 host-specific symbionts are found, most of them being rare, whereas over 50 species are native to Central Europe. Only six *Larix* symbionts occur in eastern Germany and The Netherlands (Kreisel 1985). In this case, dispersal is inhibited not by geographical barriers, but by environmental conditions. Local strains of fungi with a broad host range are apparently better adapted to these conditions and occupy their niches.

In the case of *Pseudotsuga*, only very few specific symbionts have found their way to Europe, which may at first sight indicate the effectiveness of the Atlantic as a barrier for spore dispersal. However, all (introduced?) *Pseudotsuga* symbionts remain local and rare. Apparently, environmental conditions also play a role in the establishment of these fungi. It would be interesting to study the effects of differences in host assemblages between America and Europe for properties of the tree, such as vitality and potential age.

Some ectomycorrhizal trees have been introduced in areas where suited native symbiotic fungi are lacking, for instance *Pinus* spp. in New Zealand, South Africa and many tropical countries. Successful forestry appeared to depend on either accidental import of ectomycorrhizal fungi or artificial inoculation of nursery plants with selected strains (Mikola 1973).

B. Introductions of Fungi

Many pathogenic fungi have been introduced accidentally to new areas, sometimes with detrimental effects on valuable crops. A well-known example is the introduction of *Phytophtora infestans* (Mont.) de Bary, the cause of late blight in potato, in Europe around 1845, resulting in the notorious famine in Ireland, where 1 000 000 people died. The coffee rust, *Hemileia vastatrix* Berk. & Br. arrived in Ceylon in 1869 and destroyed 200 000 ha of coffee plantation in 20 years; it turned the British from coffee drinkers into tea drinkers. It is sometimes suggested that such epidemics are typical of monocultures. However, the introduction of *Cryptonectria parasitica* (Murr.) Barr from East Asia into North America around 1900 almost exterminated the American chestnut, *Castanea dentata*, Borkh. an important component of native forests. Almost equally detrimental was the introduction from China into The Netherlands around 1915 of *Ceratocystis ulmi* (Buisman) C. Moreau, causing wilting of elm trees (Dutch elm disease) (Campbell and Madden 1990).

These catastrophic introductions have led on the one hand to quarantaine procedures to prevent unwanted introductions, on the other to the use of some fungi for biological control of introduced weeds. Successful examples are the suppression of *Chondrilla juncea* L. by the rust fungus *Puccinia chondrillina* Bubák & Syd. in Australia and the USA, and of *Ageratina riparia* on Hawaii by an unidentified *Cercosporella*-like fungus (Evans and Ellison 1990).

Little is known about accidental introductions of saprotrophic and mycorrhizal fungi (see also Sect. IV.A). The exchange of such fungi between North America and Europe is surprisingly small compared to the enormous exchange of vascular plants in the past centuries. This may be caused by (1) natural occurrence of many species on the two continents, so that some possible introductions are not noticed; (2) lack of intentional transportation of fungi, whereas many plants were introduced by purpose; (3) most successful plants being ruderals of man-made environments, which are poor in macrofungi; (4) the niches of soil-inhabiting fungi being possibly completely saturated so that colonization by new species may be difficult.

Most known cases of successful introductions concern Phallales, which combine a striking appearance with limited spore-dispersal capacity (spread by insects) and a preference for disturbed habitats, for instance in Europe *Mutinus ravenelii* (Berk. & Curt.) E. Fischer and *M. elegans* (Mot.) E. Fischer from North America and *Anthurus archeri* (Berk.) E. Fischer from the southern hemisphere (Kreisel 1985). The extension of the latter species was described in detail by Parent and Thoen (1986).

C. Decline and Extinction of Fungi

The dramatic loss of natural and seminatural habitats on earth, caused by human activities, is reflected in the rarification or extinction of many vascular plants and animals. It is inevitable that fungi are subject to the same environmental changes. Nevertheless, to my knowledge, not a single species has been reported to be completely extinct on a global scale. This is certainly in part an artefact caused by limited research efforts in this field and the methodological complications outlined above. How do we know that a rare species has disappeared when its ephemeral sporocarps have been collected at intervals of many years in scattered localities? (see also Sect. III.B). However, it is probably true that fungi are less vulnerable to extinction than vascular plants because (1) there are few species endemic to small areas; (2) their dispersal capacity is, in general, larger, at least on a continental scale; (3) it is hardly possible to exterminate fungi on purpose, e.g. by harvesting of sporocarps, as has been the case for some plants and larger animals.

However, decline and extinction of regional populations of macrofungi were observed in several parts of Europe. These conclusions are based on an analysis of distribution patterns in past and present, comparison of foray reports and some other methods mentioned in Section III.F. Evidence of a considerable decline in some ecological and taxonomic groups of fungi has stimulated efforts for fungal conservation. Examples of decreasing fungi are treated in the following section. Here, only one example of a regionally nearly extinct fungus is described in some detail in order to illustrate some of the methodological problems encountered. It concerns the decline in The Netherlands of *Sarcodon imbricatus* (L.: Fr.) P. Karst., an obligate ectomycorrhizal species of coniferous trees, producing large epigeous sporocarps. The species was reported as common in the eastern Netherlands up to the 1950s, became rare in the next decade and was recorded from only four localities in the period 1973–1985 (Arnolds 1989; Fig. 8). Nowadays it is found in only one locality. The maps are only a weak reflection of its true decline, because of the enormous increase in mycofloristic research in The Netherlands: from the decade 1980–1989, over 200000 records on macrofungi are available, from all the years before 1950 only 11000. The decrease is better expressed in two other parameters: (1) the proportion of records of *Sarcodon imbricatus* with regard to the total number of fungal records, decreasing from 0.21% before 1949 to 0.09% between 1950 and 1959, 0.03% between 1960 and 1969, 0.01% between 1970–1979 and less than 0.005% afterwards; (2) the proportion of forays from which the species was reported, decreasing from 31% in

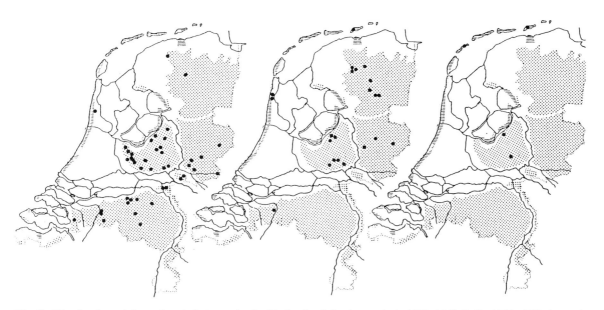

Fig. 8. Distribution of *Sarcodon imbricatus* in the Netherlands in the periods 1890–1949 (*left*), 1950–1972 (*centre*) and 1973–1985 (*right*). *Dotted* Acid, Pleistocene sands; *hatched* calcareous, Holocene sands. (Arnolds 1989)

1900–1949 to 14% in 1950–1959, 1.3% in 1960–1969, 0.8% in 1970–1979 and 0 in 1980–1989 (Arnolds and Jansen 1992).

V. Conservation of Fungi

The need of attention for the conservation of fungi has only recently become a topic of major concern (Winterhoff and Krieglsteiner 1984; Arnolds 1991a). Research in this field is mainly restricted to macrofungi, and geographically to Europe. The ninth Congress of European Mycologists in 1985 erected a permanent European Council for Conservation of Fungi (ECCF) with representatives from all European countries. The council organized three specialist meetings. In addition, the International Union for Nature Conservation (IUCN) founded a committee for fungi in 1990.

One of the main tools for the conservation of fungi is the publication of Red Lists: enumerations of species which are considered to be threatened in their long-term survival in a certain area, usually a country or other political unit. Different categories of threatened species are distinguished: (1) (presumably) extinct species, which have not been observed for several decades, in spite of intensive fieldwork; (2) endangered species, which are rare and strongly decreasing and/or inhabiting habitats which are endangered and declining; (3) vulnerable species, which are likely to become endangered in the near future if the causal agents of their decline are not reduced or removed; (4) rare or potentially threatened species, which have small populations or few sites that are not threatened at present, but may become at risk without adequate protection.

In view of the quoted criteria, it is obvious that data in Red Lists are strongly dependent on available knowledge on floristics, distribution patterns in the past and present, and ecology of the species involved. In practice, many species are included on the basis of circumstantial evidence and personal judgement rather than sound results of research. In this respect, most Red Lists are preliminary attempts to summarize available information and suppositions.

The aims of Red Lists are (1) to inform professional and amateur mycologists in order to stimulate them to collect data on threatened species and areas where many such species are present; (2) to inform nature conservationists and environmental planners to facilitate the use of mycological data for protection and management of nature areas; (3) to inform decision-makers and politicians, so that they are able to develop measures and laws in favour of fungal conservation; (4) to provide data for the selection of species for monitoring programmes; (5) to provide data for the selection of species to be protected by law; (6) to compare the situation in different countries in order to estimate the international status of a species.

A survey of available Red Lists was given by Arnolds and De Vries (1993). They were published to date in 11 European countries, and, in addition, in most states of Germany. A Red List for Europe is in preparation (Ing 1993). The number of species on each list varies from 123 in the former Czechoslovakia to 1032 in the former West Germany. These numbers are not necessarily proportional to the need of conservation efforts, and are also strongly influenced by differences in the application of the criteria. Altogether, almost 3000 species of macrofungi are considered to be (potentially) threatened in some parts of Europe, including 2760 basidiomycetes, approximately 50% of all European species. The threatened species include saprotrophic fungi on soils (35%), ectomycorrhizal species (31%), saprotrophs on wood (28%), parasitic fungi on trees (2%) and even some fungi on dung (1%).

From a comparison of Red Lists, it appears that some ecological and taxonomic groups of fungi deserve special attention in large areas of Europe. (1) Species largely restricted to pristine and old-growth forests, mainly lignicolous fungi on large logs and other wood debris, including many polypores (Høiland and Bendiksen 1993; Kotiranta and Niemelä 1993). Nowadays, most of these species are restricted to forests in remote parts of Scandinavia and the Central European mountains. They are mainly threatened by logging, in mountain regions also by forest decline due to air pollution (acid rain). (2) Fungi, characteristic of peat bogs, fens and other wetlands, such as *Armillaria ectypa* (Fr.) Herink, listed on 8 out of 11 Red Lists (Arnolds and De Vries 1993). These habitats are subject in most areas to digging of peat, drainage, reclamation and/or afforestation. Extended, undamaged bogs are at present mainly found in North Europe. (3) Saprotrophic species characteristic of coastal and inland sand dunes, sometimes also steppe, including many gasteromycetes, for instance of the genera

Geastrum Pers.: Pers., *Tulostoma* Pers.: Pers. and *Disciseda* Czern. They are threatened throughout Europe by digging of sand, recreational activities, afforestation and natural succession on remaining relics (Winterhoff and Krieglsteiner 1984). (4) Saprotrophic fungi of old, nutrient-poor, unfertilized meadows and hayfields, including many species of *Hygrocybe* (Fr.) Kumm, *Entoloma* (Fr.) Kumm., *Dermoloma* (J. Lange) Herink, Clavariaceae and Geoglossaceae, are strongly threatened in most parts of Europe. The distribution of grassland relics rich in *Hygrocybe* species in The Netherlands and Denmark (Rald 1985) is well documented. The decline of grassland fungi in Sweden was described by Nitare (1988), who used Geoglossaceae as indicator species. Grassland fungi are threatened on the one hand by intensification of grassland use (e.g. use of dung and fertilizers), on the other by abandonment and afforestation, both of which processes are strongly stimulated by EEC agricultural policy. (5) Many ectomycorrhizal fungi, associated with both frondose and coniferous trees, are strongly threatened in some densely populated and industrialized parts of Europe. Reports on decline come mainly from Czechia (Fellner 1993), parts of Germany (Derbsch and Schmitt 1987) and The Netherlands (Arnolds 1988; Nauta and Vellinga 1993). The decrease in ectomycorrhizal sporocarps is most prominent in forests on acidic soils poor in nutrients and humus, including well-known edible species such as *Cantharellus cibarius* Fr.: Fr. and *Boletus edulis* Bull.: Fr. (Jansen and Van Dobben 1987; Derbsch and Schmitt 1987). However, circumstantial and experimental evidence suggests that the influence of collecting sporocarps itself is negligible (Egli et al. 1990). The decrease in sporocarps may or may not coincide with a reduction of mycorrhizal rootlets (Dighton and Jansen 1991). The decline of ectomycorrhizal fungi is now generally attributed to the effects of air pollutants on trees (reduced photosynthesis) and soil (acidification and nitrogen accumulation) (Jansen and Van Dobben 1987; Arnolds 1991b; Dighton and Jansen 1991).

An important aspect of fungal conservation is appropriate management of habitats. For example, management of poor grasslands by cutting or grazing is necessary to prevent natural succession to forests. Removal of litter may help to restore the ectomycorrhizal flora in forests. Adequate management of roadsides may help to create refugia for endangered grassland fungi and ectomycorrhizal species (when planted with trees) in cultural landscapes. A survey of the relations between management and fungi was given by Keizer (1993).

Little is known about threatened fungi and their conservation outside West and Central Europe. One example is the polypore *Oxyporus nobilissimus* W.B. Cooke, producing enormous basidiocarps on large stumps of *Tsuga* and *Abies* in old-growth forests in North America. It seems to be endemic in Oregon and Washington, where it is regarded as very rare and endangered by forestry practices.

VI. Conclusions

Biogeographical information on fungi is becoming increasingly important for the understanding of evolutionary processes and biodiversity patterns, as well as for control of spread of crop pests and for nature conservation. However, this kind of information is still scanty, scattered over the literature and therefore often difficult to obtain. Gathering knowledge on distribution patterns of fungi is often hampered by the biological properties of these organisms and methodological problems. Relatively little is known on global distribution patterns as well as the frequency and significance of local endemism. Programmes for mapping of (macro)fungi exist in only some European countries.

Expansion of ranges of fungi has mainly been documented for plant pathogens, usually due to accidental introductions and sometimes with profound effects on ecosystem functioning or crop production. Relatively few examples exist of saprotrophic or mycorrhizal fungi invading other continents. On the other hand, in some areas, in particular densely populated parts of Europe, a drastic decline of many species has been established. These changes in the mycoflora are mainly caused by destruction of (semi)natural habitats and environmental pollution. In particular, ectomycorrhizal species appear to be sensitive to acidification and eutrophication of forest ecosystems. It is to be expected that similar losses of biodiversity are currently taking place in other parts of the world, in particular the tropics, but escaping our knowledge. The exploration and documentation of biodiversity of fungi in these areas is therefore urgently needed. Effective conservation of fungi can be achieved only in the context of integral ecosystem protection and management.

References

Arnolds E (1988) The changing macromycete flora in The Netherlands. Trans Br Mycol Soc 90:391–404

Arnolds E (1989) Former and present distribution of stipitate hydnaceous fungi (Basidiomycetes) in The Netherlands. Nova Hedwigia 48:107–142

Arnolds E (1991a) Mycologists and nature conservation. In: Hawksworth DL (ed) Frontiers in mycology. CAB International, Wallingford, pp 243–264

Arnolds E (1991b) Decline of ectomycorrhizal fungi in Europe. Agric Ecosyst Environ 35:209–244

Arnolds E (1992) Mapping and monitoring of macromycetes in relation to nature conservation. McIlvainea 10(2):4–27

Arnolds E, de Vires B (1993) Conservation of fungi in Europe. In: Pegler DN, Boddy L, Ing B, Kirk PM (eds) Fungi of Europe: investigation, recording and conservation. Royal Botanic Gardens, Kew, pp 221–230

Arnolds E, Jansen E (1992) New evidence for changes in the macromycete flora of the Netherlands. Nova Hedwigia 55:325–351

Campbell LC, Madden LV (1990) An introduction to plant disease epidemology

Corner EJH (1968) A monograph of *Thelephora* (Basidiomycetes). Beih Nova Hedwigia 27. J Cramer, Lehre

Demoulin V (1971) Le genre *Lycoperdon* en Europe et en Amérique du Nord/Etude taxonomique et phytogéographique. PhD Thesis, Université de Liège, Liège

Demoulin V (1973) Phytogeography of the fungal genus *Lycoperdon* in relation to the opening of the Atlantic. Nature 242:123–125

Demoulin V (1987) La chorologie des gastéromycètes. Mém Soc R Bot Belg 9:37–46

Dennis RL (1970) A Middle Pennsylvanian basidiomycete mycelium with clamp connections. Mycologia 62:578–584

Derbsch H, Schmitt JA (1987) Atlas der Pilze des Saarlandes Teil 2: Nachweise, Vorkommen und Beschreibungen. Minister für Umwelt, Saarbrücken

Diehl WW (1937) A basis for mycogeography. J Wash Acad Sci 27:244–254

Dighton J, Jansen AE (1991) Atmospheric pollutants and ectomycorrhizae: more question than answers? Environ Pollut 73:179–204

Egli S, Ayer F, Chatelain F (1990) Der Einfluss des Pilzsammelns auf die Pilzflora. Mycol Helv 3:417–428

Evans HC, Ellison CA (1990) Classical biological control of weeds with micro-organisms: past, present, prospects. Aspects Appl Biol 24:39–49

Fellner R (1993) Air pollution and mycorrhizal fungi in Central Europe. In: Pegler DN, Boddy L, Ing B, Kirk PM (eds) Fungi of Europe: investigation, recording and conservation. Royal Botanic Gardens, Kew, pp 239–250

Gross G, Runge A, Winterhoff W (1980) Bauchpilze (Gasteromycetes s.l.) in der Bundesrepublik und West Berlin. Beih Z Mykol 2:1–220

Guzman G (1970) Monografia del género *Scleroderma* Pers. emend. Fr. (Fungi-Basidiomycetes). Darwiniana 16:233–407

Hallenberg N (1991) Speciation and distribution in Corticiaceae (Basidiomycetes). Plant Syst Evol 177:93–110

Hesler LR, Smith AH (1963) North American Species of Hygrophorus. University of Knoxville

Høiland K, Bendiksen E (1993) Problems concerning lignicolous fungi in boreal forests in Norway. In: Arnolds E, Kreisel H (eds) Conservation of fungi in Europe. Ernst-Moritz-Arndt Universität, Greifswald, pp 51–57

Horak E (1983) Mycogeography in the South Pacific region: Agaricales, Boletales. Aust J Bot Suppl Ser 10: 1–41

Ing B (1993) Towards a Red List of endangered European macrofungi. In: Pegler DN, Boddy L, Ing B, Kirk PM (eds) Fungi of Europe: investigation, recording and conservation. Royal Botanic Gardens, Kew, pp 231–237

Jansen AE (1984) Vegetation and macrofungi of acid oakwoods in the north-east of The Netherlands. Agricultural Research Reports 923, Pudoc, Wageningen

Jansen E, van Dobben HF (1987) Is decline of *Cantharellus cibarius* in The Netherlands due to air pollution? Ambio 16:211–213

Keizer PJ (1990) The expansion of *Schizopora carneolutea* (Basidiomycetes) in Europe, in particular in The Netherlands. Persoonia 14:167–171

Keizer PJ (1993) The influence of nature management on the macromycete flora. In: Pegler DN, Boddy L, Ing B, Kirk PM (eds) Fungi of Europe: investigation, recording and conservation. Royal Botanic Gardens, Kew, pp 251–269

Kile GA, McDonald GI, Byler JW (1991) Ecology and disease in natural forests. In: Shaw CG, Kile GA (eds) *Armillaria* root disease. US Dep Agric Agric Handb 691:102–121

Kotiranta H, Niemelä T (1993) Uhanalaiset Käävät Suomessa (Threatened polypores in Finland). Vesija Ympäristöhallinnon Julkaisuja-Sarja B 17:1–116 Helsinki

Kreisel H (1967) Taxonomisch-Pflanzengeographische Monographie der Gattung *Bovista*. Beih Nova Hedwigia 25:1–244, Figs 1–70

Kreisel H (1985) Geographische Verbreitung der Pilze. In: Kreisel M-H (ed) Handbuch für Pilzfreunde, vol 4. Gustav Fischer, Jena, pp 48–66

Krieglsteiner GJ (1991) Verbreitungsatlas der Grosspilze Deutschlands (West) Band 1A, B. Ulmer, Stuttgart

Krieglsteiner GJ (1993) Verbreitungsatlas der Grosspilze Deutschlands (West) Band 2. Ulmer, Stuttgart

Lange L (1974) The distribution of macromycetes in Europe. Dan Bot Ark 30(1):1–105

Lawrynowicz M (1989) Chorology of the European hypogeous Ascomycetes, I. Elaphomycetales. Acta Mycol 25:3–41

Lawrynowicz M (1990) Chorology of the European hypogeous Ascomycetes, II. Tuberales. Acta Mycol 26:7–75

Mikola P (1973) Application of mycorrhizal symbiosis in forestry practice. In: Marks GC, Kozlowski TT (eds) Ectomycorrhiza, their ecology and physiology. Academic Press, New York, pp 383–411

Murakami Y (1987) Spatial distribution of *Russula* species in *Castanopsis cuspidata* forest. Trans Br Mycol Soc 89:187–193

Nauta M, Vellinga EC (1993) Distribution and decline of macrofungi in the Netherlands. In: Pegler DN, Boddy L, Ing B, Kirk PM (eds) Fungi of Europe: investigation, recording and conservation. Royal Botanic Gardens, Kew, pp 21–46

Nauta M, Vellinga EC (1995) Atlas van Nederlandse paddestoelen. Balkema, Rotterdam

Nitare J (1988) Jordtungor, en svampgrupp på tillbakegång i naturliga fodermarker. Sven Bot Tidskr 82:341–368

Parent GH, Thoen D (1986) Etat actuel de l'extension de l'aire de *Clathrus archeri* (Berkeley) Dring (syn.: *Anthurus archeri* (Berk.) Ed Fischer) en Europe et particulièrement en France et au Benelux. Bull Trimest Soc Mycol Fr 102:237–272

Pegler DN (1983a) The genus *Lentinus*, a world monograph. Royal Botanic Gardens, Kew

Pegler DN (1983b) Agaric flora of the lesser Antilles. Royal Botanic Gardens, Kew

Pirozynski KA (1968) Geographical distribution of fungi. In: Ainsworth GC, Sussman AS (eds) The fungi, an advanced treatise, vol 3. Academic Press, London, pp 487–504

Rald E (1985) Vokshatte som indikatorarten for mykologisk vaerdifulde overdrevs-lokaliteter. Svampe 11:1–9

Redhead SA (1989) A biogeographical overview of the Canadian mushroom flora. Can J Bot 67:3003–3062

Ryvarden L (1991) Genera of polypores, nomenclature and taxonomy. Fungiflora, Oslo

Väisänen R, Heliövaara K, Kotiranta H, Niemelä T (1992) Biogeographical analysis of Finnish polypore assemblages. Karstenia 32:17–28

Van Brummelen J (1967) A world monograph of the genera *Ascobolus* and *Saccobolus*. Persoonia Suppl 1:1–260

Volkmann-Kohlmeyer B, Kohlmeyer J (1993) Biogeographic observations on Pacific marine fungi. Mycologia 85:337–346

Wiley EO (1987) Methods in vicariance biogeography. In: Hovenkamp P (ed) Systematics and evolution: a matter of diversity. Utrecht University, Utrecht, pp 283–306

Winterhoff W, Krieglsteiner GJ (1984) Gefährdete Pilze in Baden-Württemberg. Beih Veröff Naturschutz Landschaftspflege Bad-Württ 10:1–120

Fungal Interactions and Biological Control Strategies

9 Competition and the Fungal Community

P. WIDDEN

CONTENTS

I. Introduction

The effects of competition on the distribution and abundance of species have, for a long time, been a major theme of ecology (Keddy 1989). The Lotka-Volterra equations, describing the interactions between two competing species (Lotka 1932), pre-dicted that stable coexistence between two species could occur only when competition between the two species was weak. The experiments of Gause (1934), designed to empirically test these predictions, led to the formulation of the competitive exclusion principle, stated by Hardin (1960) as "complete competitors cannot coexist." The classical study by MacArthur (1958) of sympatric warblers, which all apparently had the same, or similar, feeding preferences, and therefore should not have been able to coexist, led to the concept that, where two or more apparently similar species do coexist, the resources are partitioned in such a manner that competition is avoided. As Hardin (1960) pointed out, the ensuing studies of resource partitioning were very prone to circular argument; if two similar species can be shown to be partitioning resources, then the principle is upheld, whereas if they cannot be shown to be partitioning resources, we do not understand enough concerning their biology. However, these studies also resulted in a major discussion of the nature of resources that can be competed for, the amount of "permissible" niche overlap, a consideration of the types of environment in which we can expect competition to be a major influence, and the types of habitat in which it may not be. They have also led to a major controversy concerning the degree to which the distribution and abundance of organisms in natural communities is influenced by competition.

Hairston et al. (1960) suggest that the niche of decomposer organisms is such that their communities are likely to be controlled by competition. One would therefore think that the ecologists studying decomposer organisms, such as bacteria and fungi, should be particularly interested in the matter. A glance at the ecological literature would not, however, suggest that mycologists have been concerned with the problem of competition and its consequences for the fungal community. In fact, there has been almost no contribution by mycologists to this debate (Shearer 1995). In his book on

Department of Biology, Concordia University, 1455 de Maisonneuve Boulevard West, Montreal, Quebec, Canada H3G 1M8

The Mycota IV
Environmental and Microbial Relationships
Wicklow/Söderström (Eds.)
© Springer-Verlag Berlin Heidelberg 1997

competition, Keddy (1989) quotes only two papers by mycologists. In this chapter, I will briefly review some aspects of competition theory. I will then discuss competition in the fungi and the degree to which the importance of competition in fungal communities is understood. I will conclude with some suggestions as to why mycologists should be interested in competition, and some personal views on directions that research on competition in the fungi could or should take.

II. Competition Theory

A. Definition of Competition

Keddy (1989), while noting that there are many definitions of competition, defines it as "the negative effects which one organism has upon another by consuming, or controlling access to, a resource that is limited in availability." This definition stresses the importance of resources in the discussion of competition and also suggests the need to understand which resources are limiting in any environment. The definition also implies that there are two basic ways in which competition can take place, either by consumption of the resource or by controlling access to it.

B. Resources

The resources competed for depend on the organisms competing, and include any substance or environmental attribute that may be limited in supply. In animals, intraspecific competition is often studied in relation to food, water or space. In plants, which are sessile organisms, the resources can be below ground, such as water and nutrients, or above ground such as space, light or pollinators. For fungi, decomposable substrates are the resources usually considered, though one could consider space in the phylloplane or rhizoplane or available hosts as resources that are potentially limiting.

C. Interference Competition
vs. Exploitation Competition

Competition is usually classified as either interference competition or exploitation competition (Keddy 1989). Exploitation competition occurs when one organism, by exploiting a resource, reduces its availability to another. In this case, it is not necessary for the two organisms to come into direct contact at all. As an example, Titman (1976) described competition between the diatoms *Asterionella formosa* and *Cyclotella meneghiniana* grown in nutrient solutions where either phosphate or silicate were limiting. Depending on the concentrations of the limiting nutrients, either of these two diatoms could replace the other, without there being any direct physical or chemical interference. This is quite different from the situation that is observed, for instance, where faster-growing plants occlude the light, thus preventing competitors from developing (Harper 1977), which is a form of interference competition. In plants, it has also been suggested that the production of chemicals toxic to other plants (allelopathy) also plays a role in competition, although it has been difficult to prove this (Harper 1977; Begon et al. 1990). In sessile animals, competition for space is often severe. In barnacles, direct interference competition can occur when one species quite literally pries the other off the substrate (Connell 1961).

D. Competition as a Strategy

1. r and K Selection

Another important theme in ecological research has been the concept of ecological strategies. The underlying concept behind the idea of ecological strategies is that different environments select for different traits in the organisms that occupy them. Pianka (1970, 1988) discusses the characteristics of r- and K-selected species as two extremes of a continuum from opportunistic populations (r-selected or r-strategists) to equilibrium populations (K-selected or K-strategists). Species that are r-strategists are found in disturbed environments and are characteristically short-lived, fast-growing and put a large proportion of their energy into reproduction. On the other hand, K-strategists are found in stable, undisturbed habitats. They tend to be larger, often slower-growing, and they put less energy into reproduction and more into biomass and competitive mechanisms. It is important to stress two points in relation to ecological strategies, first, that they represent a continuum, not an either/or situation, and second, that the concept implies tradeoffs between the two extreme strategies, i.e. an organism cannot be both an r- and K-strategist at the same time.

2. Ruderal, Competitive and Stress-Tolerant Strategies

Grime (1977, 1979) argued that plants and fungi exhibited three primary strategies to deal with environmental conditions that limit their biomass. Grime considered that there were two major environmental factors which limited production, stress (factors restricting production) and disturbance (destruction of biomass). This results in four extremes of habitat type: low stress/low disturbance, low stress/high disturbance, high stress/low disturbance and high stress/high disturbance. Grime argues that plants cannot adapt to the high stress/ high disturbance extreme and thus three primary strategies have evolved as a result of selection by the other three habitat types. The three strategies are: (1) **competitive** (C-selected), adaptive for low stress/low disturbance habitats; (2) **ruderal** (R-selected), adaptive to low stress/high disturbance habitats; (3) **stress-tolerant** (S-selected), adaptive to high stress/low disturbance habitats.

3. The Nature of C-Selecting Environments

As Keddy (1989) points out, Grime's ideas create a paradox because he suggests that competition is high in resource-rich environments, yet, by definition, that competition is a result of resource limitation. Thus, according to Tilman (1982), increased resources should result in decreased competition.

Part of the problem may lie in not making a distinction between competition for water and nutrients (belowground competition in plants) and competition for light and space (aboveground competition). Thus, using Grime's system, plants may be stressed by environmental conditions such as low temperatures or low concentrations of essential nutrients, or both. One could imagine non-nutrient stress that could reduce growth to the point that neither roots nor shoots are competing. However, if the stress is relieved, then growth would result in increased nutrient demand and roots could come into competition for limited resources. At this point, root competition could limit growth, preventing shoot competition. Adding nutrients to such a system would then result in increased growth, allowing shoots to grow to the point that they would now compete for light. Wilson and Tilman (1995) examined the effects of

increased fertility (nitrogen addition) and disturbance (tilling) on eight old-field plant species, and found that competitive effects moved from roots to shoots as fertility increased and that disturbance reduced competition. This study emphasises the need to define the nature of the competition and the resources being competed for. It also is important to consider the long-term impact that the organisms have on the resources they are using (Leibold 1995). Liebold (1995) suggests that considering the long-term effects of organisms on the densities of their resources may resolve the contradictions between Grime's and Tilman's views.

E. Competition and Community Structure

It is the role of competition in determining the distribution of species and the structure of communities that forms the major importance of competition theory and also a major controversy. In their book, *The Distribution and Abundance of Animals*, Andrewartha and Birch (1954), having reviewed the concepts of density-dependent and density-independent regulation of populations, concluded that the terms competition and density dependence had become synonymous. They then went on to conclude that such density-dependent mechanisms are not necessary to account for the regulation of animal numbers. Hairston et al. (1960) expressed a view quite different to that of Andrewartha and Birch, stating that, in general, decomposers, producers and predators should be controlled by resource limitation, whereas herbivores would not be limited by their food supply.

1. Does Competition Occur in Nature?

Part of the reason for this controversy lies in the difficulties involved in demonstrating that competition actually occurs, in both experimental systems and, particularly, in natural communities (Keddy 1989; Goldberg and Barton 1992). One of the clearest early demonstrations of the role of competition in the structuring of a community was that of Connell (1961). In this study, by combining field observations and manipulations, Connell was able to determine that the barnacle *Chthamalus stellatus* was limited at its highest point on the shoreline by physical conditions (heat/desiccation), whereas, lower down the shoreline, it was limited by competition for space from the barnacle *Balanus balanoides*. In the absence of competition from *B. balanoides*, *C. stellatus* could survive much lower down the shoreline than its actual distribution would indicate. Thus, the actual distribution of barnacles observed in nature was a result of a combination of factors that included environmental tolerances, interspecific competition and, in the case of *B. balanoides*, predation.

The study by Connell, quoted above, took advantage of the sessile nature of barnacles, which allowed for the constant observation of individual animals, and for the removal of competitors in a field situation. As plants are sessile organisms, and as it has been suggested that they should be limited by resources and therefore compete, it is not surprising that an extensive literature has developed on the role of competition in plant communities (Tilman 1982, 1988; Goldberg and Barton 1992). In spite of this, though there is little doubt that competition does occur in plant communities, it is still not easy to determine how important it is in any given community or to predict the competitive ability of a given species in nature (Keddy 1989).

2. Competition and Plant Diversity

At first glance, it would seem that plants have similar resource requirements and, thus, the existence of large numbers of plant species in one location would appear to violate the competitive exclusion principle. This is particularly true of planktonic algae, and was stated by Hutchinson (1961) to be the "paradox of the plankton." Titman (1976) originally examined this problem in relation to planktonic algae and developed a model which predicted that "as many competing species can coexist as there are limiting resources". While this model alone could not explain the observed diversity, Titman suggested that environmental variation would also add to the number of species that could coexist. These ideas were later applied to and are still applicable to land plant communities (Tilman 1982).

Recently, Chesson (1994) developed a general model for competition between several species in a varying environment. He suggests three general mechanisms affecting coexistence; fluctuating-independent mechanisms, relative non-linearity and the storage effect. This model supports the idea that, in a fluctuating environment, a number of competing species may coexist, the main mechanism for this being attributed to

the storage effect, which implies that, under favourable conditions, the benefits are in some way stored to tide the organism over unfavourable periods, ". . . whether this storage can be traced to a seed bank or something else". Clearly, from Chesson's model, seeds, bulbs, rhizomes etc. can be a means of storage in plants, as can spores, resting spores and sclerotia in fungi.

Recently, Tilman (1994) described a model for the coexistence of sessile organisms based on the spatial structuring of the habitat that results from occupation of space by the organisms. An essential aspect of this model is that a sessile organism occupies a discrete space which becomes available for occupation only after the death of the occupant. This model predicts that, in such a structured habitat, many species can coexist, provided that there is a two- or three-way tradeoff between colonization ability, competitive ability and longevity.

Thus, models for competition in plants suggest that diversity will increase when the number of possible limiting resources is high, when environmental heterogeneity is high and when tradeoffs between the various strategies exist. They also suggest that exploitation competition by roots (competition for nutrients or water) will be high when nutrients are scarce, but that competition for energy (light) by shoots will be high in nutrient-rich environments.

III. Use of Fungi as Models for Competition

From a theoretical viewpoint, fungi should be useful as models of competition, because they are part of the decomposer system, and therefore, as Hairston (1960) has pointed out, must be food-limited. We might therefore expect to see many examples of competitive strategies among the fungi. There are also some very practical advantages to using fungi to model competition, because of their short life cycles and the possibility of replicating experiments extensively, and because the systems required do not take up large amounts of space. However, we should probably be very careful in choosing the models for competition that we use so that we take into account the unique biology of the fungi that we are dealing with. In the discussion of competition theory, I have, for the most part, deliberately avoided a discussion of competition in animals, as their high mobility and (usually) non-clonal nature would suggest that they may compete in very different ways.

Models of plant competition may seem more applicable, as plants and fungi are both sessile and modular. Thus, the model of competition proposed by Tilman (1994), dealing with sessile organisms in structured habitats, seems to be much more applicable to the fungi. However, this model assumes that the physical presence of one individual preempts the space from others. While this seems reasonable for a tree, does it seem reasonable for a fungus, particularly a soil hyphomycete growing as a modular hyphal system through a substrate? For the most part, fungal systems do not involve spatially separated structures for the assimilation of nutrients as opposed to carbon, so models that differentiate between root and shoot competition may not be relevant to the fungi.

A useful approach to studying plant competition is the de Wit replacement series (de Wit 1960). We have used an analogous approach with soil fungi (Widden 1984; Widden and Hsu 1987). The de Wit replacement series involves growing potential competitors at differing relative densities (but the same absolute density) in controlled conditions and comparing yields at the end of the experiment. This approach has been criticised because it does not take into account the effects of changing absolute density (see Shearer 1995), but it is a useful method for comparing the outcome of competition between two plant species under different conditions (Firbank and Watkinson 1990) and, provided that the densities are high, the results are likely to be relatively independent of overall density (Cousens and O'Niell 1993). Using a similar approach, we were able to manipulate the density of spores of competing *Trichoderma* species in a sand matrix and allow the fungi to compete for tree litter (Widden 1984; Widden and Hsu 1987). The outcome of competition was then assessed by recording the success of the two species at colonizing and regrowing from the tree litter. Although these approaches appear to be similar, there are some important differences. First, using the de Wit series with plants, it is possible to know the absolute density of plants and the biomass at the end of the experiment. In the case of the fungi, it is only possible to know the starting density of spores. We do not know the biomass at the end of the experiment, nor can we talk about density in any meaningful way. It is implicitly assumed in the fungal experiments that,

at the end of the experiment, the substrate is completely colonized. In the plant experiments, the results are assessed within a single generation, whereas in the fungal experiment, we do not know how many generations have occurred. Finally, in plant experiments, it is possible to distinguish between competition for nutrients (root competition) and competition for energy (shoot competition). In the fungal experiments, it is not possible to distinguish between competition for nutrients and for energy. In comparing the competitive abilities of fungi under different environmental conditions, the lack of biomass data is probably not important, but attempting to quantify competition coefficients from such experiments would be dangerous. There is also a danger in these experiments that we subject our fungi to a second round of competition (for the agar substrate used in their recovery) before the outcome of competition is assessed.

A recent paper by Wardle et al. (1993) proposed an experimental design to overcome two of the specific problems associated with the replacement series approach to fungal competition as we had used it. They varied the starting densities and they also monitored total fungal biomass in the system. Interestingly, their results showed that the starting density did not affect the final outcome of the competition experiment.

IV. Mechanisms of Fungal Competition

A. Substrate Utilization and Exploitation Competition

Exploitation competition occurs when two or more organisms require the same resource and, according to theory, the organism which uses the resource most efficiently will outcompete the less efficient competitor for that resource Tilman (1982). When discussing fungi, we usually refer to complex substrates, such as parts of dead plants or animals, as the resources.

Garrett (1963) suggested that the first wave of colonizers to reach a substrate, such as a dead leaf, were sugar fungi, which lacked the enzymes to use more complex carbohydrates, such as starch, cellulose or lignin. These fungi would need a high level of "competitive saprophytic ability", correlates of which are an ability to germinate and grow rapidly from spores, have a versatile enzyme system, and be capable of inhibiting other organisms by producing antibiotics. Thus, a broad repertoire of hydrolytic enzymes was considered to be important for competitive ability. Gochenaur (1978) described the soil microfungi, such as species of *Penicillium*, *Trichoderma*, *Oidiodendron* and *Mortierella*, as "opportunistic decomposers" with broad environmental tolerances, which, because they can produce large numbers of propagules, remain quiescent in the soil for long periods, and come briefly to life on the addition of a suitable energy source. These fungi would, therefore, seem to fit Garrett's definition of fungi with high competitive saprophytic ability. However, analysis of the enzymatic abilities of these fungi showed that, whereas the Penicillia and other members of the Moniliaceae have broad enzymatic abilities, the typical sugar fungi, such as members of the *Mucorales*, are comparatively poor in their capacity to degrade complex carbohydrates (Gochenaur 1984). Thus, among these soil microfungi, there appear to be two major groups, fungi such as species of *Mucor* and *Mortierella*, which are only capable of using simple carbohydrates, and other organisms, such as Penicillia and Trichodermas, which can often use a wide range of complex carbon sources.

In discussing substrate colonization, however, it is probably more useful to think of resource capture than exploitation competition. Cooke and Rayner (1984) describe primary resource capture as "the process of gaining initial access to, and influence over an available resource". Once a resource is captured by a fungus, then any organism trying to gain access will have to be capable of replacing the prior colonizer (secondary resource capture). In an interesting study by Robinson et al. (1993), primary and secondary resource capture are illustrated by the interactions between *Mucor hiemalis*, *Chaetomium globosum*, *Agrocybe gibberosa* and *Sphaerobolus stelatus* when colonizing sterilized wheat-straw. *M. hiemalis* is a fast-growing sugar fungus, which is not able to use cellulose, *C. globosum* is a slower-growing, cellulose-degrading ascomycete, whereas *A. gibberosa* and *S. stelatus* are slow-growing basidiomycetes which can use both cellulose and lignin as substrates (Robinson et al. 1993). The primary resource capture by these organisms in competition was predictable from the rates of extension, so that *M. hiemalis* always colonized the substrate ahead of its competitors. However, once the substrate had been colonized, either of the basidi-

omycetes would replace *M. hiemalis* or *C. globosum*. *M. hiemalis*, therefore, is good at primary resource capture, but is not able to control the resource in the face of competition by a more combative fungus. It is worth noting that many resources, such as dead twigs, branches and leaf-litter are already colonized by fungi before they become available to the soil fungi, so that it is normal for a successful soil fungus to have to replace those fungi already present.

From the above discussion, it should be clear that there is no indication from the work described that there is exploitation competition occurring. *Mucor hiemalis* is exhibiting a ruderal strategy, capturing a resource rapidly, and, presumably, sporulating quickly, before it is replaced by more combative fungi. In fact, I am not aware of any studies which clearly demonstrate that exploitation competition for these carbon sources is taking place, although it is possible that it does. I am inclined to agree with Cooke and Rayner (1984) that it may be artificial to divide competition into exploitation competition and interference competition except when "there is continuous renewal of resources to organisms with a determinate body form".

B. Interference Competition

There are many examples in the fungal literature, mainly from dual culture experiments (see Shearer 1995), demonstrating the ability of fungi to interfere with one another. Such interference may be physical, involving direct hyphal contact (see Rayner and Todd 1979), it may involve the production of either soluble or volatile chemicals, which are effective at a distance and prevent hyphal growth (Dix and Webster 1995) or it may involve chemical lysis of the hyphae of one fungus by another. Although these phenomena are well documented, their significance in nature is not always easy to demonstrate.

1. Chemical Interference

Because of the importance of antibiotics produced by fungi, there is an extensive literature on fungal metabolites, and many have argued that antimicrobial agents evolved as agents of competition (Wicklow 1981; Gloer 1995). Almost all ecological groups of fungi are known to produce chemicals antagonistic to other fungi. In the soil, common hyphomycetes, such as species of *Penicillium* and *Trichoderma*, are well known for their ability to produce antibiotics. In the case of *Penicillium*, many compounds produced are toxic to animals, and it has been suggested that some of these compounds may be produced to defend substrates against predators (Janzen 1977; Wicklow 1985). *Trichoderma* species produce a range of antifungal antibiotics which have been considered to be important in the biological control of plant pathogens (Dennis and Webster 1971a, 1971b; Papavisas 1985). Ectomycorrhizal basidiomycetes have also been shown to produce an extensive array of antibiotic compounds which are considered to be important in biological control (Marx 1975), as have coprophilous fungi (Wicklow 1992; Gloer 1995).

In spite of this, however, there is very little direct evidence that antifungal agents produced in the soil are involved in antagonism. The study of Wright (1956) still remains one of the most convincing demonstrations that *Trichoderma* (*Gliocladium*) *virens* could, when growing on wheat straw buried in the soil, produce gliotoxin in effective concentrations. However, in studies of competition between Trichodermas, Widden and Scattolin (1988) demonstrated the ability of five *Trichoderma* species to inhibit one another over a range of temperature conditions, but they could not relate the production of chemical inhibitors to their competitive ability.

A series of studies by Marx and Davey (1969a,b) have shown that *Leucopaxillus cerealis* mycorrhizae on white pine could produce dietrene nitryl in concentrations effective against *Phytophthora cinamomi*. However, although inhibition of the root pathogen was demonstrated, it is not clear how this might relate to competition for resources by *Leucopaxillus*.

2. Physical Interference

When grown on agar, it can be clearly demonstrated that fungi can interfere with each other in a number of ways. Rayner (1978) and Rayner and Todd (1979) have described the various interactions that occur between wood-decomposing Basidiomycota when paired together on agar. At the one extreme, there is no obvious reaction, and the two fungi may merge together, whereas, at the other extreme, they may form mutual inhibition zones, presumably as a result of chemical growth inhibitors. Physical interference can be in the form of hyphal barrages (dense hyphal proliferation at the point of contact), the overgrowth of one colony by the other, or hyphal coiling. These interactions can be seen not only in culture, but also when the fungi are naturally growing on decaying wood (Rayner and Todd 1979), where zones of interaction are evident.

In *Trichoderma*, the coiling of hyphae around those of other species has been demonstrated, as has the penetration of the hyphae of other fungi (Elad et al. 1987). These interactions are similar to those observed in wood-decaying basidiomycetes by Rayner and Todd (1979). It is therefore clear that not only can fungi compete through chemical means, but that there are many potential ways in which fungi can compete through physical interference.

V. Fungal Strategies

The three basic strategies of Grime (1977) and their correlates have been discussed in relation to fungi by Pugh (1980), Pugh and Boddy (1988) and Dix and Webster (1995). Pugh (1980) argued that fungi have, in fact, adopted strategies to deal with high stress/high disturbance environments and quotes the fungi of the phylloplane as one example. Pugh also considerably changes Grime's definition of disturbance to allow the addition of substrates as a disturbance. Later, Pugh and Boddy (1988) defined two types of disturbance, destructive disturbance (disturbance as defined by Grime) and enrichment disturbance, which results from the input of uncolonized or partially colonized material. In either case, the environment is suddenly changed in such a manner that the fungi become diluted in relation to the available substrate, creating an opportunity for colonization. The ruderals that colonize under these circumstances are fast-growing, generally non-combative and produce large numbers of spores. Pugh (1980) and Pugh and Boddy (1988) and Dix and Webster (1995) all agree that the Mucorales are an example of a group of fungi that are almost exclusively ruderal, and Pugh (1980) considers most of the common soil hyphomycetes, such as the Penicillia and Trichodermae, to be ruderals. However, the Penicillia and the Trichodermas are well known for their ability to produce antibiotics in culture, and they also tend to have a larger complement of enzymes to deal with complex polysaccharides

(Gochenaur 1984). This suggests that these fungi may, in fact, be more combative in nature.

Once a substrate is occupied, combative fungi can replace the ruderals. Thus, it is on undisturbed, precolonized substrates that we expect to see competitive fungi. Combative fungi can replace ruderals by production of antibiotics, by hyphal interference or by mycoparastitism (Dix and Webster, 1995). Possibly the best examples of combative fungi are the cord-forming basidiomycetes, which invade woody substrates (Dowson et al. 1988). These cord-forming fungi are able to grow rapidly from one substrate to another by concentrating their energies into cords once a new substrate is located and they can mobilize resources from one food base to another (Dowson et al. 1988; Boddy 1993).

Such characteristics as slow growth rates, low reproductive output and conservation of resources are associated with a stress-tolerant strategy (Grime 1979). With fungi, this certainly appears to be the case, as the major group of microfungi found in polar soils are slow-growing, non-sporulating fungi (Dowding and Widden 1974; Widden and Parkinson 1979).

It is important to recognize that fungal strategies are infinitely variable, and that members of a single group may vary in their strategies. Thus, in cord-forming basidiomycetes, *Hypholoma fasiculare* is more combative than *Phanerochaete velutina*, which is more combative than *Coriolus versicolor* (Chapela et al. 1988; Boddy 1993). In a detailed study of the ability of species of *Trichoderma* to compete for and maintain possession of Norway spruce needles, Widden and Scattolin (1988) showed that *T. polysporum* and *T. viride* had a more stress-tolerant strategy whereas *T. hamatum* and *T. koningii* had a more combative strategy, and a fifth species, similar to *T. virens*, had a more ruderal strategy.

VI. Competition in Fungal Communities

A. Does Competition Occur in Natural Fungal Communities?

When referring to fungal communities, it is first important to establish whether competition actually does occur. Although most of our evidence for the existence of competition in natural communities is indirect, the studies of decaying wood reviewed by Rayner and Todd (1979) probably represent the most elegant demonstration of competitive interactions in natural fungal communities. In decaying branches, the production of zone lines and pseudosclerotial plates as a result of both intra- and interspecific competition provides incontrovertible evidence of interference competition in a natural fungal community.

In populations of soil fungi, it has been much more difficult to demonstrate the existence of competition in nature, and most of the evidence is indirect, or comes from laboratory experiments. The study by Wardle et al. (1993) describes an approach to studying competition between pairs of fungal species, designed to overcome the major criticism of methods based on the de Wit replacement series approach. They investigated competition between *Trichoderma harzianum* and *Mucor hiemalis* in an agricultural soil, and between *M. hiemalis* and *T. polysporum* in forest litter. In this study, the effects of competition were studied at a range of total starting densities as well as a range of proportional densities between the two competitors. The proportion of the species colonizing either litter or soil (organic and mineral) particles was recorded at intervals either by dilution plating or by litter washing, and total biomass of fungi was also monitored, using the chloroform fumigation method (Jenkinson and Powlson 1976). This experiment showed that the final outcome was independent of the starting total density. It also showed that *M. hiemalis* competitively inhibited *T. harzianum* in the agricultural soil, but that *T. polysporum* competitively inhibited *M. hiemalis* in the forest litter. In both cases, the two fungi were capable of coexistence. The interpretation of these results, however, is complicated by the fact that the actual substrates being competed for are not defined. Our knowledge of these fungi would suggest that *M. hiemalis* is a true ruderal, lacking complex enzyme systems for breaking down biopolymers, whereas trichodermas generally have cellulases and are more combative. In the forest litter, with large quantities of (presumably) cellulose-containing organic matter, it is possible that *T. polysporum* was able to grow actively in the litter, and inhibition of *Mucor* was possible. In the agricultural soil, with very little organic matter, it is possible that *Trichoderma* was forced to compete with *M. hiemalis* for small quantities of available sugars. If this is so, then this may be an example of exploitation competition, as the better growth of *M. hiemalis* could have resulted from more efficient scavenging for sugars as a resource.

B. Does Competition Affect Fungal Community Structure?

Again, the strongest evidence for the effects of competition on fungal community structure is seen in the studies that have been done on decaying wood, reviewed by Rayner and Todd (1979) and more recently in Dix and Webster (1995). In these studies, the community structure can clearly be seen because of the visible barrages and discoloured zones. The competitive replacement of one species by another can also be demonstrated in inoculated logs (Rayner and Todd 1979), in wood blocks placed in soil (Dowson et al. 1988) and in felled logs (Chapela et al. 1988).

In forest litter, studies of the ecology of two basidiomycete decomposers, *Marasmius androsaceus* and *Mycena galopus*, have provided strong evidence for control of the distribution of these two species by a combination of environmental tolerances, interspecific competition and predation pressure (Frankland et al. 1995). In the spruce forest studied, *M. androsaceus* can normally be found in the upper 4 cm of the litter layer, whereas *M. galopus* is found at a depth of 4–8 cm. However, when the two fungi occur together, *M. galopus* is displaced downwards by some 4 cm (Newell, 1984a). Of the two fungi, *M. androsaceus* (which is more drought-tolerant than *M. galopus*) is the better competitor (Frankland 1984), but it is also a preferred food for the common collembolan, *Onychiurus latus* (Newell 1984a). Thus, in the field, *M. androsaceus* appears to be restricted to the upper litter layer by grazing pressure from *O. latus*, which is not tolerant of drought, and *M. galopus* appears to be restricted to lower layers by competition with *Marasmius* (Frankland 1984; Newell 1984b). This situation is remarkably similar to that described by Connell (1961) to account for the distribution of barnacle species in the intertidal zone.

C. Does Environmental Change Affect Fungal Competition?

Chesson's (1994) model of competition predicts that environmental fluctuations will enable more competing species to coexist than will a non-fluctuating environment. To microfungi, living on decaying litter on the forest floor, the environment is fluctuating both in space, due to heterogenous resources, and in time, due to diurnal and seasonal changes. If fungal competition occurs and if it is sensitive to changes in the environment, then one would expect the fungal community to be rich and diverse, which, to those of us who study fungal communities, certainly seems to be the case.

In my laboratory, we have investigated the effects of both temperature and litter type on competition between species of *Trichoderma* (Widden 1984; Widden and Hsu 1987; Widden and Scattolin 1988). These studies showed that *T. polysporum* and *T. viride*, two species known to be more prevalent in colder soils (Danielson and Davey 1973), were better competitors at low temperatures (5–10 °C), whereas *T. hamatum* and *T. koningii* were better competitors at higher temperatures (20–25 °C; Widden 1984). These studies also showed that *T. polysporum*, which is often abundant in Norway spruce forests (Söderström and Bååth 1978; Widden 1986), had a higher temperature range at which it was a good competitor for Norway spruce litter than for white pine or sugar maple litter. Carreiro and Koske (1992b) examined the effect of temperature on competition between *Geomyces pannorus*, *Mortierella hyalina* (both of which could be isolated at 0 °C; Carreiro 1992a) and *Trichoderma longibrachiatum*, which was isolated at 10 °C, for two substrates, moth wings and oak litter. Their results showed that *T. longibrachiatum* competed best for both substrates at 20 °C. At (0 °C, *G. pannorus* was a better competitor for oak leaves than *M. hyalina*, whereas *M. hyalina* competed best for moth wings.

In both the studies discussed above, the optimum growth rates for all of the fungi tested were above 20 °C and the fungi all grew well on the substrates at the higher temperatures in the absence of competition. These data therefore suggest that the low-temperature fungi are isolated from the soil during cooler months due to competitive displacement by the fungi which are abundant at warmer periods. Thus, not only does the environment affect competition, but the effects of the environment on competition can result in changes in community composition. These seasonal changes in the environment also combine with the heterogeneity that results from the existence of many different discrete substrates (leaves of different plants, insect remains etc.), to result in very complex changes in competitive balance in both space and time. This should result in microfungal communities which are also constantly changing in space and time, as is the case.

The soil environment is not only continuously changing as a result of physical and chemical heterogeneity, but the presence of other organisms can influence the outcome of competition. Grazing by animals on the fungal community can alter competitive balances by removing potential competitors (Frankland 1984; Klironomos et al. 1992; Newell 1984a). A recent study by Schoeman et al. (1996) examined the effects of temperature and the presence of *Trichoderma harzianum* metabolites on the interactions between wood decomposing basidiomycetes in culture. This study demonstrated that not only could temperature affect the interactions of these fungi, but so could the presence of *T. harzianum*. Very often, the presence of *T. harzianum* would change interactions from overgrowth of one species by another to deadlock.

D. The Basis for Competition in Nature

Although there is strong evidence from experimental studies that fungal competition in nature does occur, it is often difficult to determine the basis for that competition. Wood-decaying basidiomycetes, as I have already described, give an elegant demonstration of interference competition. In the case of cord-forming fungi, the process of invasion and maintaining control of substrates is nicely described by Boddy (1993) in terms of warfare strategies. However, in the soil, it is difficult to know the basis of competition. Although there is plenty of evidence that chemical and physical methods of eliminating competitors exist in most groups of fungi, we are usually left arguing that they must have evolved for some purpose and that competition is the most obvious candidate. This is a rather unsatisfactory state of affairs. The study of Wardle et al. (1993) is the only report I know of that even suggests that exploitation competition may occur in the soil, though there is no a priori reason to think that it does not. In the studies of fungal competition in the soil from my laboratory (Widden 1984; Widden and Hsu 1987) and in those of Carreiro and Koske (1992b), although there is strong evidence for competition, there is no evidence as to the nature of the competition.

VII. Where Do We Go from Here?

In this chapter, I have tried deliberately to relate our knowledge of fungal competition to what is known about plant competition. The evidence is clear that fungi can compete, and that they do compete in nature. Fungi clearly have evolved different strategies, and the whole range of strategies exists, from true ruderals, such as the Mucorales, to highly combative fungi, such as the cord-forming basidiomycetes, and to stress-tolerant strategies, such as those slow-growing sterile fungi that predominate in arctic soils. The properties of fungi, as evidenced from their behaviour in cluture, would suggest that many of them have evolved combative strategies, and this is consistent with the view, expressed by Hairston et al. (1960), that competition should be a major controlling force among decomposer organisms. There is also strong evidence that the constantly changing environment results in constantly changing competitive balances that will favour first one strategy, then another. The short life-span of many fungi may indeed amplify the effect of changing conditions in the soil, enabling many species to briefly flourish, sporulate and then wait for favourable conditions to return (the true ruderal strategy), and thus increasing fungal diversity. However, the reality is that virgin (uncolonized) substrates are a rarity, and most decomposer fungi have to replace an existing microflora in order to colonize new substrates, be they fallen leaves, branches or dead plant roots. Thus, whereas in plants exploitation competition may be the commonest type of competition, in fungi, interference competition is probably the norm.

Although we have this general understanding, there is very little specific information concerning competition in fungi. In comparison to the general ecological literature, there are comparatively few studies of fungal competition. In order to improve our understanding, it will be necessary first of all to have much more detailed information on the niches (Whittaker et al. 1973) of the fungi that we study. That is, what are the energy substrates and nutrients that they require, what concentrations of these substrates are required, and how do nutrients limit substrate use? We also need to develop a good understanding of the ecological strategies of the fungi we work with. What defines a stress-tolerant, combative or ruderal strategy, and what kinds of fungal environments should select for these strategies? We also need far more studies of fungal competition, either in model systems that mimic the soil, or in natural systems. We also need to identify the nature of the competition that occurs and the effects that the fungi have on the substrates available.

In theory, we may have the answers to some of these questions, particularly substrate requirements of fungi, but often our data come from laboratory studies of agar-grown fungi, which may not even closely approximate the soil system. Thus, Rayner and Todd (1979) point out that the interactions between wood-decomposing fungi growing on plates were not always the same as those seen on wood. In relation to substrate use, a good example from my own experience, is the ability of fungi to use cellulose. A number of microfungi, particularly species of *Trichoderma*, are known to produce cellulases, and are therefore assumed to be cellulolytic when growing in the soil. Even though this is so, it can sometimes be difficult to show this, using powdered cellulose in agar. However, when *Trichoderma* species are grown on native cellulose (cotton cloth) resting on moist soil, decomposition of the cellulose can be very rapid, as measured by tensile strength loss (Widden et al. 1989). However, the type of soil can have a very large effect on the results, as can the nitrogen concentration. Gillespie et al. (1988), using a similar test on 50 fungi (Hymenomycotina, Ascomycotina and Fungi Imperfecti), isolated from moorland soils, found that different moorland soils gave very different results for cotton decomposition. They also showed that there was very little correlation between results from standard agar plate tests using cellulose clearing and the tests on cotton placed on soil.

In spite of the clear difficulties involved in the study of fungal competition, there are very real reasons why more attention should be paid to competition in fungal communities. Apart from the theoretical importance of fungal competition and the wish to participate more in the broader ecological discussion, there are pressing practical reasons for more research into competition in fungal communities. There are many circumstances under which the introduction of fungi into a system is being suggested as a means to improve the success of cultured plants. These include the use of mycorrhizal inoculum to improve the growth and survival of agricultural crop plants and forest trees, and the use of fungi as biological control agents of plant pathogens. In many cases, green-

house trials are successful, but field trials fail. Often, the reasons for these failures are clearly a result of the interactions between the introduced fungi and the, much more complex, natural communities of fungi and other organisms in the soil. Competition is clearly one interaction which has the potential to reduce the abundance, and therefore the effectiveness, of introduced organisms and, therefore, needs to be fully understood if such attempts are to be successful. Without sound theoretical models, we are reduced to empirical studies for each individual case of interest.

The creation of sound predictive models for competition in the fungi is likely to be a daunting task. Peters (1991) has argued that, even though the existence of competition is not in question, the whole framework of competition theory is so weak as to lack any predictive value and he does not feel that competition theory either does, or even can, make a useful contribution to ecological understanding. Keddy (1989) also points out the general lack of predictive value to much of competition theory and makes a case for more carefully considered and clearly defined questions, planned to increase the generality of their conclusions.

VIII. Conclusions

The general impression that mycologists have not contributed in any significant way to the ongoing debate among ecologists concerning the significance of competition in structuring communities is confirmed by reading the literature. However, there is enough information to draw some general conclusions. First of all, it has clearly been demonstrated that competition does occur in fungal communities, and plays a role in their structuring. The ability of many fungi to compete using both chemical and physical interference mechanisms is well demonstrated, though there is very little evidence of exploitation competition. There is also strong evidence to suggest that not only does competition occur in nature, but that competitive interactions between fungi are strongly influenced by changes in the environment.

However, the importance of competition in determining the distribution and abundance of fungal species in natural communities has not been clearly established, nor do any useful predictive models of fungal competition exist. Because of the stress being placed on the use of introduced

fungi in agriculture and forestry, such models are
clearly needed. To develop such models, the strategies employed by the fungi need to be understood, as do their resource needs, the availability
of resources in the substrates they compete for
and the impact of the fungi on these resources.
Even with this knowledge, the development of
such models is likely to be a major challenge to
fungal ecologists, but, without them, the outcome
of attempts to use introduced fungi, either as
biocontrol agents or as beneficial symbionts, is
likely to be unpredictable, because competition
with native fungi has not been accounted for. Mycologists also need to consider the usefulness of
existing models for competition in plants or animals and to either adapt them to their own needs,
or develop new models appropriate to the fungi.

References

Andrewartha HG, Birch LC (1954) The distribution and abundance of animals. University of Chicago Press, Chicago

Begon M, Harper JL, Townsend CR (1990) Ecology, 2nd edn. Blackwell, Boston

Boddy L (1993) Saprotrophic cord-forming fungi: warfare strategies and other ecological aspects. Mycol Res 97:641–655

Boddy L, Rayner ADM (1981) Fungal communities and formation of heartwood wings in attached oak branches undergoing decay. Ann Bot 47:271–274

Carreiro MM (1993a) Effect of temperature on decomposition and development of fungal communities in leaf litter microcosms. Can J Bot 70:2177–2183

Carreiro MM, Koske RE (1993b) The effect of temperature and substratum on competition among three species of forest litter fungi. Mycol Res 96:19–24

Chapela IH, Boddy L, Rayner ADM (1988) Structure and development of fungal communities in beech logs four and a half years after felling. FEMS Microbiol Ecol 53:59–70

Chesson P (1994) Multispecies competition in variable environments. Theor Popul Biol 45:227–276

Connell J (1961) The influence of interspecific competition and other factors on the distribution of the barnacle Cthamalus stellatus. Ecology 42:710–723

Cooke RC, Rayner ADM (1984) Ecology of saprotrophic fungi. Longman, London

Cousens R, O'Niell M (1993) Density dependence of replacement series experiments. Oikos 66:347–352

Danielson RM, Davey CB (1973) Non-nutritional factors affecting the growth of Trichoderma in culture. Soil Biol Biochem 5:484–495

Dennis C, Webster J (1971a) Antagonistic properties of species groups of Trichoderma I. Production of non-volatile antibiotics. Trans Br Mycol Soc 57:25–39

Dennis C, Webster J (1971b) Antagonistic properties of species groups of Trichoderma I. Production of volatile antibiotics. Trans Br Mycol Soc 57:41–48

de Wit CT (1960) On competition. Versl Landbouwkd Onderz 66:1–82

Dix NJ, Webster J (1995) Fungal ecology. Chapman and Hall, New York

Dowding P, Widden P (1974) Some relationships between fungi and their environment in tundra regions. In: Holding AJ, Heal OW, MacLean SF, Flannigan PW (eds) Soil organisms and decomposition in the tundra. IBP Tundra Biome Steering Committee, Stockholm, pp 123–142

Dowson CG, Rayner ADM, Boddy L (1988) The form and outcome of mycelial interactions involving cord-forming decomposer basidiomycetes in homogenous and heterogenous environments. New Phytol 109:423–432

Elad Y, Sadowsky Z, Chet I (1987) Scanning electron microscopical observations of early stages of interaction of Trichoderma harzianum and Rhizoctonia solani. Trans Br Mycol Soc 88:259–263

Firbank LG, Watkinson AR (1990) On the effects of competition: from monocultures to mixtures. In: Grace JB, Tilman D (eds) Perspectives in plant competition. Academic Press, San Diego, pp 165–192

Frankland JC (1984) Autecology and the mycelium of a woodland decomposer. In: Jennings DH, Rayner ADM (eds) The ecology and physiology of the fungal mycelium. Cambridge University Press, Cambridge, pp 241–260

Frankland JC, Poskitt JM, Howard DM (1995) Spatial development of populations of a decomposer fungus, Mycena galopus. Can J Bot 73 (Suppl 1):S1399–S1406

Garrett SD (1963) Soil fungi and soil fertility. Pergamon, London

Gause GF (1934) The struggle for existence. Hafner, New York

Gillespie J, Latter PM, Widden P (1988) Cellulolysis of cotton by fungi in three upland soils. In: Harrison AF, Latter PM (eds) Cotton strip assay: an index of decomposition in soils. Institute of Terrestrial Ecology, Merlewood, Grange-over-Sands, pp 60–67

Gloer JB (1995) The chemistry of fungal antagonism and defence. Can J Bot 73 (Suppl 1):S1265–S1274

Gochenaur SE (1978) Fungi of a Long Island oak-birch forest I. Community organization and seasonal occurrence of the opportunistic decomposers of the A horizon. Mycologia 70:975–994

Gochenaur SE (1984) Fungi of a Long Island oak-birch forest II. Population dynamics and hydrolase patterns for the soil Penicillia. Mycologia 76:218–231

Goldberg DE, Barton AM (1992) Patterns and consequences of interspecific competition in natural communities: a review of field experiments with plants. Am Nat 4:776–801

Grime JP (1977) Evidence for the existence of three primary strategies in plants and its relevance to ecological and evolutionary theory. Am Nat 111:1169–1194

Grime JP (1979) Plant strategies and vegetation processes. John Wiley, New York

Hairston NG, Smith FE, Slobodkin LB (1960) Community structure, population control and competition. Am Nat 94:421–425

Hardin G (1960) The competitive exclusion principle. Science 131:1292–1297

Harper JL (1977) Population biology of plants. Academic Press, New York

Hutchinson GE (1961) The paradox of the plankton. Am Nat 95:137–145

Janzen DH (1977) Why fruits rot, seeds mold and meat spoils. Am Nat 111:691–713

Jenkinson DS, Powlson DS (1976) The effect of biocidal treatments on metabolism in soil. V. A method for measuring soil biomass. Soil Biol Biochem 8:204–213

Keddy P (1989) Competition. Chapman and Hall, New York

Klironomos JN, Widden P, Deslandes I (1992) Feeding preferences of the collembolan *Folsomia candida* in relation to microfungal successions on decaying litter. Soil Biol Biochem 24:685–692

Liebold M (1995) Revisioning the niche. Ecology 76:1371–1382

Lotka AJ (1932) The growth of mixed populations: Two species competing for a common food supply. J Wash Acad Sci 22:461–469

MacArthur RH (1958) Population ecology of some warblers of northeastern coniferous forests. Ecology 39:599–619

Marx DH (1975) The role of ectomycorrhizae in the protection of pine from root infection by *Phytophthora cinnamomi*. In: Bruehl GW (ed) Biology and control of soil-borne plant pathogens. American Phytopathological Society, St. Paul, pp 112–115

Marx DH, Davey CB (1969a) The influence of ectotrophic mycorrhizal fungi on the resistance of pine roots to pathogenic infections. III. Resistance of aseptically formed mycorrhizae to infection by *Phytophthora cinnamomi*. Phytopathology 59:549–558

Marx DH, Davey CB (1969b) The influence of ectotrophic mycorrhizal fungi on the resistance of pine roots to pathogenic infections. IV. Resistance of naturally occurring mycorrhizae to infection by *Phytophthora cinnamomi*. Phytopathology 59:559–565

Newell K (1984a) Interaction between two decomposer basidiomycetes and a collembolan under sitka spruce: distribution, abundance and selective grazing. Soil Biol Biochem 16:227–233

Newell K (1984b) Interaction between two decomposer basidiomycetes and a collembolan under sitka spruce: grazing and its potential effects on fungal distribution and litter decomposition. Soil Biol Biochem 16:235–239

Papavisas GC (1985) *Trichoderma* and *Gliocladium*: biology, ecology and potential for biocontrol. Annu Rev Phytopathol 23:23–54

Peters RH (1991) A critique for ecology. Cambridge University Press, New York

Pianka ER (1970) On r and K selection. Am Nat 102:592–597

Pianka ER (1988) Evolutionary ecology, 9th edn. Harper and Row, Philadelphia

Pugh GJF (1980) Strategies in fungal ecology. Trans Br Mycol Soc 75:1–14

Pugh GJF, Boddy L (1988) A view of disturbance and life strategies in fungi. Proc R Soc Edinb 94B:3–11

Rayner ADM (1978) Interactions between fungi colonizing hardwood stumps and their possible role in determining patterns of colonization and succession. Ann Appl Biol 89:131–134

Rayner ADM, Todd NK (1979) Population and community structure and dynamics of fungi in decaying wood. Adv Bot Res 7:333–420

Robinson CH, Dighton J, Frankland JC (1993) Resource capture by interacting fungal colonizers of straw. Mycol Res 97:547–558

Schoeman MW, Webber JF, Dickinson DJ (1996) The effect of diffusible metabolites of *Trichoderma harzianum* Rifai on in vitro interactions between basidiomycete isolates at two different temperature regimes. Mycol Res 100: (in press)

Shearer C (1995) Fungal competition. Can J Bot 73 (Suppl 1):S1259–S1264

Söderström BE, Bååth E (1978) Soil microfungi in three Swedish coniferous soils. Holarct Ecol 1:62–72

Swift MJ (1976) Species diversity and the structure of microbial communities in terrestrial habitats. In: Anderson MJ, MacFayden A (eds) The role of terrestrial and aquatic organisms in decomposition processes. Blackwell, Oxford, pp 185–222

Tilman D (1982) Resource competition and community structure. Princeton University Press, Princeton

Tilman D (1988) Plant strategies and the dynamics and structure of plant communities. Princeton University Press, Princeton

Tilman D (1994) Competition and biodiversity in spatially structured habitats. Ecology 75:2–16

Titman D (1976) Ecological competition between algae: experimental confirmation of a resource-based competition theory. Science 192:463–465

Wardle DA, Parkinson D, Waller JF (1993) Interspecific competitive interactions between pairs of fungal species in natural substrates. Oecologia 94:165–172

Whittaker RH, Levin SA, Root RB (1973) Niche, habitat and ecotope Am Nat 107:321–338

Wicklow DT (1982) Interference competition and the organization of fungal communities. In: Wicklow DT, Carroll GC (eds) The fungal community: its organization and role in the community. Marcel Dekker, New York, pp 351–375

Wicklow DT (1985) Ecological adaptation and classification in *Aspergillus* and *Penicillium*. In: Samson RA, Pitt JI (eds) Advances in *Penicillium* and *Aspergillus* systematics. Plenum, New York, pp 255–265

Wicklow DT (1992) The coprophilous fungal community: an experimental system. In: Carroll GC, Wicklow DT (eds) The fungal community: its organization and role in the community, 2nd edn. Marcel Dekker, New York, pp 715–728

Widden P (1984) The effects of temperature on competition for spruce needles between sympatric species of *Trichoderma*. Mycologia 76:873–883

Widden P (1986) Community structure of microfungi from forest soils in southern Quebec, using discriminant function and factor analysis. Can J Bot 64:1402–1412

Widden P, Abitbol JJ (1980) Seasonality of *Trichoderma* species in a spruce forest soil. Mycologia 72:775–784

Widden P, Hsu D (1987) Competition between *Trichoderma* species: effect of temperature and litter type. Soil Biol Biochem 19:89–93

Widden P, Parkinson D (1979) Populations of fungi in a high arctic ecosystem. Can J Bot 57:1324–1331

Widden P, Scattolin V (1988) Competitive interactions and ecological strategies on *Trichoderma* species colonizing spruce litter. Mycologia 80:795–803

Wilson SD, Tilman D (1995) Competitive responses of eight old-field plant species in four environments. Ecology 76:1169–1180

Widden P, Cunningham J, Breil B (1989) Decomposition of cotton by *Trichoderma* species: influence of temperature, soil type and nitrogen levels. Can J Microbiol 35:469–473

Wright JM (1956) The production of antibiotics in the soil. III. Production of gliotoxin in wheat straw buried in soil. Ann Appl Biol 44:461–466

10 Mycoparasitism

P. JEFFRIES

CONTENTS

I. Introduction

There are many ways in which different fungal species can interact such that the presence of one in some way affects the behaviour of the other. At one extreme, one living fungus serves directly as the nutrient source for another; this is mycoparasitism, the major concern in this chapter. In nature, parasitic relationships between fungi probably play an important role in the development of community structure. At the other extreme, the two mycelia grow together in the exploitation of a particular environmental resource and do not apparently effect one another (neutralism). This is unusual, however, as two organisms in the same ecological niche almost inevitably interact in some way or other, often as competitors. In between these two extremes are many other interactions in which one fungus is compromised by the presence of another, yet nutrient exchange has not been demonstrated. Interspecific interactions in this latter category have been discussed in detail elsewhere (Rayner and Webber 1984; Pearce 1990). There are also interactions in which one fungus derives an advantage from the presence of another but the latter does not derive reciprocal benefit although it is not

harmed (commensalism). The phenomenon of fungal individualism whereby different clones of fungi can recognize and respond to others antagonistically has been well described (Rayner 1991), but these interactions occur between mycelia of the same species and are not considered as mycoparasitism.

II. The Variety of Mycoparasitic Relationships

Barnett and Binder (1973) divided mycoparasites into two main categories depending on their mode of nutrition. Necrotrophic or destructive mycoparasites kill their hosts as a result of their parasitic activity, whilst biotrophic mycoparasites obtain their nutrients directly from the living mycelium of the hosts. In necrotrophic relationships the antagonistic action of the mycoparasites is strongly aggressive and the mycoparasite dominates the association. Hyphae of the parasite contact and grow in association with those of the host, sometimes coiling around them, and frequently penetrating. Secretion of hyphal wall degrading enzymes or exotoxins may cause the death of the cytoplasm of the host prior to hyphal contact, or alternatively cytoplasmic death may not occur until after contact has been established. Necrotrophic parasites tend to have a broad range of host fungi, and are relatively unspecialized in their mechanism of parasitism. For example, they often release toxins and lytic enzymes into the environment, are overtly destructive and usually lack specialized infection structures. In this way their behaviour parallels that of the necrotrophic fungi which parasitize plants.

In a biotrophic mycoparasitic relationship, the living host supports the growth of the parasite for an extended period of time, may not appear diseased, and its growth rate, sporulation and metabolism may appear overtly to be little affected,

Research School of Biosciences, University of Kent, Canterbury, Kent, CT2 7NJ, UK

The Mycota IV
Environmental and Microbial Relationships
Wicklow/Söderström (Eds.)
© Springer-Verlag Berlin Heidelberg 1997

at least in the early stages of the relationship. The parasitic relationship is physiologically balanced and the parasite appears to be highly adapted to this mode of life. Biotrophic mycoparasites tend to have more restricted host ranges than necrotrophs, and often form specialized infection structures or host-parasite interfaces. Exotoxin production has not been demonstrated in any biotrophic mycoparasitic interaction. Three types of biotrophic relationship can be distinguished by the morphology of the interaction (Table 1).

There are also different degrees of specialization within interfungal parasitic relationships. The conventional concept of parasitism applies most readily to biotrophic relationships where, by definition, the parasite exploits only living host cytoplasm as a nutrient source. Owing to the limitations of transporting material across an often complex interface between the partners (Jeffries 1987), it is usually only the simpler organic compounds such as sugars and amino acids that are likely to be translocated from the host. In contrast, necrotrophic relationships involve the utilization of dead host biomass, which has been killed as a direct result of the activity of the necrotroph itself. This can involve the degradation of the major structural polymers of the host or/and the soluble components of the cytoplasm and is closely paralleled in many saprotrophic situations. Indeed, saprotrophy, necrotrophy and biotrophy

Table 1. Types of mycoparasite grouped according to their host-parasite interface

Necrotrophs	
Contact necrotrophs	Grow in close contact with the hyphae of the host, but penetration has not been observed
Invasive necrotrophs:	Hyphae of the mycoparasite penetrate those of the host and grow within them, causing necrosis and hyphal lysis
Biotrophs	
Haustorial biotrophs	Penetrate hyphae of the host by means of short hyphal branches (haustoria). The plasmalemma of the host is invaginated around the haustorium
Intracellular biotrophs	Penetrate hyphae of the host, and the naked protoplast of the mycoparasite thallus enters the invaded cytoplasm
Fusion biotrophs	The walls of host and parasite hyphae become closely associated at a contact zone. No penetration of the hyphae of the host occurs, but intercellular channels are formed connecting the protoplasts of host and parasite

are not necessarily mutually exclusive ways of life (Cooke and Rayner 1984) and the relationships between fungi may switch from one form to another as the association develops.

The distinction between these nutritional modes may not always be clear, and it has been pointed out that the colonization of living tissues, via essentially saprotrophic behaviour involving the utilization of diffusates of dead cells, can be confused with necrotrophy and even biotrophy (Rayner et al. 1985). To overcome some of these difficulties, Hawksworth (1981) recommended the use of the term fungicolous fungi as a neutral term to embrace the broad range of associations of two fungi living together even where the biological nature of the association is obscure. This avoids the problem of determining whether nutrient exchange occurs, and can also include wider symbiotic relationships such as commensal ones. For example, there are numerous fungi which have become associated with the external surface of another fungus as their particular ecological niche. They obligately grow in this location, but they are not necessarily parasites, and may be commensals or fungicolous saprotrophs and as such have been little investigated in terms of their diversity and biology. In some cases, the fungicolous fungi are only found growing on the relatively large basidiomes of mushrooms and toadstools, but in other instances the association is between the two respective mycelia. There are many such interfungal relationships, especially in tropical ecosystems, and these may represent a large pool of potential candidates for use as biological control agents of phytopathogens. There are also many fungi that are normally only found growing in association with lichens, the lichenicolous fungi, and others which are associated with Myxomycetes, the myxomyceticolous fungi (Rogerson and Stephenson 1993).

Some of the problems in defining interfungal relationships have been addressed by Cooke and Rayner (1984). They have proposed that mycelial interactions are either competitive, neutralistic or mutualistic, depending successively upon whether the outcome is detrimental to either or both, detrimental to neither but not beneficial to both, or beneficial to both. Thus mycoparasitism is deemed competitive, and is concerned directly with nutrient acquisition. Other competitive interactions are those involving combat between species either relating to defence or primary and secondary resource capture. In this respect, antagonistic attributes can aid a fungus to colonize a substrate via exploitation competition, or restrict the access to a colonized substrate through interference competition (sensu Wicklow 1981) and may not involve mycoparasitism. Thus it can be very difficult to distinguish competitive antagonism and parasitism. The various mechanisms used by fungi to antagonize or parasitize their competitors include antibiotic production, secretion of lytic enzymes, hyphal interference, and direct penetration of the host. Any particular fungus-fungus interaction may encompass more than one of these mechanisms either individually or simultaneously. Only those relationships in which one fungus obtains some nutrients either directly or indirectly from another fungus can be termed mycoparasitic. Thus, when interference competition results in the death of the antagonized hyphae, and these are then used by the antagonist as a nutrient source, mycoparasitism results.

Competition between fungi occurs during primary resource capture when two or more mycelia are limited, in terms of growth rate or size, by a common dependence on a nutrient substrate or other environmental factor. The outcome of such interactions is determined by several factors, but relative competitive saprotrophic ability (Garrett 1956) is crucial and an ability to antagonise or parasitize a competing mycelium is an important attribute in this respect. If a fungus has already colonized a primary substrate, but is subsequently replaced by an antagonistic competitor, the interaction is then described as secondary resource capture. Mycoparasitism may play a part in this process, for example when parasitism of mycelium of the pioneer wood-colonizing genus *Coriolus* by the wood-decaying *Lenzites betulina* allows the latter fungus to gain selective access to wood occupied by the former (Rayner et al. 1987). Evidence for the mycoparasitic ability of *L. betulina* is derived from the investigation of hyphal interaction on agar plates and in experimental blocks of wood. Hyphal interactions on these substrates show that at a growth temperature of 25°C *L. betulina* displaces *Coriolus*, and that during this process the mycelium of *Coriolus* is entwined and penetrated by the hyphae of *L. betulina*. Although mycoparasitism is usually an interaction which provides a primary food source, Rayner et al. (1987) considered that in this case, the interaction was more important in gaining access to the domain initially colonized by *Coriolus*, i.e. in secondary resource capture. A related phenom-

enon is exhibited by *Phanerochaete magnoliae* (Ainsworth and Rayner 1991), another wood-inhabiting fungus, which in this case is able specifically to replace colonies of *Datronia mollis* as well as forming its own hymenial surfaces through the tubes of *D. mollis* basidiomes. The mode of displacement is unusual in that it does not involve dense zones of invasive mycelium, or overt mycoparasitism. Instead, the fungus produces very sparse, rapidly extending hyphae which grow between the *Datronia* hyphae. Contact of hyphae then results in highly destructive hyphal interference which affects both fungi. This suggestion has also been made for antagonism by *Pythium oligandrum* towards several fungal hosts within the same ecological niche; direct exploitation of antagonized hyphae is of secondary importance to its primary role in combat between competing mycelia (Lutchmeah and Cooke 1984). Thus, mycoparasitism can be incidental to competitive success.

III. Mechanisms of Mycoparasitism

Antagonistic interactions of fungi can be mediated either by direct contact or at a distance. The former involves direct physical contact between the two organisms, whilst the latter refers to those instances in which one fungus releases materials into the environment such as antibiotics and lytic enzymes and induces a negative effect on the other. The boundary between indirect parasitism and competitive interactions involving primary resource capture then becomes very difficult to draw, as parasitic individuals can benefit from the increased availability or release of nutrients consequent upon their activities, and thus alleviate competition for primary nutrients. Some necrotrophic fungal parasites, such as *Trichoderma harzianum* and *Gliocladium virens*, are also able to grow well as saprotrophic competitors of other fungi, and the relative contributions of mycoparasitism and interference competition to this success can be difficult to distinguish. This is exacerbated by the fact that the antagonistic ability of a fungus can also be determined partly by its physiological state, so that changes in physiochemical or nutritional conditions associated with resource utilization will affect the outcome of interaction of combatants (Rayner and Webber 1984).

In the narrow sense, mycoparasitism could be taken to include only direct contact of the mycoparasitic fungus with the potential host. This may involve coiling around potential host structures, possibly their penetration, the development of haustoria in biotrophs, the absorption of nutrients from the cytoplasm of the host and destruction of the cytoplasm of the host. From a practical point of view, however, the production of antagonistic metabolites which precedes the physical contact necessary for invasion of the mycelium of the potential host could also be considered to constitute part of the essential reaction which leads to overt physical parasitism of the host. The morphological events relating to the invasion of a host fungus are undoubtedly induced by biochemical interactions. For example, hyphae of the mycoparasite might show directed growth towards the potential host (Figs. 1,2) through an attraction to diffusable metabolites (Evans and Cooke 1982) or to fimbriae on the host surface (Rghei et al. 1992), or may be stimulated by the proximity of the potential host to produce hydrolytic enzymes capable of degrading the walls of the latter. Thus, mycoparasitic relationships considered in the broad sense would include the biochemical and physiological reactions which precede the microscopically visible phenomena of coiling, appressorium formation, penetration, haustorium formation and cytoplasmic degradation.

The involvement of extracellular metabolites produced by the mycoparasite in the process of parasitism is well documented. The relative importance of the secretion of hyphal wall-degrading enzymes or antibiotic activity is sometimes controversial. There is good evidence, however, that both are significant mechanisms of mycoparasitism. Cell lysis is a common feature of such interactions, which may be mediated through enzymatic action. Wall breakdown is probably due to β-1,3 glucanase, chitobiosidase and chitinase activity, and it has been often demonstrated that the production of these hydrolytic enzymes increases significantly in dual cultures of mycoparasites and their hosts in comparison with that detected in cultures of either fungus grown alone. Protease activity may also be significant in some mycoparasitic interactions (Persson and Friman 1993). It seems very likely that a combination of antifungal enzymes is utilized by fungi such as Gliocladium roseum and Trichoderma harzianum in their antagonistic or mycoparasitic activities (Lorito et al. 1993). A number of non-

enzymic antifungal metabolites have been reported from antagonistic interactions of fungi, particularly in relation to those used for biocontrol. For example, the production of chaetomin in soil has been shown to play an important role in the antagonism of Pythium ultimum by Chaetomium globosum (Di Pietro et al. 1992), whilst the use of antibiotic-minus mutants has demonstrated that gliotoxin plays a major role in the biocontrol of P. ultimum by Gliocladium virens (Straney et al. 1993), although its activity could act in synergy with endochitinases (Di Pietro et al. 1993). Antibiosis may also be associated with hyphal interference. Hyphal interference was first described by Ikediugwu and Webster (1970). It occurs when the mycelium of a fungus growing either in very close proximity to ($<50\,\mu$m) or in contact with that of another species of fungus affects the latter by reducing its rate of growth and causing cytoplasmic disruption. Hyphal interference is mediated through a non-enzymic diffusible metabolite(s), but this material does not appear to be released until interacting hyphae are in close proximity, and is not found in culture filtrates of the antagonistic fungus. In contrast, several necrotrophic interactions involve freely diffusible material which can affect hyphae some distance away. For example, Hypomyces aurantius produces a non-specific toxin which induces rapid and irreversible permeability changes in the plasmalemma of its basidiomycete hosts (Kellock and Dix 1984). Culture filtrates from Hypomyces aurantius contain this material and can also be used to inhibit mycelial extension of these fungi in agar culture. The toxin is not an enzyme and is produced equally well in either the presence or absence of a potential host. During interactions with Stereum hirsutum, this substance damages membranes, and causes cytoplasmic vacuolation and bursting of host hyphae, changes associated in some instances with the deposition of brown pigments in mycelial interaction zones.

Once contact between host and mycoparasite is established, the parasitic hyphae may become attached to the hyphae of the host through a lectin or agglutinin-carbohydrate interaction (Inbar and Chet 1992; Benhamou and Chet 1993; Manocha and Sahai 1993). This early recognition phenomenon probably triggers the sequential events of the subsequent infection process such as appressorium formation, coiling and penetration, and is thus an essential prerequisite to direct

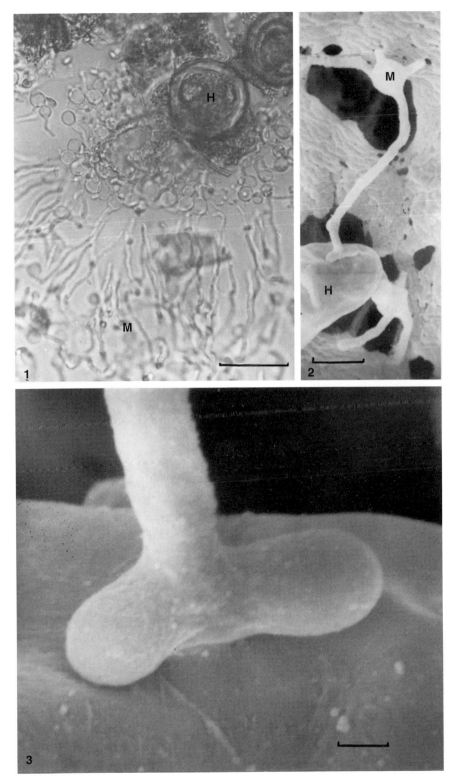

Fig. 1. Directed growth of germ tubes of the mycoparasite *Dimargaris cristalligena* (*M*) towards mycelium of the reference zygomycete host *Cokeromyces recurvatus* (*H*). *Bar 50 μm*

Fig. 2. A germinated spore of the mycoparasite *Piptocephalis unispora* (*M*) with a germ tube showing di-rected growth towards a hypha of the host *Cokeromyces recurvatus* (*H*) on which it has formed an appressorium. *Bar 10 μm*

Fig. 3. Higher magnification of the appressorium shown in Fig. 2. *Bar 1 μm*

mycoparasitism. Once host penetration has occurred, colonization and utilization of the host
metabolites is presumed to occur through secretion of catabolic enzymes by the mycoparasite and
the takeup of the products by the developing
mycelium. The mechanisms responsible for the
strict regulation of nutrient translocation in haustorial relationships remain unknown but are presumably paralleled by the mechanisms which
control haustorial function in phytopathogenic
relationships.

IV. Types of Mycoparasite

Whilst it is convenient to group mycoparasitic interactions on the basis of their physiology into
saprotrophic, necrotrophic and biotrophic categories, or into necrotrophs and biotrophs, these divisions need to be further categorized on the basis of
the morphology of the interface type (Table 1).
In some of the fungicolous and lichenicolous
saprotrophs, no nutrient exchange is implied and
thus no complex interface between the partners is
established. In necrotrophic relationships, however, there needs to be direct contact between
partners so that a channel for nutrient exchange is
established that is not easily accessible to competing microorganisms. In some necrotrophic relationships this interface is established solely by the
close contact of adjacent hyphae. These have been
referred to as the contact necrotrophs (Jeffries
and Young 1994) and they are distinctive in that
they do not involve penetration of host structures.
These examples are particularly difficult to deem
truly parasitic as there is often no evidence of
direct channels for nutrient exchange and they
may arise solely from antagonistic interactions of
mycelia, and equivalents can be seen within the
incompatibility reactions of interacting mycelia
from a single species.

The second type of necrotrophic relationship
is more suggestive of direct parasitism as it involves the penetration of one fungus by another,
accompanied almost immediately by the degeneration and death of the invaded cytoplasm. In
these cases, the mycoparasites have been termed
invasive necrotrophs (Jeffries and Young 1994)
and are very frequent in natural ecosystems. Parasitism is not axiomatic, however, and if nutrient
uptake is not demonstrable, these examples may
still represent a particularly aggressive form of
antagonism.

In contrast, the biotrophic mycoparasites
form stable interfaces with their associated host
fungi, ranging from complex multilamellate barriers between the haustorial mycoparasites and
their hosts through to the absolute fusion of the
protoplasts of the partner fungi (Jeffries 1987).
These interfaces have been categorized into three
groups based on the infection structures that are
formed (Jeffries and Young 1994). First, there are
those in which the entire thallus of the parasite
enters the hyphae of the host fungus. These are
the intracellular biotrophs and include chytridiaceous and oomycete mycoparasites. Second,
there are the haustorial biotrophs in which the
mycoparasite forms short hyphal branches from
an appressorium (Fig. 3) which penetrate the wall
of the hypha of the host and invaginate the plasma
membrane (Figs. 4, 5). Many haustorial biotrophs
have been described from the Zygomycotina, but
several other taxa also produce haustoria (Table
2). Intrahyphal growth of the branches is limited
and the haustorium differentiates an interface
which resembles that of the haustorial complexes
formed within plant cells by many biotrophic phytopathogenic fungi. The third interface type found
in biotrophic relationships is very unusual and involves the formation of channels of contact between closely appressed partner hyphae.
Specialized hyphae or buffer cells are formed
which contact the hyphae of the host but do not
penetrate them. These contact elements serve as
the interface for transfer of nutrients from the
cytoplasm of the host to that of the parasite. Electron microscope studies of the interface have
shown that the plasma membrane of the host and
that of the parasite come into direct contact in the
buffer zone and fuse; thus the cytoplasm of the
parasite becomes contiguous with that of the host.
Mycoparasites with this form of interface have

Table 2. Biotrophic mycoparasites reported to produce
haustoria. (Jeffries and Young 1994)

Caulochytrium protostelioides
Dimargaris spp.
Dispira spp.
Dispira simplex
Filobasidium floriforme
Filobasidiella neoformans
Kuzuhaea moniliformis
Piptocephalis spp.
Rhynchogastrema coronata
Syzygospora alba
Sporidesmium sclerotiovorum
Tieghemiomyces spp.
Tremella spp.
Trimorphomyces papilionaceus

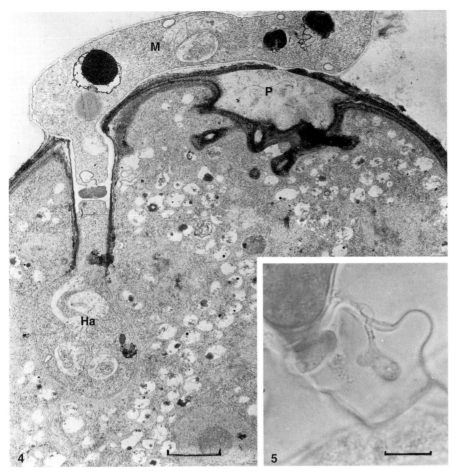

Fig. 4. A haustorium (*Ha*) of the biotrophic mycoparasite *Dimargaris cristalligena* (*M*) within a hypha of the host *Cokeromyces recurvatus*. Note the presence of a papilla (*P*) at an earlier site of attempted penetration. *Bar* 2.5 μm

Fig. 5. A haustorium of the haustorial biotroph *Dimargaris cristalligena* in a hypha of *Cokeromyces recurvatus*. *Bar* 25 μm

recently been termed fusion biotrophs (Jeffries and Young 1994) to emphasize this extraordinary phenomenon. Barnett (1964) had previously termed these the contact biotrophs, but the term fusion biotroph indicates the intimate nature of the interface more strongly. The channels between partners in these relationships can be relatively narrow, and resemble the plasmodesmata formed between adjacent plant cells, or they may be broad enough to permit the migration of organelles and other cytoplasmic components.

V. Ecological Significance of Mycoparasitism

The significance of mycoparasitism in the natural environment is certainly underrated, but there is a dearth of experimental data that reflects condi-

tions in the field. Owing to the difficulties associated with the direct observation of living fungi in the soil, most of the information concerning mycoparasitism has been obtained from *in vitro* dual culture experiments and by inference from field observations. The direct examination of microscope slides that have been previously buried in soil for a period of time, or the examination of soil smears, can show evidence of fungal activity but it is often difficult to make detailed observations, even with the use of vital fluorochromic stains. An alternative approach involves using target baits of host fungi to stimulate infection by potential mycoparasites. For example, burial in soil of oospores of *Phytophthora megasperma* var. *sojae* held between membrane filters has allowed natural infection by mycoparasites to be observed (Sneh 1977). Chytridiaceous fungi, such as *Hypochytrium catenoides*, frequently invaded the spores and produced their own spores in them within 3–4 days. Sometimes, natural mycoparasitic relationships can be inferred, as mycoparasites can cause quite distinctive growth abnormalities in the hyphae of the host, especially when such morphological alterations have also been noted in samples obtained directly from the field. Hyphal swellings in *Rhizopus oryzae* have been used in this way to infer mycoparasitic activity of *Syncephalis californica* in both naturally and artificially infested agricultural soils (Hunter et al. 1977).

It is much more difficult to make observations of mycoparasitism *in situ* than to make isolations of mycoparasites, and the importance of mycoparasitism in community dynamics can be inferred only from circumstantial observations. For example, it has been suggested that, for agaric fungi at least, necrotrophic mycoparasitism may ultimately prove to be more common than is presently realized (Rayner et al. 1985). There are, nevertheless, some clear examples where the natural occurrence of mycoparasitism is documented and some representative examples will be described here. In the case of the biotrophic haustorial mycoparasites, the highly specialized nature of their host-parasite interface, coupled with the difficulties experienced in inducing them to grow in the absence of a host, leaves little doubt that they normally grow as parasites in nature and they can be observed growing on their hosts in the field by careful observation.

In other cases, for example where potential host fungi produce large, soil-borne propagules, specialized techniques can also be used for the observation of mycoparasites. Sclerotia or large resting spores, for example, can be extracted from soil by a combination of wet and dry sieving and are then incubated in damp chambers to allow the development of any associated mycoparasites. Alternatively, laboratory-grown propagules can be buried in soils as bait for later recovery and examination in the laboratory (Adams and Ayers 1985). For example, the sclerotia of *Sclerotinia sclerotiorum* have been used in this way (Zazzerini and Tosi 1985; Ribeiro and Butler 1992). The distribution of three sclerotial mycoparasites (*Sporidesmium sclerotivorum*, *Teratosperma oligocladum* and *Laterispora brevirama*) in soils from samples obtained throughout the world has been assessed using a sclerotial bait of *Sclerotinia minor* (Adams and Ayers 1985). *Sporidesmium sclerotivorum*, for example, was found to occur in soil samples from Australia, Canada, Finland, Ja-

pan and Norway but not in those from Bermuda, Brazil, China, Egypt, France, Germany, Pakistan and The Netherlands. Although *S. sclerotivorum* was not detected in all of the soils, it should be borne in mind that the initial samples used were small and may possible not have been representative of the areas from which they were obtained. Thus it is clear that *S. sclerotivorum* is widely distributed on a global basis and could be an important natural biocontrol agent of sclerotial plant-pathogenic fungi (see Chet et al., Chap. 11, this Vol.). Australian soils screened for mycoparasites of the sclerotia of *Sclerotinia sclerotiorum* have revealed the presence of the important mycopathogens *Coniothyrium minitans* and *Dictyosporium elegans*. Both of these necrotrophs are capable of destroying the sclerotia in non-sterile soil (McCredie and Sivasithamparam 1985). The relationship between *S. sclerotiorum* and *C. minitans* was investigated in the USA during 1975 in a sunflower field naturally infested with the two organisms (Huang 1977). Sclerotia of *S. sclerotiorum* were collected biweekly from roots and basal stems of wilted plants and analyzed for infection by the mycoparasite. The results showed that *C. minitans* parasitized and killed the sclerotia produced on the root surface. This mycoparasite continued to parasitize the pathogen inside the root and upwards into the base of the stem, then infecting the sclerotia produced at these sites. By the end of the growing season, 59, 76 and 29% of sclerotia on the root surface, inside the root, and inside the stem had been killed by the mycoparasite, while 4, 9 and 68% of the sclerotia at these locations were healthy. Death of the rest of the sclerotia was due to organisms other than *C. minitans*. The data also indicated that *C. minitans* was more effective in parasitizing sclerotia produced on or inside the root than those produced in the basal stem. A second survey in 1978 showed that *C. minitans* occurred in 27% of 122 sunflower fields sampled in Manitoba and was found throughout the entire sunflower production area except the east and northeast regions, but its population density in each field generally was low. The potential for development of this fungus as an biocontrol agent is discussed by Whipps and Gerlagh (1992) and Chet et al. (Chap. 11, this Vol.).

Detailed studies of the population dynamics of mycoparasites and their hosts are necessary in order to determine where potential exists for their use in biocontrol. A promising example is *Verticil-*

lium biguttatum, which is able to infect and destroy sclerotia and hyphae of *R. solani* (Jager et al. 1979; Velvis and Jager 1983; van den Boogert and Jager 1984). Inoculation of the soil of potato fields with mycelium of *R. solani* stimulates the development of soil-borne *V. biguttatum* (van den Boogert and Jager 1983; Velvis et al. 1989). In a survey of sclerotia of *Rhizoctonia solani* collected from potato tubers from different countries, *Verticillium biguttatum* was by far the most frequently occurring mycoparasite (van den Boogert and Saat 1991). *Gliocladium roseum*, *G. solani*, other *Gliocladium* spp., *Penicillium* spp., *Trichoderma* spp., *Volutella ciliata* and *Chaetomium* spp were occasionally observed. *Verticillium biguttatum* has now also been isolated from sclerotia of *R. solani* in the United Kingdom (Morris et al. 1992). Thus, the distribution of *V. biguttatum* appears to be worldwide, and does not seem to be restricted to a particular soil type or soil pH. Apparently the fungus is able to survive in different soil types, varying from almost purely mineral soils to peat soils, at pH values between 4.8 and 7.9. Evidence has been provided for the potential regulation of *Rhizoctonia solani* by *V. biguttatum* on potato (van den Boogert and Velvis 1992).

There is a close relationship between the population dynamics of the mycoparasite and that of its host fungus. Host mycelium development on the below-ground plant appeared to be a prerequisite for *V. biguttatum* development. An increase in *R. solani* was followed by an increase in *V. biguttatum*; the increase depended on the initial soil population densities, temperature and soil type. A positive correlation was observed between host fungus population density and the parasitized fraction, suggesting that *V. biguttatum* may cause population density-dependent mortality. This may explain the decline in *Rhizoctonia* observed after growing several consecutive potato crops on the same field. In the absence of the host plant, mycoparasitism played a minor role in regulating *R. solani*, as contact between host and mycoparasite could not be established. *Verticillium biguttatum* was able to survive in plant-free soil for at least 4 years. Within a range of increasing dependence on mycoparasitism in nature, *V. biguttatum* is regarded as taking an intermediate position between the fast-growing and antibiotic-producing mycoparasites (*Gliocladium* spp. and *Trichoderma* spp.) and the slow-growing antibiotic-negative *Sporidesmium sclerotivorum*. As compared with *Verticillium biguttatum*, *S.*

sclerotivorum grows relatively poorly on various agar media (Adams and Ayers 1982), but under natural conditions the populations of these two fungi are built up on host fungi, viz. *R. solani* or *Sclerotinia* and *Sclerotium* species (Ayers and Adams 1978). *Verticillium biguttatum* has been described as an ecologically obligate mycoparasite, indicating the strictly parasitic habit under natural plant-soil conditions. This concept does not exclude the fact that the fungus can grow saprotrophically under axenic conditions, or conditions of low microbial activity.

Spores of arbuscular mycorrhizal fungi (AMF) have also been used as bait in soils and many of the mycoparasites isolated appear to be facultative with some degree of saprotrophic ability and, therefore not entirely dependent on the presence of the spores for survival (Paulitz and Menge 1986). The spores of some AMF (e.g. *Gigaspora*) are amongst the largest known within the fungal kingdom and can be extracted from field soils by a simple wet-sieving procedure. In any sample of extracted spores there is usually a proportion, sometimes the majority, which appears to be parasitized. During an assessment of the arbuscular mycorrhizal status of various crops in different agricultural soils of northern Greece, it was found that a consistent feature of all soils was the presence of large numbers of empty spores (ghosts) of AMF which usually outnumbered intact spores by a factor of 1–5 (Jeffries et al. 1988). The ghosts were always perforate, and this was taken to indicate attack by other soil fungi. Many of the ghosts had developed papillae at the infection sites, indicating that they were alive at the time of the mycoparasitic attack. The most frequently observed parasites of spores of AMF are members of the Oomycota, including *Spizellomyces* and *Pythium*-like fungi (Boyetchko and Tewari 1991). Examination of individual spores shows that the wall may be perforated by many fine radial canals. The canals are usually said to arise through penetration of the wall by the hyphae of mycoparasitic fungi and, indeed, hyphae are occasionally seen to penetrate the canals. In some cases, however, it has been suggested that amoeba-like organisms could also be responsible (Boyetchko and Tewari 1991), especially when no bacteria or fungi can be seen inside the perforated spores. In other examples, however, the hyphae of mycoparasitic fungi such as *Humicola fuscoatra* and *Anguillospora pseudolongissima* have been detected within, and isolated from, the spores of

Glomus fasciculatum and *G. versiforme* grown in glasshouse pot culture (Daniels and Menge 1980). Several zoosporic fungi are able to attack the spores of members of the *Glomales*, and sporulate either within the spores themselves, or on the outer surface. The colonization of these spores in this manner may enhance the survival and dispersal of the mycoparasites in a way analogous to that suggested for the mycoparasite *Verticillium psalliotae* infecting *Rhopalomyces elegans* (Dayal and Barron 1970). Some of the observations of spores described as being parasitized do not indicate whether the spores were viable or dead when attacked. In the absence of papillae, this is not always clear. *Spizellomyces punctatum* has been shown to invade non-viable spores of *Gigaspora margarita* (Paulitz and Menge 1984), and it was suggested that this fungus is primarily a saprotroph that attacks dead spores. This evidence suggests that some previous reports of *Phlyctochytrium* species as mycoparasites of spores may need reinterpretation. For example, treatment of agricultural soil used for growing peanuts and soybeans with the fumigant methyl bromide resulted in the enhanced development of *G. macrocarpum* in the roots and the soil during the growing season (Ross and Ruttencutter 1977). During the following season, however, a decline in the population of spores of *G. macrocarpum* appeared to be correlated with an increase in mycoparasitism by *Phlyctochytrium*. It is not clear whether this reflects an increase in necromass of the AMF that provided a suitable substrate for *Phlyctochytrium*, or whether the chytrid was directly responsible for the decline of the *Glomus*.

More detailed information is available for the antagonism of spores of *Gigaspora margarita* by *Stachybotrys chartarum* (Siqueira et al. 1984). This fungus was found as a frequent contaminant of pot cultures of inoculum of this mycorrhizal fungus. *Stachybotrys chartarum* colonizes the spores and produces conidiophores over the surface. Sometimes, hyphae were seen to penetrate the host spore, but it is not known if only dead spores were penetrated. The population dynamics of an arbuscular mycorrhizal fungus in a natural soil has been studied by Paulitz and Menge (1986), who investigated the effects of *Anguillospora pseudolongissima* parasitizing spores of *Glomus deserticola*. *Glomus deserticola* is a typical AMF and was grown in pure pot culture to induce sporulation. A preparation of soil, pieces of root

and spores was then chopped and air-dried for 7 days prior to incubation with the mycoparasite in order to kill the hyphae and leave the spores as the main source of the inoculum. Dilutions of the inoculum were mixed with various dilutions of the mycoparasite in sandy loam planted with onion seeds and the plants were harvested from 40–100 days later. The results showed that the primary effect of *A. pseudolongissima* was to reduce the number of effective propagules of *Glomus*, resulting in a delay and reduced incidence of root colonization. At low propagule densities, *A. pseudolongissima* was responsible for reducing (by up to 50%) the effective propagule density to the point at which it no longer colonized the host plant.

The cultivation of mushrooms provides a final example of an ecosystem in which mycoparasitic relationships have been documented. Antagonistic fungi can cause problems at two different stages in the process: as weed moulds colonizing the compost and inhibiting the growth of *Agaricus*, and as mycoparasites attacking *Agaricus* directly. The loss of income to mushroom growers throughout the world on an annual basis through the activity of weed moulds and mycoparasites may amount to several million pounds. As many weed moulds are competitors rather than mycoparasites they will not be considered further here.

The three most important mycoparasites of *A. bisporus* are *Verticillium fungicola*, which causes dry bubble disease, *Mycogone perniciosa*, causing wet bubble disease, and *Hypomyces rosellus*, responsible for cobweb disease. Mushrooms affected by dry or wet bubble are malformed and unsaleable. Variation in the necrotrophic reaction to these pathogens of the cultivated mushroom has been observed. With *Hypomyces rosellus*, a destructive parasite of mushrooms, and *Verticillium fungicola* var. *fungicola* the mycelium of the host is overgrown and severe necrosis occurs. The reverse reaction occurs, however, with *Mycogone perniciosa*, as the vegetative mycelium of the mushroom overgrows that of the mycoparasite. Mature mushrooms may also be attacked. It has been suggested that an inability to parasitize the actively growing vegetative mycelium of the potential host could confer an adaptive advantage on the parasite as the likelihood that the mushrooms will form is increased, thus maintaining the supply of susceptible material (Gray and Morgan-Jones 1981).

VI. Laboratory Investigations

The unequivocal demonstration of a mycoparasitic association under laboratory conditions is much easier than in the field, but cannot be taken as proof that a similar parasitic relationship occurs in nature. Such studies are often carried out in pure mixed culture under favourable environmental conditions and nutrient excess. In vitro tests of the host range provide merely a strong indication of susceptible and immune fungi in the absence of natural competitors. Under field conditions the situation will be different. Microbial competition and nutrient limitation will place a premium on those characteristics that give an advantage in competitive saprotrophic ability. Characteristics that enhance the competitiveness for carbon and nitrogen substrates, or help a fungus to withstand or reduce the antagonistic behaviour of other fungi, will increase the share of primary and secondary resources available to that fungus. Mycoparasitism is a powerful tool in this respect.

It is not difficult to isolate mycoparasitic fungi when specific methods are used involving selective media. Selectivity is often obtained by using a combination of antibiotics and other inhibitory agents. For example, a *Trichoderma*-selective medium (Elad et al. 1981; Askew and Laing 1993) and a *Pythium*-selective medium (Schmitthenner 1962) have both been used to isolate these genera from soil samples, and a chemically defined selective medium is available for the isolation of *Verticillium fungicola* from mushroom farms (Rinker et al. 1993). A combination of fungicides has been used in a similar way to isolate and enumerate *Laetisaria arvalis*, an antagonist of *Pythium ultimum* and *Rhizoctonia solani* (Papavizas et al. 1983). Such selective media can then be used to study the occurrence of the antagonistic fungi in natural substrates. For example, *Athelia bombacina*, an antagonist of the apple scab pathogen *Venturia inaequalis*, was recovered from 75% of non-inoculated apple leaves and 96% of inoculated leaves over a 6-month period (Young and Andrews 1990).

On the other hand, if the mycoparasite makes only poor axenic growth, a simple technique may be used to isolate the mycoparasite in pure dual culture by using the host fungus as a selective bait. For example, yeast-phase cells of the susceptible host *Cokeromyces recurvatus* streaked onto a plate of agar can be used to isolate species of the mycoparasites *Piptocephalis* and *Syncephalis* (Jeffries and Kirk 1976). Several mycoparasitic species of *Pythium* have also been isolated by using potential host fungi as bait. In this case, agar plates are completely precolonized by a suitable potential host fungus and a small sample of material, for example soil, suspected to contain the mycoparasite, is added. Precolonization of the plate with the potential host results in the utilization of most of the readily available nutrients and the production of compounds by the host fungus which are inhibitory to the growth of many other fungi. Thus, the only fungi capable of growing out of the substratum are those with the ability to utilize the mycelium of the precolonist as a nutrient source and are presumed to be mycoparasitic (presumptive mycoparasites sensu Mulligan and Deacon 1992).

This technique has been used successfully to selectively isolate various species of *Pythium* using *Phialophora radicicola* as a host (Deacon 1976; Deacon and Henry 1978), and for *Piptocephalis xenophila* using *Penicillium* hosts. The technique was also used to survey the occurrence of mycoparasitic *Pythium* species in a variety of soils (Foley and Deacon 1985). Samples from horticultural, grassland, arable, woodland, moorland, freshwater and coastal sites were assayed for the presence of mycoparasites by incubation on agar precolonised by *Phialophora* species. Of the total 164 samples, 84% contained one or more of the mycoparasites *Pythium oligandrum*, *P. mycoparasiticum*, *Trichoderma viride* and *Gliocladium roseum*. *Pythium oligandrum* was found in 29% of all samples, 45% of samples from "disturbed" sites (gardens, arable lands, managed grasslands), but only 11% of samples from "natural" sites subject to minimal disturbance (woodlands, moorlands, permanent pastures, coastal sites and freshwater sediments). It was commonest at pH 5.5–5.6. *Pythium mycoparasiticum* was found in 17% of all samples but was probably underestimated because of competition from *P. oligandrum* on isolation plates. Its distribution was similar to that of *P. oligandrum*, but it was much less common in horticultural and grassland sites. The mycoparasite *P. acanthicum* was found only twice, and *P. periplocum* was not found, although both could grow well on *Phialophora*-precolonized agar. Overall, mycoparasitic pythia were found in 38% of samples – at a frequency similar to *Trichoderma viride* (45%) and *Gliocladium roseum* (40%). The growth of fungi across precolonized plates indicates differences between antagonistic fungi. For example, *Trichoderma harzianum* can grow well across several bait fungi unsuitable for *P. oligandrum* (Laing and Deacon 1990) and not across others that *P. oligandrum* could colonize. *Gliocladium roseum*, on the other hand, grows across most of the fungi used as baits.

The precolonized plate technique has been further developed as a method to determine the mycoparasitic spectrum of a soil (Mulligan and Deacon 1992). Using a range of host fungi, the presumptive mycoparasites *Pythium oligandrum*, *Gliocladium roseum* (and related species), *Trichoderma* spp. and *Papulaspora* sp. were detected in, respectively, 18, 28, 24 and 21 of a total of 28 British soils. Most of the soils contained three or more mycoparasites, but the frequency of detection in experimental replicates suggested that only *P. oligandrum* and *G. roseum* were abundant in all the soils in which they occurred. The type of host fungus markedly influenced the efficiency of detection of different mycoparasites: *Fusarium culmorum* was most efficient for *P. oligandrum*, *Rhizoctonia solani* for *Trichoderma*, *Botrytis cinerea* for *Papulaspora* and *Trichoderma aureoviride*; *R. solani* and *B. cinerea* were equally efficient for detection of *G. roseum*. No single host was suitable for consistent detection of any single mycoparasite, and the authors suggested that several hosts may therefore be needed to determine the mycoparasite spectrum of a soil.

Interactions of mycoparasites and their hosts in vitro is known to vary depending on the nutrient conditions or environmental conditions. For example, Whipps (1987) investigated the interactions in vitro of six plant pathogens with four antagonistic species on a nutrient-poor substrate, tap-water agar (TWA), on soil-extract agar (SEA) with no added nutrients, and on the relatively nutrient-rich potato dextrose agar (PDA). Both the pathogens and antagonists used could be split into groups where known ecological behaviour fitted with observations obtained on the media in vitro. For instance, *Rhizoctonia solani*, *Fusarium oxysporum*, *Pyrenochaeta lycopersici* and *Phomopsis sclerotioides* are soil-borne pathogens which survive in soil as sclerotia, resistant spores or hyphae, or within plant debris, and which can actively grow through soil to infect roots or necrotic leaves touching the soil surface. In the tests, these pathogens all had the greatest growth rate on SEA but differences between this and rates on PDA and TWA were relatively small, indicating an adaptation to growth under a range of nutrient conditions likely to be found in soils and plants. *Sclerotinia sclerotiorum* and *Botrytis cinerea*, however, grew relatively poorly on TWA and SEA compared with PDA, and this again probably reflects natural ecology. These pathogens survive on plant residues, or as sclerotia in the soil, and infect aerial parts of plants where normal growth occurs under nutrient-rich conditions. In general, they show little or no ability to grow as mycelium through the soil, and have limited competitive saprotrophic ability. *Trichoderma* species and *Coniothyrium minitans* grow poorly in the low-nutrient media but at least doubled their growth rates on PDA. The former are well known for their ability to grow and sporulate profusely where large amounts of nutrients are made available in the soil, overcoming soil fungistasis (Papavizas 1985). The latter survives as spores in the soil until a suitable sclerotium is encountered and then germination, penetration and rapid growth occurs within the protected, nutrient-rich conditions of the sclerotium. *Gliocladium roseum* has growth rates similar on all media and is a ubiquitous soil organism able to grow and survive in soil (Papavizas 1985).

The in vitro study of parasitic fungi is facilitated if the fungus can be grown on laboratory media axenically and nutritional and environmental requirements for mycelial development may then be determined. This is not always possible, however, and mycoparasitic fungi, like their plant-pathogenic counterparts, differ in their nutritional flexibility with respect to being able to grow saprotrophically. The nutrient requirements of a mycoparasite grown in vitro may not be identical to those of the same fungus growing on a host. Probably a more natural system is represented when pure dual cultures are used. Complications with pure dual cultures are that the nutrient requirements of the host must also be determined in order to distinguish those of the host from those of the parasite. Secondly, a parasitized host may elaborate or modify the nutrients in the growth medium such that they are passed onto the parasite in a different form from that in which they were originally absorbed. This is likely to be the case with biotrophic mycoparasites, many of which grow well only when parasitizing the host. Thirdly, the nutrient requirements of the parasitized host may differ from those of the unparasitized host. It has thus been difficult to obtain information regarding nutritional differences between mycoparasites and their saprotrophic equivalents.

Within the genus *Pythium*, however, marked nutritional differences between the mycoparasitic species, e.g. *P. oligandrum* and *P. acanthicum*, and non-mycoparasitic species, e.g. *P. ultimum* and *P. mamillatum*, have been found (Foley and Deacon 1986). On the basis of dry weight and linear growth determinations, it was shown that the mycoparasitic strains require an organic source of nitrogen such as asparagine and thiamine. Non-mycoparasitic isolates, on the other hand, can utilize inorganic sources of nitrogen such as sodium nitrate or ammonium sulphate and most are self-sufficient for the vitamin thiamine. The mycoparasitic pythia also lack cellulolytic ability whereas the non-mycoparasites possess this enzymic capability. As the non-mycoparasitic pythia are also aggressive plant pathogens, it would seem that the ability to colonize cellulosic substrata could be of importance in this respect. Although the mycoparasitic isolates may depend on the presence of potential host fungi in the soil, the nutritional dependence could also be advantageous, as the potential hosts will have already colonized cellulosic substrata in the soil which would otherwise probably not be available to them for growth. This observation has enabled experiments to investigate the direct effects of these mycoparasites on their hosts. Non-mycoparasitic *Pythium* species were grown on cellophane films overlying agar containing nitrate as a source of nitrogen. A range of mycoparasitic *Pythium* species was compared in relation to their effects on the host fungi (Laing and Deacon 1990). The various isolates behaved differently, as evidenced by their ability to reduce cellulolytic activity of their hosts. *Pythium oligandrum* was a more aggressive mycoparasite than *P. mycoparasiticum* which, in turn, was usually more aggressive than *P. nunn*. Both *P. oligandrum* and *P. mycoparasiticum* produced oogonia in the presence of susceptible host fungi, thus deriving carbon, organic nitrogen and sterols from the host. *Pythium nunn* did not form oogonia with any host, and it was, incidentally, the only mycoparasite that could utilize nitrate as sole nitrogen source. This suggests that *P. nunn* may grow mainly as a competitor for substrates rather than a mycoparasite in nature (Laing and Deacon 1990). This is supported by a study of interactions of *P. nunn* and *P. ultimum* on bean leaves (Paulitz and Baker 1988).

Pythium nunn was able to occupy bean leaves previously colonized by *P. ultimum*, and can displace the latter fungus. It was suggested that mycoparasitism could be important for secondary resource capture, but it could also be due to competition between the fungi for primary colonization of the substrate. *Pythium mycoparasiticum* and *P. oligandrum*, on the other hand, are likely to be mycoparasitic in nature. The fact that *P. mycoparasiticum* was less aggressive in these tests might be explained by the fact that this fungus grows more slowly than *P. oligandrum*. Thus, host hyphae might outgrow those of *P. mycoparasiticum* resulting in "disease-escape" equivalent to that recorded for plant-fungus interactions (Garrett 1970). On water agar, where growth rates were more similar, *P. mycoparasiticum* was equally aggressive as *P. oligandrum*.

A number of biotrophs require the presence of host extracts for axenic culture to be successful, including *Gonatobotrys simplex*, *Gonatobotyrum fuscum*, *Hansfordia parasitica*, *Nematogonum ferrugineum* and *Stephanoma phaeospora*. The unidentified component(s) has been termed mycotrophein (Whaley and Barnett 1963; Calderone and Barnett 1972). Mycotrophein is effective in the promotion of growth even after high dilution to one part per million. Although efforts have been made to identify the compound its identity remains unknown (Hwang et al. 1985). Some non-host fungi also contain mycotrophein, which indicates that susceptibility to parasitism is also related to other factors. The nutrient requirements of these fungi when grown on mycotrophein-containing media vary, with no common factors. Some, such as *H. parasitica* and *G. fuscum*, will utilize single amino acids as nitrogen sources, whilst others, such as *G. simplex*, have more complex nitrogen requirements. *Gonatobotryum fuscum* also requires abnormally high levels of thiamine and biotin. *Stephanoma phaeospora*, formerly considered to be a typical contact biotrophic mycoparasite, has also been shown to require host extracts for growth in axenic culture (Rakvidhyasastra and Butler 1973). Some fungi, especially in the genus *Aspergillus*, released this factor into the substrate, thus supporting saprotrophic growth of *S. phaeospora* in pure dual culture of these two fungi. *Stephanoma phaeospora* is now recognized as a fusion biotroph and presumably obtains the necessary growth factors from host fungi by direct transfer through the cytoplasmic channels in the host-parasite interface.

VII. Conclusions

This chapter contains a number of examples of mycoparasitic interactions, but many more have now been described (see Jeffries and Young 1994). Few fungal genera are immune to attack, and all developmental forms (e.g. spores, sclerotia, hyphae, sporomes) are susceptible. It is evident that mycoparasitism is a widespread and significant phenomenon in fungal ecology. Whilst it is difficult to quantify the effect that mycoparasitism has on the population dynamics of the host fungi, it is now becoming obvious that it has an important influence on competitive interactions in nature. Mycoparasitism is not solely an interesting ecological occurrence, however, and there are numerous examples where it can reduce the impact of phytopathogenic fungi on their plant hosts. In this context, it will become increasingly important in the development of integrated systems of plant disease control.

References

Adams PB, Ayers WA (1982) Biological control of *Sclerotinia* lettuce drop in the field by *Sporidesmium sclerotivorum*. Phytopathology 72:485–488

Adams PB, Ayers WA (1985) The world distribution of the mycoparasites *Sporidesmium sclerotivorum*, *Teratosperma oligocladum* and *Laterispora brevirama*. Soil Biol Biochem 17:583–584

Ainsworth AM, Rayner ADM (1991) Ontogenetic stages from coenocyte to basidiome and their relation to phenoloxidase activity and colonization processes in *Phanerochaete magnoliae*. Mycol Res 12:1414–1422

Askew DJ, Laing MD (1993) An adapted selective medium for the quantitative isolation of *Trichoderma* species. Plant Pathol 42:686–690

Ayers WA, Adams PB (1978) Mycoparasitism of sclerotia of *Sclerotinia* and *Sclerotium* species by *Sclerotiorum sclerotiovorum*. Can J Microbiol 25:17–23

Barnett HL (1964) Mycoparasitism. Mycologia 56:1–19

Barnett HL, Binder FL (1973) The fungal host-parasite relationship. Annu Rev Phytopathol 11:273–292

Benhamou N, Chet I (1993) Hyphal interactions between *Trichoderma harzianum* and *Rhizoctonia solani*: ultrastructure and gold cytochemistry of the mycoparasitic process. Phytopathology 83:1062–1071

Boyetchko SM, Tewari JP (1991) Parasitism of spores of the vesicular-arbuscular mycorrhizal fungus *Glomus dimorphicum*. Phytoprotection 72:27–32

Calderone RA, Barnett HL (1972) Axenic growth and nutrition of *Gonatobotryum fuscum*. Mycologia 64:153–160

Cooke RC, Rayner ADM (1984) Ecology of saprotrophic fungi. Longman, London

Daniels BA, Menge JA (1980) Hyperparasitization of vesicular-arbuscular mycorrhizal fungi. Phytopathology 70:584–588

Dayal R, Barron GL (1970) *Verticillium psalliotae* as a parasite of *Rhopalomyces*. Mycologia 62:826–830

Deacon JW (1976) Studies on *Pythium oligandrum*, an aggressive parasite of other fungi. Trans Br Mycol Soc 66:383–391

Deacon JW, Henry CM (1978) Mycoparasitism by *Pythium oligandrum* and *Pythium acanthicum*. Soil Biol Biochem 10:409–415

Di Pietro A, Gut-Rella M, Pachlatko JP, Schwinn FJ (1992) Role of antibiotics produced by *Chaetomium globosum* in biocontrol of *Pythium ulimum*, a causal agent of damping-off. Phytopathology 82:131–135

Di Pietro A, Lorito M, Hayes CK, Broadway RM, Harman GE (1993) Endochitinase from *Gliocladium virens*: isolation, characterization and synergistic antifungal activity in combination with gliotoxin. Phytopathology 83:308–313

Elad Y, Chet I, Henis Y (1981) A selective medium for improving quantitative isolation of *Trichoderma* spp. from soil. Phytoparasitica 9:59–67

Evans GH, Cooke RC (1982) Studies on Mucoralean mycoparasites. III. Diffusible factors from *Mortierella vinacea* Dixon-Stewart that direct germ tube growth of *Piptocephalis fimbriata* Richardson & Leadbeater. New Phytol 91:245–253

Foley MF, Deacon JW (1985) Isolation of *Pythium oligandrum* and other necrotrophic mycoparasites from soil. Trans Br Mycol Soc 85:631–639

Foley MF, Deacon JW (1986) Physiological differences between mycoparasitic and plant-pathogenic *Pythium* spp. Trans Br Mycol Soc 86:225–231

Garrett SD (1956) Biology of root-infecting fungi. Cambridge University Press, Cambridge

Garrett SD (1970) Pathogenic root-infecting fungi. Cambridge University Press, Cambridge

Gray DJ, Morgan-Jones G (1981) Host-parasite relationships of *Agaricus brunnescens* and a number of mycoparasitic hyphomycetes. Mycopathologia 75:55–59

Hawksworth DL (1981) A survey of the fungicolous conidial fungi. In: Cole GT (ed) Biology of conidial fungi, vol 1. Academic Press, New York, pp 171–244

Huang HC (1977) Importance of *Coniothyrium minitans* in survival of sclerotia of *Sclerotinia sclerotiorum* in wilt of sunflower. Can J Bot 55:289–295

Hunter WE, Duniway JM, Butler EE (1977) Influence of nutrition, temperature, moisture and gas composition on parasitism of *Rhizopus oryzae* by *Syncephalis californica*. Phytopathology 67:664–669

Hwang K, Stelzig DA, Barnett HL, Roller PP, Kelsey MI (1985) Partial purification of the growth factor mycotrophein. Mycologia 77:109–113

Ikediugwu FEO, Webster J (1970) Antagonism between *Coprinus heptemerus* and other coprophilous fungi. Trans Br Mycol Soc 54:181–204

Inbar J, Chet I (1992) Biomimics of fungal cell-cell recognition by use of lectin-coated nylon fibres. J Bacteriol 174:1055–1059

Jager G, Ten Hoopen A, Velvis H (1979) Hyperparasites of *Rhizoctonia solani* in Dutch potato fields. Neth J Plant Pathol 85:253–268

Jeffries P (1987) Pathways for the exchange of materials in mycoparasitic and plant-fungal interactions. In: Ayres

PG, Pegg GF (eds) Fungal infection of plants. Cambridge University Press, Cambridge, pp 60–78

Jeffries P, Kirk PM (1976) New technique for the isolation of mycoparasitic Mucorales. Trans Br Mycol Soc 66:541–543

Jeffries P, Spyropoulos T, Vardavarkis E (1988) Vesicular-arbuscular mycorrhizal status of various crops in different agricultural soils of northern Greece. Biol Fertil Soils 5:333–337

Jeffries P, Young TWK (1994) Interfungal parasitic relationships. CAB International, Wallingford

Kellock LM, Dix NJ (1984) Antagonism by *Hypomyces aurantius*. I. Toxins and hyphal interactions. Trans Br Mycol Soc 82:327–334

Laing SAK, Deacon JW (1990) Aggressiveness and fungal host ranges of mycoparasitic *Pythium* species. Soil Biol Biochem 22:905–911

Lorito M, Harman GE, Hayes CK, Broadway RM, Tronsmo A, Woo SL, Di Pietro A (1993) Chitinolytic enzymes produced by *Trichoderma harzianum*: antifungal activity of purified endochitinase and chitobiosidase. Phytopathology 83:302–307

Lutchmeah RS, Cooke RC (1984) Aspects of antagonism by the mycoparasite *Pythium oligandrum*. Trans Br Mycol Soc 83:696–700

Manocha MS, Sahai AS (1993) Mechanisms of recognition in necrotrophic and biotrophic mycoparasites. Can J Microbiol 39:269–275

McCredie TA, Sivasithamparam K (1985) Fungi mycoparasitic on sclerotia of *Sclerotinia sclerotiorum* in some Western Australian soils. Trans Br Mycol Soc 84:736–739

Morris RAC, Coley Smith JR, Whipps JM (1992) Isolation of mycoparasite *Verticillium biguttatum* from sclerotia of *Rhizoctonia solani* in the United Kingdom. Plant Pathol 41:513–516

Mulligan DFC, Deacon JW (1992) Detection of presumptive mycoparasites in soil placed on host-colonized agar plates. Mycol Res 96:605–608

Papavizas GC (1985) *Trichoderma* and *Gliocladium*: biology, ecology, and potential for biocontrol. Annu Rev Phytopathol 23:23–54

Papavizas GC, Morris BB, Marois JJ (1983) Selective isolation and enumeration of *Laetisaria arvalis* from soil. Phytopathology 73:220–223

Paulitz TC, Baker R (1988) Interactions between *Pythium nunn* and *Pythium ultimum* on bean leaves. Can J Microbiol 34:947–951

Paulitz TC, Menge JA (1984) Is *Spizellomyces punctatum* a parasite or saprophyte of vesicular-arbuscular mycorrhizal fungi? Mycologia 76:99–107

Paulitz TC, Menge JA (1986) The effects of a mycoparasite on the mycorrhizal fungus, *Glomus deserticola*. Phytopathology 76:351–354

Pearce MH (1990) In vitro interactions between *Armillaria luteobubalina* and other wood decay fungi. Mycol Res 94:753–761

Persson Y, Friman E (1993) Intracellular proteolytic activity in mycelia of *Arthrobotrys oligospora* bearing mycoparasitic or nematode trapping structures. Exp Mycol 17:182–190

Rakvidhyasastra V, Butler EE (1973) Mycoparasitism by *Stephanoma phaeospora*. Mycologia 65:580–583

Rayner ADM (1991) The challenge of the individualistic mycelium. Mycologia 83:48–71

Rayner ADM, Webber JF (1984) Interspecific mycelial interactions – an overview. In: Jennings DH, Rayner ADM (eds) The ecology and physiology of the fungal mycelium. Cambridge University Press, Cambridge, pp 383–417

Rayner ADM, Watling R, Frankland JC (1985) Resource relations – an overview. In: Moore D, Casselton LA, Wood DA, Frankland JC (eds) Developmental biology of higher fungi. Cambridge University Press, Cambridge, pp 1–40

Rayner ADM, Boddy L, Dowson CG (1987) Temporary parasitism of *Coriolus* spp. by *Lenzites betulina*: a strategy for domain capture in wood decay fungi. FEMS Microbiol Ecol 45:53–58

Rghei NA, Castle AJ, Manocha MS (1992) Involvement of fimbriae in fungal host-mycoparasite interaction. Physiol Mol Plant Pathol 41:139–148

Ribeiro WR, Butler EE (1992) Isolation of mycoparasitic species of *Pythium* with spiny oogonia from soil in California. Mycol Res 96:857–862

Rinker DL, Bussmann S, Alm G (1993) A selective medium for *Verticillium fungicola*. Can J Plant Pathol 15:123–124

Rogerson CT, Stephenson SL (1993) Myxomyceticolous fungi. Mycologia 85:456–469

Ross JP, Ruttencutter R (1977) Population dynamics of two vesicular-arbuscular endomycorrhizal fungi and the role of hyperparasitic fungi. Phytopathology 67:490–496

Schmitthenner AF (1962) Isolation of *Pythium* from soil particles. Phytopathology 52:1133–1138

Siqueira JO, Hubbell DH, Kimbrough JM, Schenck NC (1984) *Stachybotrys chartarum* antagonistic to azygospores of *Gigaspora margarita* in vitro. Soil Biol Biochem 16:679–681

Sneh B (1977) A method for observation and study of living fungal propagules incubated in soil. Soil Biol Biochem 9:65–66

Straney DC, Wilhite SE, Lumsden RD, Ding C (1993) Mutational analysis demonstrates that gliotoxin has a major role in biocontrol of *Pythium ultimum* by *Gliocladium virens*. Phytopathology 83:1366

van den Boogert PHJF, Jager G (1983) Accumulation of hyperparasites of *Rhizoctonia solani* by addition of live mycelium of *R. solani* to soil. Neth J Plant Pathol 89:223–228

van den Boogert PHJF, Jager G (1984) Biological control of *Rhizoctonia solani* on potatoes by antagonists. 3. Inoculation of seed potatoes with different fungi. Neth J Plant Pathol 90:17–126

van den Boogert PHJF, Saat TAWM (1991) Growth of the mycoparasitic fungus *Verticillium biguttatum* from different geographical origins at near-minimum temperatures. Neth J Plant Pathol 97:115–124

van den Boogert PHJF, Velvis H (1992) Population dynamics of the mycoparasite *Verticillium biguttatum* and its host *Rhizoctonia solani*. Soil Biol Biochem 24:159–164

Velvis H, Jager G (1983) Biological control of *Rhizoctonia solani* on potatoes by antagonists. 1. Preliminary experiments with *Verticillium biguttatum*, a sclerotium-inhabiting fungus. Neth J Plant Pathol 89:113–123

Velvis H, Boogert PHJF, Jager G (1989) Role of antagonism in the decline of *Rhizoctonia solani* inoculum in soil. Soil Biol Biochem 21:125–129

Whaley JW, Barnett HL (1963) Parasitism and nutrition of *Gonatobotrys simplex*. Mycologia 55:199–210

Whipps JM (1987) Effect of media on growth and interactions between a range of soil-borne glasshouse pathogens and antagonistic fungi. New Phytol 107:127–142

Whipps JM, Gerlagh M (1992) Biology of *Coniothyrium minitans* and its potential for use in disease biocontrol. Mycol Res 96:897–907

Wicklow D (1981) Interference competition and the organization of fungal communities. In: Carroll GC, Wicklow DT (eds) The fungal community: its organization and role in the ecosystem. Dekker, New York, pp 351–375

Young CS, Andrews JH (1990) Recovery of *Athelia bombacina* from apple leaf litter. Phytopathology 80:530–535

Zazzerini A, Tosi L (1985) Antagonistic activity of fungi isolated from sclerotia of *Sclerotinia sclerotiorum*. Plant Pathol 34:415–421

11 Fungal Antagonists and Mycoparasites

I. Chet[1], J. Inbar[1,2], and Y. Hadar[1]

CONTENTS

I. Introduction

Due to the adverse environmental effects of pesticides which create health hazards for human and other nontarget organisms, including the pests' natural enemies, these chemicals have been the object of substantial criticism in recent years. The development of safer, environmentally feasible control alternatives has therefore become a top priority. In this context, biological control is becoming an urgently needed component of agriculture.

Biological control of plant pathogens is defined as the use of biological processes to lower inoculum density of the pathogen, with the aim of reducing its disease-producing activities (Baker and Cook 1974).

Biological control may be achieved by both direct and indirect strategies. Indirect strategies include the use of organic soil amendments which enhance the activity of indigenous microbial antagonists against a specific pathogen. Another indirect approach, cross-protection, involves the stimulation of plant self-defense mechanisms against a particular pathogen by prior inoculation of the plant with a nonvirulent strain. Successful cross-protection resulting in induced resistance has been documented for viruses (Costa and Muller 1980; Fletcher and Row 1975), bacterial pathogens (Sequeira and Hill 1974), and fungi (Caruso and Kuc 1977).

The direct approach involves the introduction of specific microbial antagonists into soil or plant material (Cook and Baker 1983). These antagonists have to proliferate and establish themselves in the appropriate ecological niche in order to be active against the pathogen. Antagonists are microorganisms with the potential to interfere with the growth and/or survival of plant pathogens and thereby contribute to biological control. Antagonistic interactions among microorganisms in nature include parasitism or lysis, antibiosis, and competition. These microbial interactions serve as the basic mechanisms via which biocontrol agents operate. An elucidation of the mechanisms involved in biocontrol activity is considered to be one of the key factors in developing useful biocontrol agents. Of the numerous biocontrol agents examined, only a few have been subjected to a thorough analysis of the mechanisms involved in the suppression of the pathogen. In this chapter we shall use examples to briefly demonstrate the different mechanisms, and will concentrate on mycoparasitism.

[1] Department of Plant Pathology and Microbiology, The Hebrew University of Jerusalem, Faculty of Agriculture, Rehovot 76100, Israel
[2] *Present address*: The Department of Biochemistry and Molecular biology, Israel Institute of Biological Research, P.O.B. 19, Ness Ziona 70450, Israel

The Mycota IV
Environmental and Microbial Relationships
Wicklow/Söderström (Eds.)
© Springer-Verlag Berlin Heidelberg 1997

A. Antibiosis

Handelsman and Parke (1989) restricted the definition of antibiosis to those interactions that involve a low-molecular-weight diffusible compound or an antibiotic produced by a microorganism, that inhibits the growth of another microorganism. This definition excluded proteins or enzymes that kill the target organism. Baker and Griffin (1995) extended the scope of the definition to "inhibition or destruction of an organism by the metabolic production of another." This definition includes small toxic molecules, volatiles, and lytic enzymes.

The production of inhibitory metabolites by fungal biocontrol agents has been reported in the literature over the last five decades (Bryan and McGowan 1945; Wright 1956; Dennis and Webster 1971a,b; Gnisalberti and Sivasithamparam 1991; Ordentlich et al. 1992). However, insufficient evidence exists for their contribution to pathogen suppression and disease reduction in situ. *Gliocladium virens* is a common example of the role of antibiotics in biological control by fungal antagonists. Gliovirin is a diketopiperazine antibiotic which appears to kill the fungus *Pythium ultimum* by causing coagulation of its protoplasm. Mycelium of *P. ultimum* that has been exposed to gliovirin does not grow, even after washing and transferring to fresh medium (Howell 1982; Howell et al. 1993).

Howell and Stipanovic (1983) obtained gliovirin-deficient mutants of *G. virens* via ultraviolet mutagenesis. These mutants failed to protect cotton seedlings from *P. ultimum* damping-off when applied to the seeds, whereas the normal parent strain protected the seedlings. Moreover, a gliovirin-overproducing mutant provided control similar to that of the wild type, although it exhibited a much lower growth rate. A combination of *G. virens* treatment of cotton seed with reduced levels of the fungicide metalaxyl provided disease suppression equal to that of a full fungicide treatment (Howell 1991). Antifungal, volatile alkyl pyrones produced by *Trichoderma harzianum* were identified by Claydon et al. (1987). These metabolites were inhibitory to a number of fungi in vitro and, when added to a peat soil mixture, they reduced the incidence of damping-off in lettuce caused by *Rhizoctonia solani*.

Recently, Ordentlich et al. (1992) isolated a novel inhibitory substance, 3-2(-hydroxypropyl),4-(2,4-hexadienyl),2(5H)-furanone, produced by

one isolate of *T. harzianum* that was found to suppress growth of *Fusarium oxysporoum* and may be involved in the biocontrol of *Fusarium* wilt. Nevertheless, Baker and Griffin (1995) concluded that the impact of antibiosis in biological control is uncertain. Even in cases where antifungal metabolite production by an agent reduces disease, other mechanisms may also be operating.

B. Competition

Many plant pathogens require exogenous nutrients to successfully germinate, penetrate, and infect host tissue (Baker and Griffin 1995). Garrett (1965) concluded that the most common cause of death in a microorganism is starvation. Therefore, competition for limiting nutritional factors, mainly carbon, nitrogen, and iron, may result in biological control of plant pathogens.

Research over the years has concentrated on competition by bacterial biocontrol agents, mainly for iron (Fe). Until recently, however, fungal antagonists have received very little attention. Sivan and Chet (1989b) found that a strain of *T. harzianum* (T-35) which controls *Fusarium* spp. on various crops may operate via competition for nutrients and rhizosphere colonization.

The potential of a microorganism applied as a seed treatment to proliferate and establish along the developing root system has been termed rhizosphere competence (Ahmad and Baker 1987). When T-35 conidia were applied to soil enriched with chlamydospores of *F. oxysporum* f. sp. *melonis* and f. sp. *vasinfectum* and amended with low levels of glucose and asparagine, the ability of the chlamydospores to germinate was reduced. This inhibitory effect could be reversed by adding an excess of glucose and asparagine or of seedling exudates to the soil. After its application as a seed treatment, this strain effectively colonized the rhizosphere of melon and cotton and prevented colonization of these roots by *F. oxysporum*. Thus, competition for carbon and nitrogen in the rhizosphere, as well as rhizosphere competence, may be involved in the biocontrol of *F. oxysporum* by *T. harzianum* strain T-35 (Sivan and Chet 1989b).

C. Mycoparasitism

Mycoparasitism is defined as a direct attack on a fungal thallus, followed by nutrient utilization by the parasite. The term hyperparasitism is sometimes used to describe a fungus that is parasitic on another parasitic pathogenic fungi. Barnett and Binder (1973) divided mycoparasitism into: (1) necrotrophic (destructive) parasitism, in which the relationships result in death and destruction of one or more components of the host thallus and (2) biotrophic (balanced) parasitism, in which the development of the parasite is favored by a living rather than a dead host structure.

Necrotrophic mycoparasites tend to be more aggressive, have a broad host range extending to wide taxonomic groups, and are relatively unspecialized in their mode of parasitism (Manocha 1990). The antagonistic activity of necrotrophic mycoparasites is attributed to the production of antibiotics, toxins, or hydrolytic enzymes in proportions that cause the death and destruction of their host (Manocha 1990). Biotrophic mycoparasites, on the other hand, tend to have a more restricted host range and produce specialized structures to adsorb nutritients from their host (Manocha 1990).

The parasitic relationships between fungi and their significance in biological control are the subjects of this chapter. Both types of mycoparasitism are described and discussed in the scope of their contribution to biological control. We will concentrate on the morphological, biochemical and molecular aspects of mycoparasitism in relation to biological control. The ecological aspects of this phenomenon are discussed elsewhere in this volume (see Jeffries, Chap. 10, this Vol.). Other types of interfungal relationships and other aspects of mycoparasitism are comprehensively discussed in Jeffries and Young (1994).

II. Mycoparasites as Biocontrol Agents

Many comprehensive reviews on mycoparasitism in biological control in general have been published over recent years (Baker 1986, 1987; Chet 1987, 1990; Handelsman and Parke 1989; Deacon 1991; Whipps 1991, 1992; Sivan and Chet 1992; Elad and Chet 1995; Baker and Griffin 1995; Jeffries and Youn 1994). Due to their nature, only a few examples of biotrophic mycoparasites as biocontrol agents exist (Ayers and Adams 1981; Sztejnberg et al. 1987; Adams 1990). Necrotrophic mycoparasites, being more common, saprophytic in nature, and less specialized in their mode of action, are easier to study. As a result, the majority of the mycoparasites used as biocontrol agents

in greenhouse or field trials to date have been necrotrophs (Baker 1987; Chet 1987, 1990; Adams 1990; Whipps 1991, 1992). In this chapter we will emphasize those examples in which the research that has been carried out is both applied and basic in nature.

A. Biotrophic Mycoparasites

1. *Sporidesmium sclerotivorum*

Sporidesmium sclerotivorum is a dermatiacesus hyphomycete that was isolated from field soil by Uecker et al. (1980). The fungus has been found to be an obligate parasite on sclerotia of *Sclerotinia sclerotiorum*, *S. minor*, *S. trifoliorum*, *Sclerotium cepivorum*, and *Botrytis cinerea* under natural conditions (Ayers and Adams 1981). In response to chemicals released by the host's sclerotia, macroconidia of *S. sclerotivorum* in the soil germinate and the germ tubes infect the sclerotia. The hyphae penetrate the intercellular matrix, which is composed mainly of β-glucans (Ayers et al. 1981). The production and activity of haustoria by the mycoparasite stimulate the host sclerotia to increase their glucanase, and probably other enzyme activities, resulting in the degradation of glucan into available glucose (Bullock et al. 1986). The mycoparasite establishes itself in the sclerotia, where its mycelium grows out into the surrounding soil to infect additional sclerotia and to produce new macroconidia (approx. 15 000 per sclerotium) (Adams et al. 1984). The interaction between *S. sclerotivorum* and *Sclerotinia minor* depends on both the host and parasite density (Adams 1986). The infection process is favored by soil pH (in the range of 5.5 to 7.5), water potential (>−8 bar), and temperature (20–22 °C) (Adams and Ayers 1980).

Under field conditions, a single application of an *S. sclerotivorum* preparation at a concentration of 10^2 or 10^3 macroconidia g^{-1} soil caused a 75–95% reduction in the number of sclerotia of *S. minor* per plot. Control of lettuce drop caused by *S. minor* in these plots varied from 40–83% in four consecutive lettuce crops (Adams and Ayers 1982). These results were significant but not economically important (Adams 1990). Adams (1990) concluded that "one of the biggest obstacles to practical biological control is the large quantity of the agent necessary to achieve biological control when applied directly to soil in the field." He therefore suggested two alternatives: (1) to add sclerotia of *S. minor* or a nonpathogenic *Sclerotinia* that is also a host of *S. sclerotivorum* which are already infected by the mycoparasite; (2) to apply a low dosage of the mycoparasite preparation to a diseased crop and then immediately incorporate the treated crop into the soil. This latter procedure ensures that a high percentage of the mycoparasites will be present in the soil next to the sclerotia of the pathogen (Adams 1990). Although the author assumes that these alternatives are easier and more practical, neither has been explored to any significant extent (Adams 1990).

2. *Ampelomyces quisqualis*

Ampelomyces quisqualis, a hyperparasite on Erysiphales, has been reported as a biocontrol agent of powdery mildews (Jarvis and Slingsby 1977; Sundheim 1982; Philipp et al. 1984; Sztejnberg and Mazar 1985; Sztejnberg et al. 1989). Recently, an isolate of *A. quisqualis* obtained from an *Oidium* sp. infecting *Catha edulis* in Israel proved to be infective to several powdery mildew fungi belonging to the genera *Oidium*, *Erysiphe*, *Sphaerotheca*, *Podosphaera*, *Uncinula*, and *Leveillula*. In field trials, *A. quisqualis* parasitized the powdery mildews of cucumber, carrot, and mango, and reduced the disease. *A. quisqualis* was tolerant to many fungicides used to control powdery mildews and/or other plant diseases. Treating powdery mildew of cucumber (cv. Hazera 205) with spores of *A. quisqualis* alone significantly decreased disease severity and increased cucumber yield by approximately 50%. Combining the fungicide pyrazophos with the mycoparasite resulted in a larger increase in cucumber yield (Sztejnberg et al. 1989).

In field trials, treating zucchini powdery mildews with *A. quisqualis* increased yield by 39% as compared to the untreated control, and these results were similar to those obtained using four chemical fungicide treatments. Treating powdery mildew-infected zucchini leaves with *A. quisqualis* increased the rates of photosynthesis from $3.8 \mu\text{mol } CO_2 \text{ m}^{-2}\text{s}^{-1}$ in untreated plants to 10.2 in plants treated with *A. quisqualis* as compared to 12.8 in uninfected healthy plants (Sztejnberg and Abo-Foul 1990). Electron micrographs of leaf sections of diseased cucumber plants revealed marked deterioration on the morphological organization of chloroplast membranes. Chloroplasts of *A. quisqualis*-treated plants seemed undamaged, like those of untreated plants (Abo-Foul et al. 1996). Fluorescence measurements (e.g. low-temperature fluorescence emission spectra, and room-temperature fluorescence transients) indicated a disease-correlated increase in levels of uncoupled chlorophyll (Abo-Foul et al. 1996). A simple, inexpensive medium based on potato dextrose broth (PDB) was developed for mass production of infective spores of *A. quisqualis* in fermentation for biological control (Sztejnberg et al. 1990).

The interaction between the hyperparasite *A. quisqualis* and its host fungi was studied by Hashioka and Nakai (1980) and Sundheim and Krekling (1982). The infection process of the cucumber powdery mildew *Sphaerotheca fuliginea* by *A. quisqualis* was studied by scanning electron microscopy. Within 24h after inoculation, the hyperparasite had germinated, and the germ tubes had developed appressorium-like structures at the point of contact with the powdery mildew host. Both conidia and hyphae were parasitized by penetration. Within 5 days of inoculation, the hyperparasite had developed pycnidia with conidia on the powdery mildew hyphae and conidiophores (Sunhdeim and Krekling 1982). Hashioka and Nakai (1980) used both transmission and scanning electron microscopy to study the hyphal extension and pycnidial development of the mycoparasite *A. quisqualis* Ces. inside the hyphae and conidiophores of several species of powdery mildew fungi belonging to *Microsphaera*, *Erysiphe*, and *Sphaerotheca*. The mycoparasite cells grew normally inside the host cells despite gradual degeneration of these latter cells. The invading hyphal cells of the mycoparasite migrated into the neighboring host cells by constricting themselves through the host cell's septal pore. The mycoparasite extended hyphae inside the conidiophores of the hosts and formed pycnidia which consisted of a unicellular outer layer and interior cells that later differentiated into conidiogenous structures (Hashioka and Nakai 1980).

B. Necrotrophic Mycoparasites

1. *Pythium nunn*

Pythium nunn is a mycoparasite isolated from soil suppressive to a plant parasitic *Pythium* sp. (Lifshitz et al. 1984b). When this mycoparasite was introduced into soil conducive to *Pythium* sp., the competitive saprophytic ability of this isolate was suppressed. An inverse relationship was found between propagule densities of the plant pathogen and the antagonist *P. nunn* (Lifshitz et al. 1984b).

Damping-off in cucumber caused by *P. ultimum* was suppressed by adding the antagonistic mycoparasite to steam-treated soil (Paulitz and Baker 1987). Paulitz et al. (1990) combined *P. nunn* with *T. harzianum*, isolate T-95, another mycoparasite and biocontrol agent, to control *Pythium* damping-off in cucumber.

The modes of hyphal interaction between the mycoparasite *P. nunn* and several soil fungi were studied by both phase-contrast and scanning electron microscopy (Lifshitz et al. 1984a). In the zone of interaction, *P. nunn* massively coiled around and subsequently lysed hyphae of *P. ultimum* and *P. vexans* without penetration. In contrast, *P. nunn* penetrated and eventually parasitized hyphae of *R. solani*, *P. aphanidermatum*, *Phytophthora parasitica* and *Phyto. cinnamomi* forming appressorium-like stuctures. However, *P. nunn* was not mycoparasitic against *F. oxysporum* f. sp. *cucumerinum* or *Trichoderma koningnii* and was destroyed by *T. harzianum* and *T. viride*. The authors concluded that *P. nunn* is a necrotrophic mycoparasite with a limited host range and differential modes of action among susceptible organisms (Lifshitz et al. 1984a).

Lysis and penetration of the host cell wall at the site of interaction with the mycoparasite was demonstrated by Elad et al. (1985). Calcofluor White M2R binds to the edges of polysaccharide oligomers (Kritzman et al. 1978). Using this reagent, the appearance of fluorescence indicated localized lysis of the host cell wall by *P. nunn* (Elad et al. 1985).

The cell walls of Oomycota are composed of β-glucan, cellulose, and less than 1.5% chitin. Basidiomycotina and Ascomycotina contain mainly β-glucan and chitin but no cellulose. *P. nunn* produced large amounts of β-1-3-glucanase and chitinase in liquid cultures containing cell walls of pathogenic fungi belonging to the class Basidiomycotina. This mycoparasite produced cellulase but no chitinase when grown on culture containing cell walls of two pathogens belonging to the Oomycota (Elad et al. 1985). These extracellular hydrolytic enzymes were detected in *P. nunn* when grown in dual culture with six host fungi but not with ten nonhost fungi, indicating specificity in the antagonistic activity of *P. nunn* (Baker 1987).

2. *Talaromyces flavus*

Talaromyces flavus (the perfect stage of *Penicillium dangeardii*; synonym: *P. vermiculatum*) is a mycoparasite of several soilborne plant pathogenic fungi including *R. solani* (Boosalis 1956), *S. sclerotiorum* (McLaren et al. 1986; Su and Leu 1980) and *Verticillium* spp. (Fahima and Henis 1990). In a few reports, *T. flavus* was found to be a potential biocontrol agent of both *S. sclerotiorum* and *Verticillium* wilt. By inoculating field soil with *T. flavus* together with sclerotia of *S. sclerotiorum*, McLaren et al. (1983) obtained 42–68% disease control of *Sclerotinia* wilt of sunflowers.

Laboratory investigations using light and electron microscopy indicate that *T. flavus* is a destructive hyperparasite of *S. sclerotiorum*. In dual culture, hyphae of *T. flavus* grew toward and coiled around the host hyphal cells. The coiling effect intensified as the hyphae of *T. flavus* branched repeatedly on the host surface. Tips of the hyphal branches often invaded the host by direct penetration of the cell wall without formation of appressoria. Infection of host cells by *T. flavus* resulted in granulation of the cytoplasm and collapse of the cell walls (McLaren et al. 1986).

More than 70% of the sclerotia of *S. sclerotiorum* became infected when inoculated with a conidial suspension of the hyperparasite and incubated at 24–26 °C for 3–6 weeks (Su and Leu 1980).

Direct invasion of *R. solani* hyphae via the production of penetration pegs by *T. flavus* was observed (Boosalis 1956). These pegs developed from either a mycelium coiling around the host hyphae or from a hypha in direct contact with the host. Fahima and Henis (1990) applied *T. flavus* as an ascospore suspension to soil naturally infested with *Verticillium dahliae*, the causal agent of *Verticillium* wilt in eggplant. Twelve weeks after transplanting, 77% disease reduction was achieved as compared with the untreated control.

Germinability of microsclerotia which had been incubated for 14 days in soil treated with 0.5% of a *T. flavus*-wheat bran preparation decreased from 84 to 17%, as compared with 81 and 74% in untreated soil and in soil treated with a sterilized biocontrol preparation, respectively. Scanning electron micrographs showed heavy fungal colonization and typical *T. flavus* conidia on the surface of the microsclerotia buried in the treated soil, but not in control soils. Transmission electron micrographs of microsclerotia incubated with *T. flavus* on agar revealed parasitism involving invasion of some host cells by means of small penetration pegs; the host cell walls were lysed mainly at their site of contact with the parasite hyphal tips. Further colonization of the microsclerotial cells occurred simultaneously with the degradation of the invaded host cell contents, rather than the cell walls (Fahima et al. 1992). It was suggested that mycoparasitism of *V. dahliae* microsclerotia by *T. flavus* hyphae may be involved in the biological control of *Verticillium* wilt disease. Fravel and Keinath (1991), however, claimed that *T. flavus* is known to produce compounds which mediate antibiosis, which is therefore suspected of being involved in the control of *Verticillium* wilt of eggplant and potato.

Similarly, McLaren et al. (1986) observed that hyphal cells of *S. sclerotiorum* eventually collapse as a result of infection by *T. flavus*, but host cell walls remain intact. They suggested that cell wall-degrading enzymes may not play a major role in the control of *S. sclerotiorum* by *T. flavus* and that antibiotics produced by the parasite may be involved in the deterioration of the host's hyphae (McLaren et al. 1986). Thus, the mechanism by which *T. flavus* controls plant pathogenic fungi is not always clear, and in some cases it may involve a combination of modes of actions.

3. *Coniothyrium minitans*

Coniothyrium mimitans has been found to be a natural mycoparasite of sclerotia of the plant pathogenic fungus *S. sclerotiorum*. In Canada, Huang (1977) found that sclerotia of *S. sclerotiorum* in roots and stems of sunflower, at the end of the season, became infected with the parasite *C. minitans*. This infection actually provided natural biological control of this pathogen in the field. Applying *C. minitans* to the seed furrow in field trials, in soil naturally or artificially infested with *S. sclerotiorum*, produced 42–78% disease control of sunflower wilts over 2 successive years (Huang 1980). Ahmed and Tribe (1977) obtained up to 60% disease control of white rot caused by *Sclerotium cepivorum* in onion by a combination of seed coating and seed furrow treatments. A preplanting application of *C. minitans* to greenhouse lettuce reduced both the disease caused by *S. sclerotiorum* and the number of sclerotia (Whipps 1991). In a series of experiments performed over 3 separate years in the same greenhouse, 28–90% disease reduction and over 85% sclerotial infection was obtained.

C. minitans could be recovered from the soil 13 months after its' application, and during this time it spread throughout the greenhouse indicating its' potential for long-term control (Whipps 1991). *C. minitans* is a destructive parasite that kills both hyphae and sclerotia of *S. sclerotiorum*.

By using scanning electron microscopy, it was shown that hyphae of *C. minitans* grow intracellularly in the infected sclerotia (Phillips and Price 1983; Tu 1984). Phillips and Price (1983), based on transmission electron microscopic studies, concluded that penetration of the rind cells of *S. sclerotiorum* sclerotia by *C. minitans* is due to physical pressure rather than enzymatic lysis of the cell wall. In a later study, Huang and Kokko (1987) found, by transmission electron microscopy, that there was destruction and disintegration of the sclerotial tissues, caused by penetration of the parasitic hyphae. Evidence from cell-wall etching at the penetration site suggests that chemical activity is indeed required for hyphae of *C. minitans* to penetrate the thick, melanized rind walls. The medullary tissue infected by *C. minitans* showed signs of plasmolysis, aggregation and vacuolization of the cytoplasm, and dissolution of the cell walls. The authors concluded that cell wall-lysing enzymes, responsible for the degradation of *S. sclerotiorum* hyphae (Jones and Watson 1969), may also play a significant role in the dissolution and degradation of the sclerotial rind wall at the penetration site and other affected areas (Huang and Kokko 1987).

Infection of *S. sclerotiorum* hyphae by the hyperparasite *C. minitans* has been reported by several workers (Huang and Hoes 1976; Trutmann et al. 1982; Tu 1984). However, researchers are not in complete agreement on the mode of hyperparasitism. Using light microscopy, Huang and Hoes (1976) observed that hyphal tips of *C. minitans* invade hyphae of *S. sclerotiorum* by direct penetration, without forming any special structure. Host cytoplasm disintegrates and cell walls collapse as a result of infection. Microconidia and intrahyphal hyphae were produced by *S. sclerotiorum* in infected colonies (Huang and Hoes 1976).

Production of appressoria by *C. minitans* when it comes into contact with the undamaged hyphae of *S. sclerotiorum* in dual culture on potato dextrose agar (PDA) was observed by Tu (1984).

He stated that hyphal penetration by the hyperparasite without the formation of appressoria sometimes occurs, but only on damaged host cells.

Huang and Kokko (1988), using scanning electron microscopy, confirmed previous reports from light microscopic studies that hyphal tips of *C. minitans* invade the host hyphae by direct penetration, without developing appressoria, and that indentation of the host cell wall at the point of penetration is often evident. No functional distinction between main branch and side branch hyphae of the hyperparasite was found, and tips of either type of hypha are capable of invading host hyphae by direct penetration (Huang and Kokko 1988).

4. *Gliocladium* spp.

Several species of *Gliocladium* have been reported to be hyperparasites of many fungi. The biology, ecology, and potential of this genus for biological control of plant pathogens have been extensively reviewed in a comprehensive treatise by Papavizas (1985). Huang (1978) reported that *G. catenulatum* parasitizes *S. sclerotiorum* and *Fusarium* spp. It kills the host by direct hyphal contact, causing the affected cells to collapse or disintegrate. Pseudoappressoria are formed by the hyperparasite, but hyphae derived from them do not penetrate the host cell walls. Vegetative hyphae of all species tested and macroconidia of *Fusarium* spp. are susceptible to this hyperparasite, but chlamydospores of *Fusarium equiseti* are resistant.

Phillips (1986) studied aspects of the biology of *G. virens* and its parasitism of sclerotia of *S. sclerotiorum* in soil. *G. virens* parasitized and decayed sclerotia of *S. sclerotiorum*, *S. minor*, *Botrytis cinerea*, *Sclerotium rolfsii* and *Macrophomina phaseolina* on laboratory media and caused a reduction in the survival of sclerotia of *S. sclerotiorum* in soil. However, parasitism of the mycelium was not detected.

A strain of *G. virens* isolated from the parasitized hyphae of *R. solani* by Howell (1982) significantly suppressed damping-off incidence in cotton seedlings by this pathogen and by *Pythium ultimum*. Treatment with *G. virens* more than doubled the number of surviving cotton seedlings grown in soil infested with either pathogen. *G. virens* parasitized *R. solani* by coiling around and penetrating the hyphae. *P. ultimum* was not parasitized by *G. virens*, but was strongly inhibited by antibiosis. Treatment of soil infested with propagules of *R. solani* or *P.* ultimum with *G. virens* resulted in a 63% reduction in the number of viable *R. solani* sclerotia after 3 weeks of incubation, whereas oospores of *P. ultimum* were unaffected (Howell 1982). In a recent work, strains of *G. virens* were separated into two distinct groups, P and Q, on the basis of secondary metabolite production in vitro (Howell et al. 1993).

Gliorvirin was very inhibitory to *P. ultimum*, but exhibited no activity against *R. solani*, and strains that produced it (P group) were more effective seed treatment biocontrol agents of disease incited by *P. ultimum*. Conversely, gliotoxin was more active against *R. solani* than against *P. ultimum*, and strains that produced it (Q group) were more effective seed treatments for controlling disease incited by *R. solani*. According to these results, the authors suggested that it may be necessary to treat seeds with a combination of strains in order to broaden the disease control spectrum (Howell et al. 1993).

Howell (1987) isolated mutants of *G. virens*, obtained by irradiation with ultraviolet light, that showed no mycoparasitic activity. The selected mutants retained the same antibiotic complement as the parent strains. Peat moss-Czapek's broth cultures of parent and mutant strains were similarly effective as biocontrol agents of cotton seedling disease induced by *R. solani* and as antagonists of *R. solani* sclerotia in soil. In the light of these results, Howell (1987) concluded that mycoparasitism is not a major mechanism in the biological control of *R. solani*-incited seedling disease by *G. virens* (Howell 1987).

In addition, Pachenari and Dix (1980) concluded that *G. virens* need not make intimate contact with *Botrytis allili* to cause severe internal disorganization of host cells, coagulation of cytoplasm, vacuolation, and loss of contents from organelles. Cultures of *B. allili* parasitized by *G. roseum* contained considerable β-(1-3)-glucanase and chitinase, and the cytoplasm coagulated without physical contact.

Recently, *G. virens* isolate Gl-21 was grown on various solid and liquid media: wheat bran and peanut hull meal (PHM), as well as spent glucose tartrate broth (GTB), Czapek-Dox broth (CDB), and potato dextrose broth (PDB) (Lewis et al. 1991). Aqueous extracts of these media caused leakage of carbohydrates and electrolytes from hyphae of the soilborne plant pathogen *R. solani*, and its mycelial weight was reduced. Size fraction-ation experiments indicated that a combination of factors produced by *G. virens*, rather than a single one, induced this phenomenon. Gliotoxin was de-tected in culture filtrates from *G. virens* grown on bran and PHM media. Gliotoxin preparations in-duced leakage of carbohydrates and electrolytes from *R. solani* and caused a concomitant reduc-tion in mycelial weight, which suggests that it is a leakage factor (Lewis et al. 1991). The authors speculated that hydrolytic enzymes such as β-(1-3)-glucanase, β-(1-4)-glucanase, chitinase, and protease, shown to be produced by isolates of *G. virens* (Erbeznik et al. 1986; Roberts and Lumsden 1990), have the potential to act on *R. solani* cell walls and membranes and may there-fore also be involved in inducing cytoplasmic leak-age from *R. solani* (Lewis et al. 1991).

5. *Trichoderma* spp.

More than 60 years ago, Weindling (1932) was the first to demonstrate the mycoparasitic nature of fungi from the genus *Trichoderma*. He suggested their potential use as biocontrol agents of plant pathogenic fungi (Weindling 1932). However, the first report on a biological control experiment us-ing *Trichoderma* spp. under natural field condi-tions came 40 years later by Wells et al. (1972), who used *T. harzianum* grown on an autoclaved mixture of ryegrass seeds and soil to control *Scle-rotium rolfsii* Sacc. Since then, more *Trichoderma* isolates have been obtained from natural habitats and used in biocontrol trials against several soil-borne plant pathogenic fungi under both green-house and field conditions (Chet 1987, 1990; Harman and Lumsden 1990).

Hadar et al. (1979) reported an isolate of *T. harzianum* that directly attacks the mycelium of *R. solani* in dual culture. This isolate, applied in the form of wheat bran culture to *R. solani*-infested soil, effectively controlled damping-off of bean, tomato, and eggplant seedlings under greenhouse conditions. Another isolate of *T. harzianum* ca-pable of lysing mycelia of *S. rolfsii* and *R. solani* was isolated by Elad et al. (1980) from soil natu-rally infested with those pathogens. Under green-house conditions, incorporation of a wheat bran preparation of *T. harzianum* in pathogen-infested soil significantly reduced bean diseases caused by *S. rolfsii*, *R. solani*, or both. The wheat bran prepa-ration inoculum controlled *S. rolfsii* more effi-ciently than a conidial suspension of the same antagonist. It increased growth of bean plants in noninfested soil, whereas an uninoculated wheat bran preparation increased disease incidence. In naturally infested soils, wheat bran preparations of *T. harzianum* significantly decreased diseases caused by *S. rolfsii* or *R. solani* in three field ex-periments with bean, cotton, or tomato, and they significantly increased bean yield (Elad et al. 1980).

A seed treatment was developed by Harman et al. (1980) to reduce the amount of *Trichoderma* added to the soil to control soilborne plant patho-genic fungi. *T. hamatum* conidia applied in the laboratory to seeds of pea and radish as a Methocel slurry provided protection to seeds and seedlings from *Pythium* spp. and *R. solani*, respec-tively, almost as effectively as fungicide seed treat-ment. Establishment of the mycoparasite and long-term action were demonstrated, as the propagules of *T. hamatum* increased approxi-mately 100-fold in soils planted with treated seeds. Population densities of *R. solani* and *Pythium* spp. were lower in soils containing *T. hamatum* than in soils lacking this antagonist. Replanting these soils once or even twice with untreated seeds yielded lower disease incidence than in soils originally planted with untreated seeds (Harman et al. 1980).

Addition of chitin or *R. solani* cell walls to the coating of seeds previously treated with a conidial suspension increased both the ability of *T. hamatum* to protect the seeds against *Pythium* spp. or *R. solani* and the population density of *Trichoderma* in the soil. *T. hamatum* with chitin, but without *R. solani* cell walls, effectively re-duced damping-off caused by *Pythium* spp., as compared to seed treatment containing only *T. hamatum* (Harman et al. 1980).

Sivan et al. (1984) applied a peat-bran mixture (1:1 v/v) preparation of *T. harzianum* (isolate 315) to either soil or rooting mixture and efficiently controlled damping-off induced by *Pythium aphanidermatum* in pea, cucumber, tomato, pep-per, and gypsophila. Disease reduction of up to 85% was obtained in tomatoes. Extracellular filtrate from cultures of *T. harzianum* added to a synthetic medium inhibited linear growth of *P. aphanidermatum* by 83%, whereas substances ex-creted by this fungus into the growth medium en-hanced the linear growth of *T. harzianum* by 34% (Sivan et al. 1984).

An isolate of *T. harzianum* was found by Strashnov et al. (1985a) to be tolerant of up to

20000 ppm methyl bromide (MB), whereas the plant pathogen *R. solani* was susceptible to a dose of less than 9000 ppm. Exposure to sublethal concentrations of MB had no effect on the in vitro antagonistic ability of *T. harzianum*. Under field conditions, a combination of 200 kg ha^{-1} MB and *T. harzianum* exhibited a significant synergistic effect on damping-off of carrot seedlings caused by *R. solani* and its effect on growth, yield, and disease control was similar to that of the recommended MB dosage (Strashnov et al. 1985a). *T. harzianum* also significantly reduced fruit rot of tomato caused by *R. solani*, by 21–51% (Strashnov et al. 1985b).

Several isolates of *T. harzianum* and *T. hamatum* were found to antagonize and control *Macrophomina phaseolina* in beans and melon (Elad et al. 1986). Isolates of *T. harzianum* and *T. hamatum* antagonized and controlled *Rosellinia necatrix* in almond seedlings (Freeman et al. 1986). Sztejnberg et al. (1987) combined sublethal soil heating with an application of *T. harzianum* to yield better control of *R. necatrix* than that achieved by either treatment alone.

Sivan and Chet (1986) isolated a new *Trichoderma harzianum* isolate (T-35) from the rhizosphere of cotton plants grown in fields infested with *Fusarium*. Applying this isolate in greenhouse trials as a peat-bran preparation, a conidial suspension, or a seed coating, significantly reduced the disease incidence caused by *F. oxysporum* f. sp. *vasinfectum*, *F. oxysporum* f. sp. *melonis*, and *F. roseum* Culmorum in cotton, melon, and wheat, respectively (Sivan and Chet 1986). In a further study, the isolate was tested in biological control trials over two successive growing seasons against *Fusarium* crown rot of tomato in fields naturally infested with *F. oxysporum* f. sp. *radici lycopersici* (Sivan et al. 1987). *T. harzianum* was applied as a seed coating or as a wheat bran-peat (1:1, v/v) preparation introduced into the tomato rooting mixture. *Trichoderma*-treated transplants were better protected against *Fusarium* crown rot than untreated controls when planted in MB-fumigated or nonfumigated infested fields. The total yield of tomatoes in the *T. harzianum*-treated plots increased as much as 26.2% over the controls (Sivan et al. 1987).

Integrated control of *Verticillium dahliae* in potato by *T. harzianum* and the fungicide Captan was reported by Ordentlich et al. (1990). Integrated treatment resulted in a reduction of *Verticillium* colonization of the stems, increasing both marketable and total potato yield of cv. Draga by 84 and 46%, respectively, and total yield of cv. Desiree by 80%.

C. Mycoparasitism in Suppressive Environments (Soils and Composts)

Suppression of soil-borne plant pathogens occurs in environments such as soils and soils amended with organic matter or compost as the organic component in media for container-grown plants.

Pathogen suppressiveness has been defined by Cook and Baker (1983) as "soils in which the pathogen does not establish or persist, establishes but causes little or no damage, or establishes and causes disease for a while but thereafter the disease is less important, although the pathogen may persist in the soil."

Baker and Cook (1974) divided suppression mechanisms into the two broad categories defined as general and specific. General suppression is a result of total microbial activity. In contrast, specific suppression applies when bacteria or fungi, individually or as a group, are responsible for the suppression effect. Mycoparasitism, the focus of this chapter, is a major mechanism of specific suppression. For example, Chet and Baker (1981) reported on a soil which was suppressive to *R. solani* of carnation near Bogota, Colombia. This soil contained high levels of organic matter (35%), was highly acidic (pH 5.1), and its main microbiological component was the antagonistic fungus *T. hamatum* at a population density of 8×10^5 propagules g^{-1}. The level of this mycoparasite in mineral-conducive soil was four orders of magnitude lower.

In another, related study, Henis et al. (1978, 1979) showed the effect of successive plantings on the development of suppression. A soil sown with radish every week became suppressive to *R. solani* by the fourth sowing (Henis et al. 1978) and was even more suppressive by the fifth and subsequent sowings. The population of *T. harzianum*, antagonistic to *R. solani*, increased with successive sowings of radish (Liu and Baker 1980), possibly in response to increases in the amount of *R. solani* in the soil resulting from its parasitism of the radish seedlings. The addition of *T. harzianum* spores to a conducive soil at the same density found in the suppressive soil caused the conducive soil to become suppressive. *Trichoderma* spp. were also reported to be responsible for the suppression of the take-all disease caused by *Gaeumannomyces graminis*. Both *Trichoderma* population levels and suppression were enhanced by fertilization with ammonium sulfate, but reduced by liming (Simon and Sivasithamparam 1990). Again, low pH conditions were found to be favorable to *Trichoderma* and to enhance suppression.

A practical approach to utilizing suppression in agriculture is the use of suppressive composts, mainly in container media. Composting is the breakdown of organic waste material by a succession of mixed populations of microorganisms in a thermophilic aerobic environment. The final product is compost or humus, which is the stabilized organic matter populated by microorganisms capable of suppressing soilborne plant pathogens.

The use of suppressive compost-amended container media provides effective biological control of plant pathogens (Hoitink and Fahy 1986; Hoitink et al. 1993). Therefore, compost of a wide variety of waste materials (hardwood or pine bark, municipal sludge, grape marc, or cattle manure) is an economically and ecologically sound alternative to pesticides. The mechanisms of suppression in composts do not differ substantially from those described for soils, and can be either general or specific.

For example, Nelson et al. (1983) identified specific strains of four *Trichoderma* spp. and isolates of *Gliocladium virens* as the most effective fungal hyperparasites of *R. solani* present in bark compost. A few of the 230 other fungal species also showed activity, but most were ineffective. Kwok et al. (1987) described synergistic interactions between *T. rhamatum* and *Flavobacterium balustinum*. Several other bacterial strains, including *Enterobacter*, *Pseudomonas*, and *Xanthomonas* spp., also interacted with the *Trichoderma* isolate in suppression of *Rhizoctonia* damping-off (Kwok et al. 1987).

Composted grape marc was effective in suppressing disease caused by *S. rolfsii* in beans and chickpeas (Gorodecki and Hadar 1990). Hadar and Gorodecki (1991) placed sclerotia of *S. rolfsii* on composted grape marc to isolate hyperparasites of this pathogen. Viability of sclerotia decreased from 100% to less than 10% within 40 h. It remained close to 100% for sclerotia placed on a conducive peat mix. *Penicillium* spp. and *Fusarium* spp. were observed by scanning electron microscopy to colonize the sclerotia. *Trichoderma* populations in the grape marc compost were at very low levels (10^2 cfu g^{-1} dry weight). The hyper-

parasites present in this compost are therefore quite different from those isolated from tree bark compost, where *Trichodema* and *Gliocladium* isolates predominate.

In conclusion, suppression of soil-borne plant pathogens in field soil or container media is brought about by antagonistic microorganisms. Such systems could be a source for mycoparasites to be used in biocontrol or to be incorporated in integrated disease control programs.

III. Hyphal Interactions in Mycoparasitism

A. Biotrophs

Piptosephalis virginiana is a haustorial biotrophic mycoparasite that parasitizes fungi belonging to the order Mucorales exclusively (Manocha 1981). Attachment of a biotrophic mycoparasite to its host suface is considered to be an essential prerequisite step for further penetration of the host by the parasite (Manocha and Chen 1990). *P. virginiana* attaches to the surface of both the compatible *Choanephora cucurbitarum* and *Mortierella pusilla*, and the incompatible *Phascolomyces articulosus* hosts, but not to the surface of the nonhost *Mortierella candelabrum* (Manocha 1985; Manocha et al. 1986). comparative research was performed by Manocha and his coworkers in an attempt to unravel the molecular basis for specificity and recognition in this system. Cytological and biochemical investigations were carried out to study the structure and chemical composition of cell walls of host and nonhost species (Manocha 1981, 1984, 1987). The germ tubes of the biotrophic mycoparasite *P. virginiana* were found to attach to the cell-wall surface of the host but not to that of the nonhost (Manocha 1984, 1985; Manocha et al. 1986). This attachment could be specifically inhibited by chitobiose and chitotriose. The authors therefore suggested a possible involvement of carbohydrate-binding proteins in the specificity of this interaction. A comparison of protein and glycoprotein profiles of cell-wall extracts revealed marked differences between host and nonhost species. Two high-molecular-weight glycoproteins were observed only in the extract of host cell walls and were absent in that of the nonhost (Manocha 1985; Manocha et al. 1986). Further isolation and characterization of the host cell surface proteins revealed that attachment and appressorium formation by the parasite germ tubes could be inhibited by treating host cell-wall fragments with 0.1 M NaOH or pronase E (Manocha and Chen 1991). Furthermore, the two purified glycoproteins were able to agglutinate both nongerminated and germinated spores of the mycoparasite. The pure preparation was almost 35 times as active as the crude extract (Manocha and Chen 1991). Arabinose, glucose, and N-acetylglucosamine could totally inhibit this agglutination. These glycoproteins were suggested to be two subunits of a carbohydrate-binding agglutinin present on the host cell surface, and to be involved in agglutination and attachment of the mycoparasite germ tubes (Manocha and Chen 1991).

Using fluorescein isothiocyanate-labeled lectin-binding techniques, Manocha et al. (1990) were able to show differences in the distribution pattern of glycosyl residues at the level of the cell wall between fungi that are hosts and those that are nonhosts of the mycoparasite *P. virginiana*, and at the protoplast level between compatible and incompatible hosts.

The cell walls of the compatible hosts (*C. cucurbitarum* and *M. pusilla*) and the incompatible host (*P. articulosus*), as well as that of the mycoparasite itself, contain glucose and N-acetylglucosamine. In the nonhost (*M. candelabrum*), however, other sugars, such as fucose, N-acetylgalactosamine, and galactose, could also be detected. These latter sugars could be detected on both the host and the parasite surface after mild treatment with proteinase or when grown in liquid medium. The researchers speculated that the failure of the mycoparasite to attach to the host cells after proteinase treatment or in liquid culture may be due to the appearance of galactose and galactosamine at the host cell surface (Manocha et al. 1990). The idea that N-acetylglucosamine and glucose may be involved in the attachment of *P. virginiana* to its host cell surface was supported by the observation that pretreatment of the mycoparasite germ tubes with N-acetylguclosamine or glucose inhibited their attachment to the host cells. In addition, the germ tubes attached to agarose beads coated with glucose or with N-acetylglucosamine, but not with N-acetylgalactosamine (Manocha et al. 1990).

The protoplast surfaces of compatible hosts contained all of the above-listed sugars and these protoplasts could attach to the germ tube of the mycoparasite. Only lectins specific for N-acetylglucosamine and glucose were bound at the protoplast surface of the incompatible host; these protoplasts did not attach to the mycoparasite germ tubes. Indications were found for different factors responsible for attachment and for appressorium formation, as pretreatment of the mycoparasite with glucose and N-acetylglucosamine inhibited its attachment to the host cell surface, but had no obvious effect on appressorium formation. On the other hand, appressorium formation was inhibited by heat treatment of host cell-wall fragments which still permitted attachment (Manocha et al. 1990). The authors therefore suggested a model for the recognition between *P. virginiana* and its host fungi that operates at at least two levels; the cell wall and the protoplast surface. At the cell-wall level, the attachment probably involves carbohydrate-binding agglutinins that recognize specific sugar residues on the host but not on the nonhost cell wall. After the initial recognition and attachment, at the protoplast level, the parasite distinguishes compatible from incompatible hosts. The mechanism of this distinction is not clear. Yet it seems that protoplast membrane sugars are not a major factor in recognition at this level (Manocha et al. 1990).

B. Necrotrophs

As early as 1932, Weindling reported the coiling of *Trichoderma* spp. hyphae around hyphae of other fungi. These *Trichoderma* strains were later shown to belong to the genus *Gliocladium* (Webster and Lomas 1964). Dennis and Webster (1971a,b,c) published an extensive report on the antagonistic properties of species groups of *Trichoderma*. The hyphal interaction between *Trichoderma* and plant pathogenic fungi was first comprehensively studied in their work. Since then, numerous studies on the hyphal interaction and coiling phenomenon of *Trichoderma* around its host hyphae have been carried out with the use of light and electron microscopy (Chet et al. 1981; Elad et al. 1983b; Baker 1987; Chet 1987; Fig. 1).

The destructive mode of parasitism in *Trichoderma* appears to be a process consisting of several consecutive events initiated by attraction and directed growth of *Trichoderma* towards its host, probably by chemotropism. Positive chemotropism was found in *Trichoderma* (Chet et al.

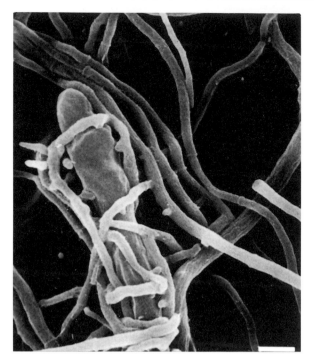

Fig. 1. A mycoparasitic relationship. Scanning electron micrograph of *T. harzianum* hyphae coiling around those of the plant pathogenic fungus *S. sclerotiorum. Bar* 10 μm

1981), as it could detect its host from a distance and begin to branch in an atypical way. These branches grew towards the pathogenic host fungi. Similar behavior was also found in *Pythium nunn* (Lifshitz et al. 1984a), *P. oligandrum* (Lewis et al. 1989) and in *Gliocladium* spp. (Huang 1978). This event is presumably a response of the antagonist to the chemical gradient of an attractant coming from the host. However, no specific stimuli other than amino acids and simple sugars have thus far been detected (R. Barak and I. Chet, unpubl. data). Hence, the specificity of the phenomenon is not clear. Apparently, it is not an essential step for mycoparasitism, although it may hold some advantage for the antagonist. Subsequently, contact is made and, in some cases, *Trichoderma* coils around or grows along the host hyphae and forms hook-like structures, presumably appressoria, that probably aid in penetrating the host hyphal cell wall (Chet et al. 1981; Elad et al. 1983c). The coiling phenomenon and appressoria formation have been reported for other mycoparasites as well (Lifshitz et al. 1984a; Tu 1984). However, Deacon (1976) concluded that in the case of *P. oligandrum*, coiling of the antagonist around its host hyphae indicates temporary host resistance rather than susceptibility. Nevertheless, in

Trichoderma, this reaction was found to be rather specific, and *Trichoderma* attacks only a few fungi. Moreover, Dennis and Webster (1971c), using plastic threads of a diameter similar to that of *P. ultimum* hyphae, concluded that the coiling of the *Trichoderma* is not merely a thigmotropic response. The *Trichoderma* hyphae never coiled around the threads but rather grew over or followed them in a straight course. This led to the idea that there is a molecular basis for the specificity. However, despite the fact that first observations and reports of this phenomenon were published decades ago, we are only now on the brink of being able to understand it. Recent reviews dealing with cellular interactions in fungi (Manocha 1990; Tunlid et al. 1992; Manocha and Sahai 1993) and the specificity of attachment of fungal parasites to their hosts (Manocha and Chen 1990) have been published. The physiology and biochemistry of biotrophic mycoparasitism in particular have been extensively reviewed by Manocha (1991).

Attachment and "recognition" between the mycoparasite and its host appears to be essential, and a crucial stage for successful continuation of the process. Lectins are sugar-binding proteins or glycoproteins of nonimmune origin which agglutinate cells and/or precipitate glycoconjugates (Goldstein et al. 1980). First discovered in plants and later in other organisms, they are involved in the interactions between the cell surface and its extracellular environment (Barondes 1981). Indeed, lectins were found to be produced by some soil-borne plant pathogenic fungi such as *R. solani* and *S. rolfsii* (Elad et al. 1983a; Barak et al. 1985, 1986) and by different members of the Sclerotiniaceae (Kellens et al. 1992).

Therefore, a role for lectins in the recognition and specificity of attachment between *Trichoderma* and its host fungi was suggested. However, until recently, no conclusive evidence to support this hypothesis was available. In an attempt to test this hypothesis, Inbar and Chet (1992) used a novel approach based on the binding of lectins to a surface of nylon fibers. This biomimetic system imitates the host hyphae and enables an examination of the role of lectins in mycoparasitism. Inert nylon fibers were chemically activated to enable the covalent binding of the lectins (Inbar and Chet 1992).

Concanavalin A (Con A), a plant lectin which is similar to the lectin of *S. rolfsii* (LSR) in its carbohydrate specificity (they are both specific to

D-glucose and D-mannose) was used first to establish the system. The *Trichoderma* recognized the LSR-treated fibers as a host, attached and coiled around them in a pattern similar to that seen with real host hyphae (Inbar and Chet 1992; Fig. 2). In contrast, in the untreated control, no interaction could be observed – the *Trichoderma* grew uninterruptedly over and along the fibers, exactly as outlined by Dennis and Webster (1971c). These findings provided the first direct evidence for the role of lectins in mycoparasitism. The researchers were able to show that inert nylon fibers coated with fungal lectins mimic the real host hyphae and can stimulate the parasite to coil around them.

Recently, a novel lectin was isolated and purified from the culture filtrate of the soilborne plant pathogenic fungus *S. rolfsii* (Inbar and Chet 1994) Agglutination of *E. coli* cells by the purified lectin

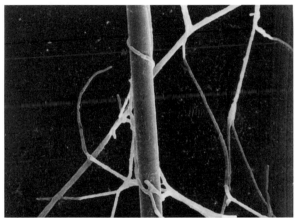

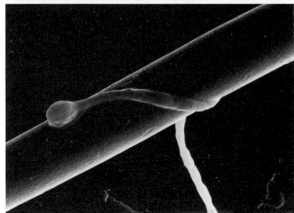

Fig. 2a,b. Biomimetic systems for simulating the interaction between *Trichoderma* and plant pathogenic fungi. **a** Scanning electron micrograph of *T. harzianum* hyphae coiling around inert mylon fibers coated with a surface lectin from the plant pathogenic fungus *S. rolfsii. Bar* 10 μm. **b** Appressorium formation by *T. harzianum* grown on nylon fibers coated with the *S. rolfsii* lectin. *Bar* 10 μm

could be inhibited by the glycoproteins mucin and asialomucin. Proteases, as well as β-1,3-glucanase, were found to be totally destructive to the agglutination activity, indicating that both protein and β-1,3-glucan are necessary for agglutination. Using the biomimetic system, it was apparent that the presence of the purified agglutinin of the surface of the fibers significantly induces mycoparasitic behavior in *T. harzianum* as compared with the untreaed ones, or with those treated with non-agglutinating extracellular proteins from *S. rolfsii* (Inbar and Chet 1994).

Subsequently, upon recognition and attachment, penetration and degradation of the host cell wall under the coiling and interaction sites were evident by visual observation, fluorescent indicators, and enzymatic studies. Using scanning electron microscopy, lysed sites and penetration holes were found in hyphae of *R. solani* and *S. rolfsii* following removal of *Trichoderma* spp. hyphae (Elad et al. 1983c). The cell walls of Basidiomycotina and Ascomycotina contain chitin and laminarin (β glucan) but no cellulose. Oomycota contain β-glucans and cellulose and relatively small amounts of chitin (<1.5%). Therefore, to penetrate the host cell wall, mycoparasites should have a system of hydrolytic enzymes that can degrade these components. Enzymatic degradation of fungal cell walls occurs mainly via the excretion of the extracellular enzymes β-1-3-glucanase and chitinase (Elad et al. 1982). Indeed, high β-1-3-glucanase and chitinase activities were detected in dual cultures when *T. harzianum* parasitized *S. rolfsii*, as compared with the low levels found with either fungus alone. Cycloheximide prevented antagonism and enzymatic activity was diminished (Elad et al. 1983c). Lynch (1987) and later Ridout et al. (1988) supported these data by finding that *T. harzianum* was able to penetrate hyphae of *R. solani*. Recently, using gold cytochemistry, *T. harzianum* hyphae were shown to coil around and penetrate cells of *R. solani*, causing extensive damage such as cell-wall alteration, plasma membrane retraction, and cytoplasm aggregation (Benhamou and Chet 1993).

Isolates of *T. harzianum* were found to differ in the levels of hydrolytic enzymes produced when the mycelium of *S. rolfsii*, *R. solani*, or *P. aphanidermatum* in soil was attacked. This phenomenon was correlated with the ability of each of the *Trichoderma* isolates to control the respective soil-borne pathogens (Elad et al. 1982). It has been found that the failure of *T. harzianum* strain

T-35, a biocontrol agent of *Fusarium* wilt, to parasitize colonies of *Fusarium oxysporum* is not due to its inability to produce and excrete fungal cell wall-degrading enzymes (Sivan and Chet 1989a). This strain, however, was shown to be an effective mycoparasite of both *R. solani* and *P. aphanidermatum* in dual cultures. Chitinase and β-1-3-glucanase were induced to higher levels in this isolate by cell walls of *R. solani* and *S. rolfsii* than by cell walls of *F. oxysporum* (Sivan and Chet 1989a). Similar findings were published by Elad et al. (1985) for *P. nunn* and its nonhost fungus *F. oxysporum* f. sp. *cucumerinum*. It was therefore suggested that the ability of a mycoparasite to produce and excrete appropriate enzymes against a given host may partially explain its specificity and potential host range (Baker 1987). However, Ordentlich et al. (1991) found no correlation between enzymatic activity, in vitro tests for antagonism, and the biological control capability of *Trichoderma* spp. against *F. oxysporum*.

Nevertheless, the involvement and importance of lytic enzymes, mainly chitinase, in the biological control of plant pathogens by both fungi and rhizobacterial agents (Chet et al. 1981; Elad et al. 1982; Ordentlich et al. 1988; Inbar and Chet 1991; Sahai and Manocha 1993), as well as their involvement in the defense of plants against pathogenic infection (Boller et al. 1983; Boller 1985; Broglie et al. 1991) is well documented.

We recently demonstrated that induction of chitinolytic enzymes in *Trichoderma* is elicited by the recognition signal (i.e., lectin-carbohydrate interactions). It was postulated that recognition is the first step in a cascade of antagonistic events triggering the parasitic response in *Trichoderma* (Inbar and Chet 1995).

IV. Molecular Aspects and Genetic Engineering in Mycoparasitism

Molecular approaches and genetic engineering techniques have recently been applied to gain a better and more basic understanding of the system, as well as to develop superior and improved strains of biocontrol agents with enhanced activity.

The chitinase gene *chi*A, encoding one of the chitinases from *Serratia marcescens*, a well-known biocontrol agent, was isolated and cloned into *E. coli* (Shapira et al. 1989). *E. coli* transformed by

the *chi*A gene under the oLpL operator and promoter of bacteriophage λ expressed and excreted the corresponding protein into the growth medium. Almost pure *S. marcescens* chitinase from *E. coli* or whole viable cells was used in greenhouse experiments against *S. rolfsii* in beans and *R. solani* in cotton. Using the chitinase preparation in the irrigation water effectively reduced the number of diseased plants. Whole viable cells of transformed *E. coli* were also effective in inhibiting *S. rolfsii*, but to a lesser degree (Shapira et al. 1989). The genetically engineered *E. coli*, a nonsoil bacterium, served here as a model system to demonstrate the role of chitinase in controlling a chitin-containing plant pathogen. It is suggested that the introduction of such engineered genes into soil bacteria will increase control efficiency by combining high expression of a gene coding for a lytic enzyme with rhizosphere competence.

Southern blot analysis of the *chi*A gene cloned from *S. marcescens* showed homology to one of the *Trichoderma* chitinase genes. Based on this, the *chi*A gene was used as a probe to isolate a chitinase gene from a cDNA library prepared from *T. harzianum* (T-35) (Chet et al. 1993). To prepare the cDNA library, *Trichoderma* (T-35) was grown with $3\,g\,l^{-1}$ chitin for 7 days, total RNA was isolated and poly (A) RNA was selected on an oligo dT column. The chitinase gene from *T. harzianum* (T-35) has recently been cloned in a Bluescript plasmid under the lac promoter. When the transformed *E. coli* was plated on LB + 0.2% chitin plates and induced by 1 mM IPTG, the bacteria showed chitinolytic activity.

In greenhouse experiments, irrigation of bean seedlings with $10^7\,cfu\,g^{-1}$ soil day^{-1} of *E. coli* XLI-Blue transformed with the *Trichoderma* chitinase gene induced by 1 mM IPTG, resulted in significant biocontrol activity. Suppression of the disease caused by *S. rolfsii* was obvious. The treated plants exhibited a better growth rate than untreated controls. The growth rate of the plants irrigated with the transformed bacteria after 18 days was similar to that of uninfected plants (Chet et al. 1993).

In an attempt to increase its effectiveness, *T. harzianum* protoplasts were cotransformed using two plasmids: pSL3chiAII (Fig. 3), containing a bacterial chitinase gene from *S. marcescens* under the control of a constitutive viral promotor, and p35SR2, a marker for selection after transformation, encoding for acetamidase. Two transformants showed increased constitutive chitinase activity (specific activity 11 and 5 times higher than the recipient, Fig. 4) and excreted a protein of ca. 58 kDa, the expected size of *S. marcescens* chitinase, when grown on synthetic medium. Antagonistic activity of the transformants was significantly higher than that of the wild-type *T. harzianum*, as evaluated by testing their ability to overgrow the plant pathogen *S. rolfsii* in dual culture (Fig. 5; Haran et al. 1993).

In another recent work, Geremia et al. (1993) identified a basic proteinase of *T. harzianum* (Prbl), which is induced in the presence of either autoclaved mycelia or a cell-wall preparation of phytopathogenic fungi. The use of synthetic oligo nucleotide probes allowed subsequent isolation of a cDNA and its corresponding genomic clone. The deduced protein sequence indicated that the proteinase is synthesized as a preproenzyme that can be classified as a subtilisin-type serine proteinase. Northern analysis showed that the induction of this enzyme is due to an increase in mRNA level (Geremia et al. 1993). In a further study, Goldman (1992) constructed transformant strains of *T. harzianum* that overexpress this alkaline protease.

This was accomplished by increasing its gene copy number by contransforming *T. harzianum* with the alkaline protease gene (prb1) and a plasmid containing a selectable marker for hygromycin B resistance. Extracellular alkaline protease activity from culture filtrates showed that some transformants produce higher alkaline proteinase activity than the wild type. One of these transformant strains produced an elevated prb1 mRNA level during mycoparasitic interactions with *R. solani* (Goldman 1992).

The major advantage of such genetic manipulations is the ability to isolate genes from one strain and introduce them into other varieties of fungi or bacteria. This enhances the potency of biocontrol agents and makes a single strain effective, stable, and consistent against more than one plant pathogenic fungus, without the hazardous effects of chemical pesticides.

This approach was taken by Broglie et al. (1991), who, in a pioneering work, produced

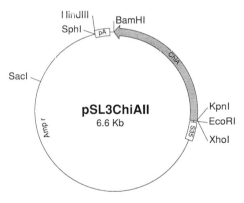

Fig. 3. Plasmid pSL3ChiAll, encoding chitinase (*chi*A) from *S. marcescens*, regulated by the constitutive 35S-CaMV promoter and terminator. (Haran et al. 1993)

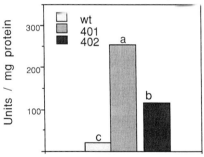

Chitinase specific activity
(Synthetic medium)

Fig. 4. Chitinase-specific activity of crude enzyme (units/mg protein) excreted by wild-type *T. harzianum* (*wt*) and transformants (*401* and *402*), after 5 days on synthetic medium. Columns *headed by different letters* are significantly different (*p* = 0.05) according to Duncan's multiple range test. (Haran et al. 1993)

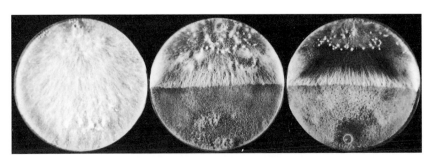

Fig. 5. From *left to right Sclerotium rolfsii*; dual cultures of *T. harzianum* wt and transformant 401, respectively (both *lower part*), with *S. rolfsii* (*upper part*), 7 days after contact. (Haran et al. 1993)

transgenic tobacco seedlings constitutively expressing a bean chitinase gene under the control of the cauliflower mosaic virus 35S promoter. The timing of the natural host defense mechanism was modified to produce fungus-resistant plants with increased ability to survive in soil infested with the fungal pathogen *R. solani* and delayed the development of disease symptoms (Broglie et al. 1991).

V. Conclusions

Mycoparasitism is a quite common and yet exciting phenomenon. It appears to play an important role in biological control, even though it should be pointed out that mycoparasitism is only one specific case in the whole complex system of parasitism.

Mycoparasitism is a complex process which includes the following steps: (1) chemotrophic growth of the antagonist towards the host; (2) recognition of the host by the mycoparasite; (3) attachment; (4) excretion of extracellular enzymes; and (5) lysis and exploitation of the host.

Mycoparasitism occurs under appropriate ecological conditions. The population and activity of the mycoparasite can be increased by relatively specific substances, such as chitin. The gene coding for chitinase is only one example of genes with mycoparasitic activity. Other potential genes are those coding for β-1-3 glucanase, protease, and lipase.

Chitinase activity was studied in both engineered microorganisms and transgenic plants. Results revealed the efficacy of the cloned chitinase in different systems, even though it did not provide a complete protection against the pathogen. Even in cases of partial protection, the delay in disease incidence may help growing seedlings overcome the relatively short period of susceptibility to damping-off diseases. Moreover, engineering various chitinases together with other genes that may act as antifungal agents may lead to better protection of plants against pathogenic fungi. It may therefore be possible to improve mycoparasitism and to enhance the plants' resistance response by integrating cloned chitinase with different lytic enzymes and other available antifungal polypeptides.

Acknowledgments. We would like to thank the Chais Family Fund for their kind support of this research.

References

Abo-Foul S, Raskin VI, Sztejnberg A, Marder JB (1996) Disruption of chlorophyll organization and function in powdery mildew-diseased cucumber leaves and its control by the hyperparasite *Ampelomyces quisqualis*. Phytopathology 86:195–199

Adams PB (1986) Effect of soil temperature, moisture, and depth on survival and activity of *Sclerotinia minor*, *Sclerotium cepivorum* and *Sporidesmium sclerotivorum*. Plant Dis 71:170

Adams PB (1990) The potential of mycoparasites for biological control of plant diseases. Annu Rev Phytopathol 28:59–72

Adams PB, Ayers WA (1980) Factors affecting parasitic activity of *Sporidesmium sclerotivorum* on sclerotia of *Sclerotinia minor* in soil. Phytopathology 70:366–368

Adams PB, Ayers WA (1982) Biological control of *Sclerotinia* lettuce drop in the field by *Sporidesmium sclerotivorum*. Phytopathology 72:485–488

Adams PB, Marois JJ, Ayers WA (1984) Population dynamics of the mycoparasite, *Sporidesmium sclerotivorum*, and its host, *Sclerotinia minor* in soil. Soil Biol Biochem 16:627–633

Ahmad JS, Baker R (1987) Rhizosphere competence of *Trichoderma harzianum*. Phytopathology 77:182–189

Ahmed AHM, Tribe HT (1977) Biological control of white rot of onion (*Sclerotium cepivorum*) by *Coniothyrium minitans*. Plant Pathol 26:75–78

Ayers WA, Adams PB (1981) Mycoparasitism and its application to biological control of plant diseases. In: Papavizas GC (ed) Biological control in crop protection, vol 5. Allanheld, Osmun Totowa, New Jersey, pp 91–105

Ayers WA, Barnett EA, Adams PB (1981) Germination of macroconidia and growth of *Sporidesmium sclerotivorum* in vitro. Can J Microbiol 27:664–669

Baker KF, Cook RJ (1974) Biological control of plant pathogens. The American Phytopathological Society, St Paul, Minnesota

Baker R (1986) Biological control: an overview. Can J Plant Pathol 8:218–221

Baker R (1987) Mycoparasitism: ecology and physiology. Can J Plant Pathol 9:370–379

Baker R, Griffin GJ (1995) Molecular strategies for biological control of fungal plant pathogens. In: Reuveni R (ed) Novel approaches to integrated pest management. Lewis, CRC Press, Boca Raton, 369pp

Barak R, Chet I (1990) Lectin of *Sclerotium rolfsii*: its purification and possible function in fungal interaction. J Appl Bacteriol 69:101–112

Barak R, Elad Y, Mirelman D, Chet I (1985) Lectins: a possible basis for specific recognition in *Trichoderma-Sclerotium rolfsii* interaction. Phytopathology 75:458–462

Barak R, Elad Y, Chet I (1986) The properties of L-fucose binding agglutinin associated with the cell wall of *Rhizoctonia solani*. Arch Microbiol 144:346–349

Barnett HL, Binder FL (1973) The fungal host-parasite relationship. Annu Rev Phytopathol 11:273–292

Barondes SH (1981) Lectins: their multiple endogenous cellular functions. Annu Rev Biochem 50:207–231

Benhamou N, Chet I (1993) Hyphal interactions between *Trichoderma harzianum* and *Rhizoctonia solani*: ultrastructure and cytochemistry of the antagonistic process. Phytopathology 83:1062–1071

Boller T (1985) Induction of hydrolases as a defence reaction against pathogens. In: Key JL, Kosuge T (eds) Cellular and molecular biology of plant stress. Alan R Liss, New York, 247pp

Boller T, Gehri A, Mauch F, Vogeli U (1983) Chitinase in bean leaves: induction by ethylene, purification, properties and possible function. Planta 157:22–31

Boosalis MG (1956) Effect of soil temperature and green-manure amendment of unsterilized soil on parasitism of *Rhizoctonia solani* by *Penicillium vermiculatum* and *Trichoderma* spp. Phytopathology 46:473–478

Broglie K, Chet I, Holliday M, Cressman R, Biddle P, Knowlton S, Mauvais CJ, Broglie R (1991) Transgenic plants with enhanced resistance to the fungal pathogen *Rhizoctonia solani*. Science 254:1194–1197

Bryan PW, McGowan JC (1945) Viridin: a highly fungistatic substance produced by *Trichoderma viride*. Nature 151:144–145

Bullock S, Adams PB, Willetts HJ, Ayers WA (1986) Production of haustoria by *Sporidesmium sclerotivorum* in sclerotia of *Sclerotinia minor*. Phytopathology 76:101–103

Caruso FL, Kuc J (1977) Protection of watermelon and muskmelon against *Colletotrichum lagenarium* by *Colletotrichum lagenarium*. Phytopathology 67:1290–1292

Chet I (1987) *Trichoderma* – application, mode of action, and potential as biocontrol agent of soilborne plant pathogenic fungi. In: Chet I (ed) Innovative approaches to plant disease control. John Wiley, New York, p 137

Chet I (1990) Mycoparasitism – recognition, physiology and ecology. In: Baker R, Dunn P (eds) New directions in biological control: alternatives for suppressing agricultural pests and diseases. Alan R Liss, New York, 725pp

Chet I, Baker R (1981) Isolation and biocontrol potential of *Trichoderma hamatum* from soil naturally suppressive to *Rhizoctonia solani*. Phytopathology 71:286–290

Chet I, Harman GE, Baker R (1981) *Trichoderma hamatum*: its hyphal interactions with *Rhizoctonia solani* and *Pythium* spp. Microb Ecol 7:29–38

Chet I, Barak Z, Oppenheim A (1993) Genetic engineering of microorganisms for improved biocontrol activity. In: Chet I (ed) Biotechnological perspectives in plant pathogen control. John Wiley, New York, 372pp

Claydon N, Allan M, Hanson JR, Avent AG (1987) Antifungal alkyl pyrons of *Trichoderma harzianum*. Trans Br Mycol Soc 88:503–513

Cook RJ, Baker KF (1983) The nature and practice of biological control of plant pathogens. The American Phytopathological Society, St Paul, Minnesota

Costa AS, Muller GW (1980) Tristeza control by cross protection: a U.S.-Brazil cooperative success. Plant Dis 64:538–541

Deacon JW (1976) Studies on *Pythium oligandrum*, an aggressive parasite of other fungi. Trans Br Mycol Soc 66:383–391

Deacon JW (1991) Significance of ecology in the development of biocontrol agents against soil-borne plant pathogens. Biocontol Sci Technol 1:5–20

Dennis C, Webster J (1971a) Antagonistic properties of species-groups of *Trichoderma*. I. Production of non-volatile antibiotics. Trans Br Mycol Soc 57:25–39

Dennis C, Webster J (1971b) Antagonistic properties of species-groups of *Trichoderma*. II. Production of volatile antibiotics. Trans Br Mycol Soc 57:41–48

Dennis C, Webster J (1971c) Antagonistic properties of species-groups of *Trichoderma*. III. Hyphal interaction. Trans Br Mycol Soc 57:363–369

Elad Y, Chet I (1995) Practical approaches for biocontrol agents implementation. In: Reuveni R (ed) Novel approaches to integrated pest management. Lewis, CRC Press, Boca Raton, 369pp

Elad Y, Chet I, Katan Y (1980) *Trichoderma harzianum*: a biocontrol agent effective against *Sclerotium rolfsii* and *Rhizoctonia solani*. Phytopathology 70:119–121

Elad Y, Chet I, Henis Y (1982) Degradation of plant pathogenic fungi by *Trichoderma harzianum*. Can J Microbiol 28:719–725

Elad Y, Barak R, Chet I (1983a) Possible role of lectins in mycoparasitism. J Bacteriol 154:1431–1435

Elad Y, Barak R, Chet I, Henis Y (1983b) Ultrastructural studies of the interaction between *Trichoderma* spp. and plant pathogenic fungi. Phytopathol Z 107:168–175

Elad Y, Chet I, Boyle P, Henis Y (1983c) Parasitism of *Trichoderma* spp. on *Rhizoctonia solani* and *Sclerotium rolfsii* – scanning electron microscopy and fluorescence microscopy. Phytopathology 73:85–88

Elad Y, Lifshitz R, Baker R (1985) Enzymatic activity of the mycoparasite *Pythium nunn* during interaction with host and non-host fungi. Physiol Plant Pathol 27:131

Elad Y, Zvieli Y, Chet I (1986) Biological control of *Macrophomina phaseolina* (Tassi) by *Trichoderma harzianum*. Crop Protect 5:288–292

Erbeznik M, Matavulj M, Stohlkovic S (1986) The effect of indole-3-acetic acid on cellulase production by *Gliocladium virens* C2-R1. Mikrobiologiya 23:39–47

Fahima T, Henis Y (1990) Interaction between pathogen, host and biocontrol agent: multiplication of *Trichoderma hamatum* and *Talaromyces flavus* on roots of diseased and healthy hosts. In: Hornby D (ed) Biological control of soil-borne plant pathogens. CAB International, Wallingford, p 165

Fahima T, Madi L, Henis Y (1992) Ultrastructure and germinability of *Verticillium dahliae* microsclerotia parsitized by *Talaromyces flavus* on agar medium and in treated soil. Biocontrol Sci Technol 2:69–78

Fletcher JT, Row JM (1975) Observations and experiments on the use of an avirulent mutant strain of tobacco mosaic virus as a means of controlling tomato mosaic. Ann Appl Biol 81:171–179

Fravel DR, Keinath AP (1991) Biocontrol of soilborne plant pathogens with fungi. In: Keister DL, Cregan PB (eds) The rhizosphere and plant growth. Kluwer, Dordrecht, p 237

Freeman S, Sztejnberg A, Chet, I (1986) Evaluation of *Trichoderma* as a biocontrol agent for *Rosellinia necatrix*. Plant Soil 94:163–170

Garrett SD (1965) Toward biological control of soil-borne plant pathogens. In: Baker KF, Synder WC (eds) Ecology of soil-borne plant pathogens. University of California Press, Berkeley, p 4

Geremia RA, Goldman GH, Jacobs D, Ardiles W, Vila SB, Van Montagu M, Herrera-Estrella A (1993) Molecular characterization of the proteinase-encoding gene, *prb* 1, related to mycoparasitism by *Trichoderma harzianum*. Mol Microbiol 8:603–613

Gnisalberti EL, Sivasithamparam K (1991) Antifungal antibiotics produced by *Trichoderma* spp. Soil Biol Biochem 23:1011–1020

Goldman GH (1992) Molecular genetic studies of mycoparasitism by *Trichoderma* spp. PhD Thesis, Ghent University, Belgium, 108 pp

Goldstein IJ, Hughes RC, Monsigny M, Osawa T, Sharon N (1980) What should be called a lectin? Nature (Lond) 285:66

Gorodecki B, Hadar Y (1990) Suppression of *Rhizoctonia solani* and *Sclerotium rolfsii* in container media containing composted separated cattle manure and composted grape marc. Crop Prot 9:271

Hadar Y, Gorodecki B (1991) Suppression of germination of sclerotia of Sclerotium rolfsii in compost. Soil Biol Biochem 23:303–306

Hadar Y, Chet I, Henis Y (1979) Biological control of Rhizoctonia solani damping-off with a wheat bran culture of Trichoderma harzianum. Phytopathology 69:1167–1172

Handelsman J, Parke JL (1989) Mechanism in biocontrol of soilborne plant pathogens. In: Kosuge T, Nester EW (eds) Plant microbe interactions, vol 3. McGraw-Hill, New York, pp 27–61

Haran S, Schickler H, Pe'er S, Logemann S, Oppenheim A, Chet I (1993) Increased constitutive chitinase activity in transformed Trichoderma harzianum. Biol Control 3:101–108

Harman GE, Lumsden RD (1990) biological disease control. In: Lynch JM (ed) The rhizosphere. Wiley, New York, p 259

Harman GE, Chet I, Baker R (1980) Trichoderma hamatum effects on seed and seedling disease induced in radish and pea by Pythium spp. or Rhizoctonia solani. Phytopathology 70:1167–1172

Hashioka Y, Nakai Y (1980) Ultrastructure of pycnidial development and mycoparasitism of Ampelomyces quisqualis parasitic on Erysiphales. Trans Mycol Soc Jpn 21:329–338

Henis Y, Ghaffar A, Baker R (1978) Integrated control of Rhizoctonia solani damping-off of radish: effect of successive plantings, PCNB, and Trichoderma harzianum on pathogen and disease. Phytopathology 68:900–907

Henis Y, Ghaffar A, Baker R (1979) Factors affecting suppressiveness to Rhizocotonia solani in soil. Phytopathology 69:1164–1169

Hoitink HAJ, Fahy PC (1986) Basis for the control of soilborne plant pathogens with composts. Annu Rev Phytopathol 24:93–114

Hoitink HAJ, Boehm MJ, Hadar Y (1993) Mechanisms of suppression of soilborne plant pathogens in compost-amended substrates. In: Hoitink HAJ, Keener HM (eds) Science and engineering of composting: design, environmental, microbiological and utilization aspects. Renaissance Publications, Worthington, Ohio, pp 601–621

Howell CR (1982) Effect of Gliocladium virens on Pythium ultimum, Rhizoctonia solani and damping-off of cotton seedling. Phytopathology 72:496–498

Howell CR (1987) Relevance of mycoparasitism in the biological control of Rhizoctonia solani by Gliocladium virens. Phytopathology 77:992–994

Howell CR (1991) Biological control of Pythium damping-off of cotton with seed coating preparation of Gliocladium virens. Phytopathology 81:738–741

Howell CR, Stipanovic RD (1983) Gliovirin, a new antibiotic from Gliocladium virens and its role in the biological control of Pythium ultimum. Can J Microbiol 29:321–324

Howell CR, Stipanovic RD, Lumsden RD (1993) Antibiotic production by strains of Gliocladium virens and its relation to the biocontrol of cotton seedling diseases. Biocontrol Sci Technol 3:435–441

Huang HC (1977) Importance of Coniothyrium minitans in survival of sclerotia of Sclerotinia sclerotiorum in wilted sunflower. Can J Bot 55:289–295

Huang HC (1978) Gliocladium catenulatum: hyperparasite of Sclerotinia sclerotiorum and Fusarium species. Can J Bot 56:2243–2246

Huang HC (1980) Control of Sclerotinia wilt of sunflower by hyperparasites. Can J Plant Pathol 2:26–32

Huang HC, Hoes JA (1976) Penetration and infection of Sclerotinia sclerotiorum by Coniothyrium minitans. Can J Bot 54:406–410

Huang HC, Kokko EG (1987) Ultrastructure of hyperparasitism of Coniothyrium minitans on sclerotia of Sclerotinia sclerotiorum. Can J Bot 65:2483–2489

Huang HC, Kokko EG (1988) Penetration of hyphae of Sclerotinia sclerotiorum by Coniothyrium minitans without the formation of appressoria. J Phytopathol 123:133–139

Inbar J, Chet I (1991) Evidence that chitinase produced by Aeromonas caviae is involved in the biological control of soil plant pathogens by this bacterium. Soil Biol Biochem 23:973–978

Inbar J, Chet I (1992) Biomimics of fungal cell-cell recognition by use of lectin-coated nylon fibers. J Bacteriol 174:1055–1059

Inbar J, Chet I (1994) A newly isolated lectin from the plant pathogenic fungus Sclerotium rolfsii: purification, characterization and its role in mycoparasitism. Microbiology 140:651–657

Inbar J, Chet I (1995) The role of recognition in the induction of specific chitinases during mycoparasitism by Trichoderma harzianum. Microbiology 141:2823–2829

Jarvis WR, Slingsby K (1977) The control of powdery mildew of greenhouse cucumber by water sprays and Ampelomyces quisqualis. Plant Dis Rep 61:726–730

Jeffries P, Young TWK (1994) Interfungal parasitic relationships. CAB International, Wallingford

Jones D, Watson D (1969) Parasitism and lysis by soil fungi of Sclerotinia sclerotiorum (Lib.) de Bary, a phytopathogenic fungus. Nature (Lond) 224:287–288

Kellens JTC, Goldstein IJ, Peumans WJ (1992) Lectins in different members of the Sclerotiniaceae. Mycol Res 96:495–502

Kritzman G, Chet I, Henis Y, Huttermann A (1978) The use of the brightener Calcofluor White M2R New in the study of fungal growth. Isr J Bot 27:138–146

Kwok OCH, Fahy PC, Hoitink HAJ, Kuter FA (1987) Interaction between bacteria and Trichoderma hamatum in suppression of Rhizoctonia damping-off in bark compost media. Phytopathology 77:1206–1212

Lewis JA, Roberts DP, Hollenbeck MD (1991) Induction of cytoplasmic leakage from Rhizoctionia solani hyphae by Gliocladium virens and partial characterization of a leakage factor. Biocontrol Sci Technol 1:21–29

Lewis K, Whipps JM, Cooke RC (1989) Mechanisms of biological disease control with special reference to the case study of Pythium oligandrum as an antagonist. In: Whipps JM, Lumsden RD (eds) Biotechnology of fungi for improving plant growth. Cambridge University Press, Cambridge, 191pp

Lifshitz R, Dupler M, Elad Y, Baker R (1984a) Hyphal interactions between Pythium nunn and several soil fungi. Can J Microbiol 30:1482–1487

Lifshitz R, Stanghellini ME, Baker R (1984b) A new species of Pythium isolated from soil in Colorado. Mycotaxon 20:373–379

Liu SY, Baker R (1980) Mechanism of biological control in soil suppressive to Rhizoctonia solani. Phytopathology 70:404–412

Lynch JM (1987) In vitro identification of Trichoderma harzianum as a potential antagonist of plant pathogens. Curr Microbiol 16:49

Manocha MS (1981) Host specificity and mechanism of resistance in a mycoparasitic system. Physiol Plant Pathol 18:257–267

Manocha MS (1984) Cell surface characteristics of *Mortierella* species and their interaction with a mycoparasite. Can J Microbiol 30:290–298

Manocha MS (1985) Specificity of mycoparasite attachment to the host cell surface. Can J Bot 63:772–778

Manocha MS (1987) Cellular and molecular aspects of fungal host-mycoparasite interaction. J Plant Dis Prot 94:431–444

Manocha MS (1990) Cell-cell interaction in fungi. J Plant Dis Prot 97:655–669

Manocha MS (1991) Physiology and biochemistry of biotrophic mycoparasitism. In: Arora Dk, Rai B, Mukerji KG, Knudsen GR (eds) Handbook of applied mycology, vol 1: soil and plants. Marcel Dekker, New York, pp 273–300

Manocha MS, Chen Y (1990) Specificity of attachment of fungal parasites to their hosts. Can J Microbiol 36:69–76

Manocha MS, Chen Y (1991) Isolation and partial characterization of host cell surface agglutinin and its role in attachment of a biotrophic mycoparasite. Can J Microbiol 37:377–383

Manocha MS, Sahai AS (1993) Mechanisms of recognition in necrotrophic and biotrophic mycoparasites. Can J Microbiol 39:269–275

Manocha MS, Balasubramanian R, Enskat (1986) Attachment of a mycoparasite with host but not with nonhost *Mortierella* species. In: Bailey J (ed) Biology and molecular biology of plant pathogen interactions. Springer, Berlin Heidelberg New York, p 59

Manocha MS, Chen Y, Rao N (1990) Involvement of cell surface sugars in recognition attachment and appressorium formation by a mycoparasite. Can J Microbiol 36:771–778

McLaren DL, Rimmer SR, Huang HC (1983) Biological control of *Sclerotinia* wilt of sunflowers by *Talaromyces flavus*. Phytopathology 73:822

McLaren DL, Huang HC, Rimmer SR (1986) Hyperparasitism of *Sclerotinia sclerotiorum* by *Talaromyces flavus*. Can J Plant Pathol 8:43–48

Nelson EB, Kuter GA, Hoitink HAJ (1983) Effect of fungal antagonists and compost age on suppression of *Rhizoctonia* damping-off in container media amended with composted hardwood bark. Phytopathology 73:1457–1462

Ordentlich A, Elad Y, Chet I (1988) The role of chitinase of *Serratia marcescens* in biocontrol of *Sclerotium rolfsii*. Phytopathology 78:84–88

Ordentlich A, Nachmias A, Chet I (1990) Integrated control of *Verticillium dahliae* in potato by *Trichoderma harzianum* and captan. Crop Prot 9:363–366

Ordentlich A, Migheli Q, Chet I (1991) biological control activity of three *Trichoderma* isolates against *Fusarium* wilts of cotton and muskmelon and lack of correlation with their lytic activity. J Phytopathol 133:177–186

Ordentlich A, Weisman Z, Gottlieb HE, Cojocaru M, Chet I (1992) New inhibitory natural product produced by the biocontrol agent *Trichoderma harzianum*. Phytochemistry 31: 485–486

Pachenari A, Dix NJ (1980) Production of toxins and wall-degrading enzymes by *Gliocladium roseum*. Trans Br Mycol Soc 74:561–566

Papavizas GC (1985) *Trichoderma* and *Gliocladium*: biology, ecology and the potential for biocontrol. Annu Rev Phytopathol 23:23–54

Paulitz TC, Baker R (1987) Biological control of *Pythium* damping-off of cucumber with *Pythium nunn*: influence of soil environment and organic amendments. Phytopathology 77:341–346

Paulitz TC, Ahmad JS, Baker R (1990) Integration of *Pythium nunn* and *Trichoderma harzianum* isolate T-95 for the biological control of *Pythium* damping-off of cucumber. Plant Soil 121: 243–250

Philipp WD, Grauer U, Grossmann F (1984) Ergänzende Untersuchungen zur biologischen und integrierten Bekämpfung von Gurkenmehltau unter Glas durch *Ampelomyces quisqualis*. Z Pflanzenkr Pflanzenschutz 91:438–443

Phillips AJL (1986) Factors affecting the parasitic activity of *Gliocladium virens* on the sclerotia of *Sclerotinia sclerotiorum* and a note on its host range. J Phytopathol 116:212–220

Phillips AJL, Price K (1983) Structural aspects of the parasitism of sclerotia of *Sclerotinia sclerotiorum* (Lib.) de Bary by *Coniothyrium minitans* Campb. Phytopathol Z 107:193–203

Ridout CJ, Coley Smith JR, Lynch JM (1988) Fractionation of extracellular enzymes from a mycoparasitic strain of *Trichoderma harzianum*. Enzyme Microb Technol 10:180–187

Roberts DP, Lumsden RD (1990) Effect of extracellular metabolites from *Gliocladium virens* on germination of sporangia and mycelial growth of *Pythium ultimum*. Phytopathology 80:461–465

Sahai AS, Manocha MS (1993) Chitinases of fungi and plants: their involvement in morphogenosis and host-parasite interaction. FEMS Microbiol Rev 11:317–338

Sequeira L, Hill LM (1974) Induced resistance in tobacco leaves: the growth of *Pseudomonas solanacearum* in protected tissues. Physiol Plant Pathol 4:447–455

Shapira R, Ordentlich A, Chet I, Oppenheim AB (1989) Control of plant diseases by chitinase expressed from cloned DNA in *Escherichia coli*. Phytopathology 79:1246–1249

Simon A, Sivasithamparam K (1990) Effect of crop rotation, nitrogenous fertilizer and lime on biological suppression of the take-all fungus. In: Hornby D (ed) Biological control of soil-borne plant pathogens. CAB International, Wallingford, pp 215–226

Sivan A, Chet I (1986) Biological control of *Fusarium* spp. in cotton, wheat and muskmelon by *Trichoderma harzianum*. Phytopathol Z 116:39–47

Sivan A, Chet I (1989a) Degradation of fungal cell walls by lytic enzymes of *Trichoderma harzianum*. J Gen Microbiol 135:675–682

Sivan A, Chet I (1989b) The possible role of competition between *Trichoderma harzianum* and *Fusarium Oxysporum* on rhizosphere colonization. Phytopathology 79:198–203

Sivan A, Chet I (1992) Microbial control of plant diseases. In: Mitchell R (ed) New concepts in environmental microbiology. Wiley-Liss, New York, 335pp

Sivan A, Elad Y, Chet I (1984) Biological control effects of a new isolate of *Trichoderma harzianum* on *Pythium aphanidermatum*. Phytopathology 74:498

Sivan A, Ucko O, Chet I (1987) Biological control of *Fusarium* crown rot of tomato by *Trichoderma harzianum* under field conditions. Plant Dis 71:587–592

Strashnov Y, Elad Y, Sivan A, Chet I (1985a) Integrated control of *Rhizoctonia solani* Kuhn on bean and carrot seedlings by methyl bromide and *Trichoderma harzianum* Rifai Aggr. Plant Pathol 34:146–151

Strashnov Y, Elad Y, Sivan A, Ruich Y, Chet I (1985b) Control of *Rhizoctonia solani* fruit rot of tomatoes by *Trichoderma harzianum* Rifai. Crop Prot 4:359–364

Su SJ, Leu LS (1980) Three parasitic fungi on *Sclerotinia sclerotiorum* (Lib.) de Bary. Plant Prot Bull 22:253–262

Sundheim L (1982) Control of cucumber powdery mildew by the hyperparasite *Ampelomyces quisqualis* and fungicides. Plant Pathol 31:209–214

Sundheim L, Krekling T (1982) Host-parasite relationships of the hyperparasite *Ampelomcyes quisqualis* and its powdery mildew host *Sphaerotheca fuliginea*. Phytopathol Z 104:202–210

Sztejnberg A, Abo-Foul S (1990) The hyperparasite *Ampelomyces quisqualis* increases yield and photosynthesis of powdery mildew-infected cucumber and zucchini. APS/CPS Annu Meet, Grand Rapids, Michigan, August, 1990

Sztejnberg A, Mazar S (1985) Biocontrol of cucumber and carrot powdry mildew by *Ampelomyces quisqualis*. Phytopathology 75:1301–1302

Sztejnberg A, Freeman S, Chet I, Katan J (1987) Control of *Rosellinia necatrix* in soil and apple orchard by solarization and *Trichoderma harzianum*. Plant Dis 71:365–369

Sztejnberg A, Galper S, Mazar S, Lisker N (1989) *Ampelomyces quisqualis* for biological and integrated control of powdery mildews in Israel. J Phytopathol 124:285–295

Sztejnberg A, Galper S, Lisker N (1990) Conditions for pycnidial production and spore formation by *Ampelomyces quisqualis*. Can J Microbiol 36:193–198

Trutmann P, Keane PJ, Merriman PR (1982) Biological control of *Sclerotinia sclerotiorum* on aerial parts of plants by the hyperparasite *Coniothyrium minitans*. Trans Br Mycol Soc 78:521–529

Tu JC (1984) Mycoparasitism by *Coniothyrium minitans* and its effects on sclerotia germination. Phytopathol Z 109:261–268

Tunlid A, Jansson H-B, Nordbring-Hertz B (1992) Fungal attachment to nematodes. Mycol Res 96:401–412

Uecker FA, Ayers WA, Adams PB (1980) *Teratosperma gliocladum* – a new hyphomycetous mycoprasite on sclerotia of *Sclerotinia sclerotiorum*, *S. trifolium* and *S. minor*. Mycotaxon 10:421–427

Webster J, Lomas N (1964) Does *Trichoderma viride* produce gliotoxin and viridin? Trans Br Mycol Soc 47:535–540

Weindling R (1932) *Trichoderma lignorum* as a parasite of other fungi. Phytopathology 22:837

Wells HD, Bell DK, Jaworski CA (1972) Efficacy of *Trichoderma harzianum* as a biocontrol agent for *Sclerotium rolfsii*. Phytopathology 62:442

Whipps JM (1991) Effects of mycoparasites on sclerotia-forming fungi. In: Beemster ABR, Bolen GJ, Berlagh M, Ruissen MA, Schippers B, Tempel A (eds) Biotic interactions and soil-borne diseases. Elsevier, Amsterdam, p 129

Whipps JM (1992) Status of biological disease control in horticulture. Biocontrol Sci Technol 2:3–24

Wright JM (1956) The production of antibiotics in soil. III. Production of gliotoxin in wheatstraw buried in soil. Ann Appl Biol 44:461

12 Entomopathogenic Fungi and Their Role in Pest Control

A.K. Charnley[1]

CONTENTS

I. Introduction

Synthetic chemical pesticides have been the mainstay of insect pest control for the past 50 years. The advent of insecticide resistance, pest resurgence and concern over the environmental impact of agricultural inputs are increasingly focussing attention on biologically based forms of pest control. The impact on insect populations of natural epizootics caused in particular by fungal and viral pathogens demonstrate the potential of microbial pest control. This fact was recognised in the latter part of the last century and culminated in the seminal attempts by Metchnikoff and Paliokov

to use the deuteromycete fungal pathogen *Metarhizium anisopliae* for pest control (in Gillespie 1988). This chapter outlines the current state of knowledge of insect fungal pathogens as it relates to their present use and future potential in crop protection.

II. Taxonomy

Relationships between fungi and insects may be mutualistic, through commensal to obligately pathogenic. The term entomogenous is often used to encompass all types of association between insects and fungi, with disease-causing fungi being referred to as entomopathogenic. A further distinction can be made between fungi which are aggressively pathogenic like *Metarhizium anisopliae* and opportunists like the wound pathogen *Mucor haemalis* (Samson et al. 1988; McCoy et al. 1988; Tanada and Kaya 1993).

Entomopathogenic fungi are to be found in most taxonomic groupings in the fungal kingdom, apart from the higher Basidiomycotina. The primitive water fungi, Mastigomycotina, have representatives with complex life cycles, e.g. *Coelomomyces psorophorae*, a mosquito pathogen with an obligate copepod secondary host. Among the Ascomycotina, *Cordyceps* spp. have fruiting structures or perithecia which can dwarf the cadavers of their insect victims. Entomophthorales are widespread members of the Zygomycotina. Mummified aphids stricken by fungi of this group are familiar features of cereal crops in temperate regions. The most widespread insect pathogenic fungi are found in the hyphomycetous Deuteromycotina. *Beauveria bassiana* and *Metarhizium anisopliae* have broad host ranges, though considerable specificity occurs among isolates.

The fungi described above are all destructively pathogenic. Laboulbeniomycetes

---

[1] School of Biology and Biochemistry University of Bath, Claverton Down, Bath, BA2 7AK, UK

The Mycota IV
Environmental and Microbial Relationships
Wicklow/Söderström (Eds.)
© Springer-Verlag Berlin Heidelberg 1997

(Ascomycotina), on the other hand, are bio-trophic. They remain largely external to their hosts, gaining nutrition via a penetrant haustoria while apparently causing little harm. Most Tricho-mycetes (Zygomycotina) have a commensal existence in the guts of their dipteran hosts.

III. Infection Process

Unique among entomopathogens, insect patho-genic fungi do not have to be ingested and can invade their hosts directly through the exoskel-eton or cuticle. Therefore they can infect non-feeding stages such as eggs and pupae. The site of invasion is often between the mouthparts, at intersegmental folds or through spiracles, where locally high humidity promotes germination and the cuticle is non-sclerotised and more easily pen-etrated (Charnley 1984, 1989; Hajek and St Leger 1994).

A. Fungal Invasion of the Host

M. anisopliae and *B. bassiana* have hydrophobic spores which appear to bind to insect cuticle by non-specific interactions, though failure to adhere to particular insect species may help to define isolate host range. Zoospores of *Lagenidium giganteum* are host-selective. Cuticle-degrading enzymes are present on the surface of conidia of *M. anisopliae* and therefore there is the potential for the fungus to modify the cuticle surface to aid attachment. Host and fungal lectins have been implicated also in the process of attachment. Ger-mination in vitro of nutrient-dependent spores of *M. anisopliae* and *B. bassiana* is consequent upon a non-specific accessible source of carbon and ni-trogen, though in vivo isolate specificity may de-pend on response to qualitative and quantitative differences in available nutrients on host cuticle. More selective pathogens appear to have more specific requirements. Ability to withstand anti-fungal compounds such as short chain fatty acids in the cuticle is a prerequisite for successful inva-sion (see Boucias and Pendland 1991). The impor-tance of signal exchange between host and pathogen is becoming increasingly clear and is first seen in the cues which cause the fungus to stop horizontal growth on the surface of the cuticle and initiate penetration. Differentiation of the germ

tube to produce the holdfast structure, or appres-sorium, is most completely understood for *M. anisopliae*. Isolate ME1 requires low concentra-tions of a complex carbon and nitrogen source and a hard surface. *Metarhizium* isolates which have come from Homoptera form appressoria in media (high concentrations of simple sugars) which are repressive for isolates from Coleoptera. This is probably an adaptation to parasitism, as the cu-ticle of plant-sucking bugs (Homoptera) is con-taminated with sugars from their copious excreta (St Leger et al. 1992a).

Numerous light and electron microscope studies on the invasion of host cuticle by entomopathogenic fungi are consistent with the involvement of both enzymes and mechanical pressure. Insect cuticle comprises between 60 and 70% protein, therefore it is perhaps not surprising that recent work has implicated proteases, in par-ticular the subtilisin-like (chymoelastase) PR1, in the penetration process (for reviews see Charnley and St Leger 1991; St Leger 1995).

Once the fungus breaks through the cuticle and underlying epidermis, it may grow profusely in the blood, in which case death is probably the result of starvation or physiological/biochemical disruption brought about by the fungus. Alterna-tively, insecticidal secondary metabolites may contribute to the demise of the insect and, in this case, extensive growth of the fungus may occur only on the cadaver of the host (Roberts 1980; Gillespie and Claydon 1989). For many fungi the reality is probably somewhere between these two extremes. Few studies have looked at the effect of fungal infection on host physiology/behaviour. This is unfortunate, because sublethal or prelethal effects of mycosis may be just as useful as the death of the host from the point of view of crop protection. Detrimental effects of mycosis on food consumption, egg laying and flight behaviour have been recorded (Nnakumusana 1985; Seyoum et al. 1995).

The life cycle is completed when the fungus sporulates on the cadaver of the host. Under the right conditions, particularly high RH, the fungus will break out through the body wall of the insect, producing aerial spores. This may allow horizontal or vertical transmission of the disease within the insect population. Resting spores produced within the dead insect will enable the fungus to survive for long periods under adverse conditions (Samson et al. 1988). Additionally, conidia may be produced in internal air spaces as the cadaver dries out under low humidity (C. Prior, pers. comm.).

B. Host Response to Fungal Invasion

The cuticle is not only the first but the major barrier to host invasion. Structural features such as sclerotisation impede penetration, while enzyme inhibitors and tyrosinases, which generate antimicrobial melanins, are front line defences against weak pathogens. Blood-borne defences seem to have little impact on virulent fungal pathogens. Phagocytosis by individual blood cells and cooperative behaviour between haemocyte subpopulations, viz. encapsulation and granuloma formation, are often not recorded. This has been attributed to a failure of the insect's non-self recognition system, in some cases brought about by toxic fungal metabolites, in others due to the removal of immunogenic components from fungal cell walls (or even the walls themselves) in the blood of infected insects (Charnley 1989; Huxham et al. 1989; Mazet et al. 1994).

IV. Epizootology of Fungal Diseases in Insects

Although epizootics of insect fungal diseases are comparatively common, the study of population level interactions between entomopathogenic fungi and their hosts is in its relative infancy (Carruthers and Soper 1987). Natural disease development and spread are affected by the characteristics of the host and pathogen populations, the environment and the impact of human activities (particularly in agroecosystems). Properties of the pathogen population which are important include virulence, dispersal, and survival in the host's environment, inoculum density and spatial distribution. Host population factors which need to be considered are susceptibility, density, movement and spatial distribution. Abiotic environmental factors such as temperature, moisture and sunlight may determine whether infection can occur. Germination and sporulation are particularly dependent on moisture. Temperature may also be limiting for disease, particularly when short generation time for the host is favoured by a temperature which is above or below the optimum for the pathogen.

Some of the more detailed epizootological studies of insect pathogenic fungi come from agroecosystems examples and include *Nomuraea rileyi* infection of *Anticarsia gemmatalis* on soybean (Ignoffo 1981) and *Entomophthora muscae* on the onion fly *Delia antiqua* (Carruthers and Haynes 1986). Forests are more diverse stable habitats than agroecosystems, consequently the insect-pathogen interactions may be more complex. The spruce budworm *Choristoneura fumiferana* is a major defoliator in balsam and spruce trees in NE America. Among a range of pathogens which attack this insect, the fungi *Erynia radicans* and *Entomophaga aulicae* produce the highest mortality (Perry and Whitfield 1984). Disease incidence depends on insect age and position in the tree canopy; the role of abiotic factors is not completely understood. Humid, tropical forests have a rich, varied entomopathogenic mycoflora (Evans 1982), including in particular *Cordyceps* spp., and it has been suggested that these fungi have a significant role in the regulation of insect populations because of the stable microclimates in such habitats.

Rangelands are more stable than agricultural systems but more uniform in structure than forest ecosystems. Grasshoppers are often the dominant phytophagous insects in such habitats. *Entomophaga grylli* mycoses cause high mortality among populations of *Camnula pellucida* and *Melanoplus bivittatus* in western North America (Pickford and Riegert 1963).

Soil is a complex habitat that harbours a large fauna and flora. *Metarhizium anisopliae* is one of the most frequent mycopathogens of soil insects in temperate regions, particularly of beetles (Keller and Zimmermann 1989). Epizootics have been found on wireworms (*Agriotes* spp.) and larvae of *Amphimallon solstitialis*. *M. anisopliae* cannot grow in normal, non-sterile soil. Therefore its ubiquity in temperate soils must reflect the broad host range of the species. In contrast, *Beauveria brongniartii* is primarily a pathogen of cockchafers, *Melolontha* spp., and other Scarabidae. Investigations of the population dynamics of cockchafers in eastern Switzerland showed the *B. brongniartii* is the main regulating factor (Zimmermann 1992). Soil can also function as a reservoir for fungi which generally infect insects on aerial parts of plants. It has been shown experimentally that spores of the lepidopteran pathogen *N. rileyi* adhere to leaves of plant seedlings as they germinate and emerge through the soil (Ignoffo et al. 1977). Fungi may persist in soil as mycelium within mummified cadavers, conidia, resting spores (e.g. Entomophthorales) or pseudosclerotia. Temperature, pH, water and organic content can affect fungal survival (Keller and Zimmermann 1989).

Aquatic ecosystems present different problems to entomopathogenic fungi. In comparison with terrestrial habitats, fluctuations in temperature and sunlight may be less important, whereas fluctuations in pH, salinity, current and dissolved solutes may affect pathogen persistence. Although some pathogens may infect aquatic insects using specialist-shaped spores or motile zoospores to aid host location, e.g. *Lagenidium giganteum*, others confine themselves to the aerial adults, thus avoiding the problems presented by the aquatic environment (Lacey and Undeen 1986).

V. Pest Control

A. Approaches to the Use of Insect Pathogenic Fungi for Pest Control

Several avenues have been taken for the use of insect pathogenic fungi in pest control. The most cost-effective is permanent introduction. This involves the establishment of a disease in a population where it is does not occur normally, in order to give long-term or permanent suppression of the pest. Although there is a history of this approach, there are comparatively few examples where detailed population studies have determined the effectiveness of introductions. Milner et al. (1982) introduced an Israeli isolate of *Zoophthora radicans* into alfalfa fields in Australia in an attempt to control an accidentally introduced pest, the spotted alfalfa aphid *Therioaphis trifolii maculata*. The fungus became established and spread from the release site in subsequent seasons, causing high percentage mortality among aphid populations. *Entomophaga maimaga* was introduced from Japan to NE North America in 1910–1911 to control the gypsy moth *Lymantria dispar*. The fungus did not appear to have become established. However, in 1989 and 1990, it caused extensive mortality among gypsy moth in ten states in the USA (Hajek et al. 1995). It is possible that it was present but unrecognised for many years, as symptoms of mycosed gypsy moths resemble those of moths infected with a baculovirus. Recently, resting spores of the fungus have been used successfully to introduce the fungus into new areas.

Promoting natural fungal epizootics by adopting appropriate cultural and crop protection practices is an alternative way of harnessing entomopathogenic fungi for pest control. A classic example is the development of early harvesting strategies for alfalfa to maximise development and spread of natural infections of *Zoophthora* sp. among alfalfa weevils (*Hypera postica*; Brown and Nordin 1986).

Epizootics of fungal pathogens on crop pests often occur too late to be of economic value. Application of an additional inoculum can accelerate the process. When this results in secondary spread of disease, the process is termed inoculative augmentation, otherwise the strategy is called inundative augmentation. Augmentation now also encompasses the application of a fungus in a situation where mycosis may not naturally occur and is referred to as the microbial insecticide or mycoinsecticide approach (Tanada and Kaya 1993). The development of mycoinsecticides has received the most attention in recent times and is focussed on here.

Mycoinsecticides may perform inconsistently, mainly due to unfavourable environmental conditions (see below). Therefore strategies have been developed to increase efficiency and accelerate kill by combining fungi with sub-/low lethal doses of chemical pesticides. This approach is based on the assumption that, weakened by another stressor, the insect will succumb more readily to mycosis. Combinations of *B. bassiana* and insecticides have proved effective for control of the Colorado beetle in Russia (Sikura and Sikura 1983). Similarly, in China, combinations of *B. bassiana* and certain insecticides are recommended for application against crop and forest pests (Feng et al. 1994). Most experimental studies with mixtures suggest additive effects of the ingredients. However, Joshi et al. (1992) established synergy between *Metarhizium* spp. and the acyl urea insecticide Teflubenzuron against the desert locust. The alternative to the use of a chemical stressor is to combine entomopathogenic fungi with other microbial pathogens (see Zimmermann 1994). Though there are few examples where this strategy has been tried, it is given credence by observations of mixed infection in field-collected insects. However, it seems unlikely that combinations will be used widely in the west, as this strategy complicates application procedures as well as increasing the costs of pest control and initial development (particularly registration).

B. Current Use

1. Status of Mycoinsecticides

Crop protection is still dominated by chemical pesticides. World insecticide sales in 1995 amounted to US $8.75 billion, of which only some 1.5–2% were biopesticides. Mycoinsecticides accounted for a small fraction of the biopesticide market, the lion's share being taken by products based on the toxicogenic bacterium *Bacillus thuringiensis*. However, the increasing cost of development of chemical pesticides, pest resistance and pressure to reduce chemical inputs into the environment have ensured a continuing interest in alternative forms of pest control including insect pathogenic fungi (Rodgers 1993; Lisansky and Coombs 1994).

Examples of current commercial scale production of mycoinsecticides are given in Table 1. *Beauveria bassiana*, produced by communes, is used on quite a large scale in China against a number of forest and crop pests. In the former USSR, *B. bassiana*, as Boverin, is employed for control of the Colorado beetle *Leptinotarsa decemlineata* and the codling moth *Laspeyresia pomonella*. *M. anisopliae* is produced and used commercially, e.g. as Metaquino, in Brazil for the control of spittle bugs, *Mahanarva posticata*, on sugar cane. Commercial production of *M. anisopliae* by cottage-style industries is also underway in Costa Rica, Columbia, Cuba and Venezuela (Jones, 1994). Betel is one of a number of fungal preparations used on a small scale for control of scarabid beetle larvae (white grubs) around the world. Betel is based on *B. bassiana*, but elsewhere *B. brongniartii* or *M. anisopliae* are employed (Table 1; Jackson and Glare 1992).

In North America and Western Europe, mycoinsecticides/mycoacaricides have not fared well. *Hirsutella thompsonii* was developed as Mycar by Abbott for the control of the citrus rust mite. However, it was a commercial failure; problems included the requirement for cold storage, lack of a suitable bioassay and inconsistent mite control in the field (probably due to inadequate

RH after application; McCoy 1986). In the UK, *Verticillium lecanii* isolates were commercialised by Microbial Resources for control of aphids (Vertalec) and whitefly (Mycotal) on protected ornamentals and vegetable crops. The products were discontinued due to the small market size and erratic performance. However, there are signs of an improvement in the commercial prospects for mycoinsecticides. The Dutch company Koppert have brought Mycotal and Vertalec back to the market using improved production methods and formulation and a number of small companies have new products on or close to the market (Table 1). The active interest of large chemical companies like Hoechst, Schering, Ciba Geigy, Bayer and Zeneca may help improve the status of mycoinsecticides, though the future may continue to lie with small companies that are more flexible and survive in smaller niche markets.

2. Constraints on Efficiency

A need for high humidity for disease initiation and spread has often been considered the major constraint on the use of fungi for insect control. The inclusion of moisture-retaining substances in aqueous formulations and the use of oil-based formulations (Section V.C.2) may help to overcome the requirement for germination of a high environmental RH. Ability of the disease to spread among a population is not necessarily an issue with an efficiently applied virulent mycoinsecticide. Development of mycosis is affected by extremes of temperature (<15 and >32 °C), but often the fungus is only delayed. In tropical regions more equitable night temperatures may provide an opportunity for a fungus to initiate disease. Fungi are sensitive to UV. Experimentally, it has been possibly to alleviate this problem with the addition of protectants to the formulation (Hunt et al. 1995). In practice, the ingredients may be too costly.

Mycoinsecticides should be compatible with other crop protection measures. Several studies have shown that fungicides, herbicides and insecticides can prevent germination and/or mycelial growth of entomopathogenic fungi in vitro (e.g. Moorhouse et al. 1992). However, pest control by fungi is often not affected by chemical pesticides, as long as there is a ca. 7-day gap between the two applications (Anderson and Roberts 1983; Moorhouse et al. 1992).

Table 1. Commercial scale production of mycoinsecticides

Product	Fungus	Pest insect	Producer
Mycotal	*Verticillium lecanii*	Whitefly and thrips	Koppert, Holland
Vertalec	"	Aphids	"
BIO 1020[a]	*Metarhizium anisopliae*	Vine weevil	Bayer, Germany
Biogreen	"	Scarab larvae on pasture	Bio-care Technology, Australia
Metaquino	"	Spittle bugs	Brazil[bc]
Bio-Path	"	Cockroaches	EcoScience, USA
Bio-Blast	"	Termites	Ecoscience, USA
Mycotrol GH	*Beauveria bassiana*	Grasshoppers, locusts	Mycotech, USA
Mycotrol WP	"	Whitefly, aphids, thrips	"
Naturalis-L	"	Cotton pests including bollworms	Troy Biosciences, USA
Conidia	"	Coffee berry borer	AgrEvo, Germany
Betel	"	Sugar cane white grub	NPP (Calliope), France
Ostrinil	"	Corn borer	"
Boverin	"	Colorado beetle	Russia[d]
Boverol	"	Colorado beetle	Czech Republic/Slovakia[e]
Boverosil	"	"	Czech Republic/Slovakia[e]
–	"	"	Poland[e]
–	"	Various forest and crop pests	China[cf]
Engerlingspilz	*B. brongniartii*	Cockchafer	Andermatt, Switzerland
PFR-97	*Paecilomyces fumosoroseus*	Whitefly	ECO-tek[g]
PreFeRal	"	Whitefly on glasshouse vegetables and ornamentals	Biobest, Belgium[h]

[a] Registered, but not on sale, other products under development.
[b] 600000ha of sugarcane treated annually (Jones 1994).
[c] Produced by cooperatives or cottage industries.
[d] Some commentators suggest quite large-scale use (>10000ha), but this has been disputed (Sikura and Sikura 1983).
[e] From Feng et al. (1994).
– Name not known.
[f] Around 10000 tons of conidial powder used to treat 0.8–1.3 million ha.
[g] Formerly owned by W.R. Grace.
[h] Under licence from ECO-tek.

Fungal pathogens act generally in a density-dependent fashion against their hosts and have relatively slow kill, thus they are not good candidates for pest control in crops with low damage thresholds. Specificity is often perceived to be an advantage of microbial pesticides generally. However, specificity can be a problem when there is a pest complex and no one pathogen can give control (Powell and Jutsum 1993; Ravensberg 1994).

Plant allelochemicals have evolved as a defence against microbial pathogens and herbivorous insects. Whilst these compounds may be sufficiently stressful to cause an increase in the susceptibility of generalist feeders to pathogens, some adapted insects can sequester these chemicals and may thereby acquire some protection from their own pathogens.

3. Integration in Pest Management Schemes

Although mycoinsecticides can provide stand-alone pest control, they may be regarded as one weapon in an armoury of techniques employed in integrated pest management schemes (Dent 1995; Lacey and Goettel 1995). However, a virulent pathogen could have indirect detrimental effects on existing natural control, e.g. reducing the availability of hosts for parasitic insects.

C. Development of a Mycoinsecticide

1. Isolate Selection

There are some 700 species of entomopathogenic fungi known from 85 genera. However, comparatively few have been investigated as potential mycoinsecticides. Natural epizootics caused in particular by fungi in the order Entomophthorales occur frequently in natural and agricultural terrestrial ecosystems (Samson et al. 1988). This has led to a number of attempts to use entomophthoralean fungi, e.g. for pest control of aphids, with varying degrees of success. Problems have included the inability to culture certain species in vitro and the fact that the most stable spore form is the resting spore, which is not infective and not produced by the most pathogenic isolates (Latgé 1986). The genera *Cordyceps* and *Torubiella* also contain some virulent but obligate insect pathogens. As a consequence, most development work has focussed on certain Deuteromycota, particularly *M. anisopliae*, *B. bassiana*, *N. rileyi*, Ascomycotina, e.g. *Aschersonia aleyrodis* and *V. lecanii*, and Oomycota, e.g. *Lagenidium giganteum*, which are more readily cultured in vitro.

Isolates are often selected on the basis of laboratory bioassay using cultured insects under optimal conditions. However, lead isolates need to be checked in a commercial setting. Immersion of test insects in a conidial suspension as part of a bioassay represents a temporary exposure to high inoculum. Prolonged exposure of insects, e.g. in soil, may provide a lethal dose from a sublethal bioassay concentration through the accumulation of spores with time (Ferron 1985).

There is no consensus about whether isolates originating from the target host (homologous isolates) or isolates from other hosts (heterologous isolates) of an established pathogen, or an exotic fungus (a species which is not present in the geographical area of application) are likely to provide the most suitable candidate for a mycoinsecticide. One aspect of the debate is whether an adapted pathogen (homologous isolate) evolves towards a balanced relationship with its host which precludes high virulence. There is evidence from work on human viral and bacterial pathogens that this may not always be the case (Ewald 1993); similarly, in a screen for fungal pathogens of the desert locust, *Schistocerca gregaria*, the majority of isolates of *Metarhizium* spp. with high virulence

came from this or related acridids (Prior 1992). However, in other studies, heterologous isolates have proved the most virulent, e.g. see Vestergaard et al. (1995). Since there are concerns over the possible non-target effects of exotic fungi (e.g. see Lockwood 1993), it may be wise to search and screen isolates from the country where the mycoinsecticide is to be deployed in order to facilitate commercial registration (Prior 1992).

The relevance of isolate virulence to pest control is clear. However, high sporulation, stability during bulk storage and epizootic potential may be equally, if not more, important commercially, leading to a compromise on virulence during strain selection.

2. Production and Formulation

Large-scale in vivo cultivation of a fungus in laboratory reared or field-derived insects is usually only employed for obligate pathogens which do not grow readily outside their hosts. This may be a viable method of inoculum production for a programme of introduction (Soper and Ward 1981). However, it is unlikely to be economic for large-scale mycoinsecticide use. The method adopted for in vitro cultivation must take into account:

1. The inoculum produced must have optimum virulence and retain viability over an extended period in storage and after application in the field.
2. Serial in vitro transfer can lead to loss of virulence. Use of single-spore isolates with Deuteromycota (which have no sexual cycle) can help to alleviate this problem and ensure uniformity of the product. However, genetic change through chromosome transformation, transposable elements, cytoplasmically transmitted genetic elements and the parasexual cycle may promote drift (Couteaudier et al. 1994). Storage of the original culture under liquid nitrogen and periodic passage of the fungus through the host may be required to maintain virulence of a product.
3. The culture medium should balance the needs of cost effectiveness in terms of yield per unit outlay with the production of a highly virulent, stable inoculum.
4. The production system may affect the propagule type and virulence, which may impact on killing power, shelf life, environmental

stability and the formulation and application strategies.

5. The production system may have to be scaled up to produce cost-effective treatment on thousands or even millions of hectares (Lisansky and Hall 1983; Jaronski 1986; Bradley et al. 1992; Jenkins and Goettel 1995).

Three types of production system have been employed; submerged (liquid) fermentation, surface cultivation and diphasic fermentation.

Submerged Liquid Fermentation. This is the preferred option because existing large-scale deep-tank fermentation equipment can be used, the process is most easily controlled and can be much faster than other methods. The major drawback is that dimorphic fungi like *M. anisopliae*, *V. lecanii*, *Paecilomyces farinosus* and *B. brongniartii* typically produce blastospores rather than true conidia in liquid culture. In vitro-produced blastospores are similar to the in vivo yeast phase that enables many fungi to develop and spread quickly in the haemolymph of the host. The wall structure of blastospores is often similar to that of mycelium and, being unpigmented, in vitro-produced blastospores are often unstable with limited shelf life and field stability. However, good yields can be obtained, e.g. $5 \times 10^9 \mathrm{ml}^{-1}$ of *V. lecanii* (Latgé et al. 1986) and, despite constraints on their use, blastospores form the basis of some commercial formulations (e.g. Vertalec produced by Koppert, see Table 1). Certain strains of *B. bassiana* (Feng et al. 1994), *M. flavoviride* (Jenkins and Prior 1993) and *Hirsutella thompsonii* (Van Winkelhoff and McCoy 1984) will conidiate in liquid culture in the right media. Yields of $1.5 \times 10^9 \mathrm{ml}^{-1}$ of submerged conidia have been achieved with *M. flavoviride*. These spores were at least as pathogenic as aerial conidia against grasshopper and locust targets (Jenkias and Thomas 1996). Resting spores but not infective spores of *Entomophthora* spp. are formed in liquid culture (Latgé 1986), as are the sexually derived oospores of *Lagenidium giganteum* (Kerwin et al. 1986).

Several methods have been developed for producing mycelia in submerged cultures. Mycelia is applied in the field, where it will sporulate producing infective conidia. In the marcescent process, first developed for Entomophthorales (McCabe and Soper 1985) and then adapted for Deuteromycota (Rombach et al. 1988), mycelium is dried in a sugar desiccation process, milled and

then stored at low temperature prior to use. Fluid bed dried mycelial granules of *M. anisopliae*, with a shelf life of at least a year at 4 °C, can provide good long-term control of the vine weevil *O. sulcatus* when incorporated into compost (Stenzel et al. 1992). Dried mycelia fragments have been successfully field-tested (Rombach et al. 1987) and a commercial product, BIO1020, based on dried mycelial granules, has been developed by Bayer AG in Germany.

Surface Cultivation. Solid substrate is the most widely used method of production, though pH, temperature, nutrient status and aeration may be more difficult to control than in submerged liquid culture. Large-scale production has been carried out on agricultural, brewing or other wastes, though such media can be too variable and of low immediate metabolic availability. The high surface area-to-volume ratio of small cereal grains such as sorgum and rice leads to better nutrient absorption, gas exchange and heat transfer. Aeration can be improved by the use of a rotating drum; however, premature germination and reduced yield can occur when conidia are dislodged from conidiophores (Lisansky and Hall 1983). The size of the initial inoculum may influence spore yield, and procedures need to be optimised to minimise hyphal growth and maximise sporulation. *M. anisopliae* is produced commercially in Brazil for control of sugarcane pests on steam-sterilised rice in autoclavable plastic bags, and large-scale production of *B. bassiana* in China is carried out on agricultural residues.

Diphasic Fermentation. Fungus is grown in fermentation tanks to the end of the log phase. The resulting mycelium is then applied to a solid, absorbent surface such as vermiculite, sponge, cloth or a shallow liquid medium to conidiate. *B. bassiana*, as Boverin, is produced in Russia by a diphasic process. Shallow layers of sterilised liquid media in polypropylene plastic bags are inoculated with mycelium. The fungus grows as a mat on the surface producing up to 10^{13} conidia/$0.1 \mathrm{m}^2$ of the surface area on 2l of media (Bradley et al. 1992). Diphasic systems are time-consuming, labour-intensive and need a lot of space, consequently tend to be disregarded by commercial enterprises in the west. However, Mycotech have developed a two-step solid state fermentation method for producing conidia of *B. bassiana*. In this system, inoculum is produced in liquid fermentors then placed in bioreactors, where it is absorbed onto a starch-based solid substrate. The product is a dry conidial powder which has a 6-month shelf life at room temperature.

Harvesting the fungus from liquid culture is usually a matter of centrifugation then rapid controlled drying to prevent bacterial growth, though excessive temperatures can reduce viability. Following solid substrate cultivation, spores can be washed off the substrate or dried in situ to a suitable moisture content, then milled. Water content has an important bearing on conidial storage characteristics and temperature tolerance; increasing the level of desiccation can increase temperature tolerance of *M. anisopliae* and *M. flavoviride* conidia (Moore et al. 1995).

The fungus needs to be formulated to help stabilise the product during storage and to facilitate delivery to the insect target in the field. An 18-month storage period is the ideal for economic use. Mechanical harvesting and dry storage of unformulated conidia can prolong viability (71%, 12 months, 4°C, *B. bassiana*; Chen et al. 1990). Blastospores are more difficult to formulate because of their instability. Freeze-drying has been used (3 months, *B. bassiana*; Belova 1978). *M. flavoviride* conidia harvested in the light petroleum fraction oil Edelex or groundnut vegetable oil, then diluted with Shellsol K, deodorised kerosene (Edelex) or an antioxidant (groundnut oil), retained 60% viability after storage at 17°C for 30 months, as long as they were dried by the addition of non-indicating silica gel to the formulation (Moore et al. 1995). In dry or wettable powder formulations, the main ingredient may dilute the "active ingredient" to a concentration that can be handled more easily. Phyllosilicates (clays) are the most commonly used, as they are relatively inert and cheap. They can promote conidial viability over extended periods (Ward 1984). Coating of blastospores of *B. bassiana* with clay helps to prevent biodegradation in soil. Additional ingredients may include stickers, humectants, UV protectants, an emulsifier for water-based spraying of hydrophobic spores and nutrients, though inclusion of nutrients may be a natural consequence of the production process. Either way, nutrients in the formulation may allow saprophytic growth and sporulation on foliage, increasing inoculum potential, enhancing the chances of secondary pickup and vertical infection.

Use of mycelial preparations presents different problems for formulation. Pure dry mycelium

when treated with maltose or sucrose produced more conidia after storage (Pereira and Roberts 1990). Bayer's BIO1020, which consists of mycelial granules, is stored vacuum-packed after fluid-bed drying and retains the ability to form conidia after rehydration following 12-month storage at 4°C. Incorporation of mycelium into alginate pellets, with or without additional nutrient sources, has been tried successfully with *B. bassiana* (Pereira and Roberts 1991), providing enhanced shelf life and environmental stability, particularly against solar radiation after application. Cornstarch or cornstarch oil formulation also enhanced conidial production by mycelium after several months' storage (Pereira and Roberts 1991).

3. Application

The timing of application may be important. Suitable weather conditions (high humidity, equitable temperature) may occur during late evening/early morning. The pest should be at the most susceptible stage (the most juvenile stages are often the most susceptible and easiest to control). The method of application depends on the nature of the inoculum and the niche of the pest insect.

Conidia or blastospores of Deuteromycota like *B. bassiana*, *M. anisopliae* and *V. lecanii* can be suspended in a liquid or mixed with a powder carrier and sprayed with conventional machinery used for the application of synthetic chemical insecticides. High-volume hydraulic sprayers are used to apply *V. lecanii* spores in water against aphids on chrysanthemums in glasshouses. Secondary spore pickup rather than just direct hit is critical to success, which makes sedentary insects like *Aphis gossypii* more difficult to control. Repeated low-dose treatments may be helpful (Helyer and Wardlow 1987) though regular use of large volumes of water on a crop may promote plant fungal disease. Koppert recommends two to three applications of Mycotal 7 days apart to control whiteflies on cucumber. Improved targeting against aphids which prefer abaxial leaf surfaces can be provided by electrostatic sprayers. The equipment imparts a charge to droplets, thereby increasing abaxial deposition and aphid control (Sopp et al. 1989).

Mist blowers and helicopters have been used successfully to apply *B. brongniartii* to swarming adults of the cockchafer *Melolontha melolontha* in Switzerland (2×10^{14} spores/ha). Ultra-low volume (ULV) application of water or mineral oil-based formulations on the ground or by aircraft has been used on >0.8 million ha in China for control of various forest and crop insects (Ying 1992). Bateman et al. (1993) have shown in laboratory experiments that mineral or vegetable oil formulations of *M. anisopliae* and *M. flavoviride* were more virulent than water-based formulations against the desert locust *Schistocerca* gregaria, and decreased reliance on high environmental RH. These formulations in ULV sprays have been used successfully in field trials against the brown locust *Locustana pardalina* in South Africa (Bateman et al. 1994) and the variegated grasshopper *Zonocerus variegatus* in West Africa (Lomer et al. 1993).

Soil-borne pests can be treated either prophylactically or curatively with fungal spores. Prior incorporation of conidia of *M. anisopliae* into compost can give year-long protection of *Impatiens wallerana* against the vine weevil (Moorhouse et al. 1993). The success of drenches of aqueous spore suspensions is influenced by the depth to which spores percolate, the volume of the drench, adsorption to soil particles and movement of the insect that will facilitate uptake of a lethal dose. Direct drilling of *M. anisopliae* conidia using existing (crop-sowing) machinery to a depth of 20–25 mm in pasture gave long-term control of the redheaded cockchafer *Adorphorus couloni* in trials in Tasmania (Rath 1992).

A novel method of pest control was suggested by the observation that *B. bassiana* exists as an endophyte in certain genotypes of maize (Vakili 1990). Bing and Lewis (1991) prepared a granular formulation of *B. bassiana* by spraying a suspension of *B. bassiana* (1.1×10^8 conidia/g) onto corn grits in a rotating drum of a Gustafon minimixer. Hand-held inoculators were used to apply 0.4 g of the granules to whorls of plants ($10^7 \times 4.55$ conidia/plant $\cong 2.5 \times 10^{12}$ conidia/ha). Conidia of *B. bassiana* applied in this way moved within the plants and provided season-long control of the European cornborer, *Ostrinia nubilalis*.

Baits and traps provide ways of bringing the insects to a source of inoculum, rather than the other way around. Application of conidia of *B. bassiana* to bran has given some success in trials against rangeland grasshoppers in Canada (Johnson and Goettel 1993; Inglis et al. 1996). Sex pheromone has been used to attract male diamond-backed moths, *Plutella xylostella*, to traps where they are infected with *Zoophthora radicans*. Fungus is carried by contaminated moths to susceptible larvae, thus initiating or enhancing an epizootic (Furlong et al. 1995). Traps laced with conidia of *M. anisopliae* are commercially available in the USA for cockroach control.

Preparations of dried mycelial pellets, which need to be rehydrated prior to the production of infective conidia, are particularly suitable for application to soil (Stenzel et al. 1992). Rombach et al. (1987) applied ~0.3-mm pieces of dried mycelium of *B. bassiana* to rice plants with a spinning disc applicator against the brown plant hopper, *Nilaparvata lugens*. In both systems, the mycelia rehydrated and then sporulated. The resulting conidia infected the insects.

4. Safety

It is sometimes assumed a priori that microbial pesticides must be considerably safer to humans and their environment than synthetic chemical insecticides. However, quite correctly, registration procedures in most countries require that mycoinsecticdes, like chemical insecticides, are safety tested (for reviews see Austwick 1980; Laird et al. 1990). The possible side-effects of entomopathogenic fungi may be summarised as infections, toxicosis and allergies in non-target animals or humans, Among non-target arthropods, fungal infections of pollinating bees, parasitoids and predatory beetles could be particularly important. The absence of natural epizootics of candidate fungi among pollinators suggests that the risks are low, while several studies have shown the compatibility of fungi and parasitoids and predators in integrated control programmes.

None of the entomopathogenic fungi currently in use or under consideration is invasively pathogenic to humans. However, immunocompromised individuals are open to opportunistic infections which could potentially include certain entomopathogens, though none of the fungi under development can grow efficiently at 37 °C.

Many fungi produce toxic secondary metabolites which have detrimental effects on cultured animal cells. The mammalian toxicity of many of these compounds has not been determined and it is difficult to assess the risks associated with the small quantities of such compounds which may be present either in the inoculum or in mycosed insect cadavers. Quality control during production is essential to ensure there is no contamination from toxicogenic spoilage fungi such as *Aspergillus flavus*.

Hypersensitive reactions to fungal antigens derived from hyphae, spores or from metabolites are perhaps the most likely health hazards to humans. Experiments with mice, rats and guinea pigs suggest that the main route of sensitisation is the respiratory system and people involved with large-scale production are at a particular risk.

VI. Future Developments

A. Potential Targets

Entomopathogenic fungi have potential for control of some, but not all, insect pests and it is important to identify appropriate targets for mycoinsecticide development. Fungi, in contrast to bacteria and viruses, invade their hosts by actively penetrating the exoskeleton (cuticle). Therefore fungi are particularly important natural pathogens of sucking insects such as aphids, whitefly, thrips and leafhoppers, since the feeding strategy of these insects tends to preclude acquisition of pathogens that are infectious per os. Larval and adult beetles are frequently hosts to fungal infections but appear to have comparatively few bacterial and viral pathogens. Thus fungi are often the pathogens of choice for bug and beetle pests.

Certain ecological niches lend themselves in particular to the deployment of mycoinsecticides (see Table 2). The habitats in question have in common that chemical control is difficult or inappropriate and the environments are conducive to fungal infection. Commercial-scale use of mycoinsecticides occurs on certain field crops which do not appear to provide optimum conditions for mycosis (see examples in Table 1). As a result, pest control does not always match that achievable with chemical insecticides. In these situations, moderate pest control is commercially acceptable, it would seem, because of cheap labour-intensive production methods. Recent developments, however, challenge pre-conceptions of where and under what conditions mycoinsecticides are likely to be successful, e.g. it has been claimed recently that new a emulsifiable formulation of *B. bassiana*, Naturalis L, from Troy Biosciences in the USA, gives effective control of the major cotton pests as well as certain other pests of vegetables and ornamentals (Wright and

Knauf 1994; Table 1). Mineral oil formulations of *Metarhizium* spp. and *B. bassiana* promote infection of grasshoppers and locusts, and good control has been achieved under arid field conditions (Bateman et al. 1994).

B. Constraints on the Commercial Use of Entomopathogenic Fungi

The low investment by multinational agrochemical companies in research on mycoinsecticides has significantly hindered development. The commercial view of insect pathogens has been that they are too specific, too expensive, difficult to formulate, too erratic, have a short shelf life and are difficult to patent (Powell and Jutsum 1993; Rodgers 1993; Lisansky and Coombs 1994; Ravensberg 1994). This has led on the whole to the development of microbial insecticides by small- to medium-sized companies for niche markets, viz. where chemicals do not work well (through resistance or the withdrawal of registration for environmental reasons of effective products, e.g. protected crops), or are environmentally unacceptable and have been banned (e.g. forests in North America). In addition, these niches have environments which promote activity of mycoinsecticides, viz. protection from temperature extremes, UV and desiccation, and crops which can sustain some damage without economic loss.

Implementation of reregistration schemes for existing chemical pesticides and government-backed use reduction schemes in a number of countries will result in fewer chemical products and may open up more niche markets for microbials including fungi. Growing public demand for food with low or no chemical residues increases the pressure for biological alternatives to synthetic pesticides.

The cost of registration is a major constraint on the development of a small market product. Some countries charge lower fees for a biological product in comparison with a chemical (Ravensberg 1994). These fees may be only a fraction of the costs incurred in providing the toxicological and non-target risk assessment data, though the data package required for the registration of biologicals is generally less than for synthetic chemicals. The USA has a tiered approach to the registration process. If the results of the first tier tests indicate no adverse effects, then data

Table 2. Ecological niches which lend themselves in particular to the deployment mycoinsecticides

Habitat	Pests	Reasons for poor control with insecticides	Fungal pathogen
Soil	Vine weevil, citrus root weevil, cockchafer, white grubs	Prophylactic control, no good chemicals Chemicals too costly Environmental concerns	*Metarhizium* spp. *Beauveria* spp.
Cryptic habitats	Stem borer, e.g. European corn-borer Bark beetles	Not enough surface feeding to pick up lethal dose	*Beauveria* spp. *Metarhizium* spp.
Glasshouses	Whiteflies, thrips, aphids	Insecticide resistance Consumer resistance to chemical use	*Verticillium lecanii* *Aschersonia aleyrodis* *Metarhizium* spp. *Paecilomyces fumosoroseus*
Rice	Plant hoppers Black bugs	Insecticide resistance Environmental concerns	*Metarhizium* spp. *B. bassiana* *Hirsutella citriformis* *P. lilacinus*
	Mosquitoes		*Lagenidium giganteum*
Nests of social insects	Termites, leaf-cutting ants, fire ants	Application problems Insecticide resistance Environmental concerns	*Metarhizium* spp.
Nuisance insects – Human habitation – Poultry houses	Cockroaches, flies	Application problems Insecticide resistance Environmental concerns	*Metarhizium* spp. *Entomophthora muscae*
Tropical environments – Sugarcane	Whitefly, thrips, froghoppers	Ineffective control with existing products Environmental concerns	*Metarhizium* spp. *Aschersonia aleyrodis* *Paecilomyces fumosoroseus*

from the other tiers are not required. This reduces the cost significantly (Plimmer 1993). There is an enormous variation in the requirements for registration of microbial insecticides between countries, a problem that has been addressed in the European Community at least in part by the adoption of a common policy on registration of pesticide active ingredients (EC directive 91/414, "plant protection products regulations"). Use of genetically engineered microorganisms may incur additional registration requirements, but to date no live genetically engineered microbial pesticides have been registered.

The rate of application of a mycoinsectide required to give adequate control in critical to commercial success; 10^{13} conidia per hectare is the current bench mark (Bradley et al. 1992). Good control of the vine weevil on glasshouse ornamentals was achieved experimentally using an equivalent dose of 1.6×10^{14} conidia/ha (Moorhouse et al. 1993). The recommended rate for the use of *V. lecanii*, as Mycotal, against glasshouse whitefly is 3×10^{13} conida/ha on cucumber. *B. bassiana*, as Boverin, is used at 6×10^{12}–2.2×10^{13} spores ha^{-1} against the Colorado beetle. Since Mycotal contains 10^{10} conidia/g and Boverin contains 6×10^{9}–1.2×10^{10} conidia/g, these rates represent as much as 10 kg of product or more per hectare. Indeed, fermentation requirements for the large-scale production of most entomopathogenic fungi are equivalent to tens to hundreds of litres per hectare (Bradley et al. 1992). Improved production efficiency or enhanced fungal virulence will be needed to ensure feasability of large-scale mycoinsecticide production (Jaronski 1986).

C. Strain Improvement

As our understanding of the epizootology of disease and the biochemical basis of pathogenicity/virulence of entomopathogenic fungi improves, and techniques are developed for their genetic manipulation, it will be possible to devise strategies for strain improvement. Characteristics which should be addressed include: enhanced shelf life and environmental stability (e.g. UV resistance, temperature tolerance), improved sporulation during mass production, ability to initiate infection at low humidity, expansion of the host range, accelerated kill (reduced LT_{50}) and increased killing power (reduced LD_{50}).

A simple approach to enhancing host range would be to include a mixture of isolates/species in the mycoinsecticide formulation. This tactic could also be used to enhance performance against a single target if isolates with differing virulence strategies or environmental requirements e.g. temperature were employed. However, incompatibility between isolates may lead to one isolate prevailing over the other and the absence of co-operativity. In any event, extra costs incurred with producing, formulating and registering several isolates in a single mycoinsecticide may prove prohibitive.

Culture conditions can influence the characteristics of fungal spores and can be manipulated to increase mycoinsecticide efficiency. Blastospores of B. bassiana from nitrogen-limited cultures had higher concentrations of carbohydrate and lipid and were significantly more virulent (lower LT_{50}) towards the rice green leafhopper than blastospores from carbon-limited cultures (Lane et al. 1991). Growth of B. bassiana, M. anisopliae and P. farinosus on agar-based media with low water activity or with a high concentration of glycerol encouraged accumulation of polyols in conidia that were more pathogenic at lower RH than those produced on control media (Hallsworth and Magan 1994, 1995).

Genetic modification of entomopathogenic fungi to improve efficiency of pest control is complicated by the fact that the leading candidates are largely Deuteromycota with no known sexual stage. Chemical mutagenesis, the parasexual cycle, protoplast fusion and direct genetic manipulation could be used. Mutants of M. anisopliae and P. farinosus have been generated which were significantly more virulent (reduced LT_{50}) at low RH than parentals (Matewele et al. 1994). Samuels et al. (1986) produced high-sporulating mutants of M. anisopliae.

The parasexual cycle and protoplast fusion have been used to cross isolates of M. anisopliae and V. lecanii (see review by Heale et al. 1989). However, rarely have the progeny had improved characteristics; indeed, the reverse can be the case. It has been suggested that disruption of clusters of pathogenicity genes is the cause (Charnley 1989). Recently, however, Riba et al. (1994) successfully crossed the nonentomopathogenic B. sulfurescens, which produces an entomotoxic glycoprotein, with

an atoxigenic pathogenic isolate of B. bassiana. Stable, partial diploid, hypervirulent, toxigenic recombinants ensued.

Prospects for the use of somatic hybridization in strain improvement have increased enormously with the advent of molecular techniques to target suitable isolates and monitor the outcome of crosses (St Leger 1994). Strain selection and parasexual crossing will be the most effective method of obtaining environmental tolerance, as these traits are probably controlled polygenically. The alternative, direct genetic manipulation, would provide enhanced targeting for single genes or gene clusters. Genetic engineering needs the establishment of transformation and cloning systems which has been achieved for M. anisopliae (Bernier et al. 1989; St Leger et al. 1992b; Smithson et al. 1995). For the future, it should be possible to increase isolate virulence and/or extend the host range by altering the timing and release of virulence factors, increasing copy number of virulence genes and introducing specificity genes from other isolates or toxin genes from other organisms. Indeed, St Leger et al. (1996) have recently inserted extra copies of the prl gene (encoding the protease PR1) from M. anisopliae into the genome of M. anisopliae such that the gene was constitutively overexpressed in the tobacco hornworm, Manduca sexta. This resulted in the activation of the host's prophenoloxidase system. The combined effects of PR1 and the reaction products of phenoloxidase caused a 25% reduction in time to death and a 40% reduction in food consumption by insects infected with engineered fungus in comparison with insects infected with wild type.

It has to be said that however exciting the prospects are for genetic engineering, there are going to be regulatory problems in getting a transgenic isolate approved for release. Perhaps in the short term it may be prudent to concentrate on the use of somatic hybridization techniques for strain improvement.

VII. Conclusions

Mycoinsecticides have had little impact on insect pest control to date. However, the current rate of progress in research on epizootology, mass production, formulation, application and mechanisms of pathogenesis suggests a promising future for fungal pathogens in pest control. It is sometimes suggested that failures of chemical insecticides due to the development of resistance are unlikely to be repeated with insects' own pathogens because of the complex multigenic interaction between host and fungus. However, given the high rate of increase that some insect species can achieve, the selection pressure that would be exerted by widespread, year-on-year application of a mycoinsecticide and the existing immune defences of insects, it would seem that the development of resistance to mycoinsecticides will occur with time. The propensity for resistance development is shown by the occurrence in a laboratory colony of a biotype of the pea aphid, *Acyrthosiphon pisum*, resistant to *Erynia neoaphidis* (Milner et al. 1982). Therefore it is to be hoped that laboratory selection experiments will be carried out before instances of resistance are reported in the field, so that strategies for managing resistance can be developed in advance. Much can be learned in this regard from research on resistance to *Bacillus thuringiensis* (Tabashnik 1994).

Acknowledgments. I would like to thank Dr. C. Prior for helpful comments on the manuscript.

References

Anderson TE, Roberts DW (1983) Compatibility of *Beauveria bassiana* isolates with insecticide formulations used in Colorado potato beetle (Coleoptera: Chrysomelidae) control. J Econ Entomol 76:1437–1441

Austwick PKC (1980) The pathogenic aspects of the use of fungi: the need for risk analysis and registration of fungi. In: Environmental protection and biological forms of control of pest organisms. Ecol Bull 31:91–102

Bateman RP, Carey M, Moore D, Prior C (1993) The enhanced infectivity of *Metarhizium flavoviride* in oil formulations to desert locusts at low humidities. Ann Appl Biol 122:145–152

Bateman RP, Price RE, Muller EJ, Brown HD (1994) Controlling brown locust hopper bands in South Africa with a mycoinsecticide spray. Brighton Crop Protection Conf, Pests and Diseases, British Crop Protection Council, Farnham, pp 609–616

Belova RN (1978) Development of the technology of Boverin production by the submersion method. In: Ignoffo CM (ed) Proc 1st joint US/USSR conf on the production, selection and standardisation of entomopathogenic fungi of the US/USSR joint working group on the production of substances by microbiological means. National Science Foundation, Washington, DC, pp 102–119

Bernier L, Cooper RM, Charnley AK, Clarkson JM (1989) Transformation of the fungus *Metarhizium anisopliae* to benomyl resistance. FEMS Microbiol Lett 60:261–266

Bing LA, Lewis LC (1991) Suppression of the European cornborer, *Ostrinia nubilalis* (Hubner) (Lepidoptera: Pyralidae) by endophytic *Beauveria bassiana* (Balsamo) Vuillemin, Environ Entomol 20:1207–1211

Boucias DG, Pendland JC (1991) Attachment of mycopathogens to cuticle: the initial event of mycosis in arthropod hosts. In: Cole GT, Hoch HC (eds) The fungal spore and disease initiation in plants and animals. Plenum, New York, pp 101–128

Bradley CA, Black WE, Kearns R, Wood P (1992) Role of production technology in mycoinsecticide development. In: Leatham GF (ed) Frontiers in industrial mycology. Chapman and Hall, New York, pp 160–173

Brown GC, Nordin GL (1986) Evaluation of an early harvest approach for induction of *Erynia* epizootics in alfalfa weevil populations, J Kans Entomol Soc 59:446–453

Carruthers RI, Haynes DL (1986) Temperature, moisture and habitat effects on *Entomophthora muscae* (Entomophthorales, Entomophthoraceae) conidial germination and survival in the onion agroecosystem. Environ Entomol 15:1154–1160

Carruthers RI, Soper RS (1987) Fungal diseases. In: Fuxa JR, Tanada Y (eds) Epizootology of insects diseases. Wiley, New York, pp 357–416

Charnley AK (1984) Physiological aspects of destructive pathogenesis in insects by fungi: a speculative review. In: Anderson JM, Rayner ADM, Walton DWH (eds) British Mycological Society Symposium No 6. Cambridge University Press, Cambridge, pp 229–270

Charnley AK (1989) Mechanisms of fungal pathogenesis in insects. In: Whipps JM, Lumsden RD (eds) Biotechnology for improving plant growth. Cambridge University Press, Cambridge, pp 85–125

Charnley AK, St Leger RJ (1991) The role of cuticle degrading enzymes in fungal pathogenesis of insects. In Cole GT, Hoch HC (eds) The fungal spore and disease initiation in plants and animals. Plenum Press, New York, pp 267–286

Chen CJ, Wu JW, Li ZZ, Wang ZX, Li YW, Chang SH, Yin XP, Dai LY, Tao L, Zhang YA, Tang J, Ding S, Ding GG, Gao XH, Tan YC (1990) Application of microbial pesticides in IPM. In: Chen CJ (ed) Integrated management of pine caterpillars in China. China Forestry Publishing House, Beijung, pp 214–308 (in Chinese; quoted in Feng et al. 1994)

Couteaudier Y, Maurer P, Viaud M, Riba G (1994) Genetic stability of fungi. In: Int Colloq Invertebrate Pathology and Microbial Control, Montpellier, France, Aug 1994, Proc vol 1, pp 343–349

Dent D (1995) Integrated pest management. Chapman and Hall, New York, 356 pp

Evans HC (1982) Entomogenous fungi in tropical forest ecosystems: an appraisal. Ecol Entomol 7:47–60

Ewald PW (1993) The evolution of virulence. Sci Am April:56–62

Feng MG, Poprawski TJ, Khachatourians GG (1994) Production, formulation and application of the entomopathogenic fungus *Beauveria bassiana* for insect control: current status. Biocontrol Sci Technol 4:3–34

Ferron P (1985) Fungal control. In: Kerkut GA, Gilbert LI (eds) Comprehensive insect physiology, biochemistry and pharmacology, vol 12. Pergamon Press, Oxford, pp 313–346

Furlong MJ, Pell JK, Choo OP, Rahman SA (1995) Field and laboratory evaluation of a sex pheromone trap for the autodissemination of the fungal entomopathogen *Zoophthora radicans* (Entomophthorales) by the diamond-back moth, *Plutella xylostella* (Lepidoptera: Yponomeutidae). Bull Entomol Res 85:331–337

Gillespie AT (1988) Use of fungi to control pests of agricultural importance. In: Burge MN (ed) Fungi in biological control systems. Manchester University Press, Manchester, pp 37–60

Gillespie AT, Claydon N (1989) The use of entomogenous fungi for pest control and the role of toxins in pathogenesis. Pestic Sci 27:203–215

Hajek AE, St Leger RJ (1994) Interactions between fungal pathogens and insect hosts. Annu Rev Entomol 39:293–322

Hajek AE, Humber RA, Elkington JS (1995) Mysterious origin of *Entomophaga maimaiga* in North America. Am Entomol Spring:31–42

Hallsworth JE, Magan N (1994) Improved biological control by changing polyols/trehalose in conidia of entomopathogens. In: Brighton Crop Protection Conf, Pests and Diseases 8D-8, British Crop Protection Council, Farnham, pp 1091–1096

Hallsworth JE, Magan N (1995) Manipulation of intracellular glycerol and erythritol enhances germination of conidia at low water activity. Microbiology 141:1109–1115

Heale JB, Isaac JE, Chandler D (1989) Prospects for strain improvement in entomopathogenic fungi. Pestic Sci 26:79–92

Helyer NL, Wardlow LR (1987) Aphid control on chrysanthemums using frequent, low-dose applications of *Verticillium lecanii*. WPRS Bull X/2:62–65

Hunt TR, Moore D, Higgins PM, Prior C (1995) Effect of sunscreens, irradiance and resting periods on the germination of *Metarhizium flavoviride* conidia. Entomophaga 39:313–322

Huxham IM, Lackie AM, McCorkindale NJ (1989) Inhibitory effects of cyclicdepsipeptide destruxins from the fungus *Metarhizium ansiopliae* on cellular immunity in insects. J Insect Physiol 35:97–105

Ignoffo CM (1981) The fungus *Nomuraea rileyi* as a microbial insecticide. In: Burges HD (ed) Microbial control of pests and plant diseases 1970–1980. Academic Press, London, pp 513–538

Ignoffo CA, Garcia C, Hostetter DL, Pinnel RE (1977) Laboratory studies of the entomopathogenic fungus *Nomuraea rileyi*: soil-borne contamination of soybean seedlings and dispersal of diseased larvae of *Trichoplusia ni*. J Invertebr Pathol 29:147–152

Inglis GD, Johnson DL, Goettel MS (1996) Effect of bait substrate and formulation on infection of grasshopper nymphs by *Beauveria bassiana*. Biocontrol Sci Technol 6:35–50

Jackson TA, Glare TR (eds) (1992) Use of pathogens in scarab pest management. Intercept, Andover. 298pp

Jaronski ST (1986) Commercial development of Deuteromycetous fungi of arthropods: a critical appraisal. In: Samson RA, Vlak JM, Peters D (eds) Fundamental and applied aspects of invertebrate pathology. Foundation of the IVth Int Colloquium of Invertebrate Pathology, Wageningen, pp 653–656

Jenkins NE, Prior C (1993) Growth and formulation of true conidia by *Metarhizium flavoviride* in a simple liquid medium. Mycol Res 7:1489–1494

Jenkins NE, Goettel MS (1995) Methods for the mass production of microbial agents of grasshoppers and locusts. In: Microbial control of grasshoppers and locusts. Memoirs of the Entomological Society of Canada (in press)

Jenkins NE, Thomas MB (1996) Effects of formulation and application method on the efficacy of aerial and submerged conidia of *Metarhizium flavoviride* for locust and grasshopper control. Pestic Sci 46:299–306

Johnson DL, Goettel MS (1993) Reduction of grasshopper populations following field application of the fungus *Beauveria bassiana*. Biocontrol Sci Technol 3:165–175

Jones KA (1994) Registration and use of microbial insecticides in developing countries. In: VIth Int Colloq on Invertebrate Pathology and Microbial Control, Montpellier, France, Aug 1994, Proc vol 1, pp 82–88

Joshi L, Charnley AK, Arnold G, Brain P, Bateman RP (1992) Synergism between entomopathogenic fungi, *Metarhizium* spp., and the benzoylphenyl urea insecticide Teflubenzuron, against the desert locust *Schistocerca gregaria*. Proc Brighton Crop Protection Conf, Pests and Diseases, British Crop Protection Council, Franham, pp 369–374

Keller S, Zimmermann G (1989) Mycopathogens of soil insects. In: Wilding N, Collins NM, Hammond PM, Webber JF (eds) Insect-fungus interactions. Academic Press, London, pp 239–270

Kerwin JL, Simmons CA, Washino RK (1986) Oosporogenesis by *Lagenidium giganteum* in liquid culture. J Invertebr Pathol 47:258–270

Lacey L, Goettel MS (1995) Current developments in microbial control of insect pests and prospects for the early 21st century. Entomophaga 40:3–27

Lacey LA, Undeen AH (1986) Microbial control of blackflies and mosquitoes. Annu Rev Entomol 31: 265–296

Laird M, Lacey LA, Davidson EW (eds) (1990) Safety of microbial insecticides. CRC Press, Boc Raton, 259pp

Lane BS, Trinci AP, Gillespie AT (1991) Influence of cultural conditions on the virulence of conidia and blastospores of *Beauveria bassiana* to the green leafhopper, *Nephotettix virescens*. Mycol Res 95:829–833

Latgé J-P (1986) The Entomophthorales after the resting spore stage. In: Samson RA, Vlak JM, Peters D (eds) Fundamental and applied aspects of invertebrate pathology. Foundation of the IVth Int Colloquium of Invertebrate Pathology, Wageningen, pp 651–652

Latgé J-P, Hall RA, Cabrera RI, Kerwin JC (1986) Liquid fermentation of entomopathogenic fungi. In: Samson RA, Vlak JM, Peters D (eds) Fundamental and applied aspects of invertebrate pathology. Foundation of the IVth Int Colloquium of Invertebrate Pathology, Wageningen, pp 603–606

Lisansky SG, Hall RA (1983) Fungal control of insects. In: Smith N (ed) Filamentous fungi, vol IV. Arnold, London, pp 325–345

Lisansky SG, Coombs J (1994) Developments in the market for biopesticides. In: Brighton Crop Protection Conf, Pests and Diseases, British Crop Protection Council, Farnham, pp 1049–1054

Lockwood JA (1993) Environmental issues involved in the biological control of rangeland grasshoppers (Orthoptera: Acrididae) with exotic agents. Environ Entomol 22:503–518

Lomer CJ, Bateman RP, Godonou I, Kpindu D, Shah PA, Paraiso A, Prior C (1993) Field infection of *Zonocerus variegatus* following application of an oil-based formulation of *Metarhizium flavoviride*. Biocontrol Sci Technol 3:337–346

Matewele P, Trinci APJ, Gillespie AT (1994) Mutants of entomopathogenic fungi that germinate and grow at reduced water activities and reduced relative humidities are more virulent to *Nephotettix virescens* (green leafhopper) than the parental strains. Mycol Res 98:1329–1333

Mazet I, Hung S-Y, Boucias DG (1994) Detection of toxic metabolites in the haemolymph of *Beauveria bassiana*-infected *Spodoptera exigua* larvae. Experientia 50:142–147

McCabe D, Soper RS (1985) Preparation of an entomopathogenic fungal insect control agent. US Patent No 4,530,834, July 23, pp 1–4

McCoy CW (1986) Factors governing the efficacy of *Hirsutella thompsonii* in the field. In: Samson RA, Vlak JM, Peters D (eds) Fundamental and applied aspects of invertebrate pathology. Foundation of the IVth Int Colloquium of Invertebrate Pathology, Wageningen, pp 171–174

McCoy CW, Samson RA, Boucias DG (1988) Entomogenous fungi. In: Ignoffo CM (ed) CRC Handbook of natural pesticides, vol V Microbial insecticides, Part A Entomogenous protozoa and fungi. CRC Press, Boca Raton, pp 151–236

Milner RJ, Soper RS, Lutton GG (1982) Field release of an Israeli strain of the fungus *Zoophthora radicans* for the biological control of *Therioaphis trifolii* f. *maculata*. J Aust Entomol Soc 21:113–118

Moore D, Bateman RP, Carey M, Prior C (1995) Long-term storage of *Metarhizium flavoviride* conidia in oil formulations for the control of locusts and grasshoppers. Biocontrol Sci Technol 5:193–199

Moorhouse ER, Gillespie AT, Sellers EK, Charnley AK (1992) Influence of fungicides and insecticides on the entomogenous fungus *Metarhizium anisopliae*, a pathogen of the vine weevil, *Otiorhynchus sulcatus*. Biocontrol Sci Technol 2:49–58

Moorhouse ER, Gillespie AT, Charnley AK (1993) Application of *Metarhizium anisopliae* (Metsch.) Sor. conidia to control *Otiorhynchus sulcatus* (F.) (Coleoptera: Curculionidae) larvae on glasshouse pot plants. Ann Appl Biol 122:623–636

Nnakumusana ES (1985) Laboratory infection of mosquito larvae by entomopathogenic fungi with particular reference to *Aspergillus parasiticus* and its effects on fecundity and longevity of mosquitoes exposed to conidial infections in larval stages. Curr Sci 54:1221–1228

Pereira RM, Roberts DW (1990) Dry mycelium preparations of the entomopathogenic fungi *Metarhizium anisopliae* and *Beauveria bassiana*. J Invertebr Pathol 56:39–46

Pereira RM, Roberts DW (1991) Alginate and cornstarch mycelial formulations of entomopathogenic fungi, *Beauveria bassiana* and *Metarhizium anisopliae*. J Econ Entomol 84:1657–1661

Perry DF, Whitfield GH (1984) The interrelationships between microbial entomopathogens and insect hosts: a system study approach with particular reference to the Entomophthorales and the Eastern spruce budworm. In: Anderson JM, Rayner ADM, Walton D (eds) Animal-microbial interactions. Cambridge University Press, Cambridge, pp 307–331

Pickford R, Riegert PW (1963) The fungus disease caused by *Entomophthora grylli* Fres and its effect on grasshopper populations in Saskatchewan in 1963. Can Entomol 96:1158–1166

Powell KA, Jutsum AR (1993) Technical and commercial aspects of biocontrol products. Pestic Sci 37:315–321

Plimmer JR (1993) Regulatory problems associated with natural products and biopesticides. Pestic Sci 39:103–108

Prior C (1992) Discovery and characterization of fungal pathogens for locust and grasshopper control. In: Lomer CJ, Prior C (eds) Biological control of locusts and grasshoppers. CAB International, Wallingford, pp 159–180

Rath AW (1992) *Metarhizium anisopliae* for control of the Tasmanian pasture scarab *Adoryphorus couloni*. In: Jackson TA, Glare TR (eds) Use of pathogens in scarab pest management. Intercept, Andover, pp 217–227

Ravensberg WJ (1994) Biological control of pests: current trends and future prospects. In: Brighton Crop Protection Conf, Pests and Diseases, Brighton Crop Protection Council Franham, pp 591–600

Riba G, Couteaudier Y, Maurer P, Neuvéglise (1994) Molecular methods offer a new challenge for fungal bioinsecticides. In: Int Colloq Invertebrate Pathology and Microbial Control, Proc Montpellier, France, Aug 1994, vol 1, pp 16–22

Roberts DW (1980) Toxins of entomopathogenic fungi. In: Burges HD (ed) Microbial control of insects, mites and plant diseases, vol 2. Academic Press, New York, pp 441–463

Rodgers PB (1993) Potential of biopesticides in agriculture. Pestic Sci 39:117–129

Rombach MC, Aguda RM, Roberts DW (1987) Biological control of the brown planthopper, *Nilaparvata lugens* (Homoptera: Delphacidae) with dry mycelium applications of *Metarhizium anisopliae* (Deuteromycotina). Philipp Entomol 613–619

Rombach MC, Aguda RM, Roberts DW (1988) Production of *Beauveria bassiana* (Deuteromycotina: Hyphomycetes) in different liquid media and subsequent conidiation of dry mycelium. Entomophaga 33:315–324

Samson RA, Evans HC, Latgé JP (1988) Atlas of entomopathogenic fungi. Springer Berlin Heidelberg, New York, 187pp

Samuels KZ, Heale JB, Llewellyn MJ (1986) Enhanced extracellular enzyme production and sporulation traits in isolates of *Metarhizium anisopliae* used for the control of *Nilauaparta lugens*. In: Samson RA, Vlak JM, Peters D (eds) Fundamental and applied aspects of invertebrate pathology. Foundation of the IVth Int Colloquium of Invertebrate Pathology, Wageningen, p 242

Seyoum E, Moore D, Charnley AK (1995) Reduction in flight activity and food consumption by the desert locust, *Schistocerca gregaria*, after infection with *Metarhizium flavoviride*. Z Angew Entomol 118:310–315

Sikura AI, Sikura LV (1983) The use of biopreparations. Zashch Rast (Mosc) 5:38–39 (in Russian) (quoted in Feng et al. 1994)

Smithson SL, Paterson IC, Bailey AM, Screen SE, Cobb B, Hunt BA, Cooper RM, Charnley AK, Clarkson JM (1995) Cloning and characterisation of a gene encoding a cuticle-degrading protease from the insect pathogenic fungus *Metarhizium anisopliae*. Gene 166:161–165

Soper RS, Ward MG (1981) Production, formulation and application of fungi for insect control. In: Papavizas GC (ed) Biological control in crop production. BARC Symp No 5. Allanheld, Osmun, Montclair, pp 161–180

Sopp P, Gillespie AT, Palmer A (1989) Application of *Verticillium lecanii* by a low-volume electrostatic rotary atomiser and a high-volume hydraulic sprayer. Entomophaga 34:417–428

Stenzel K, Hölters J, Andersch W, Smit TAM (1992) Bio 1020: granular *Metarhizium* – a new product for biocontrol of soil pests. In: Brighton Crop Protection Con, Pests and Diseases, British Crop Protection Council, Farnham, pp 363–368

St Leger RJ (1994) Mycoinsecticides: an opportunity for genetic engineering. In: Int Colloquium of Invertebrate Pathology and Microbial Control. Proc vol 1, pp 299–304

St Leger RJ (1995) The role of cuticle-degrading proteases in fungal pathogenesis of insects. Can J Bot 73 (Suppl):S1119–S1125

St Leger RJ, May B, Allee LL, Frank DC, Staples RC, Roberts DW (1992a) Genetic differences in allozymes and in formation of infection structures among isolates of the entomopathogenic fungus *Metarhizium anisopliae*. J Invertebr Pathol 60.89–101

St Leger RJ, Frank DC, Roberts DW, Staples RC (1992b) Molecular cloning and regulatory analysis of the cuticle-degrading protease structural gene from the entomopathogenic fungus *Metarhizium anisopliae*. Eur J Biochem 204:991–1001

St Leger RJ, Joshi L, Bidochka MJ, Roberts DW (1996) Construction of an improved mycoinsecticide overexpressing a toxic protease. Proc Nat Acad Sci 93:6349–6354

Tabashnik BE (1994) Evolution of resistance to *Bacillus thuringiensis*. Annu Rev Entomol 39:47–79

Tanada Y, Kaya HK (1993) Insect pathology. Academic Press, San Diego, 666pp

Vakili NG (1990) Biocontrol of stalk rot in corn. Proc 44th Annu Corn Sorghum Res Conf 44:87–105

Van Winkelhoff AJ, McCoy CW (1984) Conidiation of *Hirsutella thompsonii* var. *synnematosa* in submerged culture. J Invertebr Pathol 43:59–68

Vestergaard S, Gillespie AT, Butt TM, Schreuter G, Eilenberg (1995) Pathogenicity of the hyphomycete fungi *Verticillium lecanii* and *Metarhizium anisopliae* to the Western Flower Thrips *Frankliniella occidentalis*. Biocontrol Sci Technol 5:185–192

Ward MG (1984) Formulation of biological insecticides. In: Scher HB (ed) Advances in pesticide formulation technology, ACS Symposium series. American Chemical Society, Washington, pp 175–184

Wright JE, Knauf TA (1994) Evaluation of Naturalis-L for control of cotton insects. In: Brighton Crop Protection Conf, Pests and Diseases, British Crop Protection Council, Farnham, pp 45–52

Ying FW (1992) Current situation of *Beauveria bassiana* for pine caterpillar and its prospect in China. In: Proc XIX Int Congr of Entomology, Beijing, China, 300pp

Zimmermann (1992) Use of the fungus *Beauveria brongniartii* for control of the European cockchafer, *Melolontha melolontha* spp. in Europe. In: Jackson TA, Glare TR (eds) Use of pathogens in scarab pest management. Intercept, Andover, pp 199–208

Zimmermann G (1994) Strategies for the utilization of entomopathogenic fungi. In: VIth Int Colloq Invertebrate Pathology and Microbial Control, Montpellier, France, Aug 1994, Proc vol 1, pp 67–73

13 Fungi as Biological Control Agents for Plant Parasitic Nematodes

B.R. Kerry[1] and B.A. Jaffee[2]

CONTENTS

I. Introduction

Since Zopf (1988) observed that the nematode-trapping fungus, *Arthrobotrys oligospora*, was able to capture and colonise motile nematodes, nematophagous fungi have been considered as potential biological control agents for plant parasitic nematodes. Although most research on biological control of nematodes has concerned nematophagous fungi, few species have been studied in detail and little is understood concerning their mode of action or dynamics in soil.

[1] Department of Entomology and Nematology, IACR-Rothamsted, Harpenden, Herts AL5 2JQ, UK
[2] Department of Nematology, University of California at Davis, Davis, California 95616-8668, USA

There have been several reports of the natural control of specific nematode pests where suppressive soils have developed due to an increase in nematophagous fungi which limit nematode multiplication. Suppressive soils usually contain more than one species of fungus, but little is known of the dynamics and interactions of these fungal communities. Such information is essential if the suppression is to be managed effectively (Kerry 1988; Sikora 1992). Fungi may also kill nematodes by the production of toxins (Barron and Thorn 1987) or through the destruction of the feeding sites of sedentary nematodes in roots (Stiles and Glawe 1989). Plant pathogenic fungi often compete with nematodes in roots and significantly reduce nematode multiplication (James 1968); non-pathogenic isolates of these fungi may have potential as biological control agents for endoparasitic nematodes (Sikora 1992). Mycorrhizae improve the growth of nematode-infected plants and may also affect nematode development (Hussey and Roncadori 1982).

A few fungal species have been developed as products for the biological control of plant parasitic nematodes. None has been commercially successful and there are no biological control agents in widespread use. The problems and progress in the development of biological control strategies for nematodes have been extensively reviewed in recent years (Kerry 1987; Stirling 1991; Sikora 1992). Here, we briefly consider the types of fungi that parasitize nematodes and their biology before discussing in some detail the role of two fungi, the obligate parasite *Hirsutella rhossiliensis* and the facultative parasite *Verticillium chlamydosporium*, in the regulation of nematode populations. The need for detailed information on the biology and ecology of these fungi is stressed, as it is essential for an understanding of their potential impact on nematode populations in a range of conditions. Failures in the development of biological control agents have

The Mycota IV
Environmental and Microbial Relationships
Wicklow/Söderström (Eds.)
© Springer-Verlag Berlin Heidelberg 1997

resulted from a lack of information on the key factors that affect their efficacy.

II. Types of Nematophagous Fungi

The type of fungus used as a biological control agent against a specific nematode species depends on the mode of parasitism of the pest. Most plant parasitic nematodes kill the plant cells on which they feed and so must move to new feeding sites and remain active throughout their life cycle. Thus, fungi which parasitise these migratory nematodes must produce adhesive spores, or traps or toxins to immobilize their host before colonisation can occur. The major nematode pests such as cyst and root-knot nematodes have saccate females which are sedentary on or in roots and produce large numbers of eggs. Such sedentary stages may be colonised by hyphae in the rhizosphere or root tissue without the development of specialised structures, apart from appressoria.

Fungi which kill and colonize plant parasitic nematodes have been identifed in the Chytridiomycota, Oomycota, Zygomycotina, Deuteromycota, Basidiomycotina but not the Ascomycotina (Barron 1977). However, the telemorphic states of nematophagous species of fungi such as *Verticillium* and *Hirsutella*, when described, are likely to confirm their position in the Ascomycotina. Clearly, the nematode parasitic fungi are a diverse group, and the nematophagous habit is thought to have evolved

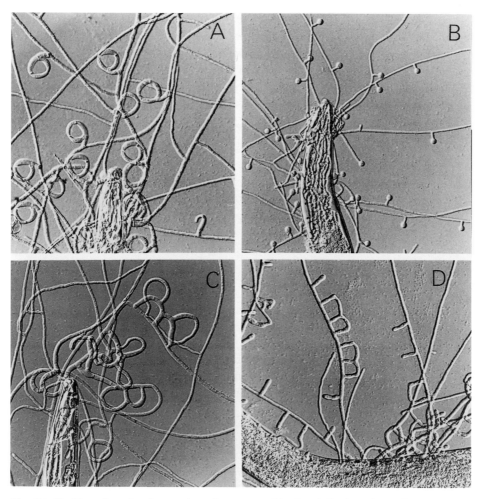

Fig. 1A–D. Trapping fungi growing from parasitised nematodes (*Steinernema glaseri*) in soil extract which contained no other nematodes. A portion of the parasitised nematode is present in each micrograph. **A** Constricting rings produced by *Arthrobotrys dactyloides*. **B** Adhesive knobs produced by *Monacrosporium ellipsosporum*. **C** Adhesive three-dimensional networks produced by *A. oligospora* (the networks appear to be two-dimensional because they have been flattened by the converslip. **D** Adhesive branches and two-dimensional networks produced by *M. cionopagum*. Magnification = 240×. (Jaffee et al. 1992a)

independently in the different fungal classes. As a consequence, the fungi are described here within three groups based on their modes of parasitism of nematodes rather than their systematic similarities.

A. Nematode-Trapping Fungi

Six different types of adhesive or non-adhesive trap are produced by species of nematode-trapping fungi on a sparse mycelium (Barron 1977). Although much work has been done on the capture and infection of nematodes by these fungi, in particular *A. oligospora* (Jansson and Nordbring-Hertz 1988), their biology and ecology are poorly understood. The simplest trapping device consists of the production of an adhesin on part or all of the hyphal surface of fungi in two genera of the Zygomycotina, *Stylopage* and *Cystopage*. Adhesive traps in the Deuteromycota are more specialized and consist of simple hyphal branches and knobs or more complex two- and three-dimensional networks (Fig. 1). Such devices are formed by *Monacrosporium cionopagum* (branches), *Monacrosporium ellipsosporum* (knobs) and *A. oligospora* (networks). Non-adhesive ring structures include the most highly developed trap, the constricting ring, produced by fungi such as *Arthrobotrys dactyloides* in which the three cells that form the ring expand rapidly in response to contact with a nematode on their inner surface; the aperture of the ring is closed and the nematode firmly held. Some fungi, such as *Dactylella candida* produce both adhesive knobs and non-constricting ring traps on the same mycelium. Trapping fungi produce large spores capable of supporting the development of a trap produced directly on the germ tube or on a limited mycelium (Barron 1977).

Nematode-trapping fungi have been found in a wide range of habitats including the bark of pine trees (Tubaki and Yamanaka 1984) and the thin soils of Antarctica (Gray 1985). They can be isolated readily from soil and cultured in vitro and, as a consequence, were the first organisms tested as biocontrol agents for nematodes.

B. Other Facultative Parasites

A wide range of Hyphomycotina have been isolated from the sedentary stages of nematodes and especially from eggs and females of cyst (*Heterodera* and *Globodera* spp.) and root-knot (*Meloidogyne* spp.) nematodes. These fungi have

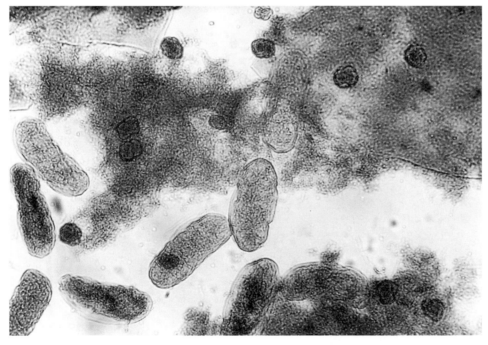

Fig. 2. Eggs of the cereal cyst nematode parasitised by *Verticillium chlamydosporium*. Eggs at all stages of embryonic development are colonised but those that are immature are parasitized more readily than those containing second-stage juveniles. Note the characteristic chlamydospores of the fungus. Magnification = 240×

been described as opportunistic because they do not produce specialized infection structures, apart from an appressorium, and they may survive and proliferate in soil in the absence of nematodes. Approximately 150 species of fungi have been isolated from saccate female nematodes, cysts and eggs from around the world, but their status as parasites is often not known (Kerry 1988). The most common fungi isolated include species of *Acremonium*, *Cylindrocarpon*, *Fusarium*, *Paecilomyces* and *Verticillium* (Fig. 2). In general, the root provides an effective barrier to colonization by nematophagous fungi, and few endoparasitic nematode stages are parasitised within roots. It is only when the egg masses of root-knot nematodes or the females of cyst-forming species emerge on the root surface that they are attacked by these fungi (Kerry et al. 1982; Kerry and de Leij 1992). Old cysts in soil contain more fungi than females on roots, but most of them are saprophytic and their role in the development and hatching of encysted eggs is unclear. Cysts are formed from the bodies of dead females and only their viable egg contents may be parasitised. *Verticillium chlamydosporium* is a typical facultative parasite and is discussed in detail below (Fig. 3).

Paecilomyces lilacinus has been tested most widely in the field for the biological control of cyst and root-knot nematodes. In many tests it is not possible to separate the effects of the organic carrier, used in considerable quantities to introduce the fungus into soil, from that of the fungus itself (Kerry 1990), but in some experiments significant biological control has been achieved (Jatala 1986). Control from applications of *P. lilacinus* has been variable and is much affected by environmental conditions. Improved methods of formulation and application may help provide more consistent levels of control.

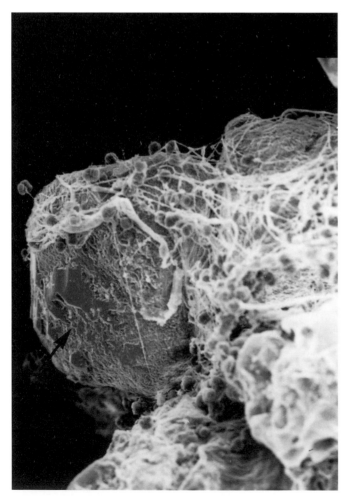

Fig. 3. Scanning electron micrograph of hyphae of *Verticillium chlamydosporium* spreading from the root surface to colonize an egg mass (*arrow*) of the root-knot nematode, *Meloidogyne incognita*. Chlamydospores are present on the surface of both root and egg mass. Magnification = 200×

C. Endoparasites of Vermiform and Saccate Nematodes

These fungi have very limited growth in soil outside the colonised nematode cadaver. Those that infect motile nematodes produce adhesive spores that attach to the cuticle of passing nematodes or motile zoospores that encyst on the nematode's surface. Chytridiomycota such as *Catenaria anguillullae* may be separated from Oomycota such as *Myzocytium* spp. and *Lagenidium* spp. in the development of the mycelium and in the number of flagella on the zoospore; chytrids have a single backward-directed flagellum, whereas the other fungi produce zoospores with two flagella which arise from a groove in their ventral surface. The movement of zoospores in soil requires water-filled pores (Duniway 1976) which do not favour the movement of nematodes (Wallace 1963). Both vermiform and sedentary nematodes are infected by representatives of these groups of fungi. Infection of females of the cereal-cyst nematode by the oomycete, *Nematophthora gynophila* (Fig. 4), was much affected by rainfall in June and July, when the nematodes were exposed to parasitism on the root surface (Kerry et al. 1982). Several different isolates of *C. anguillulae* have been compared (Voss and Wyss 1990) but

none was considered to have potential as a biological control agent; virulence appeared to be rapidly lost in vitro and the isolates tested were not virulent against the active stages of several tylenchid nematodes. Hyphomycotina such as *Hirsutella rhossiliensis* (discussed below) *Drechmeria coniospora* and some *Verticillium* spp. produce adhesive spores that attach to the cuticle of passing nematodes (Fig. 5). There have been no reports of these fungi parasitising sedentary nematodes although they will infect the motile second-stage juveniles of cyst and root-knot nematodes.

III. The Biology of Nematophagous Fungi

A. Infection Processes and Host Specificity

Before penetration and infection can occur, nematophagous fungi must contact a nematode host. This contact, or transmission, often depends on movement or growth by the fungus. Thus, the zoospores of *Nematophthora gynophila* and *Catenaria anguillulae* swim through the water films surrounding soil particles in soil pores, and hyphae of other fungi, such as *Verticillium*

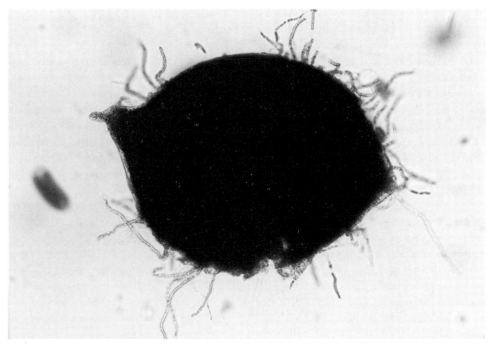

Fig. 4. Female cereal cyst nematode parasitized by *Nematophthora gynophila*. The body wall is disrupted and is penetrated by hyphae which act as exit tubes for the infective zoospores. The infected female produces no, or few viable eggs and fails to form a cyst. Magnification = 125×

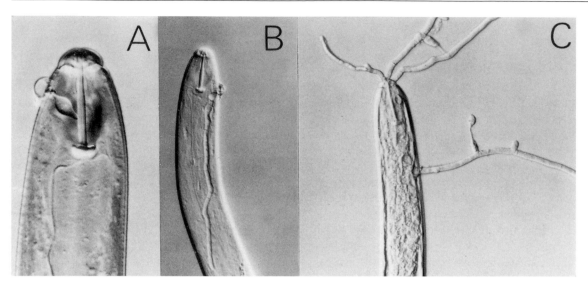

Fig. 5A–C. *Hirsutella rhossiliensis* parasitising *Heterodera schachtii* juveniles. **A** Conidium of the fungus adhered to the cuticle of the nematode. Note the post-infection bulb; magnification = 1000×. **B** An assimilative hypha extending from the postinfection bulb; the nematode host is still alive; magnification = 400×. **C** Within 3 days at 20 °C the host is packed with assimilative hyphae and is killed and the fungus has begun to sporulate. Thick-walled hyphae spread into the soil and give rise to phialides, which bear single conidia; magnification = 400×. (Jaffee 1992)

chlamydosporium and *Paecilomyces lilacinus*, grow from progagules or from a colonized substrate through soil or along the root surface to nematodes. In contrast, traps and adhesive conidia do not colonise soil; fungi that produce these structures depend on nematode movement for transmission.

Although transmission may result from random movement, there is evidence of directed movement by the fungus or the nematode. In glass capillaries, zoospores of *C. anguillulae* move towards their host (Jansson and Thiman 1992) and, on agar plates, bacterial-feeding nematodes are attracted to certain nematode-trapping and endoparasitic fungi (Field and Webster 1977; Jansson and Nordbring-Hertz 1979). Such attraction has yet to be documented in soil.

Simple physical contact between nematode and fungus does not guarantee infection, because the egg shell and cuticle of living nematodes are impenetrable to most soil fungi. Moreover, few fungi appear capable of penetrating the natural openings of living nematodes. Most nematophagous fungi penetrate nematodes directly through the egg shell or cuticle and must adhere to, or otherwise maintain contact with, the nematode as a prerequisite to penetration (Tunlid et al. 1992). Such fungi may be firmly attached to the nematode surface by an appressorium or appressorium-like structure from which a penetra-

tion tube develops (Barron 1973; Stirling and Mankau 1978; Dijksterhuis et al. 1990; Kim et al. 1992; Lopez-Llorca and Robertson 1993; Sjollema et al. 1993). Even the adhesive conidium of *H. rhossiliensis* and the rings and adhesive structures of the trapping fungi act like appressoria (Wimble and Young 1983) in that they ensure the firm attachment required for penetration (Dijksterhuis et al. 1994; Tunlid et al. 1992).

Transmission triggers directed growth of the nematophagous fungus into the nematode, and mechanisms controlling that growth were reviewed recently (Dackman et al. 1992; Tunlid et al. 1992; Dijksterhuis et al. 1994). Typically, a narrow penetration tube grows through the nematode eggshell or cuticle, which sometimes is indented just before penetration (Sturhan and Schneider 1980; Sjollema et al. 1993), suggesting mechanical pressure. Enzymes also are produced and may weaken the cuticle (Dackman et al. 1992; Dijksterhuis et al. 1994). Penetration of the egg shell by *V. chlamydosporium* is an enzymatic and physical phenomenon (Morgan-Jones et al. 1983; Segers et al. 1994). Subtilisin serine proteases are thought to be key enzymes in the infection of nematode eggs by *V. chlamydosporium* (Segers et al. 1994, 1995, 1996) and by *Paecilomyces lilacinus* (Bonants et al. 1995). It takes several days to destroy a nematode egg and eventually the mycelium lyses, leaving an empty, shrivelled egg shell;

in general, immature eggs which do not contain second-stage juveniles are more susceptible to infection, but eggs in all stages of embryonic development may be colonised. After penetration, many fungi form a spherical infection bulb inside the nematode egg or body cavity. The function of this bulb is not known, but it seems to act as a transitional structure between penetration and assimilation (Dijksterhuis et al. 1994). Perhaps the sphere is the most efficient shape (i.e. it may require the least amount of cell wall synthesis) for the limited amount of cytoplasm which has moved from the appressorium, through the penetration tube, and into the nematode.

In contrast to insects, nematodes apparently offer little resistance to infection once the egg or body cavity has been penetrated. Thin-walled hyphae grow through and digest and assimilate the host, which is killed early or late in the process, depending on the fungus and nematode. Generally, the cuticle or egg shell is not totally degraded and eventually delimits fungal rather than nematode tissue. The host ranges of nematophagous fungi often are broad, possibly because one species of nematode seldom dominates in the soil, and extreme selectivity would therefore be disadvantageous. Nevertheless, not all fungi parasitize all nematodes, perhaps because egg shell and cuticle biochemistry differ between nematode species.

B. Proliferation

Proliferation of obligate parasites of nematodes depends greatly on nematode abundance, and is likely to be density-dependent (Jaffee et al. 1992b), which is discussed in greater detail in the section on *Hirsutella rhossiliensis* (Sect. IV.B). Many species of nematophagous fungi are facultative parasites and their dependence on nematodes for their growth and survival differs greatly. For example, *V. chlamydosporium* utilizes root exudates and grows along root surfaces even in the absence of nematode hosts (see Sect. V.B). *Paecilomyces lilacinus* is a common soil saprophyte and may be an opportunistic parasite of nematode eggs, whereas species of the Basidiomycotina degrade wood and may trap nematodes in order to obtain nitrogen in the nitrogen-poor environment of rotting logs (Thorn and Barron 1984; Barron 1992). Obviously, predicting the growth and survival of facultative para-

sites requires an understanding of how they compete for and utilize both nematodes and other substrates.

The dependence of nematode-trapping fungi on nematodes and other substrates is unclear (Persson and Baath 1992; Jaffee et al. 1993). Drechsler (1937) believed that the growth of *A. oligospora*, *A. dactyloides* and related species on agar was a laboratory artefact and that these fungi functioned primarily as parasites in soil. This position was challenged by Cooke (1968), and most reviews suggest that saprophytism is important to at least some species (Dackman et al. 1992; Dijksterhuis et al. 1994). Differences in fungal growth rate and trap formation on agar plates and in competitive saprophytic ability in soil (Cooke 1963) do suggest that some species are more saprophytic than others. Moreover, some species appear to colonize organic matter rapidly when soil containing organic debris is sprinkled on agar plates (Drechsler 1937). Cooke (1963), however, examined seven species of nematode-trapping fungi and reported that none exhibited appreciable competitive saprophytic ability in soil and that none could be considered a good saprophyte. The accepted paradigm, which seems to reconcile the views that trapping fungi are both effective and ineffective saprophytes, is that capture of nematodes enables trapping fungi to compete saprophytically (Cooke 1968; Thorn and Barron 1984; Quinn 1987; Barron 1992). This explanation is reasonable but it is virtually untested. Before accepting their relative importance, saprophytism and parasitism should be quantified in the natural environment, but this task is not easy.

C. Dispersal

Nematophagous fungi lack the means for long-distance dispersal but have many active and passive ways to achieve local dispersal. Thus, zoospores swim short distances and hyphae grow through soil. Several species of nematode-trapping fungi spread their traps through soil by producing them on hyphae that radiate from the parasitised nematode (Jaffee et al. 1992a). Conidiophores of nematode-trapping fungi may function similarly, in that they are long (often >300 mm) and extend through soil pores and thus produce conidia at some distance from the host.

In all of the above examples, dispersal is active and involves expenditure of energy by the fungus. In other examples, dispersal is passive. The microconidia of *V. chlamydosporium* are small enough to move passively with water through soil pores (de Leij et al. 1993a). Conidia of *Nematoctonus* spp. and *Drechmeria coniospora* also seem adapted for passive dispersal, as they are unable to adhere to nematodes until after they have separated from the conidiophore (Barron 1977; Dijksterhuis et al. 1990).

Soil animals and especially nematodes may disperse nematophagous fungi. For example, conidia of *Drechmeria coniospora* adhere to motile nematodes, which continue to move for many hours before they die; this phenomenon has been used to introduce and spread the fungus through soil (Jansson et al. 1985). However, second-stage juveniles of cyst or root-knot nematodes with adherent conidia of *H. rhossiliensis* may penetrate roots and avoid fungal infection if they feed and moult before the fungus penetrates (Jaffee and Muldoon 1989). Trapping fungi generally restrain the host, but some kinds of traps detach readily from the parent hypha, and other traps will detach if the nematode is large or vigorous; detached traps nevertheless produce penetration pegs and nematodes are infected.

D. Survival

When nematodes or other suitable substrates are scarce, nematophagous fungi may persist in various resting structures including thick-walled conidia, clamydospores and oospores. The parasitized host also may serve as a resting structure, because the host cuticle may protect the otherwise vulnerable hyphae. *Hirsutella rhossiliensis* has been isolated from parasitized nematodes extracted from air-dried field soil stored in the laboratory (B.A. Jaffee, unpubl.). Under moist conditions, parasitized nematodes rapidly disappear from soil as the fungus converts the material and energy in the assimilative hyphae within the nematode host into new fungal structures external to the host (Kerry and Crump 1977; Jaffee 1992).

Survival may be reduced by a wide range of soil organisms that consume or otherwise antagonize fungi. These organisms include enchytraeids, collembola, nematodes, mites, amoebae, bacteria and other fungi. The significance of these natural enemies is largely unknown. In fact, surprisingly little information is available on the persistence of traps, conidia, hyphae and other structures in soil.

This information is fundamental for understanding fungal activity and for predicting the success of biological control.

IV. *Hirsutella rhossiliensis* as a Regulator of Nematode Populations

A. Taxonomy, Life History and Variability

Most species in the genus *Hirsutella* are parasites of arthropods (Minter and Brady 1980), and at present, *H. rhossiliensis* is the only one known to attack nematodes. Although this species was described by Minter and Brady (1980), it was Sturhan and Schneider (1980) who first reported and detailed its nematophagous habit. *Hirsutella rhossiliensis* is a Deuteromycota, with no sexual stage reported. Because the fungus grows slowly in culture, has little competitive saprophytic ability, and produces specialised conidia for attacking nematodes, it is assumed to be an obligate parasite in nature (Jaffee and Zehr 1985; Jaffee 1992).

The life history of *H. rhossiliensis* is simple (Jaffee et al. 1990). Transmission occurs when the conidium (which is non-motile, surrounded by an adhesin, and borne on a phialide) attaches to a passing nematode. After 12 h at 25 °C, a germ tube penetrates the cuticle directly beneath the conidium; virtually every conidium that adhered was able to penetrate the cyst nematode *Heterodera schachtii* unless temperatures were >30 °C or less than 15 °C (Tedford et al. 1995). *Hirsutella rhossiliensis* forms and infection bulb once inside the host from which thin-walled assimilative hyphae grow and ramify through the nematode, which is dead and completely filled with hyphae within about 3 days. If conditions are suitable, the fungus grows from the host and produces new conidia; during conidiation, the assimilative hyphae within the nematode become increasingly transparent because sporulation utilises their cytoplasm. The hyphae that grow through the soil are thick-walled and may extend 3 mm or more from the host and thus help disperse the conidia. The phialide of *H. rhossiliensis* is essential for transmission because conidia detached from phialides do not adhere to passing nematodes (McInnis and Jaffee 1989). This is true for certain other nematophagous fungi, and Barron (1977) discusses the role of the conidiophore in positioning the conidium in the soil pore, where it will be more likely to encounter nematodes rather than fungivores. Although only recently described, *H. rhossiliensis* has been isolated from many species of nematodes in many parts of the world. Genetic variation among isolates is substantial, but not variation in pathogenicity and morphology (Tedford et al. 1994). Apparently, the populations are isolated and undergo molecular divergence, but the genes controlling pathogenicity are either not subject to change or are under similar selection pressures, regardless of geographical location, host nematode and crop.

B. Evidence for Density-Dependent Parasitism

At any moment, the size of the conidial pool (number of conidia/cm^3 soil) largely determines the probability of transmission (the chance of a nematode encountering at least one conidium). When conidia are dense, this probability is high, and when conidia are scarce, the probability is low. The current probability of transmission, however, is not greatly affected by current nematode density because transmitted conidia, although lost from the conidial pool, usually represent a minute portion of that pool (Jaffee et al. 1992b). Current nematode density has little influence on the immediate probability of parasitism of individual nematodes, but it greatly influences the number of nematodes encountering conidia. Increased numbers of nematodes mean increased numbers of encounters. Thus, when the numbers of nematodes increase, more conidia are produced which will, in turn, increase the probability of transmission. Negative feedback is central to density-dependent parasitism; as high densities of nematodes lead to high densities of conidia, these in turn lead to reduced densities of nematodes and then of conidia.

The self-regulation implicit in density-dependent parasitism is attractive. Like a pesticide or a resistance gene, the fungus should suppress nematode density but, unlike a pesticide or resistance gene, the pressure exerted by the fungus will change with nematode density and therefore will tend to regulate nematode density in a dynamic equilibrium. Time delays are an important feature of this and similar systems and both help and hinder attempts to achieve biological control. On the one hand, the probability of parasitism may increase slowly because conidial density increases only after nematode density increases; the processes of transmission, infection and sporulation take time. On the other hand, the probability of parasitism may remain high even after numbers of nematodes decline if the conidia or other fungal structures are persistent.

Observation, experimentation and theory indicate that parasitism of nematodes by *H. rhossiliensis* is density-dependent. In the field, the percentage of nematodes parasitized was correlated with nematode density when many samples were collected at one time in a mature peach orchard (Jaffee and McInnis 1991). In soil microcosms in the laboratory, the probability of parasitism increased to nearly 100% when a high nematode density was maintained, and declined to near 0% at low densities. A mathematical model based on density-dependent parasitism closely described the laboratory dynamics (Jaffee 1992).

C. Population Regulation

The regulation provided by density-dependent processes may be strong or weak, depending on the biology of the host and parasite and on the biotic and abiotic environment. Unfortunately, *H. rhossiliensis* exerts little population control unless host densities have been high. In orchards, the probability of parasitism does not reach high levels for 7 years or more from planting (Underwood et al. 1994). In laboratory experiments and in computer models, the rates of transmission, sporulation, and conidium mortality are such that the conidial pool increases only slowly (Jaffee 1992).

There are several ways of increasing the rate at which the pool of conidia enlarges, and one was proposed almost 70 years ago by Linford et al. (1938) which focused on nematode-trapping fungi, but is also relevant to *H. rhossiliensis*. They suggested that the addition of organic matter to soil enhanced the density of bacterial-feeding nematodes, and that these non-pest nematodes acted as alternative hosts for nematode-trapping fungi, which tend to have broad host ranges. Thus, an increase in bacterial-feeding nematodes should support an increase in nematode-trapping fungi, which, in turn, should result in a greater probability of parasitism of all nematodes, including pest nematodes. This scenario is similar to the density-dependent process described for *H. rhossiliensis*, except that here the increase in the pool of traps depends on high densities of a non-damaging host. Whether this systematically operates with trapping fungi remains unclear, but recent experiments indicate that it does occur with the endoparasitic fungus *Drechmeria coniospora* (van den Boogert et al. 1994). However, several organic amendments suppressed rather than enhanced parasitism by *H. rhossiliensis*, perhaps due to fungicidal levels of ammonia (Jaffee et al. 1994).

D. Inundative Release

As an alternative to enhancing density-dependent processes, sufficient quantities of inoculum might be added to soil that the density of conidia increases rapidly. If added at or before planting, the inoculum might protect seedlings from the initial invasion of nematodes. The conidium is the most obvious form of *H. rhossiliensis* inoculum to add to soil but was found to be ineffective; as noted before, conidia must be produced in situ to adhere to nematodes (McInnis and Jaffee 1989).

Nematodes colonized by *H. rhossiliensis* have been used as fungal inoculum for experiments but cannot be produced in sufficient quantity for treatment of large volumes of soil. In some respects, the colonized nematode is analogous to a mass of assimilative hyphae. Such a mass, grown in simple shake culture, was found to support sporulation when added to soil either alone or when dried in an alginate matrix; the conidia produced from the pelletized hyphae adhered to and infected nematodes (Lackey et al. 1992, 1993, 1994). Assimilative hyphae of two nematode-trapping fungi also produce traps when added to soil as pelletized hyphae (Jaffee and Muldoon 1995).

V. The Ecology of *Verticillium chlamydosporium* and Biological Control of Cyst and Root-Knot Nematodes

A. Taxonomy, Life History and Variability

Verticillium chlamydosporium is a Deuteromycota and a facultative parasite recorded from the eggs of snails (Barron and Onions 1966), nematodes (Willcox and Tribe 1974) and fungal hyphae (Sneh et al. 1977). The fungus is widespread and has been isolated from cyst and root-knot nematodes around the world (Kerry 1988).

Verticillium chlamydosporium may be part of a complex of similar species which have been isolated from nematode eggs and there is considerable variation between different isolates in terms of their virulence, in vitro growth, and ability to colonize the rhizosphere; other species infect motile stages (Gams 1988). The teleomorph is considered to belong to the Cordiceps but has not been described (H. Evans, pers. comm.). The fungus produces thick-walled dictyochlamydospores on short pedicels and conidia on simple phialides (Barron and Onions 1966). *Verticillium chlamydosporium* extensively colonizes the rhizosphere of several plant species but does not penetrate the epidermis and spread in the root cortex (de Leij and Kerry 1991). Colonisation of soil organic matter may be limited, but the passive spread of conidia, at least down the soil profile, may be considerable.

Verticillium chlamydosporium does not produce trapping structures or adhesive spores and there is no evidence that motile nematodes are colonized in soil, although immobile second-stage juveniles of cyst and root knot nematodes may be colonized on agar. Infection occurs when a hypha contacts a nematode egg or sedentary female in the rhizosphere. An appressorium has been observed when the fungus infects eggs (see Sect. III.A). Infection of root-knot nematode eggs by an isolate of *V. chlamydosporium* was reduced at temperatures above 30°C, as many eggs hatched and the second-stage juveniles escaped before the fungus colonised the whole egg mass (de Leij et al. 1992). Infection of cyst nematode females within 2 weeks of their emergence on the root surface results in premature maturation to a small cyst, reduced fecundity and extensive parasitism of the eggs (Kerry 1990). Late infection of the female may not affect cyst size, and few eggs may be destroyed. Hyphae, chlamydospores and conidia have been observed within infected cyst nematodes and root-knot egg masses, on the root surface, and have also been extracted from soil.

B. Regulation of Nematode Populations

The fungus has been associated with soils which suppress the multiplication of cyst nematode populations (Kerry et al. 1982; Thomas 1982). In such soils, there are usually several species of parasitic fungus occurring together and it is difficult to elucidate the importance of individual species. However, *V. chlamydosporium* was significantly more abundant in soils in which cereal and beet cyst nematode multiplication was suppressed than where these nematodes multiplied (Kerry et al. 1993). Also, in some soils where *V. chlamydosporium* was the only important nematophagous fungus present, nematode populations still declined and a reduction in the activity of the fungus through partial soil sterilisation with formalin (38% formaldehyde) resulted in significant increases in nematode multiplication (B.R. Kerry, unpubl. data). Thus, the fungus appears to be an important natural regulator of some nematode populations and is active in a range of soil types. Athough the activity of nematophagous fungi may be affected by the addition of various soil amendments (Rodriguez-Kabana et al. 1987), natural control is slow to establish in soil and difficult to manipulate.

Most work on the regulation of nematode populations through the application of *V. chlamydosporium* has concerned the control of root-knot nematodes. Isolates differ in their ability to kill cyst and root-knot nematodes, but selected isolates have proved effective against several species of the latter (de Leij and Kerry 1991). In general, the control of root-knot nematodes depends greatly on factors which affect the growth of the fungus during its saprophytic phase.

The effective control of *Meloidogyne* spp. is dependent on the colonisation of the rhizosphere. Isolates of the fungus which are unable to colonise the rhizosphere extensively fail to cause significant control of these nematodes (de Leij and Kerry 1991). Also, the plant species and cultivar have marked effects on the extent of colonization, with plants such as maize supporting ten times the numbers of colony-forming units compared to sorghum (Kerry and de Leij 1992). The presence of nematodes may also increase the growth of the fungus, presumably by their influence on the supply of nutrients in the rhizosphere. In general, the more abundant the fungus is in the rhizosphere, the greater the extent of parasitism and the control of nematode multiplication. However, in plants very susceptible to *Meloidogyne* spp. or at large nematode densities, egg masses remain embedded in the large galls and are not exposed to parasitism by the fungus, which is confined to the surface of the roots. In such situations, significant numbers of nematode eggs escape infection and control is poor (Kerry 1993). Proliferation of the fungus after application to soil is less in mineral than organic soils (de Leij et al. 1993b) and is affected by temperature (de Leij et al. 1992).

Cyst nematode females are exposed on the surface of roots to parasitism by *V. chlamydosporium* for several weeks and interactions tend to be more complex than those with root-knot nematodes. As discussed, early infection of the female nematodes results in the development of few healthy eggs. There was no simple relationship between the extent of root colonisation by *V. chlamydosporium* and the multiplication of *H. schachtii* on oilseed rape and sugar beet (Clyde 1992). Although there were similar amounts of fungus in the rhizospheres of sugar beet and oilseed rape and numbers of females on untreated plants, there was much greater nematode control on oilseed rape. Presumably, there were differences in the distribution of the nematode or differences in the physiological state of the fungus on sugar beet which significantly reduced the nematode control.

C. Potential in Biological Control

Clearly, the growth of *V. chlamydosporium* in the rhizosphere and its efficacy in the regulation of nematode populations are affected by several key factors. *Verticillium chlamydosporium* can be cultured on a range of solid and liquid media. Approximately 5×10^6 chlamydospores g^{-1} medium develop on a 1:1 sand barley bran mixture in 3 weeks at $20\,^{\circ}$C, whereas few chlamydospores are produced in Czapek Dox broth and other liquid media, which support only hyphal growth and the development of conidia. Applications of the fungus require an additional nutrient source to support growth in soil if the inoculum has been produced in liquid culture but chlamydospores contain sufficient reserves and can be applied successfully in aqueous suspensions (de Leij and Kerry 1991). Little work has been done on the production, formulation and storage of the fungus. Chlamydospores are the most convenient form of inoculum and have provided consistent control of nematode multiplication in a range of laboratory tests and a small plot trial (de Leij et al. 1993b). In this trial, a broadcast application of 5000 chlamydospores g^{-1} soil incorporated at planting established the fungus throughout the growing season and controlled a small population of *M. hapla* on a tomato crop grown in microplots. However, the failure to produce such spores in liquid fermentation may severely restrict commercial development.

There is a need to test the fungus in a range of conditions, but it is unlikely that this fungus will prevent increase of nematode populations on highly susceptible crops, and it would probably be more successful if used in combination with poor hosts for the nematode (Kerry 1995). As the fungus does not reduce the initial invasion of the root system, its effect on nematode populations is akin to that of a resistant cultivar, and at large nematode infestations crop damage may be significant unless tolerant crops are grown. *Verticillium chlamydosporium* shows considerable promise, but more trials are needed to assess its potential as a biological control agent.

VI. Exploitation of Nematophagous Fungi

The potential of nematophagous fungi in the development of new management strategies for nematode pests would appear to be considerable. In much of northern Europe, farmers grow susceptible cereal crops continuously in the presence of the potentially damaging cereal cyst nematode without significant yield loss because of the buildup of nematophagous fungi, especially *N. gynophila* and *V. chlamydosporium*, in soil. The biological control of cereal cyst nematode has proved a sustainable method of nematode management in intensive agriculture and has effectively controlled this pest for at least the last 20 years in a wide range of soils. Soils which suppress the multiplication of a number of nematode pests on susceptible crops have been reported around the world. Such natural control has been slow to establish and the fungal agents involved have been difficult to manipulate in soil using methods that are practical for the grower. Green manure crops are known to increase the activity of some facultative parasites of nematode eggs (Schlang et al. 1988) but not of obligate parasites, which may be inhibited (Jaffee et al. 1994); other practices, such as partial soil sterilization, may be useful in increasing the rate of establishment of antagonistic organisms in soil. To date, the degree of control observed in naturally suppressive soils has not been achieved through the application of specific fungal agents. For effective and consistent nematode control, experience from field and laboratory experiments has highlighted the need to integrate biological agents with other measures to reduce nematode infestations. These measures include the use of nematicides, crop rotation, solarization and partial soil sterilants, as well as methods to increase the activity of nematophagous fungi such as green manures and specific soil amendments.

The stage of the nematode attacked has a marked effect on the exploitation of the agent. For example, fungi that parasitise second-stage juveniles of root-knot nematodes may kill sufficient nematodes to reduce plant damage and provide significant yield increases but are unlikely to control nematode multiplication in the long term. Reduction in the numbers of juveniles results in less intraspecific competition and the fecundity of females which survive is increased, especially if plant growth is also improved. Hence, nematode populations will only be controlled if most juvenile stages are killed (Kerry 1980); few nematicides and no agent which attacks juveniles are able to prevent nematode population growth, especially if there is more than one generation in a growing season. Generally, those fungi that kill adult females will have little effect on plant damage in the short term but may control nematode multiplication. The influence of nematode density on the activity of nematophagous fungi has been discussed. In general, obligate parasites are more likely to establish if they are applied to soils where nematodes are numerous, whereas facultative parasites may be considered as a preventive treatment to stop the buildup of damaging infestations.

The nematode target must be considered in the selection of a suitable control agent and its method of application. Nematodes with large multiplication rates or those which are damaging at very low densities, such as those which are vectors of virus diseases, are likely to be difficult to control with a biological agent alone. The introduction of biological control agents on seeds is a favoured method of application which enables large areas of crops to be treated with a minimum of inoculum (Rhodes 1993). However, few nematophagous fungi are sufficiently aggressive rhizosphere colonisers to provide prolonged root protection from such a type of inoculum, and seed application is likely to be successful in reducing nematode damage only over a relatively short period of time. Initially, the most appropriate approach to exploit nematophagous fungi may be the application of selected isolates to protect horticultural crops, as there are opportunities to add agents during transplantation in the form of bare root dips or in peat blocks. Hence, it is often easier to deliver inundative treatments to the site of action in horticultural crops than in field crops where the agent must be introduced at planting to relatively large volumes of soil. The commercial development of biological control agents for plant parasitic nematodes will depend on the careful selection of an agent to meet the specific requirements of a niche market (Powell 1993). Even when promising control agents have been identified, difficulties in mass production and formulation and limited market opportunities have restricted the development of biological control agents for plant parasitic nematodes. However, further planned restrictions on the use of nematicides, especially the ban on methyl bromide in the year 2000, are likely to change attitudes towards the use of biological control for nematodes on some crops.

Nematophagous fungi may also be exploited through the identification, characterisation and production of specific compounds, such as enzymes, toxins and adhesins, which they produce, and which could form the basis of novel nematode control strategies or other applications. Although several agrochemical companies have tested many soil microorganisms for the production of nematotoxins, the modes of action of nematophagous fungi have been little studied and few key compounds have been identified.

VII. Conclusions

Surveys made of nematophagous fungi in similar or different habitats around the world have frequently led to the identification of new species but, in general, there is a restricted microflora capable of attacking nematodes, and several species have a worldwide distribution. Hence, as several selected isolates of nematophagous fungi have shown considerable potential as biological control agents, it would seem more productive to devote limited research resources to the exploitation of known fungi rather than to continue the search for ever more effective agents. However, several countries have now banned the release of non-indigenous organisms, and so surveys remain essential for the identification and selection of local fungi; of course, fungi more effective than those already isolated may be found. In such situations,

it is important that the nematophagous fungi isolated and tested are maintained as pure cultures within international collections and are accessible for comparison with isolates obtained locally.

The potential of molecular techniques in the development of biological control agents for plant parasitic nematodes would appear to be great but it is largely unexploited. The development of sensitive diagnostic methods based on molecular techniques could be used in studies of the epidemiology and ecology of nematophagous fungi in soil. Such studies currently depend on the extraction of colonized hosts or the isolation of fungal propagules on semiselective media. Dilution plating techniques may provide information on relative changes in abundance, but it is often impossible to distinguish colonies that have originated from hyphal fragments, conidia or chlamydospores. Increases in colony counts on dilution plates may be due, therefore, to increased vegetative growth or to sporulation, and estimates of the abundance of *V. chlamydosporium* on roots may not be simply related to the levels of nematode control in different soils (de Leij et al. 1993b). There is a need to develop methods such as ELISA that may enable different fungal propagules to be quantified in soil (Bourne et al. 1994). Immunological methods are well developed for the quantification of microorganisms whereas DNA techniques remain largely qualitative, although the use of PCR is likely to enable estimates of population densities to be made in the future (Curran and Robinson 1993). Methods based on DNA techniques are usually very discriminating and have been used to separate isolates of *Verticillium* spp. (Carder et al. 1993; Arora et al. 1996). They are likely to play an important role in understanding variation within a taxon and the phylogeny of different groups of nematophagous fungi, but they are not able to distinguish between different stages of development within a species. The regulatory procedures governing the release of organisms are dependent upon the development of suitable methods of monitoring which may need to separate introduced from indigenous isolates of the same species of fungus. Such procedures are likely to depend on the development of suitable molecular markers and DNA probes. The rational selection of fungal strains and their improvement by genetic engineering or by crossing strains via the parasexual cycle may be possible (Clarkson 1992). However, such developments are dependent on a thorough understanding of the modes of action of selected fungal isolates and the characterisation of the key enzymes and toxins involved.

There is still a need for a critical evaluation of the potential biological control agents for plant parasitic nematodes in commercial cropping situations and their integration with existing production systems. Too often, agents have been released without knowledge of their biology and ecology, and their use has been limited because nematode control was often poor and inconsistent. Also, few attempts have been made to monitor the activity of agents after their release and so there is no understanding of the reasons for poor control. Detailed studies of both *H. rhossiliensis* and *V. chlamydosporium* have highlighted the need for such information before an evaluation of the potential of these fungi as biological control agents can be made. Basic research on the mode of action and epidemiology of selected fungal agents must be supported by developments in methods of mass production, formulation and application, and will frequently require inputs from appropriate commercial organisations. The development of biological control strategies usually requires much research, which has largely been done in laboratories within the public sector. If fungi are to be exploited for nematode control by growers, their selection and development must be based on sound science and their use carefully targeted to ensure commercial interest. There is little doubt that the consumers and many growers are keen to find alternatives to the use of nematicides. To respond to these challenges and develop sustainable methods of nematode management will require close collaboration between scientists in both public and private organizations.

References

Arora DK, Hirsch PR, Kerry BR (1996) PCR-based molecular discrimination of *Verticillium chlamydosporium* isolates. Mycol Res 100:801–809

Barron GL (1973) Nematophagous fungi: *Rhopalomyces elegans*. Can J Bot 51:2505–2507

Barron GL (1977) The nematode-destroying fungi. Canadian Biological Publications, Guelph, Ontario, Canada

Barron GL (1992) Lignolytic and celluloytic fungi as predators and parasites. In: Carroll GC, Wicklow DT (eds) The fungal community: its organization and role in the ecosystem. Dekker, New York, pp 331–326

Barron GL, Onions AHS (1966) *Verticillium chlamydosporium* and its relationships to *Diheterospora*, *Stemphyliopsis* and *Paecilomyces*. Can J Bot 44:861–869

Barron GL, Thorne RG (1987) Destruction of nematodes by species of *Pleurotus*. Canad. J Bot 65:774–778

Bonants PJM, Fitters PFL, Thijs H, den Belder E, Waalwijk C, Henfling JWDM (1995) A basic serine protease from *Paecilomyces lilacinus* with biological activity against *Meloidogyne hapla* eggs. Microbiology 141:775–784

Bourne JM, Kerry BR, de Leij FAAM (1994) Methods for the study of *Verticillium chlamydosporium* in the rhizosphere. J Nematol 26(S):587–591

Carder JH, Segers R, Butt TM, Barbara DJ, von Mende N, Coosemans J (1993) Taxonomy of the nematophagous fungi *Verticillium chlamydosporium* and *V. suchlasporium* based on secreted enzyme activities and RFLP analysis. J Invert Pathol 62:178–184

Clarkson JM (1992) Molecular biology of filamentous fungi used for biological control. In: Kinghorn JR, Turner G (eds) Applied molecular genetics of filamentous fungi. Blackie Academic and Professional, Glasgow, pp 175–190

Clyde J (1992) Studies on the infection of *Heterodera schachtii* by the nematophagous fungus *Verticillium chlamydosporium*. In: Jensen DF, Hockenhull J, Fokkema NH (eds) New approaches in biological control of soil-borne diseases. IOBC/WPRS Bull, pp 67–69

Cooke RC (1963) Ecological characteristics of nematode-trapping hyphomycetes. I. Preliminary studies. Ann Appl Biol 52:431–437

Cooke RC (1968) Relationships between nematode-destroying fungi and soil-borne phytonematodes. Phytopathology 58:909–913

Curran J, Robinson MP (1993) Molecular aids to nematode diagnosis. In: Evans K, Trudgill DL, Webster JM (eds) Plant parasitic nematodes in temperate agriculture. CAB International, Wallingford, pp 545–564

Dackman C, Jansson H-B, Nordbring-Hertz B (1992) Nematophagous fungi and their activities in soil. In: Stotzky G, Bollag JM (eds) Soil biochemistry, vol 7. Dekker, New York, pp 95–130

de Leij FAAM, Kerry BR (1991) The nematophagous fungus, *Verticillium chlamydosporium*, as a potential biological control agent for *Meloidogyne arenaria*. Rev Nematol 14:157–164

de Leij FAAM, Dennehy JA, Kerry BR (1992) The effect of temperature and nematode species on interactions between the nematophagous fungus *Verticillium chlamydoporium* and root-knot nematodes (*Meloidogyne* spp.). Nematologica 38:65–79

de Leij FAAN, Dennehy JA, Kerry BR (1993a) Effect of watering on the distribution of *Verticillium chlamydosporium* in soil and the colonization of egg masses by the fungus. Nematologica 39:250–265

de Leij FAAM, Kerry BR, Dennehy JA (1993b) *Verticillium chlamydosporium* as a biological control agent for *Meloidogyne incognita* and *M. hapla* in pot and microplot tests. Nematologica 39:115–126

Dijksterhuis J, Veenhuis M, Harder W (1990) Ultrastructural study of adhesion and initial stages of infection of nematodes by conidia of *Drechmeria coniospora*. Mycol Res 94:1–8

Dijksterhuis J, Veenhuis M, Harder W, Nordbring-Hertz B (1994) Nematophagous fungi: structure-function relationships and physiological aspects. Adv Microb Physiol 36:111–143

Drechsler C (1937) Some hyphomycetes that prey on free-living terricolous nematodes. Mycologia 29:447–552

Duniway (1976) Movement of zoospores of *Phytophthora cryptogen* in soils of various textures and matric potentials. Phytopathology 66:877–882

Field JI, Webster J (1977) Traps of predacious fungi attract nematodes. Trans Br Mycol Soc 68:467–469

Gams W (1988) A contribution to the knowledge of nematophagous species of *Verticillium*. Neth J Plant Pathol 94:123–148

Gray NF (1985) Nematophagous fungi from the maritime antarctic: Factors affecting distribution. Mycopathologia 90:165–176

Hussey RS, Roncadori RW (1982) Vesicular-arbuscular mycorrhizae may limit nematode activity and improve plant growth. Plant Dis 66:9–14

Jaffee BA (1992) Population biology and biological control of nematodes. Can J Microbiol 38:359–364

Jaffee BA, Muldoon AE (1989) Suppression of cyst nematode by natural infestation of nematophagous fungus. J Nematol 21:505–510

Jaffee BA, McInnis TM (1991) Sampling strategies for detection of density-dependent parasitism of soil-borne nematodes by nematophagous fungi. Rev Nematol 14:147–150

Jaffee BA, Muldoon AE (1995) Susceptibility of root-knot and cyst nematodes to the nematode-trapping fungi *Monacrosporium ellipsosporum* and *M. cionopagum*. Soil Biol Biochem 27:1083–1090

Jaffee BA, Zehr EI (1985) Parasitic and saprophytic abilities of the nematode-attacking fungus *Hirsutella rhossiliensis*. J Nematol 17:341–345

Jaffee BA, Muldoon AE, Phillips R, Mangel M (1990) Rates of spore transmission, mortality, and production for the nematophagous fungus *Hirsutella rhossiliensis*. Phytopathology 80:1083–1088

Jaffee BA, Muldoon AE, Tedford EC (1992a) Trap production by nematophagous fungi growing from parasitized nematodes. Phytopathology 82:615–620

Jaffee BA, Phillips R, Muldoon AE, Mangel M (1992b) Density-dependent host pothogen dynamics in soil microcosms. Ecology 73:495–506

Jaffee BA, Tedford EC, Muldoon AE (1993) Tests for denity-dependent parasitism of nematodes by trapping and endoparasitic fungi. Biol Contr 3:329–336

Jaffee BA, Ferris H, Stapleton JJ, Norton MVK, Muldoon AE (1994) Parasitism of nematodes by the fungus *Hirsutella rhossiliensis* as affected by certain organic amendments. J Nematol 26:152–161

James GL (1968) The interrelationships of the causal fungus of brown root rot of tomatoes and potato root eelworm, *Heterodera rostochiensis* Woll. Ann Appl Biol 61:505–510

Jansson H-B, Nordbring-Hertz B (1979) Attraction of nematodes to living mycelium of nematophagous fungi. J Gen Microbiol 112:89–93

Jansson H-B, Nordbring-Hertz B (1988) Infection events in the fungus-nematode system. In: Poinar GO, Jansson H-B (eds) Diseases of nematodes, vol II. CRC Press, Boca Raton, pp 59–72

Jansson H-B, Thiman L (1992) A preliminary study of chemotaxis of zoospores of the nematode-parasitic fungus *Catenaria anguillulae*. Mycologia 84:109–112

Jansson H-B, Jeyaprakash A, Zuckerman BM (1985) Control of root-knot nematode on tomato by the endoparasitic fungus *Meria coniospora*. J Nematol 17:327–329

Jatala P (1986) Biological control of plant parasitic nematodes. Annu Rev Phytopathol 24:453–489

Kerry BR (1980) Biocontrol: Fungal parasites of female cyst nematodes. J Nematol 26:253–259

Kerry BR (1987) Biological control. In: Brown RH, Kerry BR (eds) Principles and practice of nematode control in crops. Academic Press, New York, pp 233–263

Kerry BR (1988) Fungal parasites of cyst nematodes. Agric Ecosyst Environ 24:293–305

Kerry BR (1990) An assessment of progress toward microbial control of plant parasitic nematodes. J Nematol 22:621–631

Kerry BR (1993) The use of microbial agents for the biological control of plant parasitic nematodes. In: Gareth Jones D (ed) Exploitation of microorganisms. Chapman and Hall, London, pp 81–104

Kerry BR (1995) Ecological considerations for the use of the nematophagous fungus, *Verticillium chlamydosporium*, to control plant parasitic nematodes. Can J Bot 73:565–570

Kerry BR, Crump DH (1977) Observations on fungal parasites of females and eggs of the cereal cyst-nematode, *Heterodera avenae*, and other cyst nematodes. Nematologica 23:193–201

Kerry BR, de Leij FAAM (1992) Key factors in the development of fungal agents for the control of cyst and root-knot nematodes. In: Tjamos EC, Papavizas GC, Cook RJ (eds) Biological control of plant diseases. Plenum Press, New York, pp 139–144

Kerry BR, Crump DH, Mullen LA (1982) Studies of the cereal cyst nematode, *Hederodera avenae*, under continuous cereals, 1975–1978. II. Fungal parasitism of nematode females and eggs. Ann Appl Biol 100:489–499

Kerry BR, Kirkwood IA, Leij de FAAM, Barba J, Leijdens MB, Brookes PC (1993) Growth and survival of *Verticillum chlamydosporium* Goddard, a parasite of nematodes, in soil. Biocon Sci Tech 3:355–365

Kim DG, Riggs RD, Kim KS (1992) Ultrastructure of *Heterodera glycines* parasitized by Arkansas fungus 18. Phytopathology 82:429–433

Lackey BA, Jaffee BA, Muldoon AE (1992) Sporulation of the nematophagous fungus *Hirsutella rhossiliensis* from hyphae produced in vitro and added to soil. Phytopathology 82:1326–1330

Lackey BA, Muldoon AE, Jaffee BA (1993) Alginate pellet formulation of *Hirsutella rhossiliensis* for biological control of plant-parasitic nematodes. Biol Contr 3:155–160

Lackey BA, Jaffee BA, Muldoon AE (1994) Effect of nematode inoculum on suppression of root-knot and cyst nematodes by the nematophagous fungus *Hirsutella rhossiliensis*. Phytopathology 84:415–420

Linford MB, Yap F, Oliverira JM (1938) Reduction of soil populations of the root-knot nematode during decomposition of organic matter. Soil Sci 45:127–141

Lopez-Llorca LV, Robertson WM (1993) Ultrastructure of infection of cyst nematode eggs by the nematophagous fungus *Verticillium suchlasporium*. Nematologica 39:65–74

McInnis TM, Jaffee BA (1989) An assay for *Hirsutella rhossiliensis* spores and the importance of phialides for nematode inoculation. J Nematol 21:229–234

Minter DW, Brady BL (1980) Mononematous species of *Hirsutella*. Trans Br Mycol Soc 74:271–282

Morgan-Jones G, White JF, Rodriguez-Kabana R (1983) Phytonematode pathology: ultrastructural studies. I. Parasitism of *Meloidogyne arenaria* eggs by *Verticillium chlamydosporium*. Nematropica 13:245–260

Persson Y, Baath E (1992) Quantification of myco-parasitism by the nematode-trapping fungus *Arthrobotrys oligospora* on *Rhizoctonia solani* and the influence of nutrient levels. FEMS Microbiol Ecol 101:11–16

Powell KA (1993) The commercial exploitation of microorganisms in agriculture. In: Gareth Jones D (ed) Exploitation of microorganisms. Chapman and Hall, London, pp 441–459

Quinn MA (1987) The influence of saprophytic competition on nematode predation by nematode-trapping fungi. J Invertebr Pathol 49:170–174

Rhodes DJ (1993) Formulation of biological control agents. In: Gareth Jones D (ed) Exploitation of microorganisms. Chapman and Hall, London, pp 411–439

Rodriguez-Kabana R, Morgan-Jones G, Chet I (1987) Biological control of nematodes: soil amendments and microbial antagonists. Plant Soil 100:237–247

Schlang J, Steudel W, Muller J (1988) Influence of resistant green manure crops on the population dynamics of *Heterodera schachtii* and its fungal egg parasites. Nematologica 34:193

Segers R, Butt TM, Kerry BR, Peberdy JF (1994) The nematophagous fungus *Verticillium chlamydosporium* produces a chymoelastase-like protease that hydrolyses host nematode proteins in situ. Microbiol 140:2715–2723

Segers R, Butt TM, Keen JF, Kerry BR, Peberdy JF (1995) The subtilisins of the invertebrate mycopathogens *Verticillium chlamydosporium* and *Metarhizium anisopliae* are serologically and functionally related. FEMS Microbiol Lett 126:227–232

Segers R, Butt TM, Kerry BR, Beckett A, Peberdy JF (1996) The role of the proteinase VCP1 produced by the nematophagous *Verticillium chlamydosporium* on the infection process of nematode eggs. Mycol Res 100:421–428

Sikora RA (1992) Management of the antagonistic potential in agricultural ecosystems for the biological control of plant parasitic nematodes. Annu Rev Phytopathol 30:245–270

Sjollema KA, Dijksterhuis J, Veenhuis M, Harder W (1993) An electron microscopical study of the infection of the nematode *Panagrellus redivivus* by the endoparasitic fungus *Verticillium balanoides*. Mycol Res 97:479–484

Sneh B, Humble SJ, Lockwood JL (1977) Parasitism of oospores of *Phytophthora megasperma* var. *sohae*, *P. cactorum*, *Pythium* sp. and *Aphanomyces euteiches* in soil by oomycetes, chytridiomycetes, hyphomycetes, actinomycetes and bacteria. Phytopathology 67:622–628

Stiles CM, Glawe DA (1989) Colonization of soybean roots by fungi isolated from cysts of *Heterodera glycines*. Mycologia 81:797–799

Stirling GR (1991) Biological control of plant parasitic nematodes. CAB International, Wallingford

Stirling GR, Mankau R (1978) *Dactylella oviparasitica*, a new fungal parasite of *Meloidogyne* eggs. Mycologia 70:774–783

Sturhan D, Schneider R (1980) *Hirsutella heteroderae*, ein neuer nematodenparasitarer Pilz. Phytopathol Z 99:105–115

Tedford EC, Jaffee BA, Muldoon AE (19944) Variability among isolates of the nematophagous fungus *Hirsutella rhossiliensis*. Mycol Res 98:1127–1136

Tedford EC, Jaffee BA, Muldoon AE (1995) Effect of temperature on infection of the cyst nematode *Heterodera schachtii* by the nematophagous fungus *Hirsutella rhossiliensis*. J Invertr Pathol 66:6–10

Thomas E (1982) On the occurrence of parasitic fungi and other pathogens, and their influence on the development of populations of the beet cyst nematode (*Heterodera schachtii*) in the northern Rhineland. Gesunde Pflanzen 34:162–168

Thorn RG, Barron GL (1984) Carnivorous mushrooms. Science 224:76–78

Tubaki K, Yamanaka K (1984) An undescribed nematode trapping species of Arthrobotrys. Trans Mycol Soc Jpn 25:349

Tunlid A, Jansson H-B, Nordbring-Hertz B (1992) Fungal attachment to nematodes. Mycol Res 96:401–412

Underwood T, Jaffee BA, Verdegaal P, Norton MVK, Asai WK, Muldoon AE, McKenry MVH, Ferris H (1994) Effect of lime on *Criconemella xenoplax* and bacterial canker in two California orchards. J Nematol 26:606–611

van den Boogert PHJF, Velvis H, Ettema CH, Bouwman LA (1994) The role of organic matter in the population dynamics of endoparasitic nematophagous fungus *Drechmeria coniospora* in microcosms. Nematologica 40:249–257

Voss B, Wyss U (1990) Variation between strains of the nematophagous endoparasitic fungus *Catenaria anguillulae* Sorokin 1. Factors affecting parasitism in vitro. Z Pflanzenkr Pflanzensch 97:416–430

Wallace HR (1963) The biology of plant parasitic nematodes. Arnold, London, 280 pp

Willcox J, Tribe HT (1974) Fungal parasitism in cysts of *Heterodera* 1. Preliminary investigations. Trans Br Mycol Soc 62:585–594

Wimble DB, Young TWK (1983) Structure of adhesive knobs in *Dactylella lysipaga*. Trans Brit Mycol Soc 80:515–519

Zopf W (1988) Zur Kenntnis der Infektions – Krankheiten niederer Tiere und Pflanzen. Nova Acta Acad Leop Carol 52:314–376

14 The Impact of Phylloplane Microorganisms on Mycoherbicide Efficacy and Development

D.A. SCHISLER[1]

CONTENTS

[1] Fermentation Biochemistry Research, National Center for Agricultural Utilization Research, USDA-ARS, 1815 N. University St., Peoria, Illinois 61604, USA

I. Introduction

Microbial interactions can be advantageous or deleterious to a mycoherbicide's effectiveness and can occur both on the phylloplane of weedy plants and during the infection process. My purpose in this chapter is to provide a glimpse of the complexity of microbially related factors that potentially impact mycoherbicide agents rather than to exhaustively analyze the recently reviewed topics of biological control and microbial ecology in the phyllosphere (Andrews and Hirano 1991; Andrews 1992). In the last portion of this chapter, my intention is to stimulate further thinking regarding management of the phylloplane microbial environment to favor mycoherbicide success. Some review of terms and concepts is first in order. Weedy plants cause an economic loss of over 4 billion dollars a year in the United States alone (Bridges and Anderson 1992; Table 1). Plant pathogenic fungi can be used by man to intentionally infect weedy plants, rendering the infected plants less able to compete with the agricultural crop or managed forest that the weed pervades. Under ideal conditions for pathogenesis, the weedy plant is killed outright. When such pathogenic fungi are inundatively distributed over large areas that the weed infests and there is limited expectation of long-term pathogen persistence, the fungus is being used as a mycoherbicide.

Several mycoherbicides have been registered in North America including COLLEGO [*Colletotrichum gloeosporioides* (Penz.) Penz. & Sacc. f. sp. *aeschynomene*] for the control of northern jointvetch [*Aeschynomene virginica* (L.) B.S.P.] (Daniel et al. 1973; Smith 1986), Dr. Biosedge [*Puccinia canaliculata* (Schw.) Lagerh.] for the control of yellow nutsedge (*Cyperus esculentus* L.) (Phatak et al. 1988), DeVine [*Phytophthora palmivora* (Butler) Butler] for the control of strangler vine (*Morrenia odorata* Lindl.) (Burnett et al. 1974; Kenney 1986; Ridings 1986), and BioMal (*C. gloeosporioides* f. sp.

The Mycota IV
Environmental and Microbial Relationships
Wicklow/Söderström (Eds.)
© Springer-Verlag Berlin Heidelberg 1997

malvae) for the control of round-leaved mallow (*Malva pusilla* Sm.) (Mortensen 1988; Makowski and Mortensen 1992). Throughout this chapter, I will most often use the fungal specific term mycoherbicide instead of the more broadly applicable term bioherbicide (inundative use of any plant pathogenic microorganism to infect and control weeds) to emphasize the fact that this chapter deals specifically with **fungal**-microbial interactions. The study of biologically controlling weeds using pathogenic microorganisms is still in its beginnings, first being reviewed by Wilson (1969) and recently by Charudattan (1991), Van Dyke (1991), and TeBeest et al. (1992). The topic has also been the subject of an American Chemical Society symposium series (Hoagland 1990).

Why are bioherbicides of interest? Or perhaps more importantly, why should bioherbicides be developed and used? Several features of bioherbicides are attractive in today's socio-political environment of concern for developing environmentally safe, sustainable, renewable agricultural strategies. The continued development of bioherbicide products promises a reduction in the number and amount of chemical herbicides applied to managed plant systems. Additionally, plant pathogens can be selected for development that are highly host-specific to target weeds and are not harmful to agricultural crops or workers. However, mycoherbicides also have several impediments to development and consistency in controlling weedy plant targets. As an example, the pathogenic fungal component of a mycoherbicide invariably has an obligate requirement for an extended period of dew or free moisture to permit spore germination and subsequent host infection. A mycoherbicide faces challenges to every stage

Table 1. Estimated average annual losses due to weeds in several commodity groups in the United States, 1989–1991. (Bridges and Anderson 1992)

Commodity group	Average annual monetary losses ($ \times 1000$)[a]
Field crops	3 359 671
Noncitrus fruits	201 181
Citrus fruits	118 943
Tree nuts	46 542
Vegetables	386 418
US totals	4 112 755

[a] Estimate based on using best management practices with chemical herbicides. Total loss estimated to be $19.4 billion if chemical herbicides not available.

of the infection process from the time fungal spores are placed on a phylloplane surface that is physically, structurally, chemically, and microbiologically complex through the engagement of host plant defenses resisting fungal penetration (Kenerley and Andrews 1990; Juniper 1991). It is with good reason that "pathogenicity is the exception rather than the rule" (Baker and Cook 1974).

II. Topic Area Defined

The complexity of each aspect of mycoherbicide discovery, development, formulation, and application has been reviewed in some detail (Charudattan 1990a; Daigle and Connick 1990; Stowell 1991). These topics will be treated here only by stating that a mycoherbicide propagule arrives at the phylloplane with an inherent potential to incite disease. This potential is affected by many factors, including the genetics of the specific fungal strain (Weideman and TeBeest 1990), the fermentation environment utilized to produce propagules (Schisler et al. 1991b; Jackson and Schisler 1992), and the formulation of the pathogen (Boyette et al. 1991; Connick et al. 1991a; Schisler et al. 1995). Most of this chapter will concentrate on the impact of the phylloplane microflora on the potential of the mycoherbicide propagule to incite disease after arrival on the leaf surface. Factors which impact the initial potential of a dose of a mycoherbicide to incite disease in a target weed are schematically depicted in Fig. 1. For clarity, potential interactions between a mycoherbicide and the microbiota will be discussed in this text as isolated events. The reader should remain aware, however, that a web of overlapping microbial interactions is a better representation of reality on the phylloplane. These interactions, in turn, are influenced by such factors as available nutrients, plant-produced compounds, plant health, pesticide residues, moisture availability, temperature, and light quality, to name a few. Further complicating our ability to understand microbial-mycoherbicide interactions is the reality that the phylloplane is not physically, chemically, or microbiologically static or homogeneous (Blakeman and Atkinson 1981; Blakeman 1985; Juniper 1991; Andrews and Harris, Chap. 1, this Vol.); but rather an area in transition from the time of leaf emergence until abscission. For ex-

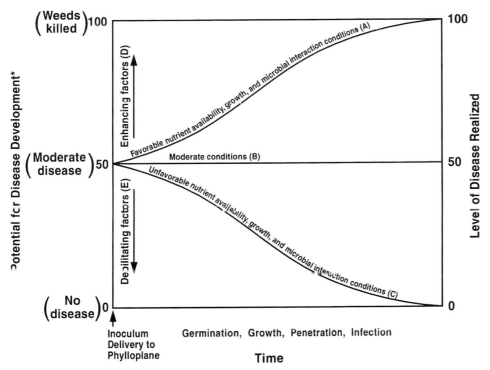

Fig. 1. Diagrammatic representation of factors that may impact the potential of a moderate dose of a mycoherbicide to incite disease in a target weed. * *Dimensionless "potential for disease development"* value arbitrarily set at *50* for an "average" mycoherbicide strain and "average" growth conditions. *A* Relative potential for disease development increases as spore germination, growth, and penetration proceed in an environment favorable to the mycoherbicide; *B* moderate conditions have little net effect on the original potential for disease development value; *C* potential for disease development decreases with time as inadequate nutrients, poor fungal growth conditions and deleterious microbial interactions cause cumulative losses in mycoherbicide inoculum potential; *D* Y-axis starting point of all curves is shifted upwards if the mycoherbicide is highly virulent or requires a short dew period for infection. Ending point may be shifted upwards if the host is predisposed to disease development; *E* Y-axis starting point of all curves is shifted downward for mycoherbicide strains that are less virulent or require long dew periods. Ending point may be shifted downward if the host has been induced for disease resistance

ample, a maturing leaf phylloplane goes through a succession of dominant microbial communities (Hudson 1971; Jensen 1971), each of which can have a differing overall influence on mycoherbicide spores that arrive. Older weeds targeted for control by mycoherbicides will have a variety of leaf ages and microbial communities present on an individual plant. Thus the overall influence of "typical" phylloplane microorganisms will differ not only from crop to crop, but from plant to plant, plant leaf to plant leaf, and quite possibly within a single leaf. The following sections on deleterious and advantageous microbial-mycoherbicide interactions should therefore be viewed in the context of taking place in a dynamic, highly interactive environment. Many of the interactions suggested are based on speculation from limited available literature outside mycoherbicide research, while space constraints and breadth of topic necessitate superficial coverage of some ar-

eas where a wealth of literature exists. The final section of this chapter will then discuss potential methods of utilizing the phylloplane microbiota and applying ecological principles to favor mycoherbicide success.

III. Mycoherbicide-Microbe Interactions

Due to the variety of microbial interactions that potentially affect a mycoherbicide on the phylloplane, I have for convenience separated the interactions into subclasses of direct vs. indirect as well as deleterious vs. advantageous. For purposes of this chapter, direct interactions are defined as mycoherbicide/microbial interactions that are not mediated by another microbe or the host plant while the converse is true for indirect interactions. A summary of these potential interactions is pre-

Table 2. Summary of selected microbial activities that can affect mycoherbicides on weed leaf surfaces

Microbial activity	Effect on mycoherbicide[a]
Direct[b]	
Production of toxic metabolites	–
Mycoparasitism	–
Prior niche possession	–
Degradation of spore mucilage	–
Biosurfactant production	+
Enhancement of spore germination	+
Enhancement of appressoria formation	+
Utilization/conversion of plant defense elicitors	+
Detoxification of phytoalexins	+
Indirect	
Induction of plant resistance	–
Predisposition of weed to disease	+
Displacement of deleterious microbes	+
Suppression of host defense	+

[a] +, Microbial activity has a positive influence on the potential efficacy of the mycoherbicide.
–, microbial activity has a negative influence on the potential efficacy of the mycoherbicide.
[b] For purposes of this chapter, direct interactions are defined as mycoherbicide/microbial interactions that are not mediated by another microbe or the host plant while the converse is true for indirect interactions.

sented in Table 2. An overriding consideration in the importance of any individual mycoherbicide-microbial interaction is the time the mycoherbicide inoculum is exposed on the phylloplane before infection is achieved. Thus, microbial interactions would generally have a higher probability of affecting potential mycoherbicide agents such as *Colletotrichum graminicola* and *C. coccodes*, which require dew periods in excess of 12h to achieve maximal infection of johnsongrass [*Sorghum halepense* (L.) Pers.; Chiang et al. 1989] and velvetleaf (*Abutilon theophrasti* Medic.; Wymore et al. 1988), respectively, than *C. truncatum* which requires only 4–6h of dew to infect the weed *Sesbania exaltata* (Boyette 1991).

A. Direct Interactions Deleterious to Mycoherbicides

1. Production of Detrimental Metabolites

An enormous number of toxic compounds are produced by bacteria and fungi that can inhabit the phylloplane (Lynch 1990). Such compounds include antibiotics, volatiles, and enzymes which have been implicated or demonstrated to be instrumental in reducing the level of disease incited by plant pathogenic fungi (Fravel 1988). Fluorescent pseudomonads, which are commonly isolated phylloplane inhabitants, can produce a variety of compounds, including phenzaines, pyoluteorin, pyrrolnitrin, and 2,4-diacetylphlorglucinol, which can be deleterious to fungal pathogens (O'Sullivan and O'Gara 1992). Microbial production of compounds with a broad spectrum of fungistatic or fungicidal activity would be especially harmful to mycoherbicide propagules in the comparatively vulnerable process of germinating and infecting target weeds.

The impact of microbially produced toxic volatiles, though considerable in the rhizosphere, would most likely be of lesser importance in the phylloplane, where microbially produced toxic volatile compounds would normally rapidly dissipate. However, one cannot exclude the possibility that low air movement and temperatures combined with the presence of free moisture could permit water-soluble, microbially produced volatiles with low vapor pressures an opportunity to reach concentrations deleterious to nearby mycoherbicide propagules. Ammonia gas, for example, is stable at physiological temperatures, is highly soluble in water especially at lower temperatures (33% by weight at 20 °C), and has a relatively low vapor pressure in unsaturated aqueous solutions (Kroschwitz 1992). A variety of bacteria can produce ammonia, which is highly active in suppressing several genera of plant pathogenic fungi (Howell et al. 1988; Fernando and Linderman 1995).

Enzymes produced by microbial inhabitants of the phylloplane of weedy plants could also negatively affect a mycoherbicide. The cell walls of basidiomycetes and ascomycetes contain chitin and glucans, and the production of cell wall-degrading enzymes by antagonists has been implicated as a mechanism of biologically controlling plant pathogenic fungi (Ordentlich et al. 1988; Urquhart et al. 1994). Interestingly, Lorito et al. (1993) have shown a synergistic interaction between chitinolytic enzymes from *Trichoderma harzianum* and cells of *Enterobacter cloacae* that resulted in inhibited spore germination and germ tube elongation for a variety of fungal pathogens in vitro. Presumably, some mycoherbicide applications would face similar synergistically antagonistic environments.

2. Mycoparasitism

Mycoparasitism, although an exploitable factor in the biological control of plant pathogenic fungi, would be expected to provide a limited barrier to mycoherbicide efficacy, primarily due to the nominal amount of time available for increases in mycoparasite populations before mycoherbicide propagules have infected host tissues. Low populations of mycoparasites could even be advantageous to a mycoherbicide in limiting the establishment of fungi antagonistic to the mycoherbicide. However, with repeated application of mycoherbicides, low populations of mycoparasites specific to the mycoherbicide could increase, in theory, to levels deleterious to the mycoherbicide. Studies of the population ecology of mycoparasites in areas of long-term treatment with the successful mycoherbicide Collego [*Colletotrichum gloeosporioides* (Penz.) Penz. and Sacc. f. sp.*aeschynomene*] (Templeton 1992) would greatly contribute to determining the importance of mycoparasites on mycoherbicide efficacy.

3. Prior Niche Possession (Biological Buffering)

In natural systems, the prior microbial occupation of phylloplane sites made favorable for colonization by the presence of nutrients from plant cells or external sources such as insect honeydew or pollen can negate or severely hinder the establishment of newly arrived microbial propagules (Lindemann and Suslow 1987). When used to treat young weed seedlings, mycoherbicide propagules would face a reduced level of competition since microbial populations on the recently emerged phylloplane surface would most likely be small (Blakeman 1985). In older seedlings, the extremely large numbers of mycoherbicide propagules combined with endogenous nutrient reserves within spores could limit the impact of prior niche possession. However, the combined effects of antibiotics, mycoparasitism, and competition for limiting available nutrients (such as iron; Loper and Buyer 1991), would be expected to contribute to attenuated viability, virulence and, under severe conditions, propagule death (Hyakumachi et al. 1992). Further research into this relatively unexplored area will undoubtedly uncover mycoherbicide-microbial interactions equally complex to those being described for bacterial strain interactions in the phylloplane (Kinkel and Lindow 1993).

4. Degradation of Spore Mucilage

Propagules of numerous fungal pathogens of plants may produce mucilage after arrival on plant surfaces. Many fungi within the genus *Colletotrichum* produce conidia in a mucilagenous matrix. This genus contains numerous mycoherbicide candidates (Charudattan 1991). Mucilage is involved in host recognition, attachment of propagules and infective structures to host surfaces, protection of spores from dehydration and toxic compounds of plant or microbial origin, and can contain enzymes that aid early stages of a pathogen's infection of a host (Nicholson et al. 1989; Nicholson 1990; Pascholati et al. 1993). Though extensive microbial degradation of mucilage would leave mycoherbicide spores severely impaired in theory, it is open to speculation as to how important this process could be to mycoherbicide efficacy in field applications. Mucilage coverage of spores produced via liquid culture is generally reduced compared to spores produced on solid substrates, since mucilage is water-soluble (Nicholson et al. 1989). Thus, microbial degradation of the limited mucilage coverage of liquid-culture-produced spores or the microbial inhibition of mucilage production by mycoherbicide propagules could be of considerable significance compared to its importance when conidia are produced in natural environments.

B. Indirect Interactions Deleterious to Mycoherbicides

1. Microbially Induced Disease Resistance

An indirect interaction with great potential for reducing the efficacy of a mycoherbicide is microbially induced resistance in the target weed. Induced disease resistance is "the process of active resistance dependent on the host plant's physical or chemical barriers, activated by biotic or abiotic agents" and can be a localized response or systemically expressed throughout the induced plant (Kloepper et al. 1992). Seedlings of the weedy plant sicklepod (*Cassia obtusifolia* L.) exposed 3 days earlier to the mycoherbicide *Alternaria cassiae* Jurair and Khan developed 43% fewer lesions when reinoculated with *A. cassiae* than control plants inoculated with *A. cassiae* for the first time (Weete 1992). Infection of *Ipomoea purpurea* by *Colletotrichum dematium* also increases resistance to subsequent infections on previously uninfected leaves (Sims and Vision 1995). In field applications of mycoherbicides, previous activity by less virulent pathogens could render the mycoherbicide less effective in controlling the weedy-plant target (Yamaguchi et al. 1992). Furthermore, infection is only one method by which weeds may acquire induced resistance. Nonpathogenic rhizosphere bacteria have been implicated in inducing resistance to plant disease (Wei et al. 1991). The list of microbially produced compounds which have been demonstrated to induce resistance is extensive and includes volatiles (VanLoon and Pennings 1993), complex carbohydrates such as the fungal cell wall components chitin and β-1,3-glucans (Keen 1990), and proteins (Ricci et al. 1993). As a general rule, with age, weed seedlings become more resistant to plant pathogens. Studies to separate plant age-based resistance from increasing induced resistance as phylloplane (and rhizosphere) microbial populations mount have not been attempted to my knowledge. However, the action of microbial colonists surely must at least contribute to the increased level of pathogen resistance seen in older seedlings.

C. Advantageous Mycoherbicide-Microbe Interactions

1. Concept of Microbial Facilitators

In this section, microbial interactions that have the potential for improving mycoherbicide efficacy in inciting disease will be presented. A summary of the potentially advantageous mycoherbicide-microbe interactions identified is found in Table 2. The term microbial facilitator is defined as a nonpathogenic microorganism that, when combined with propagules of a bioherbicide, can increase the severity of disease initiated by the pathogen (Schisler et al. 1991a). This term will be used in discussing the variety of potentially advantageous mycoherbicide-microbe interactions described below. Microbial facilitators enhance the efficacy of the mycoherbicides *Colletotrichum truncatum* (Schisler et al. 1991a; Table 3) and *C. coccodes* (Fernando et al. 1994). A variety of Gram-negative and -positive bacterial microbial facilitators were isolated by Schisler and co-workers, while Fernando et al. found several fluorescent *Pseudomonas* strains and two unidentified bacterial strains to be effective. Interestingly, though yeasts make up a preponderance of phylloplane microorganisms on mature leaves, only Gram-positive and Gram-negative bacterial

Table 3. Impact of selected bacterial microbial facilitators on disease incited by *Colletotrichum truncatum* in seedlings of *Sesbania exaltata*. (Schisler et al. 1991a)

Inoculum composition	Seeding shoot dry weight (mg)	Leaves per plant
HkCt + hkb[a]	75[b,c]	5.0
Ct + hkb[d]	50	3.4
Ct + A30[e]	40	2.5
Ct + A67	37*	2.5
Ct + A79	38*	2.7
Ct + B24	37*	2.4*
Ct + B31	35**	2.1*
Ct + B33	30**	1.9**

[a] Heat-killed conidia of *C. truncatum* (HkCt) and heat-killed mixture of bacterial microbial facilitators (hkb).
[b] Measurements were made on 27-day-old seedling of *S. exaltata* 8 days after inoculation.
[c] Within a column, values followed by one or two asterisks are significantly different from the Ct + hkb treatment ($P < 0.05$, $P < 0.01$, respectively).
[d] Conidia of *C. truncatum* (Ct) + hkb (Ct dose = 5×10^5 conidia/ml).
[e] Conidia of Ct+ a bacterial microbial facilitator strain (bacteria dose = 1×10^8 cfu/ml).

microbial facilitators have been isolated. As this area of mycoherbicide research has just begun to be explored, the discovery of yeast, actinomycete, and filamentous fungi microbial facilitators may be forthcoming.

D. Direct Microbial Interactions Advantageous to Mycoherbicides

1. Biosurfactant Activity

Recently, Bunster et al. (1989) demonstrated that 50% of *Pseudomonas* strains tested increased leaf wettability by the production of biosurfactants. Within 24 h, droplets containing active strains spread over a hydrophobic polystyrene surface while those droplets with no surface-active strains did not. Laycock et al. (1991) described a peptidolipid biosurfactant, produced by a strain of *Pseudomonas fluorescens* biovar II, that facilitated spread of the bacterium on broccoli florets. The presence of such surface-active strains naturally occurring on the phylloplane or included in a mycoherbicide formulation would lessen the difficulties a mycoherbicide would encounter in contacting and dispersing across hydrophobic weedy leaf surfaces, even if rain splash removed surfactants included in the mycoherbicide formulation. However, prior to concluding that biosurfactants are uniformly advantageous to mycoherbicides, studies to determine whether biosurfactants influence mycoherbicide spore adhesion to plant surfaces would need to be conducted.

2. Enhancement of Spore Germination and Appressoria Formation

The germination of mycoherbicide propagules on the phylloplane can be problematic, not only due to the possibility of inadequate free moisture, but, for some mycoherbicides, due to the presence of self-inhibitors of spore germination which would be particularly concentrated when applying the high concentrations of spores used in mycoherbicide applications (Leite and Nicholson 1992). However, microbial stimulation of spore germination and breaking of spore dormancy is widespread in nature (Allen 1976), a factor to the advantage of an applied mycoherbicide. Self-inhibitors of germination that are found in conidial mucilage, such as mycosporin-alanine in

Colletotrichum graminicola (Leite and Nicholson 1992) and CG-SI 1, 2 and 3 (Tsurushima et al. 1995) should be susceptible to microbial breakdown or inactivation. Siderophores produced by phylloplane bacteria can stimulate germination of *C. musae* (McCracken and Swinburne 1979). The provision of limiting growth factors (vitamins, solubilized minerals, etc.) made available via microbial activity (Fig. 2) could also stimulate conidial germination. A microbial facilitator may also benefit a mycoherbicide by utilizing or inactivating plant-produced phylloplane compounds such as terpenoids and phenolics which would normally slow conidial germination and appressoria formation (Blakeman and Atkinson 1981).

Microorganisms can also act to increase the number of appressoria formed by germinating spores and the rapidity at which this is accomplished. Microbes can produce appressoria-promoting substances such as siderophores (Slade et al. 1986) or utilize available nutrients to produce an environment nutritionally hostile to vegetative fungal growth (Blakeman and Parbery 1977) that, in turn, triggers appressoria formation. Studies by Fernando et al. (1996) indicated that the appressoria-stimulating effect of some microbial facilitators of *C. coccodes* was reversible with the addition of Fe^{3+}. Though microbial facilitators of Schisler et al. (1991a) also stimulated appressoria formation (Fig. 3), Fe^{3+} sequestering by microbial facilitators did not appear to contribute to increased appressoria formation by *C. truncatum* in the presence of these microbes (unpubl. results). There is inconclusive evidence that a microbially created, nutritionally hostile environment can account for increased appressoria formation by *Colletotrichum* fungi (Emmett and Parbery 1975). Fernando and coworkers (1996) found *C. coccodes* varied in appressoria formation in the presence of microbial facilitators and nutrients depending on the specific amino acid tested. Schisler et al. (1992) found that appressoria formation was sometimes inversely correlated with the concentration of microbial facilitator applied, though this response varied with the strain tested.

3. Utilization/Conversion of Elicitors of Plant Defenses

Many pathogens produce molecules that can induce resistance with (Darvill and Albersheim 1984) or without (Kamoun et al. 1993) prior injury

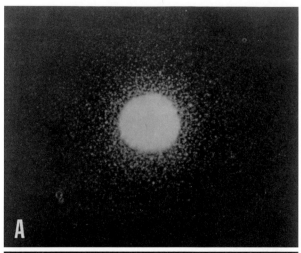

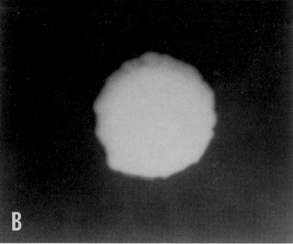

Fig. 2A,B. A cross-feeding test for thiamine excretion. **A** Excreter yeast colony surrounded by a zone of micro-colonies of a thiamine-requiring strain (cross-feeding zone). **B** Nonexcreter colony. (Haj-Ahmad et al. 1992)

of plant cells in the elicited tissue. Utilization of such compounds by microbial facilitators could conceivably lessen or delay an induced resistance response by the weedy plant host.

4. Detoxification of Phytoalexins

Many successful plant pathogens possess the ability to detoxify phytoalexins, toxic compounds synthesized by plants in response to microbial infection (VanEtten et al. 1989; Kuć 1995). It is open to speculation whether microbial facilitators could assist pathogens in detoxifying phytoalexins. Certainly, the possibility cannot be ruled out considering, for instance, that some strains of *Pseudomonas* spp. are known microbial facilita-

tors and can degrade a wide variety of compounds including xenobiotics (Golovleva et al. 1992).

E. Indirect Microbial Interactions Advantageous to Mycoherbicides

1. Microbial Activity Which Predisposes Weeds to Infection

Nongenetic, plant-stressing factors acting prior to infection can increase the susceptibility of a plant to disease. Such factors can be abiotic (water, temperature, nutrient, chemical herbicides, etc.; Schoeneweiss 1975; Altman et al. 1990) or the result of biotic activity. Biotic activity that can predispose plants to infection includes the production of enzymes and toxins. Sebastian et al. (1987) reported that a phyllosphere strain of a *Pseudomonas* sp. produced cutinase, and suggested that enzyme activity provided a carbon source to other phyllosphere bacteria. Cutin is a polyester structural component of the plant cuticle and its degradation may expose aspects of surface topography necessary for recognition of the infection court by pathogens (Nicholson 1990). According to Ruinen (1966), many commonly isolated phylloplane yeasts such as *Aureobasidium pullulans*, *Cryptococcus laurentii*, and *Rhodotorula glutinis* are lipolytic and involved in the destruction of previously formed cuticle and the prevention of its subsequent reformation. Commonly produced microbial enzymes such as proteases, cellulases, pectinases, polygalacturonases, and lipases would all have potential for predisposing a plant to pathogen attack. Kohmoto et al. (1989) presented evidence of host-specific toxins of some strains of the common mycoherbicide genus *Alternaria* inducing host susceptibility to infection without host cell necrosis. Pathogenic attack is another form of biotic activity that can predispose plants to infection. Prior infection of the weedy plant groundsel (*Senecio vulgaris*) by the rust *Puccinia lagenophorae* Cooke permitted several secondary fungal invaders to kill the host, yet these weak pathogens rarely killed rust-free plants (Hallett and Ayres 1992). Similarly, prior infection of *Xanthium occidentale* Bertol. (Noogoora burr) by *Puccinia xanthii* permitted a secondary infection by *C. orbiculare* to kill the weed (Morin et al. 1993). The concept of "induced susceptibility" (Daly 1976) and predisposition to plant disease has received considerably

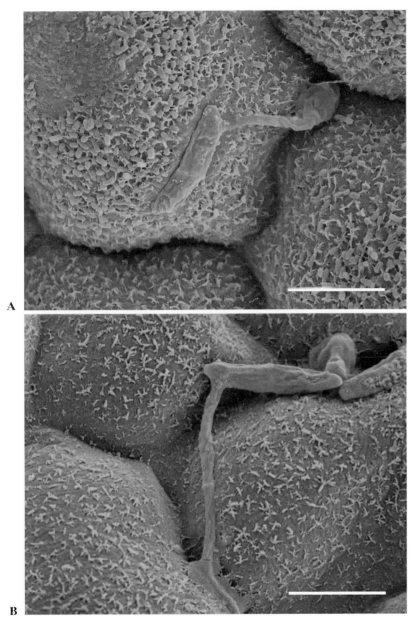

Fig. 3A,B. Behavior of *Colletotrichum truncatum* conidia: **A** coinoculated with microbial facilitator strain *Enterobacter* sp. A67 or **B** without microbial facilitator on the leaf surface of hemp sesbania (*Sesbania exaltata*) after 6 h dew at 26 °C. Note that the conidium in **A** is representative of the majority that have relatively short germ tubes with terminal appressoria while conidia in **B** are representative of the higher proportion of conidia with long germ tubes and no appressorium formation. *Bar* 10 μm. (Electronmicrographs of experimental material courtesy Lee Baker, USDA-ARS, NCAUR, Peoria, IL)

less research attention in recent years than "induced resistance", perhaps due to the more appealing potential of plant disease control applications inherent with induced resistance research. However, with the considerable contribution that understanding this phenomenon would bring to applying mycoherbicides more successfully and predictably, there is cause for optimism

that an upturn in research in this important area will be forthcoming.

2. Microbial Facilitator Displacement of Strains Deleterious to Pathogen Efficacy

Deleterious microbial interactions were previously discussed, including microbial production of antimicrobial compounds, mycoparasitism, and microbially induced host resistance. Inactivation of microorganisms deleterious to a mycoherbicide or their replacement by benign microbes would considerably improve mycoherbicide performance. The establishment of microbial populations that do not induce host plant resistance to the exclusion of those that do would also benefit mycoherbicide performance. This could occur, for instance, via the microbial production of antibacterial compounds or the establishment of microbes that block the attachment of microbials (Sequeira 1984) that would otherwise induce host resistance.

3. Suppression of Host Defense

As mentioned previously, there is an enormous array of compounds that are capable of inducing disease resistance. However, microbial facilitators could, in theory, act to keep inducing compounds from triggering a host defense response. For instance, the pathogen *Mycosphaerella pinodes* secretes a sugar-peptide inhibitor active against ATPase in the plasma membrane of pea. Suppressor activity was nonspecific in vitro, specific to pea in vivo, and temporarily suppressed the expression of all defense reactions of host plants (Oku et al. 1993). Additionally, Huang et al. (1991) demonstrated reduced phenylalanine ammonia-lyase (PAL) activity, slower would healing, and a higher susceptibility of orange fruit to *Penicillium digitatum* when a strain of *Pseudomonas* or its culture filtrate was present on wounded fruit. Could similar microbial activity allow mycoherbicide penetration to proceed while host defense systems are retarded or temporarily deactivated?

IV. Application of Ecological/ Biotechnical Precepts to Promote Favorable Microbial-Mycoherbicide Interactions

A proactive approach to favoring mycoherbicides could be adopted in the form of applying treatments that favor native phylloplane microbial facilitators or discourage microorganisms harmful to a mycoherbicide. Conversely, microbial facilitors could be selected for, characterized,

and ultimately formulated as a microbial facilitator product designed to be utilized in conjunction with a compatible mycoherbicide. Such studies would also have the potential of identifying mechanisms of microbial facilitation so that the mechanism of disease enhancement could be applied without the need of delivering cells of the actual microbial facilitator, thus saving the added expense of producing and formulating a second biological product. Although a staggering amount of basic scientific knowledge remains to be accumulated before many of the proposals outlined below can be predictably applied, it is hoped that the ideas advanced might stimulate research that begins to fill this void.

A. Enrichment of Native Microbial Facilitator Populations or Activity

With the variety of microorganisms potentially present on the phylloplane of a given weedy plant, severe difficulties can be anticipated in identifying compounds that would selectively or semiselectively enhance microbial facilitators. One approach to the problem would be to question what physiological or metabolic characteristics can be identified that are common to a set of microbial facilitators. If the production of siderophores is important, how could the phylloplane environment be modified to favor siderophore production by these microorganisms? If the production of mildly deleterious secondary metabolites that induce appressoria formation is a shared characteristic, can common metabolic pathways be identified and pathway precursors then be provided? Such precursor compounds would not have to be directly applied as a spray. Chemical herbicides frequently are a safe and effective tool for weed control, and mycoherbicides have been successfully integrated with these chemicals (Altman et al. 1990; Smith 1991). Such integration can result in increased activity of the mycoherbicide (Yu et al. 1988), possibly due to increased plant stress and levels of exudation induced by the chemical. Herbicides can differ greatly in their impact on the quality and quantity of root exudates (Youssef and Heitefuss 1983; Marks and Cerra 1991). Further studies on applying sublethal levels of different families of chemical herbicides should consider whether a directed increase in phylloplane microorganisms is possible via pairing the mycoherbicide with a chemical herbicide that induces leaf exudates favorable to the microbial facilitator.

B. Techniques for Isolating Microbial Facilitators

Only two laboratories have reported on specific research efforts directed at discovering microbial facilitators of mycoherbicides (Schisler et al. 1991a; Fernando et al. 1994). Schisler and coworkers screened a wide variety of mixed microbial populations for their overall ability to increase appressoria formation by *C. truncatum* and then obtained active individual strains from those populations that were superior in this attribute. Fernando et al. isolated microorganisms from a variety of ages of leaves taken from different ages of velvetleaf plants. Clearly, organisms superior to those reported in both laboratories and specific to individual mycoherbicide candidates remain to be discovered. Selection techniques such as those described by Schisler et al. serve as a reasonable starting point. In vitro selection techniques could be based on identifying microbial characteristics found to be prerequisite to microbial facilitator activity. Such attributes could include the production of cuticle-degrading enzymes, siderophores, or vitamins (Haj-Ahmad et al. 1992; Fig. 2). Selection techniques could also be tailored to assure that candidates are competent in osmoregulation and exhibit chemotaxis to compounds liberated by a mycoherbicide propagule, attributes that phylloplane-competent strains would be likely to exhibit in natural field environments. The utilization of enrichment techniques based on supplying nutrients that semiselectively increase the population of a desired microbial function or group prior to isolating individual strains can also prove useful (Schisler and Slininger 1994). The overall objective of screening should be to balance a desirably high rejection rate of inferior candidate organisms based on a highly selective screening process against an undesirable overly severe technique that rejects too many legitimate performers. Furthermore, the assay developed should be conducted using parameters and conditions that approximate those of the field environment, since positive microbial performance identified with such a bioassay maximizes the likelihood of similar performance in the field. Additional points in developing effective protocols for discovering and developing biocontrol agents have been recently proposed by Cook (1993). Once an initial set of potential microbial facilitators has been identified, performance under controlled in vivo conditions should be demonstrated using cells produced in liquid culture.

C. Characterization and Production of Microbial Facilitators

Cells produced in liquid culture can differ radically in efficacy from those produced on solid substrates (Wilson and Lindow 1993; Slininger et al. 1994). Since deep tank liquid fermentation represents the industry standard for biomass production, studies to insure that effective candidate strains are amenable to production in liquid culture must also be completed at an early stage of the screening process (Slininger et al. 1994). If microbial facilitator or any biological control research is to proceed beyond the level of laboratory phenomenon, it should be realized that field efficacy is a moot point if the strain cannot be economically produced in liquid culture. The nutritional environment during fermentation of a mycoherbicide agent affects not only the number of spores produced but also their subsequent efficacy (Schisler et al. 1991b) and longevity (Jackson et al. 1995). This fact must be recognized in optimizing fermentation protocols of prospective mycoherbicide agents and microbial facilitators. Knowledge gained from growth studies in liquid culture may also enable nutrient amendments to be chosen that broadly favor mycoherbicide and microbial facilitator growth while having less impact on competing microorganisms (Morris and Rouse 1985). Alternatively, possession of a competent microbial facilitator strain may justify genetically engineering the strain to enable its utilization of an uncommon carbon source that could then be supplied in a product formulation (Colbert et al. 1993). Further research on techniques of liquid culture production should also prove useful in diminishing current difficulties frequently experienced with drying Gram-negative bacteria and some mycoherbicide candidates. Adjusting the water potential of the medium during the fermentation process (Caesar and Burr 1991; Jin et al. 1991) or modifying the pH and nutrient content of the liquid medium during culture growth (Larsson and Gustafsson 1993) has shown promise in increasing the microbial product's tolerance to dehydration.

D. Formulation of Microbial Facilitators

There is also a wide variety of needs and opportunities available in the formulation of microbial inoculants. Many of these have been well documented (Walter and Paau 1993). Briefly stated, formulations that promote the maintenance of shelf life and high viability when rehydrating microbial biomass remain of great research interest, as does developing formulations that provide free moisture to mycoherbicide propagules. Formulations that permit a dried cell product to undergo a more gradual wetting and drying during the process of applying the rehydrated biological product as a spray to weedy plants would increase product viability and efficacy (Kosanke et al. 1992). Studies designed to produce alternative mycoherbicide propagules that more readily withstand dehydration are underway (Jackson and Schisler 1995; Schisler and Jackson 1996). Invert emulsions (water in oil), which combine a hydrophobic phase with highly dispersed mycoherbicide propagules in water (Connick et al. 1991b), are a promising beginning point for research directed at solving the requirement of most mycoherbicides for free moisture in order to infect. The development of formulations that provide nutrients that favor mycoherbicide and microbial facilitator adhesion to the host could greatly improve the percentage of cells delivered to the phylloplane that are effective in initiating disease (Fletcher and McEldowney 1984; Kwon and Epstein 1993).

E. Identification and Utilization of Mechanisms of Microbial Facilitation of Disease

The identification of the mechanism(s) by which successful microbial facilitators enhance disease would be a necessary first step towards applying the mechanism to enhance mycoherbicide efficacy without requiring the presence of microbial facilitator cells. Fernando et al. (1996) saw potential for increasing *C. coccodes* appressoria formation by combining conidia with the iron chelator ethylenediaminetetraacetic acid (EDTA) after determining that appressoria stimulation caused by two *Pseudomonas* strains was reversible by supplying Fe^{3+}. Two or more facilitators with differing modes of action could be combined with expectation of additive or even synergistic effects, as has been reported in the biological control of plant pathogens by using more than one antago-

nist (Janisiewicz 1988, 1996; Toyoda et al. 1993; Schisler et al. 1994). Research subsequent to mechanism determination may also require developing fermentation regimes for the maximal, economic production of microbial facilitator metabolites found to be associated with enhancement of mycoherbicide-incited disease.

V. Risk Awareness in Continued Development of Microbial Facilitator Research

The positive effect that microbial facilitators could have on mycoherbicide-incited weed disease must be tempered with caution about the impact of these same organisms on diseases of agronomic crops. Clearly, some mechanisms of increasing appressoria formation for a fungal pathogen of weeds could also apply to a fungal pathogen of an agronomic crop. However, several factors argue against significant nontarget impacts of microbial facilitators and their persistence in native environments. Most mycoherbicides are maximally effective when used to treat weed seedlings in the earliest stages of development, frequently at a six-leaf or earlier. Therefore, crop plants could be separated in time from any microbial facilitator/mycoherbicide treatment of weeds. Secondly, the introduction of large numbers of microorganisms is normally followed by a slow population decline to native or, in some cases, unrecoverable levels when microorganisms are exposed to the extreme habitats and conditions encountered in field releases (Malvick and Moore 1988; Lindow 1992; Hirano and Upper 1993). The spread of microbial cells from the inoculation site is also limited (Lindow et al. 1988). To reduce the potential risks associated with applying large numbers of microbial facilitators, promising strains could be genetically modified to produce auxotrophic mutants (Sands et al. 1990). An obligate requirement by a microbial facilitator for a compound not readily available in field environments would permit a formulation composed of the compound, microbial facilitator, and mycoherbicide to be developed so that the microbial facilitator would not survive once the supply of the compound has been depleted. Studies of a microbial facilitator's host range and phylloplane competence and the native distribution of strains similar to a specific facilitator strain should be considered mandatory prerequisites to any large-scale field evaluation trials.

Under current regulatory guidelines, it is unclear what agency regulates the field testing of microbial facilitators. Would a microbial facilitator be considered, for regulatory purposes, as a microbial pesticide although it is not pathogenic? Should it as a microbial adjuvant be exempt from regulation? Regulatory constraints on the development of bioherbicides can be considerable, involving a variety of regulatory agencies, advisory bodies, and extensive specific regulations (Charudattan 1990b; Makowski 1997). A process at least as complicated should be anticipated for conducting large-scale field tests of microbial facilitators. Since the stated objective of many regulatory agencies is to expedite the process of field testing and registration of biological agents for pest control, perhaps the path leading to a microbial facilitator product would be comparatively short and smooth. Realistically, however, the utilization of the mechanisms by which a microbial strain enhances disease or developing techniques for increasing naturally occurring microbial facilitator populations may ultimately provide the most immediate benefit of research into microbial facilitation of weed control using mycoherbicides.

VI. Conclusions

In a perceptively written article, Zorner et al. (1993) rightly conclude that: "the discovery of active bioherbicides is not the major limitation to developing this technology. Much more effort needs to be placed on developing technology to economically ferment, stabilize and deliver these organisms in a form that will provide for consistent field efficacy." A significant consideration in obtaining consistent field efficacy is to understand phylloplane microbial dynamics, and to utilize them to the advantage of the applied mycoherbicide. Accepting an axiom from the study of controlling plant pathogens, i.e., that control procedures should target the weakest point of the pathogen's life cycle, it is conversely true that an effort to augment the damage done by a pathogen should bolster the pathogen at this same weak point. In the case of mycoherbicides, the span between initiation of germination and infection of the host tissue represents the stage of mycoherbicide deployment vulnerable to the deleterious effects of the chemical, physical, and microbiological environment present on the phylloplane. Understanding how to manage this environment to the advantage of applied mycoherbicides represents the next frontier of research that must be crossed if we expect to more fully realize the potential of prospective mycoherbicides. In 1959, Baker stated, "It has been said that the genius of Rembrandt lay in his superb shadows, even though it is the illuminated areas that excite our interest. Similarly, it is increasingly clear that mastery of the soil microflora will come only when we understand the nonpathogenic organisms, even though the pathogens have seemed the logical point of attack." In the case of mycoherbicide research, mastery of the phylloplane environment in the sense of understanding how to purposefully manage the chemical, physical, and nonpathogenic microbial factors therein, represents the best opportunity for ultimate mycoherbicide success, not the "logical attack" of continuing to look for mycoherbicide candidates with higher virulence and shorter dew period requirements in hopes that such strains will succeed based on genetic potential alone.

Acknowledgments. I thank Lee Baker for assistance with electron microscopy and Marsha Ebener and Tricia Soncasie for assistance in proofing the manuscript. Helpful reviews were provided by T. Paulitz and W.G.D. Fernando.

References

Allen RJ (1976) Spore germination and its regulation: control of spore germination and infection structure formation in the fungi. In: Pirson A, Zimmermann MH (eds) Encyclopedia of plant physiology, vol 4. Springer, Berlin Heidelberg New York, pp 51–85

Altman J, Neate S, Rovira AD (1990) Herbicide-pathogen interactions and mycoherbicides as alternative strategies for weed control. In: Hoagland RE (ed) Microbes and microbial products as herbicides. American Chemical Society, Washington DC, pp 240–259

Andrews JH (1992) Biological control in the phyllosphere. Annu Rev Phytopathol 30:603–635

Andrews JH, Hirano SS (eds) (1991) Microbial ecology of leaves. Springer, Berlin Heidelberg New York

Baker KF (1959) Epilogue to part VI. In: Holton CS, Fischer GW, Fulton RW, Hart H, McCallan SEA (eds) Plant pathology, problems and progress 1908–1958. University of Wisconsin Press, Madison, pp 337–379

Baker KF, Cook RJ (1974) Biological control of plant pathogens. Freeman, San Francisco

Blakeman JP (1985) Ecological succession of leaf surface microorganisms in relation to biological control. In: Windels CE, Lindow SE (eds) Biological control on the phylloplane. American Phytopathological Society, St Paul, pp 6–30

Blakeman JP, Atkinson P (1981) Antimicrobial substances associated with the aerial surfaces of plants. In: Blakeman JP (ed) Microbial ecology of the phylloplane. Academic Press, London, pp 245–263

Blakeman JP, Parbery DG (1977) Stimulation of appressorium formation in *Colletotrichum acutatum* by phylloplane bacteria. Physiol Plant Pathol 11:313–325

Boyette CD (1991) Control of hemp sesbania with a fungal pathogen *Colletotrichum truncatum*. US patent #5,034,328. Issued 7/23/91

Boyette CD, Quimby PC Jr, Connick WJ Jr, Daigle DJ, Fulgham FE (1991) Progress in the production, formulation, and application of mycoherbicides. In: TeBeest DO (ed) Microbial control of weeds. Chapman and Hall, New York, pp 209–222

Bridges DC, Anderson RL (1992) Crop losses due to weeds in the United States, 1992. Weed Science Society of America, Champaign, Illinois

Bunster L, Fokkema NJ, Schippers B (1989) Effect of surface-active *Pseudomonas* spp. on leaf wettability. Appl Environ Microbiol 55:1340–1345

Burnett HC, Tucker DPH, Ridings WH (1974) Phytophthora root and stem rot of milkweed vine. Plant Dis Rep 58:355–357

Caesar AJ, Burr TJ (1991) Effect of conditioning, betaine, and sucrose on survival of rhizobacteria in powder formulations. Appl Environ Microbiol 57:168–172

Charudattan R (1990a) Pathogens with potential for weed control. In: Hoagland RE (ed) Microbes and microbial products as herbicides. American Chemical Society, Washington DC, pp 132–154

Charudattan R (1990b) Release of fungi: large-scale use of fungi as biological weed control agents. In: Marois JJ, Bruening G (Tech Authors) Risk assessment in agricultural biotechnology. Regents of the University of California, pp 70–84

Charudattan R (1991) The mycoherbicide approach with plant pathogens. In: TeBeest DO (ed) Microbial control of weeds. Chapman and Hall, New York, pp 24–57

Chiang MY, Van Dyke CG, Leonard KJ (1989) Evaluation of endemic foliar fungi for potential biological control of johnsongrass (*Sorghum halepense*): screening and host range tests. Plant Dis 73:459–464

Colbert SF, Hendson M, Ferri M, Schroth MN (1993) Enhanced growth and activity of a biocontrol bacterium genetically engineered to utilize salicylate. Appl Environ Microbiol 59:2071–2076

Connick WJ Jr, Boyette CD, McAlpine JR (1991a) Formulation of mycoherbicides using a pasta-like process. Biol Contr 1:281–287

Connick WJ Jr, Daigle DJ, Quimby PC (1991b) An improved invert emulsion with high water retention for mycoherbicide delivery. Weed Tech 5:442–444

Cook RJ (1993) Making greater use of introduced microorganisms for biological control of plant pathogens. Annu Rev Phytopathol 31:53–80

Daigle DJ, Connick WJ Jr (1990) Formulation and application technology for microbial weed control. In: Hoagland RE (ed) Microbes and microbial products as herbicides. American Chemical Society, Washington DC, pp 288–304

Daly JM (1976) Some aspects of host-pathogen interactions. In: Pirson A, Zimmermann MH (eds) Encyclopedia of plant physiology, vol 4. Springer, Berlin Heidelberg New York, pp 27–50

Daniel JT, Templeton GE, Smith RJ Jr, Fox WT (1973) Biological control of northern jointvetch in rice with an endemic fungal disease. Weed Sci 21:303–307

Darvill AG, Albersheim P (1984) Phytoalexins and their elicitors – a defense against microbial infection in plants. Annu Rev Plant Physiol 35:243–275

Emmett RW, Parbery DG (1975) Appressoria. Annu Rev Phytopathol 13:147–167

Fernando WGD, Linderman RG (1995) Inhibition of *Phytophthora vignae* and stem and root rot of cowpea by soil bacteria. Biol Ag Hort 12:1–14

Fernando WGD, Watson AK, Paulitz TC (1994) Phylloplane *Pseudomonas* spp. enhance disease caused by *Colletotrichum coccodes* on velvetleaf. Biol Control 4:125–131

Fernando WGD, Watson AK, Paulitz TC (1996) The role of *Pseudomonas* spp. and competition for carbon, nitrogen and iron in the enhancement of appressorium formation by *Colletotrichum coccodes* on velvetleaf. Eur J Plant Pathol 102:1–7

Fletcher M, McEldowney S (1984) Microbial attachment to nonbiological surfaces. In: Klug MJ, Reddy CA (eds) Current perspectives in microbial ecology. ASM, Washington DC, pp 124–129

Fravel DR (1988) Role of antibiosis in the biocontrol of plant diseases. Annu Rev Phytopathol 26:75–91

Golovleva LA, Maltseva OV, Solyanikova IP (1992) Metabolism of foreign compounds in *Pseudomonas* spp. In: Galli E, Silver S, Witholt B (eds) *Pseudomonas*: molecular biology and biotechnology. ASM, Washington DC, pp 231–238

Haj-Ahmad Y, Bilinski CA, Russell I, Stewart GG (1992) Thiamine secretion in yeast. Can J Microbiol 38:1156–1161

Hallett SG, Ayres PG (1992) Invasion of rust (*Puccinia lagenophorae*) aecia on groundsel (*Senecio vulgaris*) by secondary pathogens: death of the host. Mycol Res 96:142–144

Hirano SS, Upper CD (1993) Dynamics, spread, and persistence of a single genotype of *Pseudomonas syringae* relative to those of its conspecifics on populations of snap bean leaflets. Appl Environ Microbiol 59:1082–1091

Hoagland RE (1990) Microbes and microbial products as herbicides. American Chemical Society, Washington DC

Howell CR, Beier RC, Stipanovic RD (1988) Production of ammonia by *Enterobacter cloacae* and its possible role in the biological control of *Pythium* preemergence damping-off by the bacterium. Phytopathology 78:1075–1078

Huang Y, Deverall BJ, Morris SC (1991) Promotion of infection of orange fruit by *Penicillium digitatum* with a strain of *Pseudomonas cepacia*. Phytopathology 81:615–618

Hudson HJ (1971) The development of the saprophytic fungal flora as leaves senesce and fall. In: Preece TF, Dickinson CH (eds) Ecology of leaf surface microorganisms. Academic Press, New York, pp 447–455

Hyakumachi M, Kageyama K, Ikegami H (1992) Energy stress in relation to germinability and virulence of root infecting fungi. In: Tjamos EC, Papavizas GC, Cook RJ (eds) Biological control of plant diseases. Plenum Press, New York, pp 327–330

Jackson MA, Schisler DA (1992) The composition and attributes of *Colletotrichum truncatum* spores are altered by the nutritional environment. Appl Environ Microbiol 58:2260–2265

Jackson MA, Schisler DA (1995) Liquid culture production of microsclerotia of *Colletotrichum truncatum* for use as bioherbicidal propagules. Mycol Res 99:879–884

Jackson MA, Schisler DA, Bothast RJ (1995) Conidiation environment influences fitness of the potential bioherbicide, *Colletotrichum truncatum*. In: Delfosse ES, Scott RR (eds) Proc of the 8th Int Symp on Biological control of weeds. DSIR/CSIRO, Melbourne, pp 621–626

Janisiewicz WJ (1988) Biocontrol of postharvest diseases of apples with antagonist mixtures. Phytopathology 78:194–198

Janisiewicz W (1996) Ecological diversity, niche overlap, and coexistence of antagonists used in developing mixtures for biocontrol of postharvest diseases of apples. Phytopathology 86:473–479

Jensen V (1971) The bacterial flora of beech leaves. In: Preece TF, Dickinson CH (eds) Ecology of leaf surface microorganisms. Academic Press, New York, pp 463–469

Jin X, Harman GE, Taylor AG (1991) Conidial biomass and desiccation tolerance of *Trichoderma harzianum* produced at different medium water potentials. Biol Contr 1:237–243

Juniper BE (1991) The leaf from the inside and the outside: a microbe's perspective. In: Andrews JH, Hirano SS (eds) Microbial ecology of leaves. Springer, Berlin Heidelberg New York, pp 21–42

Kamoun S, Young M, Glascock CB, Tyler BM (1993) Extracellular protein elicitors from *Phytophthora*: host specificity and induction of resistance to bacterial and fungal phytopathogens. Mol Plant Microb Interact 6:15 25

Keen NT (1990) Phytoalexins and their elicitors. In: Hoagland RE (ed) Microbes and microbial products as herbicides. American Chemical Society, Washington DC, pp 114–131

Kenerley CM, Andrews JH (1990) Interactions of pathogens on plant leaf surfaces. In: Hoagland RE (ed) Microbes and microbial products as herbicides. ACS, Washington DC, pp 192–217

Kenney DS (1986) DeVine – the way it was developed – an industrialist's view. Weed Sci 34:15–16

Kinkel LL, Lindow SE (1993) Invasion and exclusion among coexisting *Pseudomonas syringae* strains on leaves. Appl Environ Microbiol 59:3447–3454

Kloepper JW, Tuzun S, Kuc JA (1992) Proposed definitions related to induced disease resistance. Biocontrol Sci Technol 2:349–351

Kohmoto K, Otani H, Kodama M, Nishimura S (1989) Host recognition: can accessibility to fungal invasion be induced by host-specific toxins without necessitating necrotic cell death? In: Graniti A, Durbin RD, Ballio A (eds) Phytotoxins and plant pathogenesis. Springer, Berlin Heidelberg New York, pp 249–265

Kosanke JW, Osburn RM, Shuppe GI, Smith RS (1992) Slow rehydration improves the recovery of dried bacterial populations. Can J Microbiol 38:520–525

Kroschwitz JI (executive ed), Howe-Grant M (ed, 4th edn) (1992) Ammonia, encyclopedia of chemical technology. Wiley, New York

Kuć J (1995) Phytoalexins, stress metabolism, and disease resistance in plants. Annu Rev Phytopathol 33:275–297

Kwon YH, Epstein L (1993) A 90-kDa glycoprotein associated with adhesion of *Nectria haematococca* macroconidia to substrata. Mol Plant Microb Interact 6:481–487

Larsson C, Gustafsson L (1993) The role of physiological state in osmotolerance of the salt-tolerant yeast *Debaryomyces hansenii*. Can J Microbiol 39:603–609

Laycock MV, Hildebrand PD, Thibault P, Walter JA, Wright JLC (1991) Viscosin, a potent peptidolipid biosurfactant and phytopathogenic mediator produced by a pectolytic strain of *Pseudomonas fluorescens*. J Agric Food Chem 39:483–489

Leite B, Nicholson RL (1992) Mycosporine-alanine: a self-inhibitor of germination from the conidial mucilage of *Colletotrichum graminicola*. Exp Mycol 16:76–86

Lindemann J, Suslow TV (1987) Competition between ice nucleation-active wild-type and ice nucleation-deficient deletion mutant strains of *Pseudomonas syringae* and *P. fluorescens* biovar I and biological control of frost injury on strawberry blossoms. Phytopathology 77:882–886

Lindow SE (1992) Environmental release of Pseudomonads: potential benefits and risks. In: Galli E, Silver S, Witholt B (eds) *Pseudomonas*: molecular biology and biotechnology. ASM, Washington DC, pp 399–407

Lindow SE, Knudson GR, Seidler RJ, Walter MV, Lambou VW, Amy PS, Schmedding D, Prince V, Hern S (1988) Aerial dispersal and epiphytic survival of *Pseudomonas syringae* during a pretest for the release of genetically engineered strains into the environment. Appl Environ Microbiol 54:1557–1563

Loper JE, Buyer JS (1991) Siderophores in microbial interactions on plant surfaces. Mol Plant Microb Int 4:5–13

Lorito M, Di Pietro A, Hayes CK, Woo SL, Harman GE (1993) Antifungal synergistic interaction between chitinolytic enzymes from *Trichoderma harzianum* and *Enterobacter cloacae*. Phytopathology 83:721–728

Lynch JM (1990) Microbial Metabolites. In: Lynch JM (ed) The rhizosphere. Wiley, New York, pp 177–206

Makowski RMD (1997) Foliar pathogens in weed biocontrol: ecological and regulatory constraints. In: Andow DA, Ragsdale DW, Nyvall RF (eds) Ecological interactions and biological control. Westview Press, Boulder (in press)

Makowski RMD, Mortensen K (1992) The first mycoherbicide in Canada: *Colletotrichum gloeosporiodes* f. sp. *malvae* for round-leaved mallow control. In: Richardson RG (compiler) Proceedings of the 1st international weed control congress, vol 2. Weed Science Society of Victoria, Melbourne, pp 298–300

Malvick DK, Moore LW (1988) Survival and dispersal of a marked strain of *Pseudomonas syringae* in a maple nursery. Plant Pathol 37:103–105

Marks GC, Cerra R (1991) Effects of propazine and chlorthal dimethyl on *Phytophthora cinnamomi* root disease of *Pinus radiata* seedlings and associated soil microflora. Soil Biol Biochem 23:157–164

McCracken AR, Swinburne TR (1979) Siderophores produced by saprophytic bacteria as stimulants of germination of conidia of *Colletotrichum musae*. Phys Plant Pathol 15:331–340

Morin L, Auld BA, Brown JF (1993) Synergy between *Puccinia xanthii* and *Colletotrichum orbiculare* on *Xanthium occidentale*. Biol Contr 3:296–310

Morris CE, Rouse DI (1985) Role of nutrients in regulating epiphytic bacterial populations. In: Windels CE, Lindow SE (eds) Biological control on the phylloplane. APS, St Paul, Minnesota, pp 63–82

Mortensen K (1988) The potential of an endemic fungus, *Colletotrichum gloeosporioides*, for biological control of round-leaved mallow (*Malva pusilla*) and velvetleaf (*Abutilon theophrasti*). Weed Sci 36:473–478

Nicholson RL (1990) Functional significance of adhesion to the preparation of the infection court by plant pathogenic fungi. In: Hoagland RE (ed) Microbes and microbial products as herbicides. American Society, Washington DC, pp 218–239

Nicholson RL, Hipskind J, Hanau RM (1989) Protection against phenol toxicity by the spore mucilage of *Colletotrichum graminicola*, an aid to secondary spread. Physiol Mol Plant Pathol 35:243–252

Oku H, Shiraishi T, Kato T, Kim HM, Saitoh K, Tahara M (1993) Structure and mode of action of supressors, pathogenicity factors of pea pathogen, *Mycosphaerella pinodes*. In: Fritig B, Legrand M (eds) Mechanisms of plant defense responses. Kluwer, Dordrecht, p 87

Ordentlich A, Elad Y, Chet I (1988) The role of chitinase of *Serratia marcescens* in biocontrol of *Sclerotium rolfsii*. Phytopathology 78:84–88

O'Sullivan DJ, O'Gara F (1992) Traits of fluorescent *Pseudomonas* spp. involved in suppression of plant root pathogens. Microbiol Rev 56:662–676

Pascholati SF, Deising H, Leite B, Anderson D, Nicholson RL (1993) Cutinase and non-specific esterase activities in the conidial mucilage of *Colletotrichum graminicola*. Physiol Mol Plt Pathol 42:37–51

Phatak SC, Sumner DR, Wells HD, Bell DK, Glaze NC (1988) Method for controlling yellow nutsedge using *Puccinia canaliculata*. US patent #4,731,104. Issued Feb 15, 1988

Ricci P, Panabieres F, Bonnet P, Maia N, Ponchet M, Devergne JC, Marais A, Cardin L, Milat ML, Blein JP (1993) Proteinaceous elicitors of plant defense responses. In: Fritig B, Legrand M (eds) Mechanisms of plant defense responses. Kluwer, Dordrecht, pp 121–135

Ridings WH (1986) Biological control of stranglervine in citrus – a researcher's view. Weed Sci 34:31–32

Ruinen J (1966) The phyllosphere. IV. Cuticle decomposition by microorganisms in the phyllosphere. Ann Inst Pasteur 111:342–346

Sands DC, Miller RV, Ford EJ (1990) Biotechnological approaches to control of weeds with pathogens. In: Hoagland RE (ed) Microbes and microbial products as herbicides. American Chemical Society, Washington DC, pp 184–191

Schisler DA, Jackson MA (1996) Germination of soil-incorporated microsclerotia of *Colletotrichum truncatum* and colonization of seedlings of the weed *Sesbania exaltata*. Can J Microbiol 42:1032–1038

Schisler DA, Slininger PJ (1994) Selection and performance of bacterial strains for biologically controlling Fusarium dry rot of potatoes incited by *Gibberella pulicaris*. Plant Dis 78:251–255

Schisler DA, Howard KM, Bothast RJ (1991a) Enhancement of disease caused by *Colletotrichum truncatum* in *Sesbania exaltata* by coinoculating with epiphytic bacteria. Biol Contr 1:261–268

Schisler DA, Jackson MA, Bothast RJ (1991b) Influence of nutrition during conidiation of *Colletotrichum truncatum* on conidial germination and efficacy in inciting disease in *Sesbania exaltata*. Phytopathology 81:587–590

Schisler DA, Howard KM, Bothast RJ (1992) Utilization of phylloplane bacteria to augment the efficacy of the mycoherbicide, *Colletotrichum truncatum*. In: Richardson RG (compiler) Proc 1st Int Weed Control Congr, vol 2. Weed Science Society of Victoria, Melbourne, pp 461–464

Schisler DA, Slininger PJ, Bothast RJ (1994) Performance of antagonist mixtures and concentration effects of bacterial strains active against *Fusarium sambucinum*. Phytopathology 84:1114 (Abstr)

Schisler DA, Jackson MA, McGuire MR, Bothast RJ (1995) Use of pregelatinized starch and casamino acids to improve the efficacy of *Colletotrichum truncatum* conidia produced in differing nutritional environments. In: Delfosse ES, Scott RR (eds) Proc 8th Int Symp Biol control weeds. DSIR/CSIRO, Melbourne, pp 659–664

Schoeneweiss DF (1975) Predisposition, stress, and plant disease. Annu Rev Phytopathol 13:193–211

Sebastian J, Chandra AK, Kolattukudy PE (1987) Discovery of a cutinase-producing *Pseudomonas* sp. cohabiting with an apparently nitrogen-fixing *Corynebacterium* sp. in the phyllosphere. J Bacteriol 169:131–136

Sequeira L (1984) Plant-bacterial interactions. In: Pirson A, Zimmermann MH (eds) Encyclopedia of plant physiology, vol 17. Springer, Berlin Heidelberg New York, pp 187–211

Simms EL, Vision TJ (1995) Pathogen-induced systemic resistance in *Ipomoea purpurea*. Oecologia 102:494–500

Slade SJ, Swinburne TR, Archer SA (1986) The role of a bacterial siderophore and of iron in the germination and appressorium formation by conidia of *Colletotrichum acutatum*. J Gen Microbiol 132:21–26

Slininger PJ, Schisler DA, Bothast RJ (1994) Two-dimensional liquid culture focusing: a method of selecting commercially promising microbial isolates with demonstrated biological control capability. In: Ryder MH, Stephens PM, Bowen GD (eds) Improving plant productivity with rhizosphere bacteria. CSIRO Division of Soils, Glen Osmond, South Australia, pp 29–32

Smith RJ Jr (1991) Integration of biological control agents with chemical pesticides. In: TeBeest DO (ed) Microbial control of weeds. Chapman and Hall, London, pp 189–208

Smith RJ Jr (1986) Biological control of northern jointvetch (*Aeschynomene virginica*) in rice (*Oryza sativa*) and soybeans (*Glycine max*) – a researcher's view. Weed Sci 34:17–23

Stowell LJ (1991) Submerged fermentation of biological herbicides. In: TeBeest DO (ed) Microbial control of weeds. Chapman and Hall, London, pp 225–261

TeBeest DO, Yang XB, Cisar CR (1992) The status of biological control of weeds with fungal pathogens. Annu Rev Phytopathol 30:637–657

Templeton GE (1992) Use of *Colletotrichum* strains as mycoherbicides. In: Bailey JA, Jeger MJ (eds) *Colletotrichum*: biology, pathology and control. CAB International, Oxon, pp 358–380

Toyoda H, Morimoto M, Kakutani K, Morikawa M, Fukamizo T, Goto S, Terada H, Ouchi S (1993) Binary microbe system for biological control of Fusarium wilt of tomato: enhanced root-colonization of an antifungal rhizoplane bacterium supported by a chitin-degrading bacterium. Ann Phytopathol Soc Jpn 59:375–386

Tsurushima T, Ueno T, Fukami H, Irie H, Inoue M (1995) Germination self-inhibitors from *Colletotrichum gloeosporiodies* f. sp. *jussiaea*. Mol Plant Microb Interact 8:652–657

Urquhart EJ, Menzies JG, Punja ZK (1994) Growth and biological control activity of *Tilletiopsis* species against powdery mildew (*Sphaerotheca fuliginea*) on greenhouse cucumber. Phytopathology 84:341–351

Van Dyke CG (1991) Biological control of weeds with fungi. In: Arora DK, Rai B, Mukerji KG, Knudsen GR (eds) Handbook of applied mycology, vol 1. Soil and plants. Dekker, New York, pp 357–376

VanEtten HD, Matthews DE, Matthews PS (1989) Phytoalexin detoxification: importance for pathogenicity and practical implications. Annu Rev Phytopathol 27:143–164

Van Loon LO, Pennings GGH (1993) Involvement of ethylene in the induction of systemic acquired resistance in tobacco. In: Fritig B, Legrand M (eds) Mechanisms of plant defense responses. Kluwer, Dordrecht, pp 156–159

Walter JF, Paau AS (1993) Microbial inoculant production and formulation. In: Metting FB Jr (ed) Soil microbial ecology: applications in agricultural and environmental management. Dekker, New York, pp 579–594

Weete JD (1992) Induced systemic resistance to *Alternaria cassiae* in sicklepod. Physiol Mol Plant Pathol 40:437–445

Wei G, Kloepper JW, Tuzun S (1991) Induction of systemic resistance of cucumber to *Colletotrichum orbiculare* by select strains of plant growth-promoting rhizobacteria. Phytopathology 81:1508–1512

Weidemann GJ, TeBeest DO (1990) Genetic variability of fungal pathogens and their weed hosts. In: Hoagland RE (ed) Microbes and microbial products as herbicides. American Chemical Society, Washington DC, pp 176–183

Wilson CL (1969) Use of plant pathogens in weed control. Annu Rev Phytopathol 7:411–434

Wilson M, Lindow SE (1993) Effect of phenotypic plasticity on epiphytic survival and colonization by *Pseudomonas syringae*. Appl Environ Microbiol 59:410–416

Wymore LA, Poirier C, Watson AK (1988) *Colletotrichum coccodes*, a potential bioherbicide for control of velvetleaf (*Abutilon theophrasti*). Plant Dis 72:534–538

Yamaguchi K, Kida M, Arita M, Takahashi M (1992) Induction of systemic resistance by *Fusarium oxysporum* MT0062 in solanaceous crops. Ann Phytopathol Soc Jpn 58:16–22

Youssef BA, Heitefuss R (1983) Side-effects of herbicides on cotton wilt caused by *Fusarium oxysporum* f. sp. *vasinfectum*. II. Effect of herbicides on the quantitative and qualitative composition of sugars and amino acids in cotton seed and root exudates. Z Pflanzenkr Pflanzensch J Plant Dis Prot 90:36–49

Yu SM, Templeton GE, Wolf DC (1988) Trifluralin concentration and the growth of *Fusarium solani* f. sp. *cucurbitae* in liquid medium and soil. Soil Biol Biochem 20:607–612

Zorner PS, Evans SL, Savage SD (1993) Perspectives on providing a realistic technical foundation for the commercialization of bioherbicides. In: Duke SO, Menn JJ, Plimmer JR (eds) Pest control with enhanced environmental safety. ACS, Washington DC, pp 79–86

15 Fungivores

T.P. McGonigle

CONTENTS

I. Introduction

This chapter considers the impact that fungivore grazing can have on communities of fungi in soil and litter. The fungi are represented by large numbers of species (Christensen 1989) and considerable biomass (Kjøller and Struwe 1982) in soil and litter systems. The soil fauna comprises 15% of the soil biomass-C (Reichle 1977), and produces 10% of the respiratory carbon dioxide released from soil (Petersen and Luxton 1982). Between 20 and 75% of the soil fauna biomass resides in the fungivore trophic category (McGonigle 1995). Based on these considerations, it seems likely that grazing by soil fauna on fungal tissues will play a significant role in shaping the fungal communities of field systems.

In order to assess the impact that grazing on fungi has on fungal community structure, it is useful first to consider fungi in relation to the principles of community ecology. The individuals within a species and in a given location comprise a single population. Discussion of the concept of the individual with respect to fungi has been made elsewhere (Rayner and Todd 1982). The indeterminate growth of many plants has led to the quantification of modules of plant growth, rather than of individuals (Harper 1977). Fungi can be treated similarly where recognizable structures such as sporophores are visible (Shaw 1985). However, fruit bodies may represent less than only 1% of the unseen but living vegetative fungal biomass (Frankland 1982). Communities are collections of populations that are present inside defined limits in space, and within specified taxonomic or functional groups (Begon et al. 1990). Communities can be described most simply in terms of species richness, which is the number of species present. Diversity is a term usually reserved for where the relative abundance of each of the species is considered (Begon et al. 1990). For a given species richness, the diversity of the community will be determined by the relative abundances of the species present; high abundance of a small number of species equates to low diversity, whereas greater evenness or equitability of the species abundances reflects higher diversity. Fungivore grazing can be expected to affect species richness by promoting the elimination or the introduction of fungal species. In addition, the grazing of fungi can be expected to affect the relative abundance of the fungi present, and by so doing affect community diversity.

Characterization of fungal communities has, in some cases, evaluated species richness only, although this is in itself a mammoth undertaking (Christensen 1981; Swift 1976). Even prolonged

Department of Land Resource Science, University of Guelph, Guelph, Ontario, Canada N1G 2W1

The Mycota IV
Environmental and Microbial Relationships
Wicklow/Söderström (Eds.)
© Springer-Verlag Berlin Heidelberg 1997

isolation programs appear to be unable to complete the species list for a given sampling location; Christensen (1989) found that after more than 1000 isolates had already been taken from one location, each increment of 100 additional isolates consistently yielded approximately 10 species new to the site. When isolation frequencies are used as a measure of relative fungal abundance, fungal communities follow a lognormal distribution (Lussenhop 1981); this is consistent with communities for other types of organisms in other kingdoms (Begon et al. 1990). In theory, the lognormal distribution reflects the way in which species abundances are the result of the interplay of a variety of independent factors (May 1975).

The question of niche diversification among sympatric fungal species was considered by Swift (1976, 1982). Although many fungi show a surprisingly broad range of occurrence, there is some specialization for different resource types; these include various plant parts like leaves and twigs, and the different plant species they come from. There is also the opportunity for specialization among the different microenvironments within resource types, such as between the vein and mesophyll portions of a leaf, and among the various carbon substrates therein. Swift (1976) proposed the idea of the unit community, whereby each resource type is present in a number of discrete units within the ecosystem, each one with its particular combination of species from the pool available in the ecosystem as a whole.

Previous reviews of fungivore grazing have considered the processes of comminution of substrate, dispersal of fungi, and the direct impact of feeding itself, on fungal community structure (Visser 1985), and on nutrient cycling (Ingham 1992). Comminution of substrate stimulates microbial activity and speeds decomposition (Swift et al. 1979). The dispersal of fungi through the movements of grazing animals has been clearly established (Brasier 1978; Wiggins and Curl 1979). Enhanced dispersal will encourage a community to fulfil its intrinsic potential for a poor or rich species richness, playing an important role in the establishment of microbial communities. However, in this chapter it is the biological interactions which occur after dispersal, and which are involved in the determination of the fungal community structure, that are considered. The effect on fungi of the disruptive physical action of grazing animals, such as by the process of trampling, is not easy to distinguish from the effects resulting from

ingestion of the fungal material. The breaking-up of fungal material by animal body movements has been suggested (Wicklow and Yocum 1982) to be one of the reasons why fungal communities change in response to grazing.

This chapter will first outline the relevant features of the various types of invertebrate fungivores in soil and litter, and then proceed to consider some limitations encountered in the study of grazing on fungi. Examples will then be given to show the ways in which grazing can affect fungal communities through effects on species richness, community diversity, and replacement of some species by others. In the last section before the conclusion, various features of the fungivores and fungi which can modify the outcome of grazing will be discussed. Specifically, these modifying features are the selectivity and intensity of grazing, and the ability of the fungus being grazed to respond to that grazing.

II. The Fungivores in Soil and Litter Systems

A. General Considerations

The soil fauna is composed of a great diversity of species distributed among many phyla. Of particular importance with regard to fungivory are three phyla: the Annelida, the Arthropoda, and the Nematoda. Within the Annelida the group which displays extensive fungivory is the Enchytraeidae. The arthropod classes Arachnida and Insecta contain the groups Acari and Collembola, respectively, each of which has many fungivorous members. Two of the 12 orders of nematodes, the Dorylaimida and Tylenchida, contain an abundance of fungivores. Features of the Acari, Enchyraeidae, Collembola, and Nematoda which relate to fungal feeding are outlined below, along with brief mention of other fungus-feeding groups.

Using the terminology of Moore et al. (1988), animals feeding on hyphae can be divided into two categories: engulfing fungivores and fluid-feeding fungivores. These feeding modes are mostly consistent for the fungivores within defined taxonomic boundaries. The relative importance of these feeding types varies among ecosystems. An extensive review of the biomass and population densities of all the different types of soil animals across a range of ecosystems can be found in Petersen and Luxton (1982).

B. A Survey of Relevant Groups

1. Acari

Mites are very abundant in soil and litter, and many of them consume fungi. Soil mites comprise four orders. Most members of the order Cryptostigmata, or oribatid mites, are generalist feeders (Swift et al. 1979) and will ingest decaying plant material, fungal hyphae, and algae. However, there is some specialization within the order: members of the family Phthiracaridae feed only on plant residues and are able to digest cellulose but not the fungal carbohydrate trehalose; in contrast, the Oppiidae and Eremaeidae specialize on fungi and cannot digest cellulose (Luxton 1972). Some Cryptostigmata readily feed on nematodes (Rockett 1980). Members of the Gamasina in the order Mesostigmata are predatory, with different genera showing varying degrees of specialization in their choice of prey (Moore et al. 1988). The order Astigmata show a diverse range of feeding activity: members of the family Acaridae feed by engulfing hyphae, as well as taking live nematodes and protozoa, whereas some members of the family Histiostomatidae specialize in the ingestion of a slurry of decaying residues mixed with microbes (Walter and Kaplan 1990). The order Prostigmata has fluid-feeding fungivores in several families, e.g., Tydaeidae and Tarsonemidae; however, this order also contains families of predatory mites, such as the Bdellidae and Stigmaeidae (Moore et al. 1988). The group of fungivorous mites which predominates in the field depends on the type of biome under consideration. In forest systems, the Cryptostigmata are the most abundant (Mitchell and Parkinson 1976; Hogervost et al. 1993), whereas in desert systems the fungivorous Prostigmata are the more important group (Santos and Whitford 1981).

2. Collembola

The majority of Collembola are primarily fungivorous (Petersen and Luxton 1982). The springtails of the soil can be divided (Wallwork 1976) into two distinctive life-forms as follows: those at the soil surface or in litter layers are usually large, pigmented individuals with well-developed eyes. The Collembola of deeper layers are more often small, weakly pigmented, and with reduced eyes. Based on analysis of gut contents, the two life-form groups show little difference in the extent to which they engage in fungal feeding (Takeda and Ichimura 1983). One-half or more of the Collembola collected from the field typically have no gut contents, which may be due to intermittent cessation of feeding caused by the molting cycle of the animals (Anderson and Healey 1972). Collembola can molt from a few to in excess of 50 times (Christensen 1990). For springtails with some gut contents, those in a coniferous plantation were found by Poole (1959) to be filled up with fungal hyphae.

3. Enchytraeidae

Gut contents of enchytraeids contain a mixture of fungal hyphae, plant residues, and soil materials, with fungal material typically representing one third of these gut contents (Dash et al. 1980). When offered fungal baits, the gut content of enchytraeids with fungal material can increase to between 50 and 70% of the total (Dash and Cragg 1972). The gut contents of some enchytraeid species can contain twice as much fungal material as would be expected on the basis of random ingestion of substrate (O'Connor 1967). However, there is some uncertainty about the extent to which enchytraeids in the field are fungivores, and, in turn, the degree to which they rely on bacteria as food or on the saprotrophic digestion of plant residues. Estimates of fungivory in enchytraeids range from 25% (Persson et al. 1980) to as much as 80% (Whitfield 1977). In a recent review of the ecology of Enchytraeidae, Didden (1993) concluded that much work is needed on the feeding biology of the various worm species, before the role of this group in the functioning of soil systems is adequately appraised.

4. Nematoda

Free-living soil nematodes can be identified as fungivores, bacterivores, predators, or omnivores, based on their anterior morphology (Twinn 1974). The diet of omnivorous nematodes can consist of algae, fungal spores, protozoa, and other nematodes (Swift et al. 1979). In addition, there are also some free-living soil nematodes which feed on root epidermal cells and root hairs (Yeates et al. 1993). Fungivorous nematodes pierce hyphae with their stylets and feed on the fluid protoplasm of the fungus, using a pumping action (Freckman and Baldwin 1990). Empty hyphal walls remain behind. Under culture conditions the impact of grazing by fungivorous nematodes can be severe, killing all aerial hyphae and reducing growth on agar relative to that seen in the absence of the nematodes (Shafer et al. 1981).

5. Others

Various other groups in the soil fauna also eat fungi. Among the larvae of dipterous flies, those in the family Sciaridae graze fungi in dung deposited on the soil surface (Wicklow and Yocum 1982), while larvae of the families Chironomidae and Mycetophilidae are mainly fungivorous (Swift et al. 1979). The isopod *Oniscus asellus* L. was able to reduce fungal standing crop to one third of that in control leaf-fungus microcosms (Hanlon and Anderson 1980). A selection of genera of soil amoebae are able to feed on hyphae (Bamforth 1988), but the impact of this on populations of soil fungi is unclear (Chakraborty et al. 1983).

III. Some Limitations to the Study of Fungivore Grazing

Investigations into the extent to which grazing occurs can adopt several approaches: direct observation, examination of gut contents, and monitoring through time the quantities of hyphae present. Food preference tests can be used to ascertain the palatability of different fungi to the faunal taxa of interest.

A. Gut Contents

Examination of gut contents can provide valuable information on dietary choices of fauna among the available foods. However, one shortcoming of this approach is that some food types more easily retain their structural integrity, and are recognizable in the gut for a longer time. Walter (1987) found that nematodes, which were seen to be consumed by a selection of mites that are normally considered mycophagous, were not detectable in gut boluses because of the lack of sclerotization of the nematode body. Gut contents of most soil

mesofauna are typically reported as having a certain portion in the category "amorphous material" (Anderson 1975) or "unidentified" (Broady 1979).

B. Food Preference Tests

There is some uncertainty as to whether food choices seen under controlled conditions give a reliable indication of which foods are being eaten in the field. For oribatid mites, feeding choices in the laboratory are not consistent with the gut contents of field-collected individuals (Mitchell and Parkinson 1976). Variability in nitrogen content can influence the palatability of mycelium to Collembola (McMillan 1976). Further, springtails have been seen during feeding evaluation tests to switch from feeding predominately on one palatable fungus, to feeding almost exclusively on a simultaneously offered but different palatable fungus, in the course of a few days (Visser and Whittaker 1977). In the field, the diet of oribatids can change through the seasons of the year (Anderson 1975).

C. Enumeration of Hyphae

Measurement of fungal biomass through time presents difficulties because estimates of the proportion of hyphal length that is active vary according to the staining method used (Hamel et al. 1990; Schubert and Mazzitelli 1989). Fluorecein diacetate (FDA) reveals activity of cytoplasmic proteinase and acetyl-cholinase enzymes of low specificity (Rotman and Papermaster 1966), while nitro-blue tetrazolium detects activity of mitochondrial membrane-bound succinate dehydrogenase (Schaffer and Peterson 1993). There is a correlation between the presence of cytoplasm within hyphae and the quality of staining of hyphae by phenolic aniline blue (Jones and Mollison 1948); this correlation can be used to estimate the percentage of live hyphae (Nagel-de Boois and Jansen 1971). It might be assumed from direct observations of feeding that engulfing fungivores eliminate the hyphae they consume, whereas fluid-feeders leave behind dead hyphae, with hollow but otherwise intact walls. However, the feces of dipteran larvae have been found to contain many short fragments of hyphae which are without cytoplasmic contents (Swift 1982). The study of grazing by monitoring the temporal pattern of hyphal densities in soil and litter would be aided if there were a way to discern if hyphae without contents have already been grazed, or if they became empty by the relocation of cytoplasm within the fungus to growing regions of the mycelium (Dowding 1976).

IV. Grazing and Community Structure

In various sections below, studies on grazing of plants are referred to in order to provide a theoretical framework in which we can consider grazing on fungi. Comparisons have been drawn previously between grazing in fungivore-fungus and herbivore-plant systems (Wicklow 1981; Visser 1985; Shaw 1992).

A. Changes in Species Richness and Diversity

1. Grazing on a Dominant Fungus

Darwin (1859) reported that when a previously mown 90 × 120-cm turf plot was left to grow, the number of plant species within it was reduced from 20 to 11 because of the proliferation of some species. Thus, mowing can maintain increased species richness. Crawley (1983) argues that both mowing and grazing of herbaceous systems function in an essentially similar way, by the selective removal of more of some species than of others. Tansley and Adamson (1925) monitored vegetation inside and outside exclosures on chalk grassland, which was otherwise kept to between 2.5 and 5 cm in height by rabbit grazing. Although species richness was unaffected, there was increased proliferation of a few herb species inside the exclosures, showing that greater diversity and more equitability among species had previously been maintained by grazing. However, under very intense grazing which fell just short of uncovering bare soil, plant species richness itself was reduced (Tansley and Adamson 1925). These studies (Darwin 1859; Tansley and Adamson 1925) established that grazing can increase species richness or increase community diversity by the suppression of species which would otherwise be more dominating. Alternatively, when the intensity of grazing is sufficiently high, then species richness can be reduced.

When one fungus is more productive than another, selective grazing on the more abundant fungus can suppress what would otherwise have been the dominant member of the community. The outcome of grazing in this situation will be a community of greater diversity. An example from (Newell 1984a,b) was the suppression of the otherwise dominant *Marasmius androsaceus* (L. ex Fr.) Fr. under grazing pressure from the collembolan *Onychiurus latus* Gisin, leading to more extensive development of the sympatric species *Mycena galopus* (Pers ex Fr.) Kummer (Fig. 1). In the study of Newell (1984a,b) two species of fungi were considered. However, the principles involved can be applied to larger communities. Where increased representation of several otherwise suppressed species occurs at the expense of one dominant, the result will be greater equitability among species. An analogous effect may occur with the redistribution of nutrient resources among members of herbaceous communities through arbuscular mycorrhizal plant-to-plant connections (Grime et al. 1987).

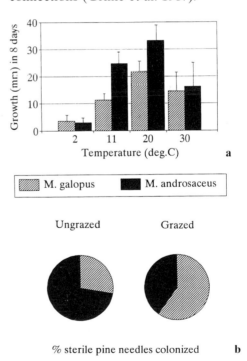

Fig. 1. a Growth of *Mycena galopus* and *Marasmius androsaceus* in culture as a function of temperature. *M. galopus* is the slower-growing species (data of Newell 1984a). b Effect of collembolan grazing on the percentage of pine needles colonized by *M. galopus* and *M. androsaceus* in microcosms. Grazing is seen to suppress *M. androsaceus* and allow greater proliferation of *M. galopus*. (Data of Newell 1984b)

2. Grazing on a Fungus That Is not Dominant

Selective grazing on a fungus that is of equal aggressiveness compared to its neighbor can act to polarize the community, making the ungrazed fungus more abundant relative to its grazed neighbor. Fungivore selection of a fungus already showing relatively low production within the community will act to reinforce or further exaggerate the dominance of the ungrazed fungus. In both cases, the trend will be toward a community of reduced diversity. One example of selective grazing on seemingly codominant fungi can be seen in Parkinson et al. (1979), where the collembolan *Onychiurus subtenuis* Folsom was introduced into aspen leaf microcosms which had been inoculated with two fungi isolated from an aspen litter system. When the two fungal isolates were kept in separate microcosms, grazing by Collembola had little impact (Fig. 2). However, when the fungi were grown together, grazing by the Collembola acted to polarize the system by the suppression of isolate sterile dark 298 and promotion of isolate basidiomycete 290 (Fig. 2). It was noted (Parkinson et al. 1979) that basidiomycete 290 was completely unpalatable to *Onychiurus subtenuis* in feeding-choice tests. More recently, Ek et al. (1994) demonstrated that grazing by *Onychiurus armatus* Tullb. was able to suppress the development of saprotrophic fungi in the genus *Paecilomyces* much more strongly when the *Pinus contorta* Dougl. ex Loud. seedlings in the microcosm were mycorrhizal with *Paxillus involutus*

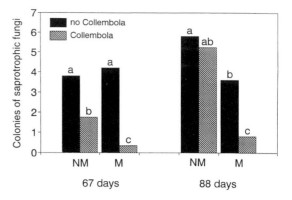

Fig. 3. The number of sporulating colonies of saprotrophic fungi seen per microcosm after 67 and 88 days of development of *Pinus contorta* seedlings which were either nonmycorrhizal (*NM*), or mycorrhizal (*M*) with *Paxillus involutus*, with or without 50 Collembola added per microcosm. The majority of the saprotrophic fungi belonged to the genus *Paecilomyces*. Bars with different letters (*a,b,c*) are significantly ($P < 0.05$) different. (Data of Ek et al. 1994)

(Batsch) Fr., than when they were nonmycorrhizal (Fig. 3). This example underlines the importance of the interplay between any direct impact of grazing, and the interactions between the various fungi present. Presumably, in response to grazing in the mycorrhizal system, deletion competition for resources swung in favor of *Paxillus* and away from *Paecilomyces*, leading to almost complete suppression of the latter. However, in the nonmycorrhizal situation, grazing was only able to cause a delay in the development of *Paecilomyces* (Fig. 3).

B. Replacement of Some Species by Others

Selective grazing can speed the transition between successional stages. Lubchenco (1983) showed that the intertidal mollusk *Littorina littorea* L. preferentially grazed on early-successional-stage seaweed species, thereby making the change to later-stage seaweed species occur more swiftly. An example of the action of grazing to promote the transition from early-stage to later-stage successional species of fungi was given by Klironomos et al. (1992). Sterile *Picea abies* (L.) Karst. needles were inoculated with one of three primary saprotrophs from the genera *Cladosporium*, *Epicoccum*, and *Phoma* (Klironomos et al. 1992), which typically invade senescent needles in the canopy. On falling to the floor, these primary saprotrophs are replaced

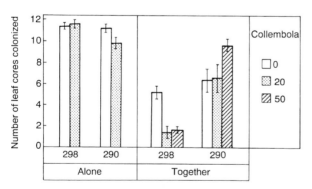

Fig. 2. Effect of collembolan grazing on the numbers of leaf cores colonized by the isolates sterile dark 298 and basidiomycete 290. In one series the fungi were inoculated onto leaf cores so as to keep the fungi in separate microcosms, while in the other series the leaf cores were inoculated with the fungi mixed in the same microcosms. Collembola were added at 0, 20, and 50 per microcosm and left for 10 days. Grazing is seen to promote the competitive success of basidiomycete 290 over sterile dark 298. (Data of Parkinson et al. 1979)

by secondary saprotrophs from the community of soil fungi. Klironomos et al. (1992) added tagged needles colonized exclusively by primary saprotrophs to similar cultures of needles colonized only by secondary saprotrophs from the genera *Penicillium* and *Trichoderma*. Treatments with or without a population of the collembolan *Folsomia candida* Willem. were set-up, and needles added were sampled over 6 weeks. Grazing by the Collembola on the primary saprotrophs, which were known from feeding choice tests to be preferred over the secondary saprotrophs, almost eliminated the primary saprotrophs and facilitated their replacement by the secondary saprotrophs (Fig. 4).

C. Modifying Factors

Irrespective of the relative dominance of any grazed fungus, and of the position of the fungal community along a successional sequence, the impact of grazing on community structure will be influenced for the most part by three modifying influences: how selective the grazing is, how intense the grazing is, and the ability of the fungi to respond to grazing.

1. Selectivity of Grazing

When a selection of fungi are offered to Collembola in culture, feeding preferences have been clearly demonstrated amongst pathogenic (Curl et al. 1988) and ectomycorrhizal (Shaw

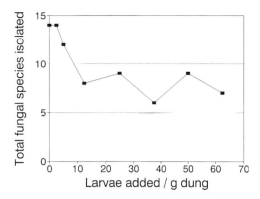

Fig. 5. Number of fungal species recovered from rabbit dung microcosms in response to the number of fungivorous dipteran fly larvae added. Increasing grazing intensity reduces the species richness, although no further consistent reduction is seen above 12.5 larvae g^{-1} dung. (Data of Wicklow and Yocum 1982)

1988) fungi. Selectivity in feeding has also been shown for mites (Harding and Stuttard 1974). Both Collembola and mites preferred *Alternaria* and *Trichoderma* over an arbuscular mycorrhizal fungus (Klironomos 1994). Collembola are attracted by the quality of fungal odors (Bengtsson et al. 1988), and the reproductive output of the animals is influenced by the nutritional quality of the fungus eaten (Booth and Anderson 1979). Crawley (1983) explains that removal of species in proportion to their abundance will not be able to affect the structure of a community; for species richness or community diversity to be influenced requires selective removal of greater quantities of some species. It seems that grazing of fungi is almost always found to involve selective feeding, so it can therefore be expected to affect fungal community structure.

2. Intensity of Grazing

Generally, intense levels of grazing lead to elimination of some species. Wicklow and Yocum (1982) imposed a range of grazing intensities by varying the number of *Lycoriella mali* Fitch. dipteran fly larvae added to 2 g dry mass-equivalent samples of rabbit dung. In the absence of grazing, 14 fungal species were recovered from each sample. However, this fell to between 6 and 9 species per sample when the grazing intensity was above 10 larvae g^{-1} dung (Fig. 5). This reduction in species richness was due to the loss of what appeared to be grazing-sensitive species. Over and above the impact of grazing seen in Fig. 5, some fungal species had a reduced frequency of occur-

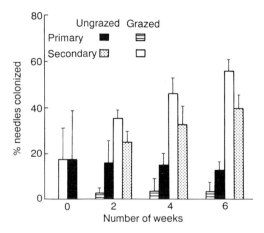

Fig. 4. Effect of collembolan grazing in microcosms on the percentage of *Picea abies* needles colonized by primary saprotroph fungi and secondary saprotroph fungi through time. Grazing is seen to encourage the replacement of primary saprotrophs by the less palatable secondary saprotrophs. (Data of Klironomos et al. 1992)

rence in the more highly grazed microcosms, although other species were unaffected (Wicklow and Yocum 1982). Reduction of species richness in response to high intensities of grazing has been seen in the fungi of aquatic systems (Bärlocher 1980). *Folsomia candida* reduced the standing crop of *Coriolus versicolor* L. ex Fr. in proportion to animal density, and bacterial biomass expanded to take the place of the fungi removed (Hanlon and Anderson 1979).

3. Responses of the Grazed Fungi

Any negative impact of grazing on the vigor of fungi through loss of biomass or damage to tissues might be offset by compensatory growth produced in response to that grazing. Indeed, where senescent parts of the mycelium are removed, and this, in turn, is able to permit regrowth from an otherwise functionally dormant hyphal system, grazing can have a positive rather than a negative impact on the grazed fungus. There are examples where compensatory responses of the fungi have been observed. Grazing of *Botrytis cinerea* Pers ex Fr. by *Folsomia candida* doubled the respiration of the fungus (Hanlon 1981). This was attributed to removal of senescent parts of the colony. Removal of old hyphae allowed new fungal growth where nutrient availability permitted (Hanlon 1981). A similar effect on respiration was shown for grazing of *Onychiurus guadriocellatus* Gisin on the fungi growing on diplopod fecal pellets (van der Drift and Jansen 1977). One collembolan per pellet decreased fungal standing crop, yet increased fungal respiration by 10% over controls (van der Drift and Jansen 1977).

The impact of grazing on fungal biomass is nonlinear. Low grazing intensities can stimulate fungal production, whereas higher levels reduce that production. *Folsomia candida* at a density of 250 animals kg^{-1} dry soil increased FDA-staining

hyphae of *Fusarium oxysporum* Schlect. to 4.8 mg^{-1} dry soil (Moore 1988) from 3.2 mg^{-1} dry soil in ungrazed controls. However, at a density of 1000 kg^{-1} dry soil these Collembola reduced the hyphae to 1.2 mg^{-1} dry soil (Moore 1988). In a leaf-litter microcosm study (Ineson et al. 1982), the number of *Folsomia candida* per microcosm increased from 20 to 300 over the first 8 weeks, falling slightly to 240 per microcosm over the remaining 4 weeks of the experiment. The standing crop of hyphae increased to 32 mg^{-1} over the initial 6 weeks, compared to a rise to only 20 mg^{-1} in ungrazed controls. However, under the more intense grazing pressure in the second half of the experiment, the fungal standing crop decreased sharply to 6 mg^{-1}, whereas in the ungrazed controls it was sustained at 20 mg^{-1} (Ineson et al. 1982).

Based on a modeling approach, Bengtsson et al. (1993) concluded that fragmentation of fungal thalli into patches, such as might be caused by grazing, will promote compensatory growth by the fungi. The effect of grazing on fungal respiration appears to consist of two phases of fungal response. Grazing of *Onychiurus armatus* reduced respiration of *Mortierella isabellina* Oudem during the 2-day period in which the Collembola were in contact with the mycelium, but fungal respiration was stimulated relative to ungrazed controls in a subsequent 5-day grazing-free interval (Bengtsson and Rundgren 1983). *Mortierella isabellina* adopts an alternate and faster-growing mode of growth in response to grazing (Hedlund et al. 1991). This alternate mode is possibly a response which helps the fungus to escape the grazer (Hedlund et al. 1991).

Following introduction of low densities of the fungivorous nematode *Aphelenchoides* into agar cultures of various fungi, Riffle (1971) noted a variety of responses by both fungus and grazer (Table 1). Fungal characteristics are clearly im-

Table 1. A summary of responses noted by Riffle (1971) for fungi cultured individually and grazed by the nematode *Aphelenchoides*. The 43 fungi studied were divided by Riffle (1971) into five groups (I to V) according to the response of fungus and nematode to grazing

Group	Fungal linear growth rate	Subculture viability	Nematode density developed	No. of fungal species	Example
I	Unchanged	Good	High	11	*Flammulina velutipes* (Curt. ex Fr.) Karst
II	Unchanged	Good	Low	12	*Suillus variegatus* (Fr.) O. Kuntze
III	Reduced	Reduced	High	5	*Amanita rubescens* ([Pers] Fr.) S.F. Grey
IV	Reduced	Lost	Low	7	*Boletus edulis* Bull. ex Fr.
V	Reduced	Lost	High	8	*Armillaria mellea* (Vahl ex Fr.) Kummer

portant in determining the outcome seen: Riffle (1971) noted that members of group I (Table 1) were fast growers whose rate of production of mycelium exceeded the rate of consumption of it by nematodes, while those in group IV were slow-growing fungi; group II were suspected of producing anti-feeding metabolites (Riffle 1971). The diversity of responses seen in Table 1 for fungal productivity and viability after imposing a standardized grazing regime emphasizes the complexity of ways in which grazing can modify fungal community structure. Depending on the type of fungus concerned, grazing can be seen to have no effect, or alternatively, a strong negative effect on the grazed fungi. Simultaneously, for both the unaffected and grazing-damaged fungi, examples are to be seen where the nematode population is stimulated to reach high densities, and in other cases not. These different types of responses of the grazer population can, in turn, be expected to feed back in different ways to modify fungal community structure through further grazing, or the lack of it.

V. Conclusions

Grazing of fungi has been shown under controlled conditions to have the capacity to change fungal community structure. If feeding is on fungi which would be dominant in the absence of grazing, diversity can be enhanced. Alternatively, grazing on codominant or subordinate fungi will act to polarize the fungal community, and diversity will be depressed. Grazing can facilitate the transition from one successional stage of fungi to the next. The impact of grazing is modified by three factors: (1) grazing must be selective to have any impact at all on community structure, (2) high intensities of grazing lead to the loss of fungal species from the community, and (3) responses of the fungi to grazing can in some cases negate any negative impact of grazing.

There is an abundance of animals in soil and litter which eat fungi. As might be expected, fungivores are found to be associated with mycelia in the field. For example, Cromack et al. (1988) reported the biomass per unit soil mass of Collembola, oribatid mites, and nematodes, to be 2.0, 1.5, and 1.2 times greater in ectomycorrhizal fungal mats than in adjacent nonmat soil. Seasonal patterns for the abundance of fungi and fungivores

are similar, with maxima in spring and fall, and a minimum in midsummer (Whittaker 1974).

The enormous taxonomic diversity among fungivores in soil and litter is of unknown significance. Much of the diversity of the soil fauna might be thought of in terms of many different species doing essentially similar things. However, there are some clues which have emerged recently which point to some degree of specialization in function. For example, fungivores in different families of oribatid mites were interpreted by Mueller et al. (1990) as playing different roles in the field. The Oppiidae occurred almost exclusively on buried litter, whereas the Tectocepheidae were more abundant on surface residues. Further, whereas application of the fungicide Captan decreased numbers of Oppiidae, it increased the density of Tectocepheidae. The hypothesis suggested (Mueller et al. 1990) to explain this observation was that the Tectocepheidae specialize on feeding of senescent hyphae.

It is clear that the potential exists for grazing by fungivores in field systems to modify the fungal communities therein. For a forest system (Persson 1983) and for a grassland system (Hunt et al. 1987), the mass of nitrogen in the fungivore biomass corresponds to 1.6 and 6.1%, respectively, of that in the fungal biomass. Kowal and Crossly (1971) estimated that soil fungivores consume 18 times their own body mass year^{-1}. Mitchell and Parkinson (1976) estimated that the oribatids alone in a forest system eat 6 g fungal hyphae m^{-2} year^{-1}, compared to a fungal standing crop of up to 280 g m^{-2}. However, due to the intractability of soil and litter systems, clear demonstrations that grazing does, in fact, play a significant role in the determination of fungal community structure in the field remain elusive.

References

Anderson JM (1975) Succession, diversity and trophic relationships of some soil animals in decomposing leaf litter. J Anim Ecol 44:475–495

Anderson JM, Healey IN (1972) Seasonal and interspecific variation in major components of the gut contents of some woodland Collembola. J Anim Ecol 41:359–368

Bamforth SS (1988) Interactions between Protozoa and other organisms. Agric Ecosyt Environ 24:229–234

Bärlocher F (1980) Leaf-eating invertebrates as competitors of aquatic hyphomycetes. Oecologia 47:303–306

Begon M, Harper JL, Townsend CR (1990) Ecology: individuals populations and communities. 2nd edition. Blackwell Scientific, Boston

Bengtsson G, Rundgren S (1983) Respiration and growth of a fungus, *Mortierella isabellina*, in response to grazing by *Onychiurus armatus* (Collembola). Soil Biol Biochem 15:469–473

Bengtsson G, Erlandsson A, Rundgren S (1988) Fungal odour attracts soil Collembola. Soil Biol Biochem 20:25–30

Bengtsson G, Hedlund K, Rundgren S (1993) Patchiness and compensatory growth in a fungus-Collembola system. Oecologia 93:296–302

Booth R, Anderson JM (1979) The influence of fungal food quality on the growth and fecundity of *Folsomia candida* (Collembola: Isotomidae). Oecologia 38:317–323

Brasier CM (1978) Mites and reproduction in *Ceratocystis ulmi* and other fungi. Trans Br Mycol Soc 70:81–89

Broady PA (1979) Feeding studies on the collembolan *Cryptopygus antarcticus* Willem at Signy Island, South Orkney Islands. Br Anarct Surv Bull 48:37–46

Chakraborty S, Old KM, Warcup JH (1983) Amoebae from a take-all suppressive soil which feed on *Gaeumannomyces graminis tricici* and other soil fungi. Soil Biol Biochem 15:17–24

Christensen K (1990) Insecta: collembola. In: Dindal DL (ed) Soil biology guide. Wiley, New York, pp 965–995

Christensen M (1981) Species diversity and dominance in fungal communities. In: Wicklow DT, Carroll GC (eds) The fungal community. Dekker, New York, pp 201–232

Christensen M (1989) A view of fungal ecology. Mycologia 81:1–19

Crawley MJ (1983) Herbivory. Blackwell Scientific, Oxford

Cromack K, Ficher BL, Moldenke AM, Entry JA, Ingham ER (1988) Interactions between soil animals and ectomycorrhizal fungal mats. Agric Ecosyst Environ 24:161–168

Curl EA, Lartey R, Peterson CM (1988) Interactions between root pathogens and soil microarthropods. Agric Ecosyst Environ 24:249–261

Darwin C (1859) The origin of species. Murray, London

Dash MC, Cragg JB (1972) Selection of microfungi by Enchytraeidae (Oligochaeta) and other members of the soil fauna. Pedobiologia 12:282–286

Dash MC, Nanda B, Behera N (1980) Fungal feeding by Enchytraeidae (Oligochaeta) in a tropical woodland in Orissa, India. Oikos 34:202–205

Didden WAM (1993) Ecology of terrestrial Enchytraeidae. Pedobiologia 37:2–29

Dowding P (1976) Allocation of resources; nutrient uptake and release by decomposer organisms. In: Anderson JM, Macfadyen A (eds) The role of terrestrial and aquatic organisms in decomposition processes. Blackwell Scientific, Oxford, pp 169–183

Ek H, Sjögren M, Arnebrant K, Söderström B (1994) Extramatrical mycelial growth, biomass allocation and nitrogen uptake in ectomycorrhizal systems in response to collembolan grazing. Appl Soil Ecol 1:155–169

Frankland JC (1982) Biomass and nutrient cycling by decomposer basidiomycetes. In: Frankland JC, Hedger JN, Swift MJ (eds) Decomposer basidiomycetes. Cambridge University Press, Cambridge, pp 241–261

Freckman DW, Baldwin JG (1990) Nematoda. In: Dindal DL (ed) Soil biology guide. Wiley, New York, pp 155–200

Grime JP, Mackay JML, Hillier SH, Read DJ (1987) Floristic diversity in a model system using experimental microcosms. Nature 328:420–422

Hamel C, Fyles H, Smith DL (1990) Measurement of development of endomycorrhizal mycelium using three different vital stains. New Phytol 115:297–302

Hanlon RDG (1981) Influence of grazing by Collembola on the activity of senescent fungal colonies grown on media of different nutrient concentration. Oikos 36:362–367

Hanlon RDG, Anderson JM (1979) The effects of Collembola grazing on microbial activity in decomposing leaf litter. Oecologia 38:93–99

Hanlon RDG, Anderson JM (1980) Influence of macroarthropod feeding activities on microflora in decomposing oak leaves. Soil Biol Biochem 12:255–261

Harper JL (1977) Population biology of plants. Academic Press, London

Harding DJL, Stuttard RA (1974) Microarthropods. In: Dickinson CH, Pugh GJF (eds) Biology of plant litter decomposition. Academic Press, London, pp 489–532

Hedlund K, Boddy L, Preston CM (1991) Mycelial responses of the soil fungus, *Mortierella isabellina*, to grazing by *Onychiurus armatus* (Collembola). Soil Biol Biochem 23:361–366

Hogervorst RF, Verhoef HA, Van Straalen NM (1993) Five-year trends in soil arthropod densities in pine forests with various levels of vitality. Biol Fertil Soils 15:189–195

Hunt HW, Coleman DC, Ingham ER, Ingham RE, Moore JC, Rose SL, Reid CPP, Morley CR (1987) The detrital food web in a shortgrass prairie. Biol Fertil Soils 3:57–68

Ineson P, Leonard MA, Anderson JM (1982) Effect of collembolan grazing upon nitrogen and cation leaching from decomposing leaf litter. Soil Biol Biochem 14:601–605

Ingham RE (1992) Interactions between invertebrates and fungi: effects on nutrient availability. In: Carroll GC, Wicklow DT (eds) The fungal community, 2nd edn. Dekker, New York, pp 669–690

Jones PCT, Mollison JE (1948) A technique for the quantitative estimation of soil micro-organisms. J Gen Microbiol 2:54–69

Kjøller A, Struwe S (1982) Microfungi in ecosystems: fungal occurrence and activity in litter and soil. Oikos 39:389–422

Klironomos JN (1994) Influences of microarthropods on the functioning of endomycorrhizal associations. PhD Thesis, University of Waterloo, Ontario, Canada

Klironomos JN, Widden P, Deslandes I (1992) Feeding preferences of the collembolan *Folsomia candida* in relation to microfungal successions on decaying litter. Soil Biol Biochem 24:685–692

Kowal NE, Crossley DA (1971) The ingestion rates of microarthropods in pine mor, estimated with radioactive calcium. Ecology 52:444–452

Leonard MA (1984) Observations on the influence of culture conditions on the fungal feeding preferences of *Folsomia candida* (Collembola: Isotomidae). Pedobiologia 26:361–367

Lubchenco J (1983) *Littorina* and *Fucus*: effects of herbivores, substratum heterogeneity, and plant escapes during succession. Ecology 64:1116–1123

Lussenhop J (1981) Analysis of microfungal component communities. In: Wicklow DT, Carroll GC (eds) The fungal community. Dekker, new York, pp 37–45

Luxton M (1972) Studies on the oribatid mites of a Danish beech wood soil. I. Nutritional biology. Pedobiologia 12:434–463

May RM (1975) Patterns of species abundance and diversity. In: Cody ML, Diamond JM (eds) Ecology and evolution of communities. Harvard University Press, Cambridge, Massachusetts, pp 81–121

McGonigle TP (1995) The significance of grazing on fungi in nutrient cycling. Can J Bot (Suppl 1):S1370–S1376

McMillan JH (1976) Laboratory observations on the food preference of *Onychiurus armatus* (Tullb.) Gisin (Collembola, Family Onychiuridae). Rev Ecol Biol Sol 13:353 364

Mitchell MJ, Parkinson D (1976) Fungal feeding of oribatid mites (Acari: Cryptostigmata) in an aspen woodland soil. Ecology 57:302–312

Moore JC (1988) The influence of microarthropods on symbiotic and non-symbiotic mutualism in detrital-based below-ground food webs. Agric Ecosyst Environ 24:147–159

Moore JC, Walter DE, Hunt HW (1988) Arthropod regulation of micro- and mesobiota in below-ground detrital food webs. Annu Rev Entomol 33:419–439

Mueller BR, Beare MH, Crossley DA (1990) Soil mites in detrital food webs of conventional and no-tillage systems. Pedobiologia 34:389 401

Nagel-de Boois HM, Jansen E (1971) The growth of fungal mycelium in forest soil layers. Rev Ecol Biol Sol 8:509–520

Newell K (1984a) Interaction between two decomposer basidiomycetes and a collembolan under sitka spruce: distribution, abundance and selective grazing. Soil Biol Biochem 16:227–233

Newell K (1984b) Interaction between two decomposer basidiomycetes and a collembolan under sitka spruce: grazing and its potential effects on fungal distribution and litter decomposition. Soil Biol Biochem 16:235–239

O'Connor FB (1967) The Enchytraeidae. In: Burges A, Raw F (eds) Soil biology. Academic Press, New York, pp 213–257

Parkinson D, Visser S, Whittaker JB (1979) Effects of collembolan grazing on fungal colonization of leaf litter. Soil Biol Biochem 11:529–535

Persson T (1983) Influence of soil animals on nitrogen mineralization in a northern scots pine forest. In: Lebrun P, Andre M, de Medts A, Gregoire-Wibo C, Wanthy G (eds) New trends in biology. Dien-Brichart, Louvain-la-Neuve, Belgium, pp 117–126

Persson T, Bååth E, Clarholm M, Lundkvist H, Söderström BE, Sohlenius B (1980) Trophic structure, biomass dynamics and carbon metabolism of soil organisms in a scots pine forest. Ecol Bull 32:419–459

Petersen H, Luxton M (1982) A comparative analysis of soil fauna populations and their role in decomposition processes. Oikos 39:287–388

Poole TB (1959) Studies on the food of Collembola in a Douglas-fir plantation. Proc Zool Soc Lond 132:71–82

Rayner ADM, Todd NK (1982) Population structure in wood-decomposing basidiomycetes. In: Frankland JC, Hedger JN, Swift MJ (eds) Decomposer basidiomycetes. Cambridge University Press, Cambridge, pp 109–128

Reichle DE (1977) The role of soil invertebrates in nutrient cycling. Ecol Bull 25:145–156

Riffle JW (1971) Effect of nematodes on root-inhabiting fungi. In: Hacskaylo E (ed) Mycorrhizae. United States Department of Agriculture Forest Service miscellaneous publication 1189. United States Government Printing Office, Washington, pp 97–113

Rockett CL (1980) Nematode predation by oribatid mites (Acari: Oribatida). Internat. J Acarol 6:219–224

Rotman B, Papermaster BW (1966) Membrane properties of living mammalian cells as studied by enzymatic hydrolysis of fluorgenic esters. Proc Natl Acad Sci USA 55:134–141

Santos PF, Whitford WG (1981) The effects of microarthropods on litter decomposition in a Chihuahuan desert ecosystem. Ecology 62:654–663

Schaffer GF, Peterson RL (1993) Modifications to clearing methods used in combination with vital staining of roots colonized with vesicular-arbuscular mycorrhizal fungi. Mycorrhiza 4:29–35

Schubert A, Mazzatelli M (1989) Enzymatic activities of VAM extraradical mycelium at different stages of development. Agric Ecosyst Environ 29:349–353

Shafer SR, Rhodes LH, Riedel RM (1981) In vitro parasitism of endomycorrhizal fungi of ericaceous plants by the mycophagous nematode *Aphelenchoides bicaudatus*. Mycologia 73:141–149

Shaw PJA (1985) Grazing preferences of *Onychiurus armatus* (Insecta: Collembola) for mycorrhizal and saprophytic fungi of pine plantations. In: Fitter AH, Atkinson D, Read DJ, Usher MB (eds) Ecological interactions in soil. Blackwell Scientific, Oxford, pp 333–337

Shaw PJA (1988) A consistent hierarchy in the fungal feeding preferences of the Collembola *Onychiurus armatus* Pedobiologia 31:179–187

Shaw PJA (1992) Fungi, fungivores, and fungal food webs. In: Carroll GC, Wicklow DT (eds) The fungal community, 2nd edn. Dekker, New York, pp 295–310

Swift MJ (1976) Species diversity and the structure of microbial communities in terrestrial habitats. In: Anderson JM, Macfadyen A (eds) The role of terrestrial and aquatic organisms in decomposition processes. Blackwell Scientific, Oxford, pp 185–222

Swift MJ (1982) Basidiomycetes as components of forest ecosystems. In: Frankland JC, Hedger JN, Swift MJ (eds) Decomposer basidiomycetes, their biology and ecology. Cambridge University Press, Cambridge, pp 307–333

Swift MJ, Heal OW, Anderson JM (1979) Decomposition in terrestrial ecosystems. University of California Press, Berkeley

Takeda H, Ichimura T (1983) Feeding attributes of four species of Collembola in a pine forest soil. Pedobiologia 25:373–381

Tansley AG, Adamson RS (1925) Studies on the vegetation of the English chalk. III. The chalk grasslands of the Hampshire-Sussex border. J Ecol 13:177–223

Twinn DC (1974) Nematodes. In: Dickinson CH, Pugh GJF (eds) Biology of plant litter decomposition. Academic Press, London, pp 421–465

Van der Drift J, Jansen E (1977) Grazing of springtails on hyphal mats and its influence on fungal growth and respiration. Ecol Bull 25:203–209

Visser S (1985) Role of the soil invertebrates in determining the composition of soil microbial communities. In: Fitter AH, Atkinson D, Read DJ, Usher MB (eds) Ecological interactions in soil. Blackwell Scientific, Oxford, pp 297–317

Visser S, Whittaker JB (1977) Feeding preferences for certain litter fungi by *Onychiurus subtenuis* (Collembola). Oikos 29:320–325

Wallwork JA (1976) The distribution and diversity of soil fauna. Academic Press, London

Walter DE (1987) Trophic behaviour of "mycophagous" microarthropods. Ecology 68:226–299

Walter DE, Kaplan DT (1990) Feeding observations on two astigmatic mites, *Schwiebea rocketti* (Acaridae) and *Histiostoma bakeri* (Histiostomatidae) associated with *Citrus* feeder roots. Pedobiologia 34:281–286

Whitfield DWA (1977) Energy budgets and ecological efficiencies on Truelove Lowland. In: Bliss LC (ed) Truelove Lowland, Devon Island, Canada: a high arctic ecosystem. University of Alberta Press, Edmonton, pp 607–620

Whittaker JB (1974) Interactions between fauna and microflora at tundra sites. In: Holding A, Heal OW, Maclean SJ, Flanagan P (eds) Soil organisms and decomposition in tundra. Tundra Biome Steering Committee, Stockholm, pp 183–196

Wicklow DT (1981) The coprophagous fungal community: a mycological system for examining ecological ideas. In: Wicklow DT, Carroll GC (eds) The fungal community. Dekker, New York pp 47–76

Wicklow DT, Yocum DH (1982) Effect of larval grazing by *Lycoriella mali* (Diptera: Sciaridae) on species abundance of coprophilous fungi. Trans Br Mycol Soc 78:29–32

Wiggins EA, Curl EA (1979) Interactions of Collembola and microflora of cotton rhizosphere. Phytopathology 69:244–249

Yeates GW, Bongers T, Goede RGM de, Freckman DW, Georgieva SS (1993) Feeding habits in soil nematode families and genera – an outline for soil ecologists. J Nematol 25:315–331

16 Applications of Fungal Ecology in the Search for New Bioactive Natural Products

J.B. GLOER[1]

CONTENTS

I. Introduction

A. Fungal Natural Products: from Mycotoxins[2] to Antibiotics

Among the most fascinating and important properties of fungi is their ability to produce a tremendous variety of so-called secondary metabolites that display a broad range of biological activities. Fungi are widely known for the production of compounds that have a deservedly negative reputation due to their activities as carcinogens or mammalian toxins (mycotoxins). Such compounds include aflatoxins, ochratoxins, citreoviridin, trichothecenes, fumonisins, and vari-

ous indole-derived tremorgenics (Miller and Trenholm 1994). In general, mycotoxins are not inherently more toxic than natural products from bacteria, plants, or other sources. However, mycotoxins are much more problematic because of their widespread occurrence as contaminants of food for humans and livestock. Knowledge of mycotoxin chemistry is essential to efforts to monitor and reduce the levels of such compounds in human and animal diets.

On the other hand, numerous important pharmaceuticals have also been discovered through studies of fungal chemistry (Masurekar 1992). This dichotomy is indicative of the diversity of bioactive compounds that fungi can produce. In fact, at least 6 of the top 20 ethical pharmaceuticals prescribed in 1994, representing combined sales of ca. US$6.7 billion (Czarnik 1996), are based on fungal natural product chemistry. Antibacterial agents such as penicillins and cephalosporins are perhaps the best-known examples, but many other compounds with various pharmacological activities have also been discovered as fungal metabolities. A notable example is mevinolin (= mevacor = lovastatin), a cholesterol-lowering agent that has annual gross sales of over $1 billion. Other medically important compounds include cyclosporin and the ergot alkaloids. Numerous other fungal metabolites have been discovered as potential pharmaceuticals or leads thereto, including asperlicin (cholecystokinin antagonist), echinocandins/pneumocandins (antifungal), papulacandins (antifungal), and zaragozic acids/squalestatins (squalene synthase inhibitors) (Masurekar 1992; Caporale 1995). Other fungal products show potential as natural agrochemicals (Gardner and McCoy 1992), with examples including destruxins (insecticides), strobilurins (fungicides), and various phytotoxins (herbicides).

The advantages of fungi as sources of useful natural products are well documented. Fungal metabolites are renewable resources, and meth-

---

[1] Department of Chemistry, University of Iowa, Iowa City, Iowa 52242, USA
[2] See also Chapter 12, Volume VI.

The Mycota IV
Environmental and Microbial Relationships
Wicklow/Söderström (Eds.)
© Springer-Verlag Berlin Heidelberg 1997

ods for large-scale production of important fungal metabolites can be developed using established techniques. Modification of metabolite structure and dramatic improvements in metabolite production efficiency can be accomplished through strain mutation, medium variation, and optimization of culture conditions. For example, manipulation of penicillin-producing cultures eventually resulted in a 6000-fold improvement in penicillin production (Demain 1992). Thus, although these steps are seldom trivial, promising metabolites isolated from fungi can be made available on a practical scale through application of existing technology.

The odds against finding a truly useful agent are daunting. For example, it requires approximately 12 years and US$350 million to develop a new pharmaceutical product, and 3600 active candidate compounds are dropped for every one that is marketed (Anonymous 1995). Even so, the potential payoff (see above), and the track record of fungi as sources of useful compounds have fostered continued industrial interest in fungal natural products chemistry within many screening programs. In recent years, it has become increasingly difficult to find new bioactive natural products from microbial sources because of the extensive screening efforts that have already taken place. In fact, the need to dereplicate cultures (i.e., weed out well-known metabolites that are responsible for positive results in a bioassay) is a source of tremendous expense and frustration (Corley and Durley 1994). Part of this problem stems from long-term heavy reliance on actinomycetes and common soil fungi as sources of bioactive metabolites. Less widely occurring fungi have been largely neglected. This is presumably due to inattention to the specific types of organisms chosen for screening and the habitats from which they are isolated, but also to disinterest in fungi that are slow-growing, difficult to isolate, and/or difficult to adapt to standard liquid fermentation protocols. Interestingly, many taxonomic and ecological groups of fungi have not been systematically explored for useful secondary metabolites, despite evidence in the literature which indirectly indicates their potential in this area.

The fungi provide almost limitless potential for metabolic variation. Fungi rank second only to the insects in estimated species biodiversity. Conservative estimates suggest that there are likely to be over 1.5 million fungal species, of which less than 5% have been described. This is over 5 times the predicted number of plant species and 50

times the estimated number of bacterial species (Hawksworth 1991). In recent years, an element of urgency has been conferred upon studies of the chemistry of certain fungi for the same reasons often cited to rationalize appeals for accelerated studies of plant chemistry, i.e., concerns about the loss of biodiversity (Balandrin et al. 1993). Many endangered plant and insect species are associated with specific fungal flora, and loss of those species would result in a concommitant loss of fungal species.

The importance of seeking isolates for industrial mass screening programs from relatively unexplored niche groups or substrates has been recognized (Monaghan and Tkacz 1990; Miller 1991; Dreyfuss and Chapela 1994; Bills 1995; Caporale 1995). However, the objective of this chapter is to provide specific examples of applications of fungal ecology to the search for bioactive natural products that may be useful in medicine or agriculture. Some examples from the literature will be cited, but emphasis will be placed on projects from our own research program that demonstrate the potential value of this approach as a complement to random screening programs.

B. Chemical Ecology: a General Guide to Natural Product Discovery

Knowledge of the ecology of plants and higher organisms has frequently led to the discovery of bioactive natural products. For example, it is now generally accepted that plants produce compounds that serve as chemical defenses against insects (Harborne 1993), herbivorous vertebrates (Rhodes 1985), or microbial attack (phytoalexins; Harborne 1987). Certain marine animals are known to produce or accumulate antifeedant/ichthyotoxic agents that protect them from attack by predators (Paul 1992). By contrast, relatively little is known of the chemical ecology of the fungi, despite the importance of fungal metabolites in modern medicine. As suggested by a number of researchers, there is ample evidence to suspect that the search for bioactive agents from fungi may be aided by application of analogous types of ecological rationale (Wicklow 1981; Demain 1992; Monaghan and Tkacz 1990; Dreyfuss and Chapela 1994; Anke et al. 1995; Caporale 1995; Gloer 1995a,b). Fungi commonly thrive in competitive environments, and it is often hypothesized that some of their secondary metabolic capabilities may be influenced by selection pressures exerted by other organisms. However, the majority of known bioactive fungal natural products have been discovered through industrial programs that involve mass random screening of uncharacterized isolates using artificial laboratory fermentation conditions. Choices of isolates for chemical investigation are based strictly on bioassay results, as are metabolite isolation procedures. Most modern industrial programs utilize an extensive array of specialized, often proprietary, biological assays for activity against specific molecular or cellular targets. Advances in assay miniaturization and robotics often permit extremely high throughput in these screens. Because of the sheer numbers of samples needed to satisfy the capacity of these assays, such programs can rarely provide support for ecological or taxonomic investigations of the organisms being screened, let alone studies of the possible roles of their metabolites.

The most important discoveries in recent years have been due to the development of novel bioassays, rather than to chemical studies of rare or specially selected types of fungi. For example, once an assay for inhibitors of HMG-CoA reductase was selected and developed, many fungi were found which produced mevinolin, or close analogs. The same is true with regard to squalene synthase and zaragozic acid-type inhibitors. Cyclosporin A is also now a commonly encountered fungal product. Interestingly, many other architecturally complex fungal metabolite-types (e.g., trichothecenes, emerimicins, destruxins, cytochalasins, paspalinines) have been found to occur as metabolites of disparate fungal taxa. On the other hand, some metabolites with potent bioactivity are known to be produced by one or more isolates of a single species. Secondary metabolite profiles can be chemotaxonomically useful in either instance (Frisvad 1989), but the latter case provides a strong argument for studies targeted toward rare fungal taxa or relatively unexplored ecosystems.

Examples of fungal chemicals likely to be associated with ecological phenomena include toxic mushroom metabolites that are presumed to play roles in defending fruiting bodies from fungivores, phytotoxins from plant pathogenic fungi that play roles in disease processes, insecticidal metabolites responsible for the toxic effects of entomopathogenic fungi, and the ergot alkaloids; chemical defenses found in the ergot of *Claviceps*

spp. and in certain grasses as metabolites of fungal endophytes. Chemical investigations of such sources could be viewed as "ecology-based" approaches leading to the discovery of fungal metabolites with somewhat predictable types of bioactivities. Studies of such phenomena are complementary to random screening because they incorporate rationale into the process of selecting fungi for chemical study. This kind of approach is particularly well suited to academic laboratories because resources for efficient mass screening on site in an appropriately wide variety of assays are not typically available in academia. Collaborations with industry are essential if compounds are to be fully evaluated. Realistically, however, throughput limitations in most academic labs, coupled with longer turnaround time for results and a somewhat more diverse (e.g., educational) mission, preclude true direct competitiveness with industrial in-house microbial chemistry programs. In any event, application of rationale at the beginning of the process limits the number of organisms that need to be investigated, reduces costs, facilitates dereplication, and permits evaluation of a hypothesis. Our own work has illustrated that such rationale can be very effective in leading to the isolation of new, patentable fungal metabolites with antifungal or antiinsectan activity. Although we have not yet demonstrated that such agents can lead to marketable products, application of ecological rationale in the search for novel bioactive fungal products may well result in findings with practical utility.

II. Observations in Fungal Ecology Associated with the Production of Bioactive Metabolites

A. Fungal Diseases

The use of biological control agents is an increasingly popular concept for the development of crop protection strategies (Gardner and McCoy 1992; Powell 1993). Such strategies involve the use of certain fungi as mycoherbicides, mycoinsecticides, or mycoparasites to control weeds, insects, or fungal pathogens, respectively. Thus, such approaches involve the deployment of a microbial disease agent effective against the pest, rather than (or as an adjunct to) the application of a chemical pesticide of some sort. Ultimately, damage to the pest is often caused by toxin(s) produced by the pathogen, and the use of measured and properly formulated quantities of the natural products themselves could provide an alternative control strategy. Use of the microorganism itself is appealing because it serves to selectively direct the toxin(s) to the target. Some toxin-producing microorganisms, most notably the bacterium *Bacillus thuringiensis*, have proven to be particularly effective in such applications. However, there are many specialized hurdles that must be overcome to implement an effective biocontrol strategy (Powell 1993), and the use of biocontrol agents will not necessarily eliminate the problem of resistance (Gardner and McCoy 1992). From a chemistry standpoint, knowledge of any metabolities involved in biocontrol effects is important as a means of avoiding unwanted side-effects, and precedents indicate that studies of fungi with mycopesticidal properties are likely to lead to discovery of pesticidal compounds. Some examples of such results are provided below.

1. Plant Pathogens

Plant pathogenic fungi are well known as producers of diverse compounds with phytotoxic effects on host plants (Turner and Aldridge 1983; Harborne 1993). The effects of the toxins are often principal causes of symptoms associated with the corresponding plant disease. In many cases, fungal phytotoxins produce damage that helps enable the fungus to invade and colonize the plant. Typically, the compounds are general phytotoxins (Ballio 1991), but in some cases, a coevolutionary process has led to at least some degree of host selectivity (Walton and Panaccione 1993). The host-selective toxins are particularly interesting from a chemical standpoint because they tend to have structural features that distinguish them from the most commonly encountered fungal metabolites. Notable examples include victorin, HC-toxin, and certain *Alternaria* toxins (Walton and Panaccione 1993). Plant pathogenic fungi as a group could be viewed as logical sources of bioactive compounds in a general sense, since they have already demonstrated the capacity to produce bioactive compounds with distinctive chemical structures. Some fungal metabolites with phytotoxic effects are known to exhibit medically relevant activities as well, including antitumor and antibiotic effects. It is interesting to note that mevinolin is a rather potent herbicide (Hoagland 1990).

Fungi pathogenic to weeds have been proposed as rational sources of herbicides. Indeed, phytotoxins with novel and unusual structures have been isolated from weed pathogens, and subsequently shown to display herbicidal activity toward weeds (Hoagland 1990). This would seem to be a particularly worthy avenue of investigation in view of the fact that other microbial natural products such as bialaphos (produced by a *Streptomyces* sp.) have already been used successfully as commercial herbicides.

2. Entomopathogenic and Nematopathogenic Fungi

The physiological effects associated with some fungal diseases of insects and nematodes have been linked to fungal toxins (Wicklow 1988). Such toxins could serve as leads to the development of new insectides or nematocides. In view of this possibility, together with the association of a significant component of predicted fungal diversity with estimates of insect diversity, it is surprising that relatively few studies of entomopathogenic fungi as sources of bioactive metabolites have been reported. A recent report (Gardner and McCoy 1992) suggests that only a small number of metabolites (ca. 20 at that time) have actually been identified from entomopathogenic fungi. A few distinctive classes of insecticidal metabolites have been discovered through studies of such species, including the beauverolides, destruxins, and viridoxins (Turner and Aldridge 1983; Gupta et al. 1993; Krasnoff et al. 1996). Some of these compounds display both dietary and topical activity against insects. Metabolites from nematopathogenic fungi are even more uncommon, however, a related approach involves investigation of nematode-trapping (nematophagous) fungi as sources of nematocidal toxins that cause paralysis or mortality of the nematode prey (Kwok et al. 1992; Stadler and Anke 1995).

3. Mycoparasites

Mycoparasitic fungi act as parasites of others, and the invaded organism often suffers negative effects from this interaction, which are likely to be caused in at least some cases by fungal toxins. Fungi isolated as colonists of others (fungicolous species) may also produce antifungal metabolites, even though a parasitic relationship may not have been demonstrated. Some mycoparasitic fungi that have been used or proposed as biocontrol agents because of their antifungal effects (e.g., *Trichoderma*, *Verticillium* spp.) have been shown to produce agents that inhibit the growth of other fungi. Examples include peptide antibiotics, phenolics, and terpenoids (Huang et al. 1995a,b; R.A.C. Morris et al. 1995). Antifungal metabolites have also been reported from other fungicolous and mycoparasitic species (Ayer et al. 1980; Choudhury et al. 1994).

B. Resistance of Key Fungal Structures to Fungivory or Microbial Attack

The vast majority of known fungal metabolites are produced in liquid fermentation cultures under conditions very different from those encountered by the fungi in nature. Although little is known about possible functions these compounds may have within the producing organisms, there is frequently a close correlation between secondary metabolite production and morphological differentiation in liquid culture (Bennett 1983). Interestingly, many fungi produce morphological structures under natural conditions (or on solid substrates in the laboratory) which are not generally formed in liquid cultures. Such structures include various fruiting bodies, sclerotia, and stromata. If there is secondary chemistry associated with such structures, it is reasonable to expect that such chemistry may not manifest itself under conditions that do not lead to formation of these bodies.

Our studies in this area have focused on important fungal survival structures (sclerotia and ascostromata) that are exposed to potential predators (fungivorous insects) under natural conditions. Our results indicate that fungal sclerotia and ascostromata often contain unique antiinsectan metabolites that may help to protect them from predation (see below). This work has been characterized by a particularly high incidence of previously undescribed natural products, even in cases where the producing fungi have long histories of prior chemical investigation.

A considerable body of direct and circumstantial evidence for the production by fungi of chemical defenses against predation has been previously compiled and reviewed (Wicklow 1988). Fungal toxins are often implicated as possible defenses in a general sense, but relatively few detailed studies of the chemistry involved have been reported.

There are many reports of grazing preferences among fungivores. For example, mycelia or fruiting structures of certain fungi are known to be avoided by fungivorous insects or arthropods (Curl et al. 1985; Shaw 1992; Wicklow 1992a). The presence of bioactive secondary metabolites is often invoked or implicated in these phenomena, and agents responsible for the observed effects have occasionally been identified (Wicklow 1988; Bernillon et al. 1989; Koshino et al. 1989). Some fungivores appear to have evolved detoxification mechanisms allowing them to consume, and sometimes specialize in consuming, toxin-producing fungi (Wicklow 1988; Dowd and VanMiddlesworth 1989; Shaw 1992). There is also significant evidence that some advantages may be conferred on certain fungi upon moderate grazing by fungivores (e.g., inoculum dispersal) (Stevenson and Dindal 1987; Shaw 1992). Many other instances of selective grazing or refusal of fungi by individual mycophagists could be cited.

1. *Claviceps* Ergot

Considering the observation that the concentration of plant defensive metabolites is often highest in reproductively important plant parts (Rhodes 1985), it seems logical to initiate discussion of fungal "chemical defenses" by considering physiological structures that are particularly important to fungal survival. One precedent provides a particularly useful introduction. The ergot alkaloids (e.g., **1**; Fig. 1) comprise a class of medicinally useful compounds that were originally isolated from the ergot (sclerotia) of *Claviceps purpurea* (Mantle 1978; Masurekar 1992), a parasitic fungus found on many species of cereal plants. This class of compound exhibits a wide array of physiological activities, and many ergot alkaloids have found medicinal uses (Floss 1976; Stadler and Giger 1984). Chemical studies of *Claviceps* were not based on a general interest in sclerotial metabolites, but instead were stimulated by the long-term implication of ergots in poisonings of humans and livestock. It is especially significant that the ergot alkaloids were originally found only in the sclerotia of the fungus (Mantle 1978). *Claviceps* sclerotia are not formed in liquid fermentation cultures, and many of the fungal ergot alkaloids would not have been discovered through screening of liquid cultures alone. The medicinal importance of the ergot alkaloids has led to the gradual development by the pharmaceutical industry of *Claviceps*

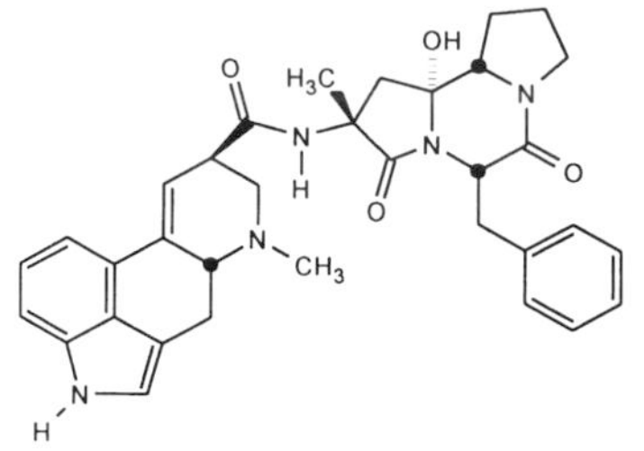

Fig. 1. Structures of compounds **1–9**

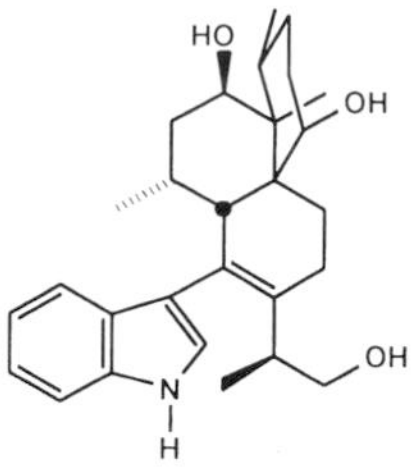

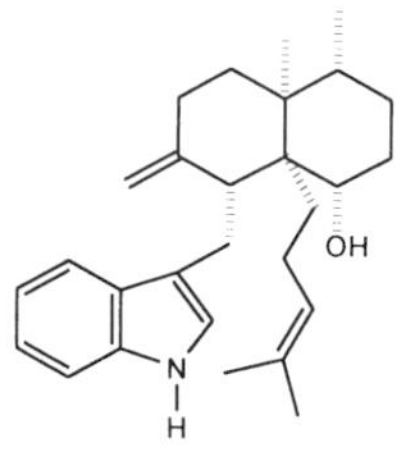

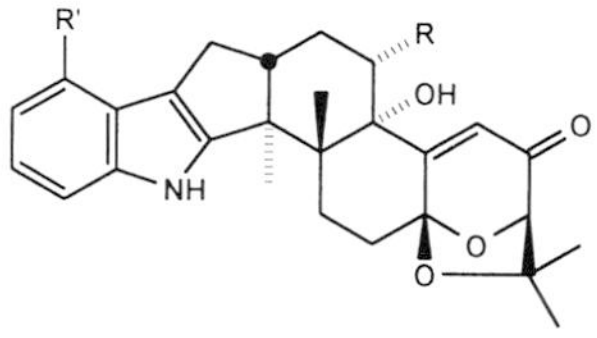

strains which produce some of the compounds in liquid fermentations, but even in these cultures, alkaloid production is associated with "sclerotial-like" cells (Mantle 1978). It has been proposed that the evolutionary development of the ergot alkaloids may have been at least partly guided by selection pressures exerted by herbivores (Kendrick 1986; Wicklow 1988).

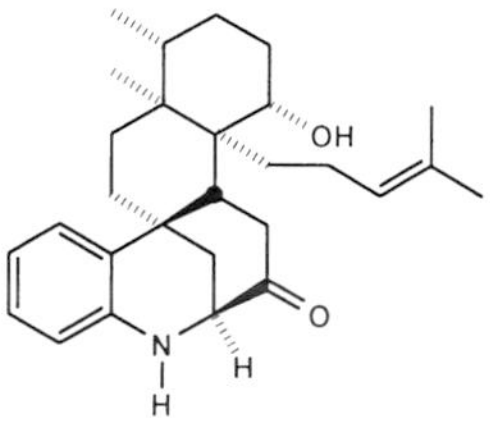

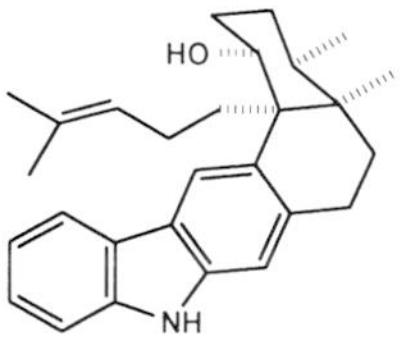

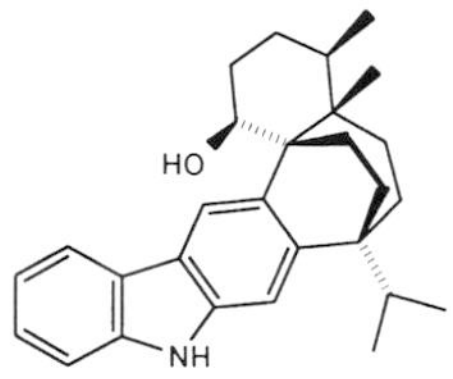

2. Sclerotia of Other Fungi

There is, of course, no reason to believe that production of bioactive sclerotial metabolities is limited to *Claviceps*. Several literature reports afford circumstantial evidence that sclerotia produced by other fungi contain biologically active metabolites. These fungi include species of *Sclerotinia*, *Sclerotium*, *Verticillium*, *Macrophomina*, and *Aspergillus* (Tanda et al. 1968; Morrall et al. 1978; Wicklow and Cole 1982; Wicklow et al. 1988).

Sclerotia, in general, are specially adapted, multicellular structures produced by certain fungi as a survival mechanism (Coley-Smith and Cooke 1971; Willets 1971, 1978). These durable resting bodies can survive periods of dry, nutrient-poor conditions that other fungal parts cannot withstand. They then serve as vital sources of primary inoculum for the fungi when conditions again become favorable for growth. Generally, sclerotia survive under more severe conditions and for

longer periods than any other kinds of fungal bodies, sometimes remaining viable in soil for periods of several years. Sclerotia are by far the largest fungal propagules, ranging in size from $30\,\mu m$ to several cm, depending on species, and their germination frequently results in the generation of very large quantities of inoculum. Thus, sclerotia represent a substantial metabolic investment for the producing fungi.

It is interesting to consider how the presence of bioactive metabolites could impact on sclerotial survival (Wicklow 1988). Sclerotia typically form in fungus-infected plant tissues, and are separately dispersed onto the soil surface or remain attached to decaying plant parts. Soil is heavily populated with insects and other invertebrates (Kevan 1965), many of which are known to consume fungi (Wicklow et al. 1988). A dormant (or germinating) sclerotium would represent a substantial nutrient reward for an insect predator, especially since sclerotia possess a much higher nutrient content than the unorganized mycelium (Willets 1971). Sclerotia damaged by insect larvae are much more susceptible to microbial decay than are undamaged sclerotia (Baker and Cook 1974). Thus, if sclerotia commonly contain metabolites which somehow limit feeding by insects, this property could clearly influence the longevity of these important fungal bodies. Based on parallel observations in plant chemical ecology (e.g., allocation of defensive metabolites to seeds; Rhodes 1985), there is reason to suspect that selection pressures exerted by insect predation may have led to the evolution of sclerotial chemical defenses (Wicklow and Cole 1982; Wicklow et al. 1988).

Our project in this area was initially stimulated by the fact that the sclerotia of *Aspergillus flavus* are avoided by the common detritivorous beetle *Carpophilus hemipterus*, an insect which feeds on the conidia and mycelia of the same fungus (Wicklow et al. 1988). In light of this observation, we investigated the secondary metabolites of *A. flavus* sclerotia, and encountered a number of antiinsectan compounds not found in the conidia or mycelia of the fungus (Gloer et al. 1988; Wicklow et al. 1988). These compounds were not produced in simple liquid shake cultures of *A. flavus*, and most of them had not been previously reported (Gloer et al. 1988). Furthermore, the most potent antiinsectan metabolite (**2**) is nontoxic to vertebrates at 300 mg kg^{-1} (Cole et al. 1981). These findings led us to initiate general studies of the chemistry of *Aspergillus* sclerotia as sources of new antiinsectan natural products (Gloer 1995a,b). Assays for activity against *C. hemipterus* and the important crop pest *Helicoverpa zea* (corn earworm) were employed to guide isolation procedures in this project. *H. zea* is unlikely to be ecologically relevant to the sclerotia of *Aspergillus* spp., but the discovery of agents with potent activity against *H. zea* could be particularly important from a practical perspective (Georghiou 1986).

A. nomius (NRRL 13137), another member of the *A. flavus* taxonomic group, produced nominine (**3**) (Rinderknecht et al. 1989), 14-hydroxypaspalinine (**4**) and 14-(N,N-dimethylvalyloxy)paspalinine (**5**) (Staub et al. 1993), and the unusual compound aspernomine (**6**) (Staub et al. 1992), all of which exhibit activity against *H. zea* larvae in dietary assays at 100 ppm. Nominine (**3**) is the most potent of these, causing 40% mortality and 96% reduction in weight gain among survivors relative to controls. These values are within an order of magnitude of the potency of commercial insecticide standards (e.g., permethrin) as insect stomach poisons. Nominine also shows topical activity, although it is not competitive with current commercial insecticides as a contact poison. Interestingly, the new paspalinine derivatives isolated also caused ca. 90% reduction in weight gain by *H. zea*, while paspalinine itself (**7**), a known tremorgen, caused essentially no ef-

fect at the same dietary concentration (Staub et al. 1993). All of these compounds were again found to be concentrated in the sclerotia of *A. nomius* (Rinderknecht et al. 1989; Staub et al. 1993).

Several additional new aflavinine analogs, as well as two new compounds with previously undescribed ring systems (tubingensins A and B, **8**, **9**), were isolated from the sclerotia of *A. tubingensis* (NRRL 4700) (TePaske et al. 1989a–c). The same compounds were also detected in sclerotia produced by strains of a more common relative, *A. niger* (TePaske 1991). These compounds are not generally as effective against the insects mentioned above, but some of them do show significant cytotoxicity. *A. tubingensis* and *A. niger* sclerotia also contain considerable amounts of a series of (mostly known) members of the aurasperone/fonsecinone class (e.g., **10**; Fig. 2) (Turner and Aldridge 1983), which are responsible for most of the effects of the sclerotial extracts on the fungivorous beetle *C. hemipterus* (TePaske 1991). Recent studies of an additional member of the *A. niger* group, *A. carbonarius* (NRRL 369), again afforded aurasperone and fonsecinone derivatives, but also provided members of a new class of aromatic compounds called carbonarins (e.g., **11**; Alfatafta 1994). In contrast to the results described above for *A. flavus*, *A. parasiticus*, and *A. nomius*, the aurasperones and fonsecinones could be detected in other fungal parts of each of these three A. higher group species, albeit in somewhat lower concentrations than are found in the sclerotia (TePaske 1991; Alfatafta 1994).

Representatives of the *A. ochraceus* taxonomic group have also afforded promising results (DeGuzman et al. 1992; Laakso 1992). For example, sclerotia of *A. sulphureus* (NRRL 4077) contained a variety of additional new indole-derived antiinsectan compounds, including radarins (e.g., **12**), sulpinines (e.g., **13**), a ring-opened analog of penitrem B (**14**), and several other related metabolites (Laakso 1992; Laakso et al. 1992a,b, 1993). All of these metabolites cause reductions in weight gain in *H. zea* comparable to that caused by nominine, although they do not induce significant mortality. Biogenetically related compounds have been isolated from liquid cultures of other fungi (Turner and Aldridge 1983). However, as was the case with members of the *A. flavus* group, detailed analysis of *A. sulphureus* cultures indicated that the active compounds are concentrated in the sclerotia, and are

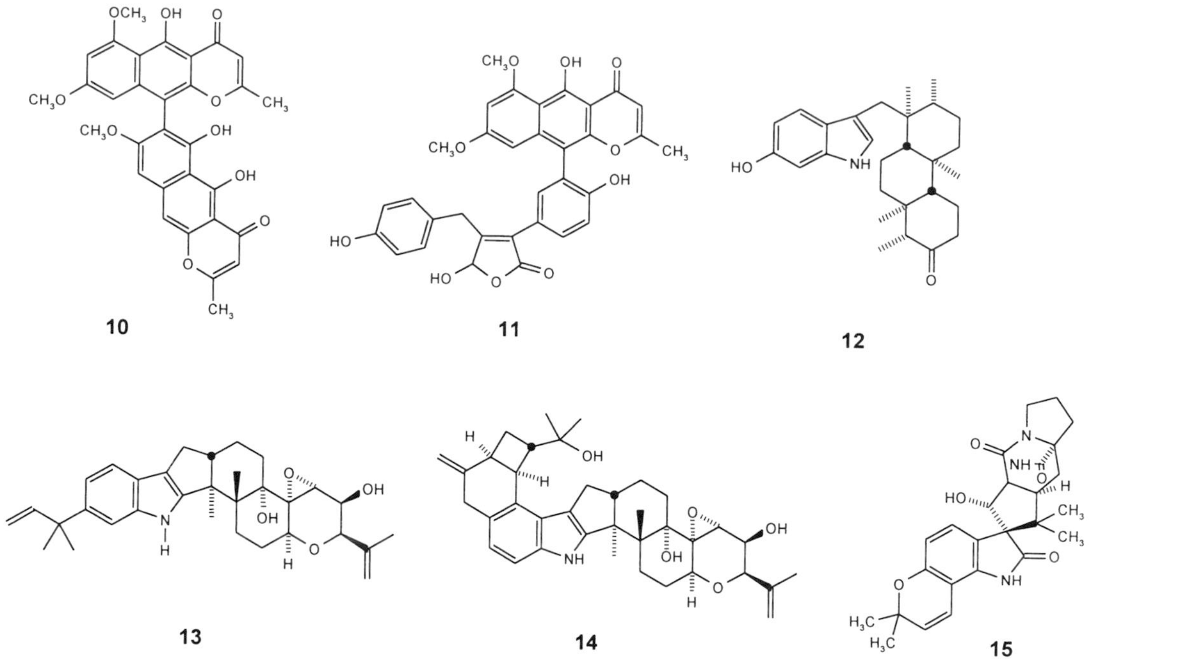

Fig. 2. Structures of compounds 10–15

relatively scarce or absent in other fungal parts or liquid cultures (Laakso 1992). More recently, a novel compound of the paraherquamide class with potent antiinsectan activity was obtained from the sclerotia of *A. sclerotiorum* (NRRL 5167). This compound, sclerotiamide (15), exhibited potent activity against *H. zea*, causing a 46% mortality rate and 98% reduction in weight gain among survivors at the 100 ppm dietary level (Whyte et al. 1996b). This level of dietary activity is comparable to that observed for azadirachtin in the same assay. Some unusual physiological effects on the larvae were also noted.

While many of the compounds isolated would not themselves completely deter feeding at sclerotial concentrations by the fungivorous insect used in our assays, other active metabolites are generally present. Indeed, the natural levels of the cocktails of compounds found in the sclerotia of most of the aspergilli we have screened would undoubtedly cause considerable feeding reduction by *C. hemipterus*. Few of the compounds have been tested in combinations, but there is also precedent for synergistic effects against insects caused by fungal metabolite mixtures (Dowd 1988). In addition, there are a great many other potential insect and arthropod predators that would most likely be affected by the presence of these metabolites, some more than others. Only two insects are used in our assays, and many compounds that show no activity against one cause considerable

effects against the other at the same dietary concentrations.

Sclerotia of a limited number of *Penicillium* spp. have also been investigated, and many of these were also found to contain antiinsectan metabolites, including griseofulvin, candidusin, and penitrem analogs (Belofsky 1996). However, most of the compounds we have encountered from *Penicillium* sclerotia to date have been previously reported from other sources.

In seeming contrast to these results, exploratory studies of a relatively small number of sclerotium-producing plant pathogens (e.g., *Sclerotinia*, *Sclerotium*, *Typhula* spp.) have afforded very few sclerotial extracts with potent antiinsectan activity in our assays. There are many possible explanations, ranging from irrelevance of the assay organisms being used, to the presence of alternative defense mechanisms or survival strategies, but the information available at this stage is insufficient to permit informed speculation about these results.

To date, our bioassay-guided chemical studies of *Aspergillus* sclerotia alone have afforded over 50 new metabolites and approximately 20 known compounds, nearly all of which display some degree of activity against insects (Gloer 1995b). The structural relationship of many of these metabolites to compounds known to display physiological effects in mammals suggests that they may have additional activities as well. Some particularly well-known bioactive compounds have been encountered, including ochratoxin A, penicillic acid, and tremorgenic mycotoxins of the paspalinine family, but the success rate in finding **new** compounds has been exceptionally high, despite extensive prior studies of the chemistry of *Aspergillus* spp. Interestingly, although sclerotia of some *A. flavus* and *A. parasiticus* isolates are known to contain aflatoxins (Wicklow and Shotwell 1983), our bioassay-guided studies have not led us to isolate a single representative of the aflatoxin class.

The existence of antiinsectan sclerotial metabolites does not conclusively prove that they have a role in chemical defense. However, the success of these studies argues for chemical studies of other fungal structures that serve similar functions. These results also demonstrate that assumptions about the identity of compounds causing ecological effects should not be based on prior knowledge of the chemistry of a fungal species. Given the vast prior knowledge about *Aspergillus* metabolites, such assumptions might have seemed warranted in the case of *Aspergillus* sclerotia, but would clearly not have been valid based on the results described above.

3. Ascostromata

Other fungi produce morphological structures that are analogous to sclerotia in function. Certain ascomycetes, for example, produce ascostromata, which are proposed to play ecological roles similar to those of sclerotia (Wicklow and Cole 1984; Horn and Wicklow 1986). Studies of such structures have shown that they also can contain bioactive metabolites. Examples include ascomata of *Epichloe*, *Petromyces*, *Eurotium*, and *Eupenicillium* spp. (Wicklow 1988; Koshino et al. 1989; Nozawa et al. 1994; Belofsky et al. 1995; Wang et al. 1995a).

Recently, we undertook what we viewed as a logical extension of the project described above by expanding our studies to include the sclerotioid ascostromata produced by certain species of *Eupenicillium*. Initial bioassays showed an incidence of antiinsectan activity comparable to that observed among *Aspergillus* sclerotial extracts, although the potency of activity was somewhat lower on average. Studies of several *Eupenicillium* spp. have already afforded interesting results, including several more new compounds with antiinsectan activity. Particularly intriguing is the finding that the ascostromata of *E. crustaceum*

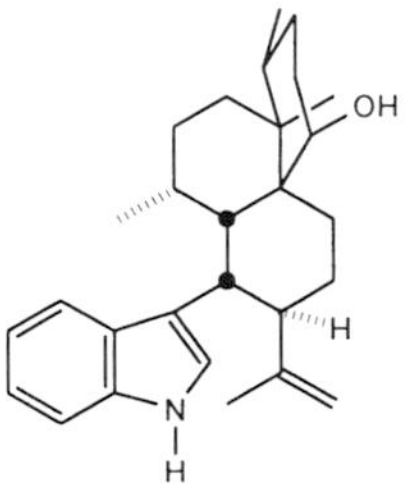

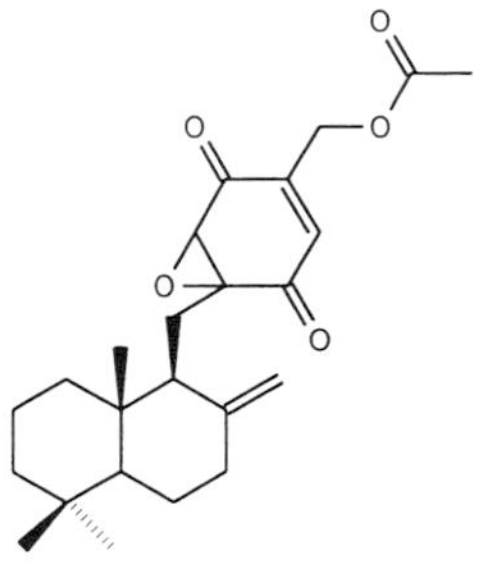

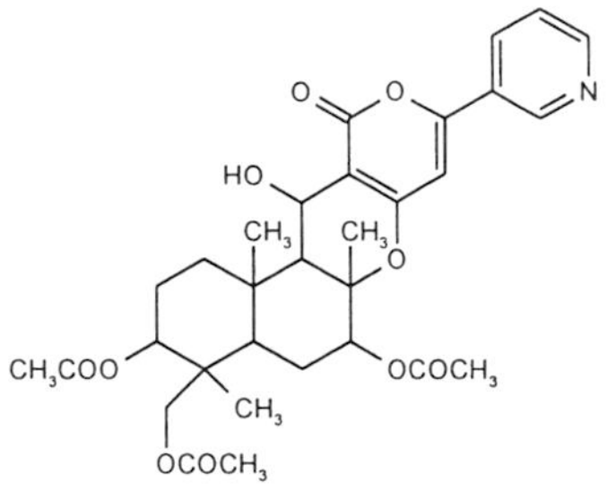

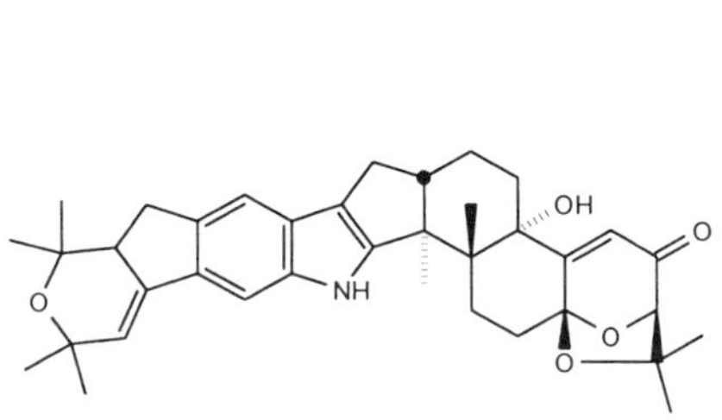

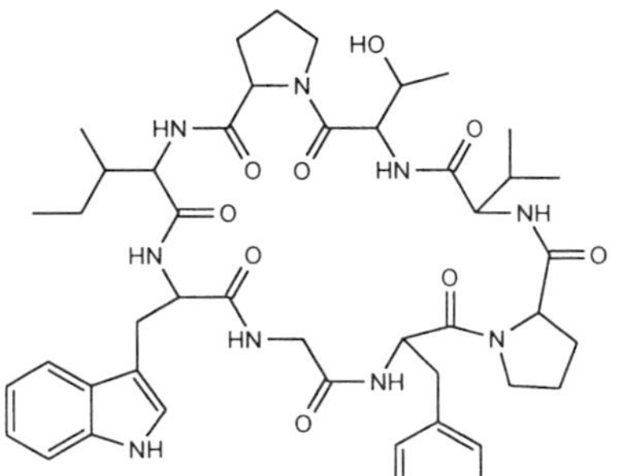

Fig. 3. Structures of compounds **16–20**

NRRL 3332 contain approximately 0.3% by weight of an aflavinine derivative (**16**; Fig. 3) (Wang et al. 1995a). Prior to this discovery, the only known sources of aflavinines had been the sclerotia of *Aspergillus* spp. Further investigation revealed that the compounds are heavily concentrated in the ascostromata. Ascostromata from several other isolates of *E. crustaceum* and related species were also found to contain aflavinines. Based on this finding, and the isolation of several known *Aspergillus* indole alkloids from *E. shearii* (Belofsky et al. 1995) it appears that at least some of the ascostroma-producing *Eupenicillium* spp. have evolved (or retained) chemical defense systems similar to those of *Aspergillus*.

On the other hand, another isolate identified as *E. crustaceum* (NRRL 22307) did not produce aflavinines, but instead was found to produce a major antiinsectan metabolite representative of a different structural class. The major component, **17**, is closely related to a known compound called macrophorin, which had been previously reported from two other fungal sources. Compound **17** exhibited significant antiinsectan activity in dietary assays at its ascostromatal concentration (*ca.* 2300 ppm), causing 63% reduction in weight gain among *H. zea* larvae and 69% reduction in feeding rate among *C. hemipterus* larvae (Wang et al. 1995a).

Studies of *E. reticulisporum* (NRRL 3446) afforded the unusual terpenoid-substituted pyridine **18**. However, compound **18** (pyripyropene A) had been reported by other researchers from an isolate of the nonsclerotium-producing species *Aspergillus fumigatus* as a potent inhibitor of acyl-CoA-cholesterol acyltransferase (Kim et al. 1994). *E. shearii* ascostromata contain a host of antiinsectan compounds that resemble metabolites isolated from sclerotium-producing *Aspergillus* spp. (e.g., shearinine A; **19**) (Belofsky et al. 1995). These metabolites exhibit potent antiinsectan effects against *H. zea* and *Spodoptera frugiperda* in dietary, topical, and/or leaf-disk bioassays. Other compounds isolated from *E. shearii* include two new types of tryptophan-containing cyclic octapeptides (shearamides; e.g., **20**) (Belofsky 1996).

4. Other Fungal Structures

Other fungal structures such as spores, cleistothecia, or fruiting bodies may also be chemically defended. Selective chemical studies of very small structures are complicated by the difficulty in separating them in quantity from other fungal material, while studies of fruiting bodies are often hindered by an inability to form them in laboratory cultures, or to predict reliably when and where they will occur in nature. Some researchers have proposed defensive roles for metabolites found in fruiting bodies of basidiomycetes (Sterner et al. 1985; Shaw 1992). The amanitins and phalloidins, well-known toxic peptides found in fruiting bodies of the genus *Amanita*, are often cited as examples of possible chemical defenses. These compounds are toxic to a variety of insects, although some insect species, either coincidentally or through selection processes, have evolved the capacity to detoxify or tolerate these metabolites (Jaenike et al. 1983; Shaw 1992). Metabolites of edible mushrooms may also display potentially useful biological activities. The strobilurins are produced by a common edible mushroom, and have proven to be important as fungicides and leads thereto (Anke 1995). These compounds have also been proposed to play an ecological role, and are reportedly produced on their natural substrate (Anke 1995), although most, if not all, of the strobilurins have been isolated from laboratory cultures. Other researchers have discussed the potential value of mycophagy to the fungi because of its importance as a mechanism of spore dispersal (Shaw 1992), citing examples of metabolites that attract fungivores. These reports are illustrative of the complexity of interpreting aspects of fungus-fungivore interactions as they might relate to secondary metabolite production. Sporadic reports suggest that smaller structures may also contain unique bioactive metabolites. Brevianamide A was found to be localized in the penicillus of *Penicillium brevicompactum* and was proposed to deter feeding in fungivorous arthropods (Wicklow 1986). Ascomata of *Chaetomium bostrychodes* were reportedly avoided by certain fly larvae, and this behavior was presumed to be due to the presence of secondary metabolites (Wicklow 1988).

C. Fungi That Confer Host Resistance to Herbivory or Disease

It has been proposed that the capability of fungi to produce mycotoxins evolved and has been retained partly because they render fungal substrates (e.g., fruits, seeds, etc.) unpalatable to herbivores (Janzen 1977; Kendrick 1986; Wicklow 1988). While this could be viewed as a defense of substrate resources from competing consumers, it is too general a concept to be especially helpful in targeting fungi for chemical investigation, and does not lend itself to field investigation. On the other hand, there are documented instances where secondary metabolites play roles in mutualisms between certain fungi and their hosts. One of the best-known examples of such a system is the production by grass endophytes of fungal ergot alkaloids that influence feeding by herbivores (Clay 1988; Carroll 1992). Evidence for parallel effects in other systems is also available. For example, fungal endophytes found in conifer needles produce metabolites that confer some level of resistance to herbivory by insects (Miller, 1991). Some industrial and academic groups have targeted endophytic fungi in general for inclusion in their screening programs (Caporale 1995; J.C. Lee et al. 1995).

D. Competitive or Antagonistic Interactions

Antagonism (interference competition) between species of naturally competing fungi has been reported in virtually every fungal ecosystem. Examples include coprophilous (Wicklow 1992a,b), carbonicolous (Wicklow and Hirschfield 1979a), lignicolous (Strunz et al. 1972), phylloplane (Fokkema 1976), rhizosphere (Carroll 1992), marine (Strongman et al. 1987), and aquatic fungi (Shearer and Zare-Maivan 1988). It has been proposed that such interactions are important factors in determining the organization, composition, and pattern of succession within these ecosystems (Webster 1970; Wicklow 1981). In many cases, the mechanism of this antagonism has been shown to involve the production of a chemical agent by one species which inhibits the growth of another, but few such reports have been followed up by studies of the chemistry associated with these phenomena. This is surprising, since the metabolites responsible for these effects are essentially natural antifungal agents.

The need for new antifungal agents is assuming a growing priority (Koltin 1990). Although effective topical antifungal agents are relatively abundant, many types of topical fungal infections recur after cessation of treatment. More importantly, very few drugs are available which are therapeutically useful in the treatment of systemic fungal infections (Richardson and Marriott 1987). Fungal diseases have become increasingly common, and there are several risk groups of growing population (e.g., AIDS and chemotherapy patients) that are especially susceptible to opportunistic fungal infections (Koltin 1990). New agriculturally useful fungicides are also being continually sought. It can be argued that fungi are logical sources to explore in search of agents that regulate fungal growth, interact with important fungal receptors, or modulate the activities of key fungal enzymes. Indeed, several compounds considered as particularly promising antifungal leads have been isolated from fungal sources, including echinocandins, pneumocandins, and papulacandins.

Our studies of one fungal niche group exemplify the potential value of exploring interspecies antagonism. Coprophilous fungi are those which colonize the dung of herbivorous vertebrates (Webster 1970; Wicklow 1981, 1992b). Studies of the composition of dung mycoflora over time have shown that many of the early colonists are eventually eliminated from the substrate and completely replaced by slower-growing, later-occurring colonists (Ikediugwu and Webster 1970a,b; Lodha 1974; Wicklow and Hirschfield 1979b; Angel and Wicklow 1983). Although nutritional factors are likely to be partly responsible for this phenomenon (Lockwood 1992), it has been shown that the nutrient levels present after early colonists have been eliminated would still support their growth in the laboratory (Webster 1970). From an evolutionary perspective, it seems logical that slower-growing colonists might evolve mechanisms which help to eliminate fast-growing, nutrient-consuming competitors from the local substrate. The composition and successional patterns for many coprophilous fungal ecosystems are well documented in the literature (Harper and Webster 1964; Webster 1970; Angel and Wicklow 1983) and several reports provide suggestions of specific candidates which appear to produce antagonistic (i.e., antifungal) agents (Harper and Webster 1964; Ikediugwu and Webster 1970a,b; Singh and Webster 1973; Wicklow and Hirschfield 1979b).

Many coprophilous fungi are somewhat taxonomically distinctive. There are a number of genera that contain mainly gut-passage coprophilous species, and most of these reside in the sordariaceae, lasiosphaeriaceae, and sporormiaceae (Ainsworth 1983; Bell 1983). Perhaps more importantly, coprophilous fungi are morphologically distinctive. Gut-passage fungi have evolved unique physiological adaptations to optimize spore discharge (e.g., phototrophic perithecia, exploding asci), and for gut-passage stimulation and survival (Malloch 1981; Bell 1983; Wicklow 1992a). Insect-dispersed fungi have evolved mechanisms by which their spores can become attached to insects or arthropods, often with selectivity for certain types of hosts (Stevenson and Dindal 1987). Such physiological adaptations must be associated with biochemical adaptations, suggesting that some unusual secondary metabolic characteristics may also be expected.

A variety of other factors, such as exploitation competition, nutritional factors, and environmental considerations, are undoubtedly involved in determining the complex patterns of fungal successions. As with other organisms, fungi are unlikely to rely on only one strategy for competition and survival. Thus, fungal successions are not caused solely by antifungal antagonism. However, antagonism appears to be an important contributing factor in this process. In any event, exploration of the secondary metabolites of antagonistic species would seem to form the basis for an ecologically rational approach to the discovery of new antifungal agents.

Based on these considerations, we targeted certain types of mid- to late-successional coprophilous fungi as potential sources of antifungal agents. In petri plate competition assays of approximately 250 different coprophilous fungal cultures representing 80 different genera carried out in our laboratory to date, over 60% have displayed unambiguous inhibitory effects toward competitor fungi at a distance. The vast majority of these cultures represent genera and/or species for which no chemistry has been previously reported. More than half of these active cultures

also displayed antifungal activity when grown in liquid culture with only minimal effort to vary the fermentation conditions. Our work thus far has focused on species that display obvious antagonistic effects at a distance, indicating the involvement of diffusible substances. However, other documented phenomena, such as hyphal interference and replacement after colony contact, may also be associated with the production of antifungal agents.

Our initial work was based on published suggestions that *Poronia punctata* (NRRL 6457), *Preussia fleischhakii* (NRRL A-24068), and *Podospora decipiens* (NRRL 6461) produce metabolites that inhibit the growth of the competitor fungi *Ascobolus furfuraceus* (NRRL 6460) and *Sordaria fimicola* (NRRL 6459) (Wicklow and Hirschfield 1979b). The antagonistic components of the first two species were found to be known compounds (Gloer and Truckenbrod 1988; Weber and Gloer 1988), but *P. decipiens* afforded a new metabolite of mixed biogenesis with an unusual ring system, which we named podosporin A (**21**; Fig. 4) (Weber et al. 1988). Podosporin A was so named because it was the first new secondary metabolite ever reported from a member of the genus *Podospora*. Podosporin A showed anti-*Candida* activity in disk assays, and caused significant inhibition against *Trichophyton mentagrophytes* at levels down to $1 \mu g ml^{-1}$. Subsequently, studies of *Preussia isomera* (CBS 415.82) afforded a group of compounds called preussomerins (e.g., **22**), which contain a ring system that had not been previously described (Weber and Gloer 1991). Discoveries of novel ring systems are relatively rare in natural products chemistry, so such findings are considered especially noteworthy. Some of these compounds showed rather potent activity against competitor fungi (MIC values $<5 \mu g ml^{-1}$). More interestingly, a subsequent report by another re-

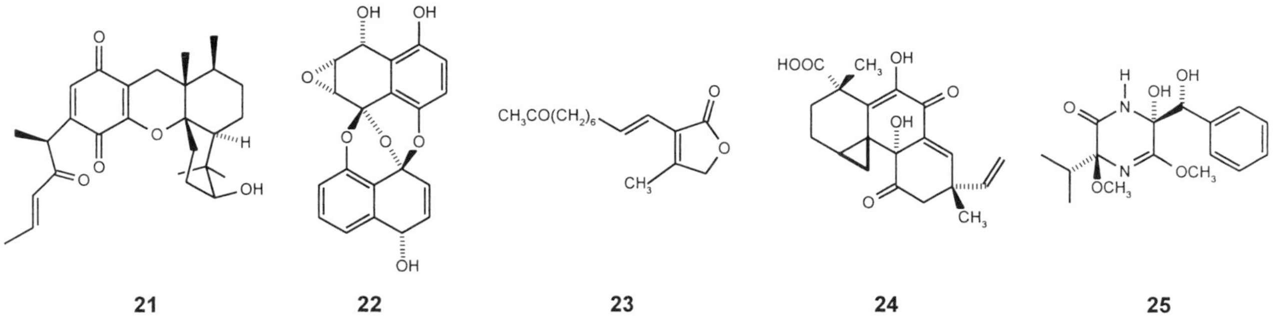

Fig. 4. Structures of compounds **21–25**

search group described several additional preussomerin analogs from an unidentified dung isolate that cause significant inhibition of Ras farnesyl-protein transferase (Singh et al. 1994). Encountering this set of fascinating, unprecedented structures at a very early stage in this work provided a compelling argument for further studies of coprophilous fungi. Other new antifungal compounds isolated early in this project include appenolides (e.g., **23**) from *Podospora appendiculata*, gliocladins (e.g., **24**) from *Gliocladium* sp., and terezines (e.g., **25**) from *Sporormiella teretispora* (Wang et al. 1993, 1995; Alfatafta 1994).

More recent studies have continued to show promise. *Mariannaea elegans* (JS 141, from a collection held by D. Malloch at the U of Toronto) forms a complex mixture of antifungal cyclic octapeptides (mariannins A–C; e.g., **26**; Fig. 5). Naturally occurring cyclic octapeptides are relatively rare, and the mariannins have no especially close known analogs. *Petriella sordida* (JS 154) forms another group of antifungal peptides. One metabolite was identified as the known cytotoxic tetrapeptide WF-1161 (**27**) which is, interestingly, the only other compound ever reported from any *Petriella* sp. (Umehara et al. 1983). However, four other peptides with quite different structures were

also isolated from this species. These compounds (petriellins A–D; e.g., **28**) contain a considerable number of unusual residues. The lead compound is a cyclic nonadecadepsipeptide containing MeVal, Pro, Pip, Val, Ala, Ile, N-MeIle, and N-MeThr residues, plus a 3-phenyllactic acid unit (K.K. Lee et al. 1995). The structures of the petriellins were assigned through analysis of 2D-NMR and FABMS-MS data. Petriellin A (**28**) exhibits potent activity toward competitor fungi, with MIC values in the sub-μg ml^{-1} range, but is considerably less active against *C. albicans* and *A. fumigatus*.

Dung isolates of *Stilbella fimetaria*, *Microascus manginii*, and *Acremonium* sp. all produced mixtures of peptides that exhibit antifungal effects. The major component in each case is another peptide, antiamoebin I (**29**), a member of the peptaibophol class of fungal antibiotics (Turner and Aldridge 1983). Several new analogs have been isolated from these fungi, though they differ from **29** only by conservative replacement of amino acid residues.

Two novel antibacterial cyclic depsipeptides were isolated from the crude fermentation broth of *Nigrosabulum globosum* (JS 176) and named nigrosabulins (e.g., **30**). *Apiospora montagnei* (JS 140) afforded an antifungal compound of mixed

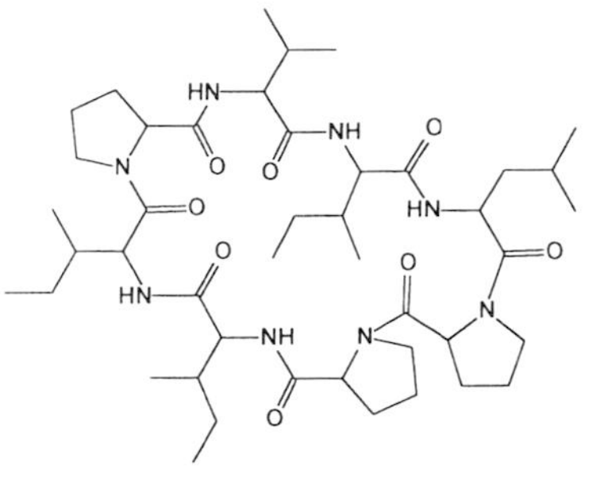

26

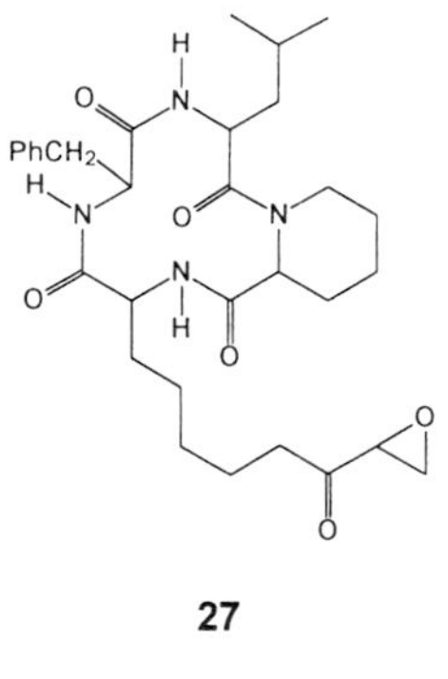

27

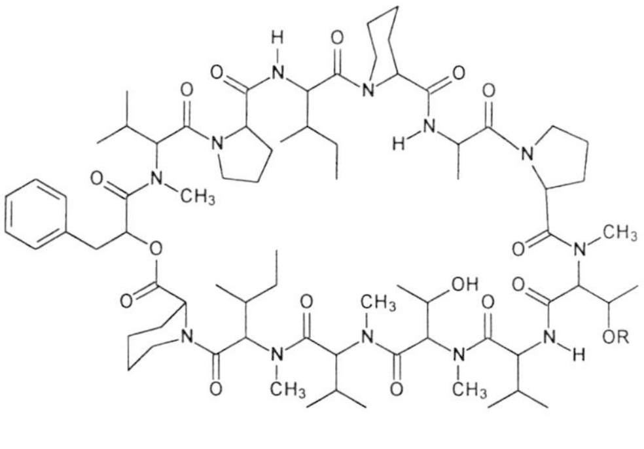

28

Aib--->Aib--->Hyp--->Gln--->Iva--->Hyp--->Aib--->Pro--->Phol

Leu<---Gly<---Iva<---Aib<---Aib<---Aib<---Phe<---Ac

29

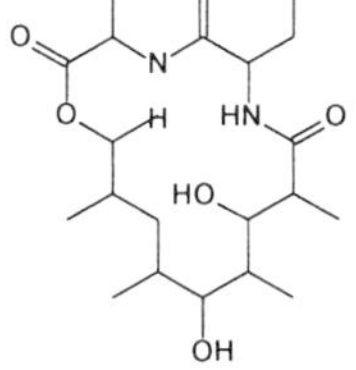

30

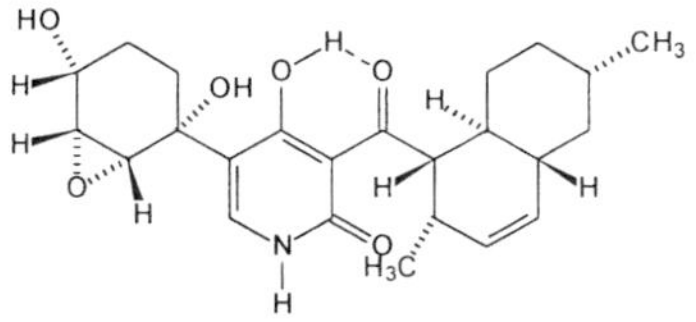

31

Fig. 5. Structures of compounds **26–31**

biogenetic origin (apiosporamide; **31**). This compound showed some unexpected NMR properties, but the structure was ultimately assigned through application of 2D NMR techniques (Alfatafta et al. 1994).

These results demonstrate that coprophilous fungi are prolific producers of antifungal peptides and amino acid-derived metabolites. However, antifungal metabolites representing other biosynthetic pathways have also been isolated from these sources. Studies of *Cercophora areolata* (JS 166) led to the isolation of several metabolites. The most potent antifungal was identified as the known trichothecene roridin E (**32**; Fig. 6), but several new compounds were also encountered, including cercophorins A–C (e.g., **33**) (Whyte et al. 1996). *Coniochaeta saccardoi* (JS 223) produced a pair of new antifungal polyketides (coniochaetones A and B; e.g., **34**) with a relatively simple, yet previously unreported cyclopentabenzopyranone ring skeleton (Wang et al. 1995). Compound **34** showed activity against *C. albicans* at 50 µg per disk and against *Aspergillus*

at 10 µg ml⁻¹. Studies of another *Coniochaeta* species (*C. hansenii*, JS 235) afforded a new prenylated analog of the viridiofungins (**35**). This finding is encouraging because the previously known viridiofungins, which show potent broad spectrum antifungal activity against *Aspergillus*, *Candida*, and *Cryptococcus* spp., were originally isolated from another fungal source by industrial researchers via a mechanism-based, mass random screening program as inhibitors of squalene synthase (Harris et al. 1993).

The compounds described above are representative of the diverse biogenetic origins of the natural products produced by coprophilous fungi. Thus far, at least 73 compounds have been isolated in this project through detailed studies of 35 coprophilous isolates that afforded liquid cultures displaying antifungal activity. Of these, 49 had not been previously reported. Nearly all of the compounds were obtained through isolation procedures guided by antifungal bioassays. Some of them are closely related to compounds that were previously known, but others contain novel struc-

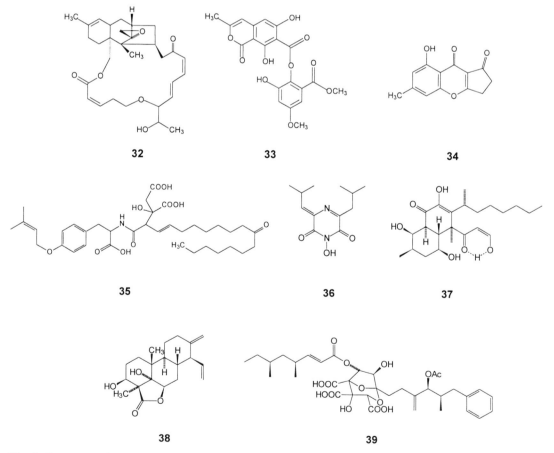

Fig. 6. Structures of compounds **32–39**

tural features. Unfortunately, while the compounds isolated in this project often display potent effects against other coprophilous fungi, few of them have displayed activity against medically relevant organisms at concentrations that would suggest significant promise for utility as pharmaceuticals. However, these results have arisen from studies of a very small subset of organisms isolated from only one of many niches in which interference competition among fungal competitors has been reported. Given the high incidence of antifungal activity and new chemistry among these isolates, the probability of encountering more promising activity or useful lead structures through further studies of the chemistry of fungal antagonism seems high.

Although the results above demonstrate that antifungal activity is a logical primary target assay, agents with antifungal activity often exhibit other potentially useful biological effects. For example, cyclosporin was originally isolated from a *Tolypocladium* sp. as an antifungal agent (Masurekar 1992). Cyclosporin is now a vitally important immunosuppressive drug, and a valuable pharmacological tool. There are also evolutionary arguments for suspecting the presence of metabolites with other bioactivities. Dung pats are often heavily colonized by insects that consume fungi (Stevenson and Dindal 1987). The characteristic fruiting bodies produced by many coprophilous fungi would have obvious potential as food sources for dung insects. Thus, coprophilous fungi may also be under evolutionary pressure to produce metabolites which can defend their important reproductive structures from insect predation. Such compounds could have potential as insecticides, but could also have other valuable biological activities.

The selection of coprophilous species has also proven useful for the purpose of enhancing biodiversity within a large-scale screening program. One industrial group has been particularly active in this area, and these researchers have described several novel bioactive lead compounds from coprophilous isolates. Examples include the selective viral endonuclease inhibitor flutimide (**36**) from *Delitschia confertaspora* (Hensens et al. 1995a) and several broad-spectrum antifungal agents, such as australifungin (**37**) from *Sporormiella australis* (Hensens et al. 1995b) and sonomolides (e.g., **38**) from an unidentified dung isolate (S.A. Morris et al. 1995). Perhaps the most encouraging chemistry report from a coprophilous

fungus to date is the same group's isolation of zaragozic acids (e.g., **39**) from *Sporormiella intermedia*. The zaragozic acids are potent broad-spectrum antifungals that act by inhibition of squalene synthase (Caporale 1995). These compounds were originally obtained from a soil isolate, but later studies showed that multiple isolates of *S. intermedia*, a commonly encountered coprophilous species, produce zaragozic acids (Bills et al. 1994).

Fungi from other ecosystems also exhibit antagonistic effects, and ventures into these areas have also met with some success. Our own preliminary investigations of antagonistic aquatic isolates have afforded several new bioactive natural products, including the stachybotrins (Xu et al. 1992), kirschsteinin (Poch et al. 1992), and euroticin (Laakso 1992). Studies of wood-decay fungi that show antagonistic effects have afforded a number of agents that could be useful in controlling economically important wood-rotting fungi (Strunz et al. 1972; Ayer and Miao 1993; Ayer and Kawahara 1995). A final example is provided by a 1992 report of several new pneumocandins with anti *Candida* and anti-*Pneumocystis* activity obtained from a random soil isolate of *Zalerion arboricola* (Schwartz et al. 1992). These compounds were subsequently suggested to be the causative agents of antagonistic effects exerted by *Z. arboricola* against a competing fungus on its natural substrate originally reported in 1973 (Buczacki 1973). While this hypothesis was not investigated further, it is interesting to consider the possibility that these potent antifungal agents might have been discovered much earlier had this documented case of fungal antagonism been followed up by chemical studies.

III. Conclusions

The results reviewed here do not prove that fungi have evolved chemical defenses, or that the antagonistic agents produced by certain fungi render a competitive advantage. Carefully controlled studies are needed to determine whether these compounds are truly significant in the life histories of the producing species. Even so, it is clear that observations in fungal ecology can be employed as leads to the discovery of novel bioactive fungal metabolites. From a mycological perspective, these findings help to validate the application of fundamental principles of chemical ecology to studies of fungi, provide additional tools with which to study the ecology of the fungi involved, foster the discovery of new taxa, and provide tools with which to investigate genetic relationships.

Investigations based on observations in fungal ecology are especially appealing from the standpoint of chemistry because it is more attractive to seek bioactive metabolites with a hypothesis in mind than to screen organisms at random. Such studies can clearly lead to the discovery of fascinating, new chemistry, and often provide unexpected analytical or structural challenges. Although unambiguous identification of natural products, even previously reported metabolites, can be a lengthy process, continuing technical improvements in methodology for isolation and structure determination now allow identification of milligram- or even sub-milligram quantities of complex, unknown natural products in many cases.

The percentage of novel compounds isolated through our own studies to date has been unusually high. This is due in part to the assay systems employed. The use of "ecologically relevant" assays, in addition to more medically and agriculturally relevant test organisms, would understandably lead to some discoveries not likely to be made by those employing only the latter assays. However, the nonrandom selction of organisms for study and the relatively unexplored nature of many of these fungi and their interactions are also contributing factors. Strategies of this type cannot replace random screening programs in the search for new bioactive natural products, but can offer insights that could increase the diversity of compounds encountered in such programs. For example, these results argue strongly for employment of solid substrate fermentation in studies of fungal metabolites, for consideration of habitat and ecological characteristics when selecting fungi for study, and for allowing longer fermentation times to access metabolites produced by slower-growing species.

There is no shortage of current and future challenges that fungal natural products chemistry can help to confront. Adequate treatments are still conspicuously lacking for many viral diseases, fungal infections, and human cancers. The long-term failure to find effective therapeutic agents in these areas does not mean that no such agents can be found. Discovery of suitable drugs may simply await the development of appropriate assays or

the discovery of appropriate molecular targets. A newer, developing challenge is posed by the increasing occurrence of antibiotic-resistance among bacterial infections, as well as the emergence of drug-resistant tuberculosis and malaria. The development of new agents effective against such diseases, particularly those having novel modes of action, is becoming an urgent priority. On the agricultural front, the demand for new insecticides is compounded by the fact that problematic insect pests are developing resistance to many commonly used commercial pesticides (Georghiou 1986). In addition, the use of some effective pesticides has been curtailed due to concerns about undesirable environmental impact. Similar concerns about other agrochemicals imply that new alternatives are needed, and that products of natural origin would be particularly appealing.

Another promising avenue of investigation can make use of advances in molecular biology and biochemistry. Knowledge of metabolite biosynthesis has been extended to an awareness of genes involved in production of the enzymes associated with the biosynthesis of some important fungal metabolites. Some fungal genes involved in antibiotic production are now known to be clustered, as is the case with many actinomycetes (Demain 1992). Potential applications of this knowledge include the cloning of genes involved in metabolite biosynthesis, improvement in metabolite production techniques, production of novel hybrid metabolites, and the use of fungal DNA libraries to aid in the discovery process.

The rapidly increasing capacity of biological screening systems has outpaced the capability of even the most prolific microbiology programs to provide adequate numbers of samples for assay. The pressure for short-term results, together with competition from other approaches to discovery of bioactive lead compounds (e.g., combinatorial chemistry, molecular modeling) has played a role in the recent termination of several industrial natural products programs in the US. Considering the diversity of fungi (and other sources) that remain unexplored, it is unfortunate that some institutions have had to abandon a line of research that has been instrumental to their success. Obviously, such occurrences increase the pressure on remaining natural products research programs to come up with promising leads, while decreasing the total effort underway in the field. At this stage, there is still no substitute for natural products chemistry as a source of truly novel lead structures, and it is likely that recognition of this fact will continue to stimulate a significant level of interest in this field. However, practical successes are more important than ever to continued support for fundamental studies of fungal taxonomy, ecology, and natural products chemistry. In this climate, the importance of cooperation between mycologists and chemists cannot be overemphasized. The results described and cited in this chapter are illustrative of the potential rewards of such collaborative efforts.

Acknowledgments. Support for our work from the National Institutes of Health (CA 46612, AI 27436, and CA 01571), the National Science Foundation (CHE-9211252 and CHE-8905894), Biotechnology Research and Development Corporation (24-1-078), the Herman Frasch Foundation (351-HF92), and the Alfred P. Sloan Foundation is gratefully acknowledged. The author also wishes to express his appreciation to Drs. D.T. Wicklow, P.F. Dowd, D. Malloch, and C.A. Shearer, as well as the chemistry graduate students and other research associates whose collaboration on projects described here has been both enjoyable and invaluable.

References

Ainsworth GC (1983) Ainsworth and Bisby's dictionary of the fungi. Commonwealth Mycological Institute, Kew, Surrey

Alfatafta AA (1994) New antifungal and antiinsectan metabolites from coprophilous and sclerotium-producing fungi. PhD Thesis, University of Iowa, Iowa City, Iowa

Alfatafta AA, Gloer JB, Scott JA, Malloch D (1994) Apiosporamide: a new antifungal agent from the coprophilous fungus *Apiospora Montagnei.* J Nat Prod 57:1696

Angel K, Wicklow DT (1983) Coprophilous fungal communities in semiarid to mesic grasslands. Can J Bot 61:594–602

Anke H, Stadler M, Mayer A, Sterner O (1995) Secondary metabolites with nematicidal and antimicrobial activity from nematophagous fungi and ascomycetes. Can J Bot 73[Suppl 1]:S932–S939

Anke T (1995) The antifungal strobilurins and their possible ecological role. Can J Bot 73[Suppl 1]:S940–S945

Anonymous (1995) SCRIP World pharmaceutical news review of 1995. PJB, London

Ayer WA, Kawahara, N (1995) Lecythophorin, a potent inhibitor of blue-stain fungi, from the hyphomycetous fungus *Lecythophora hoffmannii.* Tetrahedron Lett 36:7953–7956

Ayer WA, Miao S (1993) Secondary metabolites of the aspen fungus *Stachybotrys cylindrospora.* Can J Chem 71:487–493

Ayer WA, Lee SP, Tsuneda A, Hiratsuka Y (1980) The isolation, identification, and bioassay of the antifungal metabolites produced by *Monocillium nordinii.* Can J Microbiol 26:766–773

Baker KF, Cook RJ (1974) Biological control of plant pathogens. Freeman, San Francisco

Balandrin MF, Kinghorn AD, Farnsworth NR (1993) Plant-derived natural products in drug discovery and development: An overview. In: Balandrin MF, Kinghorn AD (eds) Human medicinal agents from plants. American Chemical Society, Washington, DC, pp 2–12

Ballio A (1991) Non-host-selective fungal phytotoxins: biochemical aspects of their mode of action. Experientia 47:783–790

Bell A (1983) Dung fungi: an illustrated guide to coprophilous fungi in New Zealand. Victoria University Press, Wellington, New Zealand, 88pp

Belofsky GN (1996) New biologically active secondary metabolites from *Penicillium* and *Eupenicillium* species. PhD Thesis, University of Iowa, Iowa City, Iowa

Belofsky GN, Gloer JB, Wicklow DT, Dowd PF (1995) Antiinsectan alkaloids: shearinines A–C and a new paxilline derivative from the ascostromata of *Eupenicillium shearii.* Tetrahedron 51:3959–3968

Bennett JW (1983) Differentiation and secondary metabolism in mycelial fungi. In: Bennett JW, Ciegler A (eds) Secondary metabolism and differentiation in fungi. Dekker, New York, pp 1–35

Bernillon J, Favre-Bonvin J, Pommier MT, Arpin N (1989) First isolation of (+)-epipentenomycin I from *Peziza* sp. carpophores. J Antibiotics 42:1430–1432

Bills GF (1995) Analysis of microfungal diversity from a user's perspective. Can J Bot 73[Suppl 1]:S33–S41

Bills GF, Pelaez F, Polishook JD, Diez-Matas MT, Harris GH, Clapp WH, Dufresne C, Byrne KM, Nallin-Omstead M, Jenkins RG, Mojena M, Huang L, Bergstrom JD (1994) Distribution of zaragozic acids (squalestatins) among filamentous ascomycetes. Mycol Res 98:733–739

Buczacki ST (1973) A microecological approach to larch canker biology. Trans Br Mycol Soc 61:315–329

Caporale LH (1995) Chemical ecology: a view from the pharmaceutical industry. Proc Natl Acad Sci USA 92:75–82

Carroll GC (1992) Fungal mutualism. In: Carroll GC, Wicklow DT (eds) The fungal community: its organization and role in the ecosystem, 2nd edn. Dekker, New York, pp 327–354

Choudhury SR, Traquair JA, Jarvis WR (1994) 4-Methyl-7,11-heptadecadienal and 4-methyl-7,11-heptadecadienoic acid: new antibiotics from *Sporothrix flocculosa* and *Sporothrix rugulosa.* J Nat Prod 57:700–704

Clay K (1988) Fungal endophytes of grasses: a defensive mutualism between plants and fungi. Ecology 69:10–16

Cole RJ, Dorner JW, Springer JP, Cox RH (1981) Indole metabolites from a strain of *A. flavus.* J Agric Food Chem 29:293

Coley-Smith JR, Cooke RC (1971) Survival and germination of fungal sclerotia. Annu Rev Phytopathol 9:65–92

Corley DG, Durley RC (1994) Strategies for database dereplication of natural products. J Nat Prod 57:1484–1490

Curl EA, Gudauskas RT, Peterson CM (1985) Effects of soil insects on populations and germination of fungal propagules. In: Parker CA, Rovira AD, Moore KJ,

Wong PTW, Kollmorgen JF (eds) Ecology and management of soil-borne plant pathogens. American Phytopathological Society, St. Paul, pp 20–23

Czarnik AW (1996) Guest editorial. Accts Chem Res 29:112–113

DeGuzman FS, Gloer JB, Dowd PF, Wicklow DT (1992) New diketopiperazine metabolites from the sclerotia of an isolate of *Aspergillus ochraceus*. J Nat Prod 55:931–939

Demain AL (1992) Regulation of Secondary Metabolism. In: Finkelstein DB, Ball C (eds) Biotechnology of filamentous fungi: technology and products. Butterworth-Heinemann, Boston, pp 89–112

Dowd PF (1988) Synergism of aflatoxin B_1 with the co-occurring fungal metabolite kojic acid to two caterpillars. Entomol Exp Appl 47:69–71

Dowd PF, VanMiddlesworth FL (1989) In vitro metabolism of the trichothecene 4-monoacetoxyscirpenol by fungus- and non-fungus-feeding insects. Experientia 45:393–395

Dreyfus MM, Chapela IH (1994) Potential of fungi in the discovery of novel low-molecular weight pharmaceuticals. In: Gullo VP (ed) The discovery of natural products with therapeutic potential. Butterworth-Heinemann, Boston, pp 49–80

Floss HG (1976) Biosynthesis of ergot alkaloids and related compounds. Tetrahedron 32:873–912

Fokkema NJ (1976) Antagonism between fungal saprophytes and pathogens on aerial plant surfaces. In: Dickinson CH, Preece TF (eds) Microbiology of aerial plant surfaces. Academic Press, New York, pp 487–506

Frisvad JC (1989) The connection between the penicillia and aspergilli and mycotoxins with special emphasis on misidentified isolates. Arch Environ Contam Toxicol 18:452–467

Gardner WA, McCoy CW (1992) Insecticides and Herbicides. In: Finkelstein DB, Ball C (eds) Biotechnology of filamentous fungi: technology and products. Butterworth-Heinemann, Boston, pp 335–359

Georghiou GP (1986) The magnitude of the resistance problem. In: Committee on strategies for the management of pesticide-resistant pest populations (ed) Pesticide resistance: strategies and tactics for management. National Academy Press, Washington, DC, pp 14–43

Gloer JB (1995a) The chemistry of fungal antagonism and defense. Can J Bot 73[Suppl 1]:S1265–S1274

Gloer JB (1995b) Antiinsectan natural products from fungal sclerotia. Accts Chem Res 28:343–350

Gloer JB, Truckenbrod SM (1988) Interference competition among coprophilous fungi: production of (+)-isoepoxydon by *Poronia punctata*. Appl Environ Microbiol 54:861–864

Gloer JB, TePaske MR, Sima JS, Dowd PF, Wicklow DT (1988) Antiinsectan aflavinine derivatives from the sclerotia of *Aspergillus flavus*. J Org Chem 53:5457–5460

Gupta S, Krasnoff SB, Renwick JAA, Roberts DW, Steiner JR, Clardy J (1993) Viridoxins A and B: novel toxins from the fungus *Metarhizium flavoviride*. J Org Chem 58:1062–1067

Harborne JB (1987) Natural fungitoxins. In: Hostettmann K, Lea PJ (eds) Biologically active natural products. Oxford University Press, Oxford, pp 195–211

Harborne JB (1993) Recent advances in chemical ecology. Nat Prod Rep 10:327–348

Harper JE, Webster J (1964) An experimental analysis of the coprophilous fungal succession. Trans Br Mycol Soc 47:511–530

Harris GH, Turner Jones ET, Meinz MS, Nallin-Omstead M, Helms GL, Bills GF, Zink D, Wilson KE (1993) Isolation and structure of viridiofungins A, B, and C. Tetrahedron Lett 34:5235–5238

Hawksworth DL (1991) The fungal dimension of biodiversity: magnitude, significance, and conservation. Mycol Res 95:641–655

Hensens OD, Goetz MA, Liesch JM, Zink DL, Raghoobar SL, Helms GL, Singh SB (1995a) Isolation and structure of flutimide, a novel endonuclease inhibitor of influenza virus. Tetrahedron Lett 36:2005–2008

Hensens OD, Helms GL, Turner Jones ET, Harris GH (1995b) Structure elucidation of australifungin, a potent inhibitor of sphinganine N-acyltransferase in sphingolipid biosynthesis from *Sporormiella australis*. J Org Chem 60:1772–1776

Hoagland RE (1990) Microbes and microbial products as herbicides: an overview. In: Hoagland RE (ed) Microbes and microbial products as herbicides. American Chemical Society, Washington, pp 2–52

Horn BW, Wicklow DT (1986) Ripening of *Eupenicillium ochrosalmoneum* ascostromata on soil. Mycologia 78:248–252

Huang Q, Tezuka Y, Hatanaka Y, Kikuchi T, Nishi A, Tubaki K (1995a) Studies on metabolites of mycoparasitic fungi. III. New sesquiterpene alcohol from *Trichoderma koningii*. Chem Pharm Bull 43:1035–1038

Huang Q, Tezuka Y, Hatanaka Y, Kikuchi T, Nishi A, Tubaki K (1995b) Studies on metabolites of mycoparasitic fungi. IV. Minor peptaibols of *Trichoderma koningii*. Chem Pharm Bull 43:1663–1667

Ikediugwu FEO, Webster J (1970a) Antagonism between *Coprinus heptemerus* and other coprophilous fungi. Trans Br Mycol Soc 54:181–204

Ikediugwu FEO, Webster J (1970b) Hyphal interference in a range of coprophilous fungi. Trans Br Mycol Soc 54:205–210

Jaenike J, Grimaldi D, Shuder AE, Greenleaf AL (1983) α-Amanitin tolerance in mycophagous *Drosophila*. Science 221:165–167

Janzen DH (1977) Why fruits rot, seeds mold, and meat spoils. Am Nat 111:691–713

Kendrick B (1986) Biology of toxigenic anamorphs. Pure Appl Chem 58:211–218

Kevan DK (1965) The soil fauna: its nature and biology. In: Baker KF, Snyder WC (eds) Ecology of soil-borne plant pathogens. University of California Press, Berkeley, pp 33–50

Kim YK, Tomoda H, Nishida H, Sunazuka T, Obata R, Omura S (1994) Pyripyropenes, novel inhibitors of acyl-Co-A: cholesterol acyltransferase produced by *Aspergillus fumigatus*. II. Structure elucidation of pyripyropenes A, B, C, and D. J Antibiot 47:154–162

Koltin Y (1990) Targets for antifungal drug discovery. Ann Rep Med Chem 25:141–148

Koshino H, Togiya S, Terada S, Yoshihara T, Sakamura S, Shimanuki T, Sato T, Tajimi A (1989) New fungitoxic sesquiterpenoids, chokols A–G, from stromata of *Epichloe typhina* and the absolute configuration of chokol E. Agric Biol Chem 53:789–796

Krasnoff SB, Gibson DM, Belofsky GN, Gloer KB, Gloer JB (1996) New destruxins from the entomopathogenic fungus *Aschersonia* sp. J Nat Prod 59:485–489

Kwok OCH, Plattner R, Wiesleder D, Wicklow DT (1992) A nematocidal toxin from *Pleurotus ostreatus* NRRL 3526. J Chem Ecol 18:127–130

Laakso JA (1992) New biologically active metabolites from aquatic and sclerotium-producing fungi. PhD Thesis, University of Iowa, Iowa City, Iowa

Laakso JA, Gloer JB, Dowd PF, Wicklow DT (1992a) Radarins A–D: new antiinsectan and cytotoxic indole diterpenoids from the sclerotia of Aspergillus sulphureus. J Org Chem 57:138–141

Laakso JA, Gloer JB, Dowd PF, Wicklow DT (1992b) Sulpinines A–C and secopenitrem B: new antiinsectan metabolites from the sclerotia of Aspergillus sulphureus. J Org Chem 57:2066–2071

Laakso JA, Gloer JB, Dowd PF, Wicklow DT (1993) A new penitrem analog with antiinsectan activity from the sclerotia of Aspergillus sulphureus. J Agric Food Chem 41:973–975

Lee JC, Lobkovsky E, Pliam NB, Strobel G, Clardy J (1995) Subglutinols A and B: immunosuppressive compounds from the endophytic fungus Fusarium subglutinans. J Org Chem 60:7076–7078

Lee KK, Gloer JB, Scott JA, Malloch D (1995) Petriellin A: a novel antifungal depsipeptide from the coprophilous fungus Petriella sordida. J Org Chem 60:5384–5385

Lockwood JL (1992) Exploitation competition. In: Carroll GC, Wicklow DT (eds) The fungal community: its organization and role in the ecosystem, 2nd edn. Dekker, New York, pp 243–263

Lodha BC (1974) Decomposition of digested litter. In: Dickinson CH, Pugh GJF (eds) Biology of plant litter decomposition, vol 1. Academic Press, New York, pp 213–241

Malloch D (1981) Moulds. Their isolation, cultivation, and identification. University of Toronto Press, Toronto, pp 19–20

Mantle PG (1978) Industrial exploitation of ergot fungi. In: Smith JE, Berry DR (eds) The filamentous fungi: industrial mycology, vol 1. Wiley Press, New York, pp 281–300

Masurekar PS (1992) Therapeutic metabolites. In: Finkelstein DB, Ball C (eds) Biotechnology of filamentous fungi: technology and products. Butterworth-Heinemann, Boston, pp 241–301

Miller JD (1991) Mycology, mycologists, and biotechnology. In: Hawksworth DL, Regensburg CA (eds) Frontiers in mycology. Honorary lectures from the 4th International Mycological Congr, pp 225–240

Miller JD, Trenholm HL (eds) (1994) Mycotoxins in grain: compounds other than aflatoxin. Eagan Press, Minnesota, 552 pp

Monaghan RL, Tkacz JS (1990) Bioactive microbial products: focus upon mechanism of action. Annu Rev Microbiol 44:271–301

Morrall RAA, Loew FM, Hayes MA (1978) Subacute toxicological evaluation of sclerotia of Sclerotinia sclerotiorum in rats. Can J Comp Med 42:473–477

Morris RAC, Ewing DF, Whipps JM, Coley-Smith JR (1995) Antifungal hydroxymethyl-phenols from the mycoparasite Verticillium biguttatum. Phytochemistry 39:1043–1048

Morris SA, Curotto JE, Zink DL, Dreikorn S, Jenkins R, Bills GF, Thompson JR, Vicente F, Basilio A, Liesch JM, Schwartz RE (1995) Sonomolides A and B, new broad spectrum antifungal agents isolated from a coprophilous fungus. Tetrahedron Lett 36:9101–9104

Nozawa K, Nakajima S, Kawai K, Udagawa S, Miyaji M (1994) Bicoumarins from ascostromata of Petromyces alliaceus. Phytochemistry 35:1049–1051

Paul VJ (ed) (1992) Ecological roles of marine natural products. Cornell University Press, Ithaca, New York

Poch GK, Gloer JB, Shearer CA (1992) New bioactive metabolites from a freshwater isolate of the fungus Kirschsteiniothelia sp. J Nat Prod 55:1093–1099

Powell KA (1993) The commercial exploitation of microorganisms in agriculture. In: Jones DG (ed) Exploitation of microorganisms. Chapman and Hall, London, pp 441–459

Rhodes DF (1985) Offensive-defensive interactions between herbivores and plants: their relevance in herbivore population dynamics and ecological theory. Am Nat 125:205–238

Richardson K, Marriott MS (1987) Antifungal agents. Annu Rep Med Chem 22:159–167

Rinderknecht BL, Gloer JB, Dowd PF, Wicklow DT (1989) Nominine: a new insecticidal indole diterpene from the sclerotia of Aspergillus nomius. J Org Chem 54:2530–2532

Schwartz RE, Sesin DF, Joshua H, Wilson KE, Kempf AJ, Goklen KA, Kuehner D, Gailliot P, Gleason C, White R, Inamine E, Bills G, Salmon P, Zitano L (1992) Pneumocandins from Zalerion arboricola: discovery and isolation. J Antibiot 45:1853–1866

Shaw PJA (1992) Fungi, fungivores, and food webs. In: Carroll GC, Wicklow DT (eds) The fungal community: its organization and role in the ecosystem, 2nd edn. Dekker, New York, pp 295–310

Shearer CA, Zare-Maivan H (1988) In vitro hyphal interactions among wood- and leaf-inhabiting ascomycetes and fungi imperfecti from freshwater habitats. Mycologia 80:31–37

Singh N, Webster J (1973) Antagonism between Stilbella erythrocephala and other coprophilous fungi. Trans Br Mycol Soc 61:487–495

Singh SB, Zink DL, Liesch JL, Ball RG, Goetz MA, Bolessa EA, Giacobbe RA, Silverman KC, Bills GF, Pelaez F, Cascales C, Gibbs JB, Lingham RB (1994) Preussomerins and deoxypreussomerins: novel inhibitors of Ras farnesyl-protein transferase. J Org Chem 59:6296–6302

Stadler M, Anke H (1995) Metabolites with nematicidal and antimicrobial activities from the ascomycete Lachnum papyraceum (Karst.) Karst. V. Production, isolation and biological activities of bromine-containing and lachnumon derivatives and four additional new bioactive metabolites. J Antibiot 48:149–153

Stadler PA, Giger RKA (1984) Ergot alkaloids and their derivatives in medicinal chemistry and therapy. In: Krogsgaard-Larsen P, Christensen SB, Kofod H (eds) Natural products and drug development. proc 20th Alfred Benzon Symp, Munksgaard, Copenhagen, pp 463–485

Staub GM, Gloer JB, Dowd PF, Wicklow DT (1992) Aspernomine: a cytotoxic antiinsectan metabolite with a novel ring system from the sclerotia of Aspergillus nomius. J Am Chem Soc 114:1015–1017

Staub GM, Gloer KB, Gloer JB, Dowd PF, Wicklow DT (1993) New paspalinine derivatives with antiinsectan activity from the sclerotia of Aspergillus nomius. Tetrahedron Lett 34:2569–2572

Sterner O, Bergman R, Kihlberg J, Wickberg B (1985) The sesquiterpenes of Lactarius vellereus and their role in a proposed chemical defense system. J Nat Prod 48:279–288

Stevenson BG, Dindal DL (1987) Functional ecology of coprophagous insects: a review. Pedobiologia 30:285–298

Strongman DB, Miller JD, Calhoun L, Findlay JA, Whitney NJ (1987) The biochemical basis for interference competition among some lignicolous marine fungi. Bot Mar 30:21–26

Strunz GM, Kakushima M, Stillwell MA (1972) Scytalidin: a new fungitoxic metabolite produced by a *Scytalidium* species. J Chem Soc Perkin Trans I 2280–2283

Tanda S, Tadakuma Y, Matsunami Y (1968) Fundamental studies on ergotial fungi. VIII. Poisonous experiments of ergot to mouse. J Agric Sci (Tokyo) 13:55–60

TePaske MR (1991) Isolation and structure determination of antiinsectan metabolites from the sclerotia of *Aspergillus* species. PhD thesis, University of Iowa, Iowa City, Iowa

TePaske MR, Gloer JB, Dowd PF, Wicklow DT (1989a) Three new aflavinines from the sclerotia of *Aspergillus tubingensis*. Tetrahedron 45:4961–4968

TePaske MR, Gloer JB, Dowd PF, Wicklow DT (1989b) Tubingensin A: a novel antiviral carbazole alkaloid from the sclerotia of *Aspergillus tubingensis*. J Org Chem 54:4743–4746

TePaske MR, Gloer JB, Dowd PF, Wicklow DT (1989c) The structure of tubingensin B: a new carbazole metabolite from the sclerotia of *Aspergillus tubingensis*. Tetrahedron Lett 30:5965–5968

Tezuka Y, Huang Q, Kikuchi T, Nishi A, Tubaki K (1994) Studies on the metabolites of mycoparasitic fungi: metabolites of *Cladobotryum varium*. Chem Pharm Bull 42:2612–2617

Turner WB, Aldridge DC (1983) Fungal metabolites II. Academic Press, New York

Umehara K, Nakahara K, Kiyoto S, Iwami M, Okamoto M, Tanaka H, Kohasaka M, Aoki H, Imanaka H (1983) Studies on WF-3161, a new antitumor antibiotic. J Antibiot 36:478–483

Walton JD, Panaccione DG (1993) Host-selective toxins and disease specificity: perspectives and progress. Annu Rev Phytopathol 31:275–303

Wang H-j, Gloer JB, Wicklow DT, Dowd PF (1995a) Aflavinines and other antiinsectan metabolites from ascostomata of *Eupenicillium crustaceum* and related species. Appl Environ Microbiol 61:4429–4435

Wang H-j, Gloer JB, Scott JA, Malloch D (1995b) Coniochaetones A and B: new antifungal benzopyranones from the coprophilous fungus *Coniochaeta saccardoi*. Tetrahedron Lett 36:5847

Wang Y (1994) Isolation and structure determination of new antifungal metabolites from coprophilous fungi. PhD Thesis, University of Iowa, Iowa City, Iowa

Wang Y, Gloer JB, Scott JA, Malloch D (1993) Appenolides A–C: new antifungal furanones from the coprophilous fungus *Podospora appendiculata*. J Nat Prod 56:341–344

Wang Y, Gloer JB, Scott JA, Malloch D (1995) Terezines A–D: new amino acid-derived bioactive metabolites from the coprophilous fungus *Sporormiella teretispora*. J Nat Prod 58:93

Weber HA, Baenziger NC, Gloer JB (1988) Podosporin A: a novel antifungal metabolite from the coprophilous fungus *Podospora decipiens* (Wint.) Niessl. J Org Chem 53:4567–4569

Weber HA, Gloer JB (1988) Interference competition among natural fungal competitors: an antifungal metabolite from the coprophilous fungus *Preussia fleishhakii* (Auerswald) Cain. J Nat Prod 51:879–883

Weber HA, Gloer JB (1991) Preussomerins A–F: novel antifungal metabolites from the coprophilous fungus *Preussia isomera* Cain. J Org Chem 56:4355–4360

Webster J (1970) Coprophilous fungi. Trans Br Mycol Soc 54:161–180

Whyte AC, Gloer JB, Scott JA, Malloch D (1996a) Cercophorins A–C: novel antifungal and cytotoxic metabolites from the coprophilous fungus *Cercophora areolata*. J Nat Prod 59:765–769

Whyte AC, Gloer JB, Wicklow DT, Dowd, PF (1996b) Sclerotiamide: a new member of the paraherquamide class with potent antiinsectan activity from the sclerotia of *Aspergillus sclerotiorum*. J Nat Prod 59:1093–1095

Wicklow DT (1981) The coprophilous fungal community: a mycological system for examining ecological ideas. In: Wicklow DT, Carroll GC (eds) The fungal community: its organization and role in the ecosystem. Dekker, New York, pp 47–76, 351–385

Wicklow DT (1986) Ecological adaptations and classification in *Aspergillus* and *Penicillium*. In: Sampson RA, Pitt JI (eds) Advances in *Penicillium* and *Aspergillus* systematics. Plenum, New York, pp 255–265

Wicklow DT (1988) Metabolites in the coevolution of fungal chemical defence systems. In: Pirozynski KA, Hawksworth D (eds) Coevolution of fungi with plants and animals. Academic Press, New York, pp 174–201

Wicklow DT (1992a) The coprophilous fungal community: an experimental system. In: Carroll GC, Wicklow DT (eds) The fungal community: its organization and role in the ecosystem, 2nd edn. Dekker, New York, pp 715–728

Wicklow DT (1992b) Interference competition. In: Carroll GD, Wicklow DT (eds) The fungal community: its organization and role in the ecosystem, 2nd edn. Dekker, New York, pp 265–274

Wicklow DT, Cole RJ (1982) Tremorgenic indole metabolites and aflatoxins in sclerotia of *Aspergillus flavus* Link: an evolutionary perspective. Can J Bot 60:525–528

Wicklow DT, Cole RJ (1984) Citreoviridin in standing corn infested by *Eupenicillium ochrosalmoneum*. Mycologia 76:959–961

Wicklow DT, Hirschfield BJ (1979a) Competitive hierarchy in post-fire ascomycetes. Mycologia 71:47–54

Wicklow DT, Hirschfield BJ (1979b) Evidence of a competitive hierarchy among coprophilous fungal populations. Can J Microbiol 25:855–858

Wicklow DT, Shotwell OL (1983) Intrafungal distribution of aflatoxins among conidia and sclerotia of *Aspergillus flavus* and *Aspergillus parasiticus*. Can J Microbiol 29:1–5

Wicklow DT, Dowd PF, TePaske MR, Gloer JB (1988) Sclerotial metabolites of *Aspergillus flavus* toxic to a detritivorous maize insect (*Carpophilus hemipterus*, Nitidulidae). Trans Br Mycol Soc 91:433–438

Willets HJ (1971) The survival of fungal sclerotia under adverse environmental conditions. Biol Rev 46:387–407

Willets HJ (1978) Sclerotium formation. In: Smith JE, Berry DR (eds) The filamentous fungi, vol III. developmental mycology. Wiley Press, New York, pp 197–213

Xu X, DeGuzman FS, Gloer JB, Shearer CA (1992) Stachybotrins A and B: novel antifungal metabolites from a brackish water isolate of the fungus *Stachybotrys* sp. J Org Chem 57:6700–6703

Decomposition, Biomass,
and Industrial Applications

17 Nutrient Cycling by Saprotrophic Fungi in Terrestrial Habitats

J. DIGHTON

CONTENTS

I. Introduction

Terrestrial ecosystems present a wide diversity of habitats ranging from polar arctic conditions, through temperate coniferous and deciduous forest and grasslands, heathland, tropical forests, and savannahs to deserts. In this short chapter, it is not possible to compare and contrast the role of fungi in nutrient cycling and bioaccumulation in each ecosystem or habitat (see Dighton 1995), but the general principles which are common to all terrestrial ecosystems will be outlined.

Primary productivity (plant growth) is dependent on adequate sunlight, moisture, temperature, and essential nutrients in soil solution. Dead plant parts (above- and below-ground) are returned to soil where the activities of bacteria, saprotrophic fungi, and soil fauna degrade the complex organic components, using the carbon skeletons for energy and some mineral elements for biomass development. However, as a result of the rapid turnover of these organisms, nutrient elements are released into soil solution as simple inorganic compounds (the process of **mineralization**). The balance between rates of decomposition and mineralization and input of dead plant parts to soil determine the type of soil profile developed over time. Where decomposition is very slow, organic matter accumulates as peat. Where decomposition

Division of Pinelands Research, Institute of Marine and Coastal Sciences, Department of Biology, Rutgers University, Camden, New Jersey 08102, USA

is rapid, as in agricultural soils, a mineral soil profile is developed with low organic matter content.

During the course of decomposition, mineral nutrients are sequestered by the decomposer soil organisms, being incorporated into the organism's biomass. The residence time of these elements is usually equivalent to the turnover time (lifespan) of that organism. During this period, the element is not in a soluble form in the soil solution, but is **immobilized** in microbial tissue. The amount of accumulation within the fungal component varies between ecosystems, depending on the chemical composition of the plant parts available for decomposition and the main fungal groups involved in the process. Thus shorter-lived, ephemeral molds, utilizing simple carbohydrates, have lower investment in biomass than longer-lived basidiomycetes, growing on woody resources; thus the potential accumulation in basidiomycetes is greater. Unlike bacteria, fungi are larger organisms and their rate of turnover is lower, particularly in the long-lived Basidiomycotina. A discussion of the role of basidiomyctes in decomposition is given by Frankland, Hedger and Swift (1982). Different nutrient elements and, in particular, metal ions may be immobilized for long periods (**bioaccumulation**) in fungi. Fungi are nondiscrete organisms (having an extending hyphal network) and are able to translocate elements within the fungal thallus (Cairney, 1992). This could account for spatial redistribution of elements. For example, if an element were always translocated away from dying regions, it would increase the length of time of immobilization into fungal components.

Figure 1 gives a diagrammatic representation of nutrient cycling in forest ecosystems. The principle components are the same in any terrestrial ecosystem and the tree (plant) could be substituted by any other plant form (grass, forb, herbaceous shrub, cactus, etc.).

It is important to remember at the outset that saprotrophic fungi involved with decomposition

The Mycota IV
Environmental and Microbial Relationships
Wicklow/Söderström (Eds.)
© Springer-Verlag Berlin Heidelberg 1997

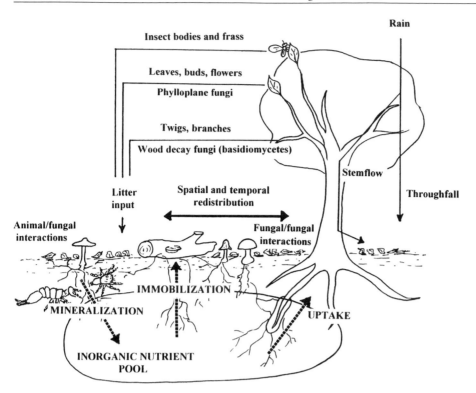

Fig. 1. Diagramatic representation of nutrient cycling in woodland ecosystems and the role of fungi in that process. (After Dighton and Boddy 1989)

and nutrient cycling in soil do not perform that function in isolation. Plant and animal remains may be comminuted and utilized by soil fauna, bacteria, and actinomycetes. Interactions between these organisms are important in determining the rate of decomposition and diversity of soil biota. The process is also dynamic so, for example, the same suite of organisms is not present on the plant or animal remains (**resource**) for the duration of the process of decomposition. It will be seen that different fungi have different enzymatic capabilities, so their appearance on a resource will be dictated by (1) their ability to utilize the resource, (2) their rate of arrival at the resource either by growth or by transport as spores etc., and (3) their ability to compete against other fungal species with similar physiological competence.

II. Decomposition

The rate at which a resource is decomposed is dependent on its chemical composition, edaphic

factors (available moisture and temperature), and the colonization of the resource by appropriate saprotrophic organisms. Many of these factors are discussed in the book by Cooke and Rayner (1984). The input of different types (chemical composition and, hence, resource quality) of plant litter varies with ecosystem type. Old, but still pertinent, data from Rodin and Bazilevich (1967) show is that the rate of plant litter input into soil of different ecosystems is partly related to plant biomass, but the amount of litter resident on the soil surface is related to the rate of decomposition (resource quality × environmental factors). Thus in cold steppe regions, plant biomass is low, litter fall is proportionately high, but the rate of decomposition is low due to climatic limitations. This leads to accumulation of organic components of the soil and evolution of peaty soil profiles. In coniferous forests, plant biomass is high, litter fall is low, but accumulation occurs mainly due to the recalcitrance of the litter to decomposition (low resource quality). In tropical forests, plant biomass is very high, litter fall is high, but the litter on the soil is low, indicating a combination of climatic

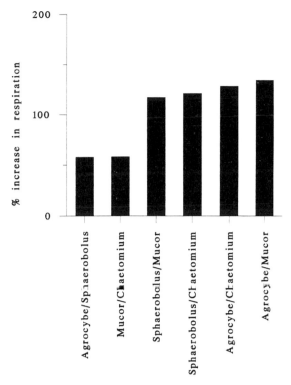

Fig. 2. Percentage increase in respiration of competing pairs of fungi colonizing straw over the mean respiration of each fungal species alone. (After Robinson et al. 1993b)

fungi). Only then is there net conversion of organic nutrient to inorganic nutrient (net mineralization).

Decomposition is a product of enzyme activity. The types of enzymes required are dependent on the substrates (chemical constituents) of the resource. Fungal species differ in the range of enzymes they are capable of producing. This, in part, dictates the succession of fungi colonizing resources (see Sect. III). Linkins et al. (1984) discussed some of the factors affecting the activity of extracellular cellulase, particularly the positive influence of temperature and the cellulose to lignin ratio. Cellulose appears to become unavailable for microbial use when the cellulose:lignin ratio declines below 0.5. In a later paper, Sinsabaugh et al. (1993) related the rate of mass loss from birch sticks to the activity of extracellular acid phosphatase, N-acetylglucosaminidase and sulphatase enzymes, giving rise to their model that microbial communities maximize production by optimizing resource allocation between macronutrient activities. The process of decomposition is governed by the production of enzymes, which are, in turn, regulated by the availability of nitrogen or phosphorus. Thus, where nutrient elements are less available, the fungi expend greater amounts of energy to produce enzymes to sequester the nutrients from organic sources.

III. Colonization

The colonization of resources by fungi is a function of the quality of the resource, rate of arrival of the fungal propagule (spore or hyphum), and the competitive interaction between fungal species on the resource. The role of plant litter quality on the pattern of fungal colonization of resources has been discussed in Dickinson and Pugh (1974). Here, many examples of the change in species composition of fungal communities as different plant substrates undergo the cascade of decay are desribed. In general, there appears to be a succession of fungi utilizing different resources within the litter. The classic assumption is that the initial colonizers used soluble carbohydrate sources (sugars) and were later replaced with fungal species having greater enzymatic competence, which are able to break down organic sources of carbon such as cellulose, and lignin. A general scheme for fungal succession in relation to resource quality is

conditions conducive to decomposition and high resource quality. One of the reasons for reduced resource quality in forested ecosystems is the diversity of litter types. Quantities cited by Dighton and Boddy (1989) suggest a forest litter composition of some 55% leaves, 10% fruits, buds, and flowers, 20% twigs, 10% branches, and 5% insect frass, etc. The wood component may be underestimated and may be nearer 40%. This woody component has a high lignin content and high C: nutrient ratio making it much less degradable by fungi. The carbon: nitrogen and lignin:nitrogen ratios are known to be determinants of the resistance of resources of decomposition and ultimate mineralization of nutrients (Melillo et al. 1982).

The model proposed by Swift et al. (1979) of mineralization of nutrients from a resource degraded by fungal activity suggests that during initial decomposition the carbohydrate component is used as an energy source until such time as the C: nutrient ratio approaches that of the decomposer organism (around 15:1 for P and 6:1 for N in

given in Table 1. However, there are few clear distinctions in the succession and, in fact, many of the species overlap in time and space. This has been described in more detail for decomposition of litter of the fern *Pteridium aquilinum* by Frankland (1994) in her discussion of fungal successions. She describes changes from lesion-forming *Rhizographus* and *Aureobasidium* on standing dead litter, through the colonization by basidiomycetes in relation to the rate of loss of cellulose and lignin and the consequential decrease in C:N ratio from some 200:1 to 30:1. In a model of Swift et al. (1979) the changes from "sugar" fungi to basidiomycetes in relation to the changes in available resources and the influence of climatic stresses are presented. In general, initial resource structure is chemically heterogenous, thus supporting a variety of fungal species. As decomposition proceeds, recalcitrant chemicals are left which can only be degraded by a restricted fungal flora; hence diversity is reduced. From these studies, the windows of opportunity for decomposition may be determined and the rate of substrate decomposition mapped.

The interaction between adjacent colonies of competing fungi have been elegantly mapped in three dimensions using wood as a resource (Rayner 1978; Rayner and Boddy 1988a,b). Clear demarcation zones are set up when genetically incompatible strains or species meet in a relatively homogenous resource. In an environment where resources are patchily distributed, such as mixed litter on the forest floor, the colonization of individual resource units is more difficult to map. The colonization pattern of individual straw resource units by a range of fungal species was found to be correlated to relative growth rates of the fungi on agar (Robinson et al. 1993a). These rates of growth allowed four species to be ranked in combative order. Mixtures of fungal species caused significant reductions in the rate of growth of less combative fungal species in the presence of combative species. Thus the cascade of decomposition is related to colonization of a substrate by fungi based on their enzymatic competence in relation to chemical resources available and also by the outcome of interaction with other potential colonizers of that resource. In a companion paper on straw decomposition, Robinson et al. (1993b) showed that where fungal interactions were taking place on straw, respiration was greater than where only one fungal species was present. This indicates that the maintenance of combative activities is energy-demanding (Fig. 2) and may affect the rate of decomposition. There is evidence that both ectomycorrhizal and ericaceous mycorrhizal fungi are able to access organic forms of nutrient (N and P) and, thus may compete with saprotrophic fungi

Table 1. A summary of trends in ecosystem and substrate (resource) succession, fungal succession on substrate and interaction between fungi and fauna in the decomposition of plant litter. (After Heal and Dighton 1986)

Ecosystem succession Increasing contribution of component to litter				
	Lower plant	Herbaceous plant	Angiosperm leaves	Coniferous leaves	Wood
Cellulose (%)	16–35	20–37	6–22	20–31	36–63
Lignin (%)	7–36	3–30	9–42	20–58	17–35
C:N ratio	13–150	29–160	21–71	63–327	294–327
Decay (% year^{-1})	20	30–70	40–60	3–50	1–90

Fungal succession				
	Sugar fungi	Yeasts	Ascomycotina	Ascomycotina
	Ascomycotina	Sugar fungi	Fungi	Fungi
	Fungi imperfecti	Ascomycotina	imperfecti	imperfecti
		Fungi imperfecti	Basidio-mycotina	Basidio-mycotina
		Basidio mycotina		

Fauna less important	Fauna important			
	Enchytraeids	Enchytraeids	Oligochaetes	Acari	Insecta
		Oligochaetes	Collembola	Collembola	Other
		Diptera	Acari	Oligochaetes	Arthropoda

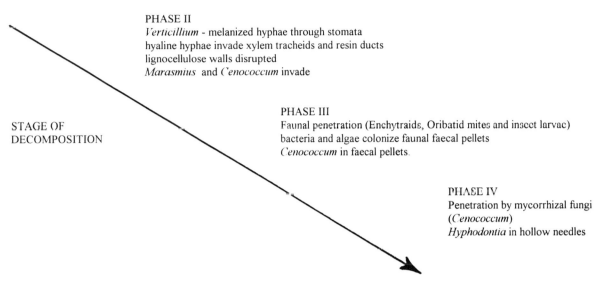

PHASE I
Lophodermium & *Ceuthespora* fruiting
hyphae in free space of mesophyll
browning of cellulose cell walls
cytoplasm missing, replaced by hyphae

PHASE II
Verticillium - melanized hyphae through stomata
hyaline hyphae invade xylem tracheids and resin ducts
lignocellulose walls disrupted
Marasmius and *Cenococcum* invade

STAGE OF
DECOMPOSITION

PHASE III
Faunal penetration (Enchytraids, Oribatid mites and insect larvae)
bacteria and algae colonize faunal faecal pellets
Cenococcum in faecal pellets.

PHASE IV
Penetration by mycorrhizal fungi
(*Cenococcum*)
Hyphodontia in hollow needles

Fig. 3. Schematic colonization pattern of fungal colonization and interaction with soil fauna during Scots pine needle decomposition as identified by the microscopic study of small volumes of forest floor material. (After Ponge 1991)

for resources in forested ecosystems (see Leake and Read, Chap. 18, this Vol.). The importance of this interaction is not well understood, although negative interactions between saprotrophic and ectomycorrhizal fungi, in terms of mycorrhizal colonization of roots, have been found (Shaw et al. 1995).

By microscopic observation of small samples of forest floor, Ponge (1991) characterized the colonization of *Pinus sylvestris* needles into four stages. His examination allowed him to identify both fungal and faunal components and their interactions (Fig. 3). This work also suggested that phylloplane fungal species are present and act as saprotrophs on freshly fallen leaf litter. In forested systems, much dead wood remains in the canopy prior to recruitment to the forest floor. This standing dead material may have a different fungal community than wood on the forest floor. The work of Boddy and Rayner (1983) on oak wood in canopies showed that twelve basidiomycete fungal species dominated in the community. Of these, *Phellinus ferreus*, *Sterium gausapatum* and *Vuilleminia comendens* were pioneer species of partially living branches, *Phlebia adiata* and *Coriolus versicolor* were secondary colonizers and *Hyphoderma setigerum* and *Sterium hirsutum* related to insect activity.

In tropical forests, a large amount of leaf litter is colonized by fungi and decomposed before dropping to the forest floor (Hedger et al. 1993). This colonization and decomposition is, in part, carried out by rhizomorphic species such as *Marasmius* and *Marasmiellus*, which tend to tie leaves together by rhizomorphs. The decomposition in the canopy results in mineralization and leaching of nutrients in the stemflow. This nutrient film is an ideal resource for epiphytic plants and lichens. Similarly, on the forest floor, Lodge and Asbury (1988) have shown that the actions of rhizomorphic basidiomycete fungi aggregate leaf litter on the forest floor of montane tropical rain forests, preventing downslope movement of organic matter and underlying soil, thus preventing soil erosion.

Other soil fauna interactions have been shown to affect both colonization of resources (litter) and rates of nutrient mineralization. Newell (1984a,b) showed that the effect of fungal grazing Collembola altered the vertical distribution of

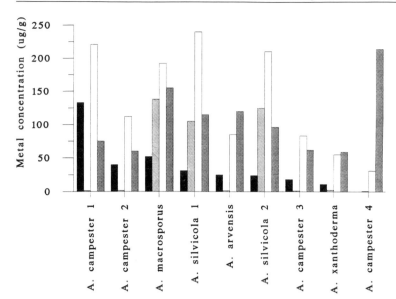

Fig. 4. Concentrations of the heavy metals, silver (*solid*), cadmium (*hatched*), copper (*open*), and zinc (*cross-hatched*) in basidiocarps of *Agaricus* species. (After Byrne 1979)

competing mycelia of *Mycena galopus* and *Marasmius androcaceous* in a spruce forest floor. Preferential grazing of *Marasmius* restricted its growth to lower depths (*Mycena* dominated in the A soil horizon). When Collembola withdrew to deeper soil layers during periods of drought, the rhizomorphic nature of *Marasmius* gave it a competitive advantage and allowed it to outgrow *Mycena* at the surface. Coleman et al. (1990) showed that reduction in microbial predators in ecosystems with high densities of soil fauna (forests) led to increased decomposition of litter (relief of grazing pressure). In contrast, systems with low densities of soil fauna (agricultural soils) the effect of faunal reduction was to reduce decomposition (suggesting a synergistic interaction).

IV. Immobilization

Where the C:nutrient ratio of a resource is very high, as in wood, the model of Swift et al. (1979) proposes initial immobilization and import of free nutrient into organic form (fungal thallus) during initial stages of decomposition until the fungal resource C:nutrient content is equivalent to that of the fungus. This immobilization of nutrients can be considerable, Stark (1972) showed that hyphae had 193 to 272% greater N content and 104 to 223% greater P content than the pine needle litter on which they were found, suggesting immobiliza-

tion. In practical forestry (particularly fast-growing trees) the potential competition between fungi decomposing woody logging residues and second-rotation tree transplants for essential nutrients has led to the establishment of burning protocols to rid the site of both woody debris and, incidentally, leaf litter and the nutrients they contain (Dighton 1995). Again, the interaction between saprotrophic fungi and soil fauna may be important in temporary nutrient immobilization. In a comparison of tilled and nontilled agricultural soils, Beare et al. (1992) showed that the exclusion of fungiverous soil arthropods reduced litter dry mass loss by only 5%, but significantly altered nitrogen dynamics in surface litter of no-till soil. Saprotrophic fungi were responsible for as much as 86% of net nitrogen immobilization ($1.8 \, \mathrm{g \, m^{-2}}$) into surface litters with exclusion of fungiverous microarthropods.

V. Bioaccumulation

The translocation of solutes in fungal tissues has been measured by tracer studies and has been shown to be of importance in allowing the colonization of low resource quality substrates. Wells and Boddy (1990) showed that 75% (*Phanerochaete velutina*) and 13% (*Phallus impudicus*) of the phosphorus added to a decomposed wood resource is translocated to newly colonized wood resources through mycelial cord systems. Maximum rates of P translocation are given as 7225 nmol P cm^{-2} day^{-1} through cords. The method of translocation of ^{14}C and ^{32}P through hyphal systems of *Rhizopus*, *Trichoderma*, and *Stemphylium* was demonstrated to be by diffusion (Olsson and Jennings 1991). The rate of translocation of carbon within the fungal thallus has been shown to react, in real time, to provide directional flow to the building phases of the hyphae (Olsson 1995). In contrast to the diffusion of C and P, translocation of ^{137}Cs through hyphae of *Schizophyllum commune*, however, has been shown to be slower than diffusion, suggesting a possible mechanism for accumulation (Gray et al. 1995, 1996). In addition, there is a suggestion from this work that there is potential preferential transport to sites of basidiocarp primordium production. This observation may support the finding of Dighton and Horrill (1988) and others (data from Yoshida and Muramatsu 1994) that radiocaesium accumulation in basidiomycete fungi could be high and long-lived. They found that up to 92% of the radiocaesium in mycorrhizal basidocarps in the UK was derived from pre-Chernobyl sources of fallout. Measures of influx and accumulation of radiocaesium into hyphae of a range of fungal species suggested that saprotrophic species had higher accumulation than mycorrhizal species (on a weight basis) (Clint et al. 1991). This rate of radiocaesium immobilization in UK upland grassland saprotrophic fungal mycelia could have accounted for a high proportion of the immobilization of Chernobyl fallout radiocaesium (Dighton et al. 1991; Dighton and Terry 1996).

Other elements, particularly heavy metals, have been shown to be adsorbed by microbial tissue. Tobin et al. (1984) demonstrated that dead mycelium of *Rhizopus arrhizus* was efficient at adsorbing a range of metal ions, but not alkali metals (Table 2). However, uptake and accumulation of metals by living fungi is proving to be of considerable importance. Byrne et al. (1979) listed values of accumulation of 9 metal elements in basidiocarps of 32 fungal species. Accumulation of silver, cadmium, copper and zinc in basidiocarps of *Agaricus* species are given in Fig. 4. It can be seen from these figures that the accumulation of each element is not the same in each species and that the pattern of accumulation in the same species may differ markedly between individual basidiocarps. It is, therefore difficult to make gen-

Table 2. Absorption of metal ions by dead hyphal mass of *Rhizopus arrhizus*. (Data from Tobin et al. 1984)

Metal ion	Absorption (mM l^{-1})
Cr	0.59
La	0.35
Mn	0.22
Cu	0.25
Zn	0.30
Cd	0.27
Ba	0.41
Hg	0.29
Pb	0.50
UO$_2$	0.82
Na	0.0
K	0.0
Rb	0.0
Ag	0.50
Cs	0.0

eralizations about fungi as a whole in their ability to concentrate metal ions.

Metal ion accumulation results from both metabolic uptake and cation exchange on appropriate binding sites. McKnight et al. (1990) showed that the concentration of, particularly, Mg was positively related to the cation exchange capacity of stipe tissue of 18 basidiomycete fungal species. Starling and Ross (1991) discussed the uptake of zinc by the fungus *Penicillium notatum* showing that it was an essential element with different uptake kinetics at low and high solution concentrations. Zinc uptake is also shown to be competitively inhibited by Cd and noncompetitively by Cu, indicating different interactions between metals and the fungal physiology. The wood decaying fungus *Phanerochaete chrysosporium* has shown the ability in field trials to degrade toxic organic compounds by converting chlorine bound in an organic form to harmless inorganic forms and to degrade aromatic hydrocarbons to CO_2 and water (Coghlan 1994). From both the aspects of metal ion accumulation and enzymatic competence of fungi, it is thought that fungi could be used to detoxify contaminated land.

VI. Conclusions

Fungi are a diverse group of organisms. In the process of decomposition and mineralization of nutrients from organic sources, a range of taxonomic groups are involved. It has been shown above that these fungi differ in their enzymatic capabilities (their ability to decompose certain resources), their rate of growth (competitiveness), and their interactions with other organisms in the ecosystem. In a world heavily influenced by human activities, fungi have been shown to possess potentially useful properties in their abilities to accumulate metal ions. It appears possible to use fungal tissues to effect detoxification of sites, although a large range of fungal species and situations have not been rigorously tested. All of these functions may be influenced by the presence of atmospheric pollutants. Evidence from the work of Wookey et al. (1991) and Newsham et al. (1992a,b) has shown that the community composition can be altered and decompositional potential of fungal communities in soil reduced in the presence of sulfur dioxide and acid precipitation. In an extreme situation of high levels of radioactivity around Chernobyl, Zhdanova et al. (1994) have shown that the effect of radiation dose can cause alteration of fungal community structure in soil.

Thus, although we can make generalized statements about the role of fungi in decomposition, nutrient cycling, and nutrient accumulation, and the similarity of action in different ecosystems, it must be remembered that there are other external influences on communities and function. In addition, Hawksworth (1991) has suggested that we may be able to identify some 5% of the possible total number of fungal species in the world. Of these, we may have isolated and investigated the physiology of a mere handful. We must ask the question: how confident are we in extrapolating these findings to fungi as a whole? Soil is an opaque medium where most of the fungal-mediated nutrient cycling processes occur. It is not an easy habitat to study and is complex in terms of the number and diversity of organisms present and their interactions. Our level of knowledge is constantly growing, but is far from complete.

References

Beare MH, Parmelee RW, Hendrix PF, Cheng W, Coleman DC, Crossley DA (1992) Microbial and faunal interactions and effects on litter nitrogen and decomposition in agroecosystems. Ecol Monogr 62:569–591

Boddy L, Rayner ADM (1983) Ecological roles of basidiomycetes forming decay communities in attached oak branches. New Phytol 93:77–88

Byrne AR, Dermelj M, Vakselj T (1979) Silver accumulation by fungi. Chemosphere 10:815–821

Cairney JWG (1992) Translocation of solutes in ectomycorrhizal and saprotrophic rhizomporphs. Mycol Res 96:135–141

Clint GM, Dighton J, Rees S (1991) Influx of ^{137}Cs into hyphae of basidiomycete fungi. Mycol Res 95:1047–1051

Coghlan A (1994) Fungi makes a meal of toxic waste. New Sci 142:18

Coleman DC, Ingham ER, Hunt HW, Elliott ET, Reid CPP, Moore JC (1990) Seasonal and faunal effects on decomposition in semiarid prarie, meadow and lodgepole pine forest. Pedobiologia 34:207–219

Cooke RC, Rayner ADM (1984) Ecology of saprotrophic fungi. Longman, London

Dickinson CH, Pugh GJF (eds) (1974) Biology of plant litter decomposition. Academic Press, New York

Dighton J (1995) Nutrient cycling in different terrestrial ecosystems in relation to fungi. Can J Bot 73(suppl): S1349–S1360

Dighton J, Boddy L (1989) Role of fungi in nitrogen, phosphorus and sulphur cycling in temperate forest ecosystems. In: Boddy L, Marchant R, Read DJ (eds) Nitrogen, phosphorus and sulphur utilization by fungi. Cambridge University Press, Cambridge, pp 267–29

Dighton J, Horrill AD (1988) Radiocaesium accumulation in the mycorrhizal fungi Lactarius rufus and Inocybe longicystis, in upland Britain, following the Chernobyl accident. Trans Br Mycol Soc 91:335–357

Dighton J, Terry GM (1996) Uptake and immobilization of caesium in UK grassland and forest soils by fungi following the Chernobyl accident. In: Frankland JC, Magan N, Gadd GM (eds) Fungi and pollution. Cambridge University Press, Cambridge, pp 184–200

Dighton J, Clint GM, Poskitt JM (1991) Uptake and accumulation of ^{137}Cs by upland grassland soil fungi: a potential pool of Cs immobilization. Mycol Res 95:1052–1056

Frankland JC, Hedger JN, Swift MJ (1982) Decomposer basidiomycetes: their biology and ecology. Cambridge University Press, Cambridge

Frankland JC (1994) Mechanisms in fungal succession. In: Carroll GC, Wicklow DT (eds) The fungal community; its organization and role in the ecosystem. Dekker, New York, pp 383–401

Gray SN, Dighton J, Olsson S, Jennings DH (1995) Real-time measurement of uptake and translocation of ^{137}Cs within mycelium of Schizophyllum commune by autoradiography followed by quantitative image analysis. New Phytol 129:449–465

Gray SN, Dighton J, Jennings DH (1996) The physiology of basidiomycete linear organs. III. Uptake and translocation of radiocaesium within differentiated mycelia of Armillaria spp. growing in microcosms and in the field. New Phytol 132:471–482

Hawksworth DL (1991) The fungal dimension of biodiversity: magnitude, significance and conservation. Mycol Res 95:641–655

Heal OW, Dighton J (1986) Nutrient cycling and decomposition in natural terrestrial ecosystems. In: Mitchell MJ, Nakas JP (eds) Microfloral and faunal interactions. Martinus Nijhoff/Dr. W. Junk, Dordrecht, pp 14–73

Hedger J, Lewis P, Gitay H (1993) Litter-trapping by fungi in moist tropical forest. In: Isaac S, Frankland JC, Watling R, Whalley AJS (eds) Aspects of tropical mycology. Cambridge Univesity Press, Cambridge, pp 15–35

Linkins AE, Melillo JM, Sisabaugh RL (1984) Factors affecting cellulase activity in terrestrial and aquatic ecosystems. In: Klug MJ, Reddy CA (eds) Current perspectives in microbial ecology. American Society for Microbiology, Washington, pp 572–579

Lodge DJ, Asbury CE (1988) Basidiomycetes reduce export of organic matter from forest slopes. Mycologia 80:888–890

McKnight KB, McKnight KH, Harper KT (1990) Cation exchange capacities and mineral element concentrations of macrofungal stipe tissue. Mycologia 82:91–98

Melillo JM, Aber JD, Muratore JF (1982) Nitrogen and lignin control of hardwood litter decomposition dynamics. Ecology 63:621–626

Newell K (1984a) Interactions between two decomposer basidiomycets and a collembolan under Sitka spruce: distribution, abundance and selective grazing. Soil Biol Biochem 16:227–233

Newell K (1984b) Interactions between two decomposer basidiomycets and a collembolan under Sitka spruce: grazing and the potential effects of fungal distribution and litter decomposition. Soil Biol Biochem 16:235–239

Newsham KK, Frankland JC, Boddy L, Ineson P (1992a) Effects of dry deposited sulphur dioxide on fungal decomposition of angiosperm tree leaf litter. I. Changes in communities of fungal saprotrophs. New Pytol 122:97–110

Newsham KK, Boddy L, Frankland JC, Ineson P (1992b) Effects of dry deposited sulphur dioxide on fungal decomposition of angiosperm tree leaf litter. III. Decomposition rates and fungal respiration. New Pytol 122:127–140

Olsson S (1995) Mycelial density profiles of fungi on heterogenous media and their interpretation in terms of nutrient allocation patterns. Mycol Res 99:143–153

Olsson S, Jennings DH (1991) Evidence for diffusion being the mechanism of translocation in the hyphae of three moulds. Exp Mycol 15:302–309

Ponge JF (1991) Succession of fungi and fauna during decomposition of needles in a small area of Scots pine litter. Plant Soil 138:99–113

Rayner ADM (1978) Interactions between fungi colonizing hardwood stumps and their possible role in determining patterns of colonization and succession. Ann Appl Biol 89:131–134

Rayner ADM, Boddy L (1988a) Fungal decomposition of wood: its biology and ecology. Wiley, New York

Rayner ADM, Boddy L (1988b) Fungal communities in the decay of wood. Adv Microb Ecol 10:115–136

Robinson CH, Dighton J, Frankland JC (1993a) Resource capture by interacting fungal colonizers of straw. Mycol Res 97:547–558

Robinson CH, Dighton J, Frankland JC, Coward PA (1993b) Nutrient and carbon dioxide release by interacting species of straw-decomposing fungi. Plant Soil 151:139–142

Rodin LE, Bazilevich NI (1967) Production and mineral cycling in terrestrial vegetation. Oliver and Boyd, Edinburgh

Shaw TM, Dighton J, Sanders FE (1995) Interactions between ectomycorrhizal and saprotrophic fungi on agar and in association with seedlings of lodgepole pine (*Pinus contorta*). Mycol Res 99:159–165

Sinsabaugh RL, Antibus RK, Linkins AE, McClaugherty CA, Rayburn L, Repert D, Weiland T (1993) Wood decomposition: nitrogen and phosphorus dynamics in relation to extracellular enzyme activity. Ecology 74:1586–1593

Stark N (1972) Nutrient cycling pathways and litter fungi. Bioscience 22:355–360

Starling AP, Ross IS (1991) Uptake of zinc by *Penicillium notatum*. Mycol Res 95:712–714

Swift MJ, Heal OW, Anderson JM (1979) Decomposition in terrestrial ecosystems. Blackwell Scientific, Oxford

Tobin JM, Cooper DG, Neufeld RJ (1984) Uptake of metal ions by *Rhizopus arrhizus* biomass. Appl Environ Microbiol 47:821–82

Wells JM, Boddy L (1990) Wood decay and phosphorus and fungal biomass allocation in mycelial cord systems. New Phytol 116:285–295

Wookey PA, Ineson P, Mansfield TA (1991) Effects of atmospheric sulphur dioxide on microbial activity in decomposing forest litter. Agric Ecosyst Environ 33:263–280

Yoshida S, Muramatsu Y (1994) Accumulation of radiocaesium in basidiomycetes collected from Japanese forests. Sci Total Environ 157:197–205

Zhdanova NN, Vasilevskaya AI, Artyshkova LV, Lashko TN, Gavrilyuk VI, Dighton J (1994) Changes in micromycete communities in soil in response to pollution by long lived radionuclides emitted in the Chernobyl accident. Mycol Res 98:789–795

18 Mycorrhizal Fungi in Terrestrial Habitats

J.R. LEAKE and D.J. READ

CONTENTS

I. Introduction

The past decade has seen increased interest in, and evidence for, direct recycling of nutrients from organic matter to plants by mycorrhizal fungi. It is now established that there are some categories of mycorrhizas which, by means of extracellular, cell wall-bound and intracellular enzymes, can hydrolyse, oxidise, assimilate and transform many of the major nitrogen- and phosphorus-containing organic molecules in plant, microbial and animal detritus. These nutrient elements are then passed to the host. Hitherto, such essential steps in biogeochemical cycling have been attributed exclusively to free-living

Department of Animal and Plant Sciences, The University of Sheffield, Western Bank Sheffield S10 2TN, UK

saprotrophic organisms and the involvement of mycorrhizas discounted or ignored.

The misconception that mycorrhizal fungi are unimportant in nutrient mobilisation from organic matter has arisen partly because one of the most important types, that forming arbuscular mycorrhizas, has an obligate requirement for carbon supply from a host plant. The inability of zygomycetous fungi of arbuscular mycorrhizas to sustain growth apart from a host plant, even in the presence of exogenous carbohydrates and organic nutrients, has been widely interpreted as evidence for their complete lack of saprotrophic capabilities, athough the fungi have been shown to produce plant cell wall-degrading enzymes, enabling penetration and growth within the cells of their host roots (Garcia-Romera et al. 1991; Garcia-Garrido et al. 1992).

The view that saprotrophy is generally absent in mycorrhizal fungi was reinforced by many early studies of ectomycorrhizal species, in which polymeric substrates of particularly high C:N ratio such as cellulose (Norkrans 1950; Lyr 1963; Palmer and Hacskaylo 1970) and recalcitrant humic compounds (Lundeberg 1970) were employed as substrates. The failure of the fungi to derive sufficient carbon from polymers of this kind to sustain their growth in pure culture has been interpreted as evidence that they lack nutrient-mobilising activity and must, as a consequence, be dependent upon the mineralising activities of the saprotrophic population. However, although this is consistent with the observation that, with the exception of orchidaceous endophytes (Hadley 1969), most mycorrhizal fungi have poorly developed cellulolytic capabilities (see, for example Richter and Bruhn 1989) and do not compete successfully with specialist decomposer organisms for complex sources of carbon in nature (Paul and Clark 1989), it fails to recognise the unique physiology of the mycorrhizal symbiosis. The fact that carbon is supplied by the host simply places a premium on access to other nutrients. It is the low

The Mycota IV
Environmental and Microbial Relationships
Wicklow/Söderström (Eds.)
© Springer-Verlag Berlin Heidelberg 1997

availability of these, in particular nitrogen (N) and phosphorus (P), which exerts primary control over ecosystem activity and determines the effectiveness of the mycorrhizal symbiosis.

While selection may have favoured the provision of access for saprophytes to carbon-rich components of litter such as carbohydrates, lipids and phenolic polymers, it would be expected, conversely, to favour effective exploitation of the major sources of N- and P-rich compounds by mycorrhizal fungi. Amongst these, proteins, amino acids, chitin, nucleic acids, phospholipids and sugar phosphates are likely to be the most important.

This chapter summarises and evaluates the evidence for the view that such selection has taken place and that in both structural and functional terms some groups of mycorrhizal fungi, notably those of the ecto and ericoid type that predominate in ecosystems characterised by accumulation of N and P in organic residues, are well adapted to compete successfully with saprotrophs for the labile nutrient pools of the rooting environment.

II. Location and Activity of Mycorrhizal Roots and Mycelium in Relation to Substrate

A. The Association Between Mycorrhizal Roots and Decomposing Organic Matter

The earliest studies of the symbiosis (Frank 1894) recognised the close association between ectomycorrhizal roots and surface organic horizons of the soil. Subsequently, the fact that this type of root is preferentially located in a specific well-defined zone of the soil profile, namely the decomposition or fermentation (FH) horizon, has been demonstrated in temperate forests of the major deciduous (e.g. *Fagus*: Meyer 1973) and coniferous types (e.g. *Pinus*: Mikola and Laiho 1962; Harvey et al. 1976; Persson 1978, 1983; and *Picea*: Mikola and Laiho 1962). A similar distribution has also been recorded in tropical ectomycorrhizal trees which are particularly associated with soils with surface organic matter accumulation (Singer 1984; Singer and Araujo 1979; Alexander 1989), and in more typical tropical forest (St. John et al. 1983a).

In plants with ericoid mycorrhiza, prominent amongst which in heathland ecosystems of the northern hemisphere are *Calluna vulgaris* and members of the genera *Erica*, *Gaultheria*, *Rhododendron* and *Vaccinium*, the fine hair roots that are heavily colonised by the fungal endophyte are characteristically confined to the top 10 cm of the soil profile, being most abundant in the litter (Reiners 1965; Gimingham 1972; Persson 1978). The use of the isotopes ^{32}P applied to the soil as phosphate (Boggie et al. 1958) and ^{14}C, fed to the shoots as CO_2 (Wallén 1983) has confirmed that the active roots are confined to the top few centimetres of the heathland soil profile.

It is thus clear that the pattern of distribution of mycorrhizal roots is comparable in both ecto- and ericoid-dominated systems. Each type of community is characteristic of nutrient-impoverished soils and is species-poor. Consequently, the roots of the dominant plants proliferate predominantly in residues produced by themselves, their close relatives and their mycorrhizal fungal partners. Plants with arbuscular mycorrhizal associations are typically found on soils with much higher rates of mineralisation, so there is less evidence of close associations between organic matter and roots with this kind of mycorrhiza. Nonetheless, using non-radiactive tracers Fitter (1986) and Mamolos et al. (1995) have found highest root activities in grassland to occur in the top 5 cm of the soil, in which the greatest inputs of organic matter will occur.

B. The Association Between Mycorrhizal Mycelium and Decomposing Organic Matter

Careful excavation of forest soil has revealed that the vegetative mycelium of ectomycorrhizal fungi spreads to occupy a major part of the volume and surface of the fermenting litter (Ogawa 1977; Cromack et al. 1979; Soma and Saito 1979). The use of transparent observation chambers enabling non-destructive analysis of foraging patterns of ectomycorrhizal mycelium (Finlay and Read 1986) has shown that many of these fungi produce a lax fan-like growth, comparable with that described by Ogawa in the field. However, there were regions of dense mycelial proliferation, termed patches, in some discrete areas. While addition of mineral sources of N or P as ammonium and phosphate ions, respectively, failed to stimulate patch formation, introduction of organic residues, collected in the FH horizon of forest soil otherwise homogeneously humified peat, did elicit these types of proliferation (Read 1991; Bending and Read 1995a), as did senescing shoots of feather-mosses, which are the forest-floor dominants in many boreal forests (Carleton and Read 1990). Unestam (1991) found that hyphal growth was stimulated under decomposing leaves added to petri-dish rhizoscopes. Within a few weeks, intensive ramification and rhizomorph formation was induced in five out of six species of ectomycorrhizal fungus. Ponge (1990) provided circumstantial evidence for the involvement of ectomycorrhizal fungi in litter decomposition by observing under a light microscope the penetration of pine needles and bracken leaflets, as well as animal corpses and exoskeletons, by hyphae which were believed to be of the ectomycorrhizal fungi *Cenococcum geophilum* and *Hyphodontia* spp.

Despite the widely held view that arbuscular mycorrhizas have no saprotrophic capabilities, observations on the absorbtive mycelium of these fungi have demonstrated that they show affinity for, and proliferate preferentially in decomposing organic materials in soil (Nicolson 1959; St. John et al. 1983a,b). Their ability to survive independently of live host plants for many weeks in dead roots (Tommerup and Abbott 1981) and in peat (Warner 1984), where they penetrated dead *Sphagnum* cells, may indicate saprotrophy. Mosse (1959) found that decomposition of peat was increased in the presence of an arbuscular fungus, and detailed observations on its extra-radical mycelium revealed the formation of fine haustoria-like branches firmly attaching to particles of soil organic matter, the distal, and finest, portions of which could not be separated undamaged from the peat. Further evidence of some saprotrophic capability was provided by Warner and Moss (1980), Hepper and Warner (1983).

This kind of morphological plasticity has also been observed in wood-decomposing fungi, where lax fans and intensively branched mycelia were described respectively as representing exploratory and exploitative modes of growth (Rayner and Boddy 1988). Clearly, there is a possibility that the exploitative mycelial growth of ectomycorrhizal mycelium, the dense proliferation of ericoid hair roots, and the branching, penetration and formation of haustorial-like connections by arbuscular mycorrhizas in organic residues, in each case, represent parallel structural and functional adaptations to permit direct exploitation of organic nutrient sources or to efficiently capture mineral

nutrients released by saprophytes. The extent to which experimental evidence supports the direct exploitation route of nutrient cycling is examined in Section III.

C. Seasonal Patterns of Activity of Roots and Mycorrhizas

Periods of greatest root growth and mycorrhizal activity in boreal and temperate forests and heathlands are typically correlated with times of greatest nutrient supply rather than nutrient demand. Mycelial growth and production of mycorrhizal fruit bodies are maximal when carbon allocation below ground by plants reaches its peak in autumn. In deciduous forests, this corresponds to the stage of leaf senescence, when fresh organic substrates are deposited in the litter layer. In mediterranean and many boreal forests it occurs as the autumn rains rewet the summer-dry organic layers. Late autumn brings the peak activity of mycorrhizal roots of conifers like *Pinus sylvestris* and ericaceous plants like *Calluna vulgaris* (Persson 1983), at the time when labile forms of organic nitrogen like amino acids reach their zenith in soil (Abuarghub and Read 1988). In contrast, plants with arbuscular mycorrhizas typically have their main root growth and mycorrhizal activity in spring and summer (Brundrett and Kendrick 1988), although some geophytes have greatest root and mycorrhizal activity in autumn and winter (Brundrett and Kendrick 1990).

III. Nutrient Mobilisation and Export from Soil Colonised by Mycorrhizal Mycelium

A. Arbuscular Mycorrhizas

There is increasing direct and indirect evidence that at least some arbuscular mycorrhizal associations may have a direct role in mineralization and uptake of phosphorus from organic compounds. Phosphatase activity has been found associated with the cell walls of hyphae of arbuscular mycorrhizal fungi (MacDonald and Lewis 1978) and arbuscular mycorrhiza-specific phosphatases have been found in host roots (Pacovsky et al. 1991). Acid phosphatase activity is greater in the roots and mycorrhizosphere of infected plants compared to non-mycorrhizal plants (Dodd et al. 1987; Mohandas 1992; Tarafdar and Marschner 1994). In the latter study, phosphatase activity and arbuscular infection were increased in the presence of sodium phytate, and the enzyme activity was strongly correlated with hyphal length. Mycorrhiza accounted for 24–33% of total P uptake from an inorganic source and 48–59% of uptake from the phytate. Similarly, Jayachandran et al. (1992) reported that phytic acid, glycerophosphate, RNA, ATP and CMP were effective phosphorus sources for *Andropogon gerardii* mycorrhizal with *Glomus etunicatum*, but only slightly available to uninfected plants. However, since this experiment was not conducted under axenic conditions, the possibility remains that the mycorrhizal plants were simply more effective at uptake of P mineralized by saprophytic microorganisms.

These studies appear to contradict the results obtained from experiments in which ^{32}P-labelled orthophosphate is added to soil, its specific activity subsequently being compared with that of ^{31}P in mycorrhizal and non-mycorrhizal plants (Sanders and Tinker 1971; Hayman and Mosse 1972; Mosse et al. 1973; Powell 1975; Bolan 1991). In these experiments, significant differences between the specific activities in the two groups of plants are normally absent, suggesting that mycorrhizal and non-mycorrhizal plants use the same labile P sources, but that infected plants are more effective at scavenging such P. Bolan (1991) reviewed the use of ^{32}P for determining the sources of P used by mycorrhizal and non-mycorrhizal plants, and concluded both that the process of isotopic exchange between the labelled and unlabelled P in soil is complex, and that the assumption that ^{32}P only exchanges with the labile inorganic fraction of soil P may not be valid. In addition, it should be noted that in most cases the soils used for these experiments had a very low organic matter (and hence low organic P) content, most also being subjected to sterilisation treatments which mineralise most of the labile organic P, rendering it available to both mycorrhizal and non-mycorrhizal plants.

Whereas the potential involvement of arbuscular mycorrhizal fungi in mobilising organic phosphorus sources has received some attention, their ability to utilise organic sources of nitrogen has been little studied. Ames et al. (1983) provided one of the first indications that these mycorrhizas may mineralise organic nitrogen, but their experiments were also not conducted axenically.

In a study using aseptically reared *Festuca* plants mycorrhizal with axenic cultures of *Glomus caledonium*, Francis (1985) showed increased uptake of the amino acids glutamine, alanine and asparagine by infected plants. Jayaratne (1992) found that *Gigaspora margarita* increased the uptake of nitrogen from glutamine and asparagine by aseptically reared *Plantago lanceolata*, glutamine being as effective a source of nitrogen as nitrate but only in mycorrhizal plants. However, in both of these studies the increase in amino acid uptake was more limited in quantity and chemical forms than has been found in ericoid and ectomycorrhizas (see Bajwa and Read 1986; Abuzinadah and Read 1988), which supports the view that the mobilization of nutrients from organic matter is better developed in the latter associations.

Arbuscular mycorrhizal fungi are known to produce cell wall-degrading enzymes including pectinase, cellulase and hemicellulase (Garcia-Romera et al. 1990) and their growth may be stimulated by the protein-containing fraction of soil extracts (Siqueira and Hubbell 1986). However, further studies of utilization of organic nutrients by axenically reared arbuscular mycorrhizal plants are urgently required to properly establish their potential role in nutrient mobilization. Attention should be focussed on plants and mycorrhizal fungi from soils rich in organic sources of nitrogen and phosphorus, where selection pressures for mobilisation of organic nutrients will be highest. One such endophyte is *Glomus tenue*, which occurs frequently in seminatural grasslands (Rabatin 1979; McGonigle and Fitter 1990; Blaszkowski 1994), and is often present in ecosystems in which mineralization rates are retarded and the main pools of nutrients are present in organic matter, for example in peaty soils (Ali 1969), alpine pastures (Crush 1973; Blaschke 1991), particularly in the nival zone (Read and Haselwandter 1981), wetlands (Tanner and Clayton 1985) and in tropical cloud-forest epiphytes (Rabatin et al. 1993). Of particular note is the reported growth of this endophyte in decomposing organic material such as senescent moss stems (Rabatin 1980).

B. Ectomycorrhizas

Although there have been many studies of mobilisation of nutrients from some organic materials by ectomycorrhizal fungi, Dighton (1991) has noted:

"Most work to date has been carried out under laboratory conditions, often using the fungal symbiont in isolation from its host root. There is a need to demonstrate degradation of organic substrates in the field by mycorrhizas and also to look at the intact mycorrhizal system . . . the intact system needs to be studied to evaluate the relative role of each component and to look at the physiology of the distal end of the hyphae where the relevant interaction between the fungus and substrate occurs."

Two recent studies which satisfy these criteria have reported nutrient depletion from natural substrates (freshly senesced and partially decomposed conifer litter) by intact ectomycorrhizal systems. Bending and Read (1995a) grew *Pinus sylvestris* in the mycorrhizal condition with *Suillus bovinus* or *Thelephora terrestris* in root chambers containing non-sterile peat substrate and determined nutrient release from small patches of pine litter. Once the plants and their mycorrhizas were well established, replicate plastic trays containing weighed amounts of partially decomposed pine litter were introduced. The initial nitrogen and phosphorus content of litter was determined in parallel replicate samples. The mycorrhizal fungi grew intensively into the litter patches, and then gradually senesced after a few months. Control trays of litter were placed in microcosms containing peat but no plants, and incubated in the same conditions as the experimental microcosms. After 120 days, the total nitrogen and phosphorus content of the litter samples were measured and the quantities of nutrients released during incubation determined (Table 1a). There was no significant loss of N or P in the controls compared to their initial concentration in litter ($p > 0.05$); however, there was a significant depletion of N in litter colonised by *Thelephora*, ($p < 0.05$) and by *Suillus* ($p < 0.01$). The phosphorus content of the litter was also unchanged in the control samples, but was significantly decreased in litter colonised by *Suillus* ($p < 0.01$) but not *Thelephora* ($p > 0.05$). The differing abilities of *Suillus bovinus* and *Thelephora terrestris* to obtain nutrients from litter is consistent with their differing patterns of mycelial growth and enzymatic capabilities. *Suillus* produces particularly intensive mycelial growth on litter whereas the growth of *Thelephora* mycelium was not markedly stimulated by leaf litter (Unestam 1991; Bending and Read 1995a). *Suillus* has much more active proteolytic capabilities than *Thelephora* (J.R. Leake, unpubl.).

The hypothesis that these ectomycorrhizal systems are adapted for selective exploitation of N and P compounds was supported by the fact that with the exception of the plant macronutrient potassium, the other major cations (Ca and Mg) increased in concentration and content in the litter patches, and the C:N ratio of litter was increased significantly ($p < 0.01$) from 44:1 in the control to 52:1 in patches colonised by both ectomycorrhizal fungi.

In a similar study conducted in the field, Entry et al. (1991) reported decomposition and nutrient release from freshly fallen litter of *Pseudosuga menziezii* in mesh bags incubated for a year in soil containing perennial mats of mycorrhizal mycelium in which the fungal tissue comprises 45–55% of the total soil biomass (Fogel and Hunt 1979), and in adjacent soil with smaller, and more typical

Table 1. Nitrogen and phosphorus release

a Release of N and P from *Pinus* litter incubated in non-sterile laboratory microcosms colonised by two ectomycorrhizal fungi for 120 days (Bending and Read 1995a). Significant release of nutrients compared to the initial concentration in litter at the start of the experiment is indicated.

b Release of N and P from *Pseudotsuga menziesii* needles after 1 year in soil containing ectomycorrhizal mat or adjacent non-mat soil. (Entry et al. 1991). The differences between the mean percentage N and P release for the mat and non-mat soils are significant ($P < 0.05$)

Treatment	Nitrogen release from litter		Phosphorus release from litter	
a Microcosm study	μg N/g^{-1} litter	Release (%)	μg P/g^{-1} litter	Release (%)
Control (no mycorrhizal fungus)	661	5 ns	0	0 ns
Thelephora terrestris	1800	13**	22	3 ns
Suillus bovinus	3134	23***	164	23***
b Field study				
Control (no ectomycorrhizal mat)	2334	16	318	19
Colonised by ectomycorrhizal mat	4674	32*	566	33*

ns = $P > 0.05$, * = $P < 0.05$, ** = $P < 0.01$, *** = $P < 0.001$.

quantities of mycelium. Rates of release of N and P from the litter were significantly higher in the mat soils than in the adjacent non-mat soil (Table 1b). The quantities of nutrients released in the microcosm study of Bending and Read (1995a) and in the field study of Entry et al. (1991) are remarkably similar. The most intensive exploitation by rhizomorph-forming ectomycorrhizal fungi caused approximately 25% of the litter nutrient content to be released over a period of 120 days in controlled environment conditions and approximately 33% after 365 days in the field. Since in both cases the litter contained only trace amounts of mineral N, we must assume that it was mobilised from organic compounds. Aguilera et al. (1993) also found evidence of depletion of labile organic N in ectomycorrhizal mycelial mats in soil and Durall et al. (1994) showed in a microcosm study the decomposition of ^{14}C-labelled substrates by mycorrhizal fungi growing in association with Douglas fir. Increased activities of key nutrient-mobilizing enzymes in association with ectomycorrhizal mycelium will discussed in Section IV below.

C. Ericoid Mycorrhizas

Indirect evidence that ericoid mycorrhizas obtain access to nitrogen contained in organic matter was provided by Stribley and Read (1980). They added ^{15}NH$_4$ to heathland soil and measured both ^{15}N enrichment and total N content of mycorrhizal and non-mycorrhizal *Vaccinium* plants after growth in the substrate. Plants colonised by the

mycorrhizal fungus were found to have a much higher total N content than their uncolonised counterparts but a smaller amount of ^{15}N enrichment, suggesting that they were acquiring the element from alternative, presumably organic, sources.

Rates of mineralisation of nitrogen under ericaceous plants like *Calluna* are typically exceptionally low, and when ammonium is released it may be rapidly immobilised (Adams 1986). The efficiency of capture of labile nitrogen, including organic compounds, by ericoid mycorrhizas is most dramatically illustrated by spruce-check phenomenon, in which *Picea sitchensis* planted into heathland goes into growth check and suffers severe nitrogen deficiency in competition with *Calluna* (Taylor and Tabbush 1990). Application of fertiliser or herbicide treatment of *Calluna* cures the problem. The direct utilisation of organic N by ericoid mycorrhizas effectively short-circuits the nitrogen cycle by preventing ammonification of organic N, thereby restricting the supply of nitrogen to potential competitor species whose roots or mycorrhizas are less effective at uptake of these organic sources.

IV. Oxidative, Hydrolytic and Other Nutrient-Mobilising Enzymes Associated with Mycorrhizal Colonisation of Litter

Studies of mycelial mats on the forest floor (Griffiths and Caldwell 1992) and of patches formed in laboratory observation chambers

Table 2. Changes in activities of key nutrient mobilizing enzymes following colonisation by ectomycorrhizal mycelium and in ectomycorrhizal mycelial mats compared to adjacent non-mat soil

A: Activity in litter colonized by *Paxillus involutus* as % of that in non-sterile control litter without active mycorrhiza (Bending and Read 1995b)
B: Activity in soils colonized by ectomycorrhizal mycelial mat as a % of that in adjacent non-mat soil (Griffiths and Caldwell 1992). Data for *G. monticola* are for each of two sites

Enzyme	A: Bending and Read (1995b)	B: Griffiths and Caldwell (1992)			
	Paxillus involutus	*Hysterangium setchellii*	*Hysterangium coriaceum*	*Hysterangium garneri*	*Gautieria monticola*
Acid proteinase	232*	189*	254*	213*	74 ns, 130 ns
Acid phosphatase	124**	262*	216*	358*	218*, 244*
Polyphenol oxidase	300***	ND	ND	ND	ND
Peroxidase	122 ns	2800*	7400*	ND	3300*, 12000*

ND = not determined.
ns = no significant difference to control ($P > 0.05$).
Significant differences to control are indicated with an asterisk: * = $P < 0.05$, ** $P < 0.01$, *** $P < 0.001$.

(Bending and Read 1995b), where ectomycorrhizal mycelium is the dominant component of the microbial biomass, have demonstrated that increases in amounts and activities of key nutrient-mobilising enzymes occur synchronously with intensive colonisation of substrates by ectomycorrhizal fungi (Table 2). In the latter study, the use of non-sterile litter provided strong indirect evidence that such enzymes were derived from the mycorrhizal fungi, as does the fact that many mycorrhizal fungi produce these enzymes in pure culture studies (see Sect. V). The enhanced activities of proteinase and phosphatase activities are especially noteworthy, since they will mobilise nutrients from relatively labile substrates. The higher peroxidase and polyphenol oxidase activities suggest potential to achieve ring cleavage of complex phenolic compounds, and enhance release of peptides and amino acids from protein-polyphenol complexes within mycorrhizal mycelial mats.

Two fundamental questions arise from these studies: To what extent are the mycorrhizal fungi directly involved in organic matter breakdown and nutrient mobilization? and: what are the nutrient sources which are being removed from litter which is intensively colonised by mycelium of ectomycorrhizal fungi? These questions are addressed in the following two sections.

V. Production of Organic Matter-Degrading Enzymes by Mycorrhizal Fungi in Vitro

The failure of arbuscular mycorrhizal fungi to sustain growth in pure cultures and the difficulties in establishing axenic mycorrhizal synthesies with these fungi have, with few exceptions (see Sect. III.A), prevented investigations into the potential production of nutrient-mobilising enzymes by these fungi. However, all of the major hydrolytic enzymes involved in mobilisation of nitrogen and phosphorus from organic compounds by saprophytic microorganisms have also been recorded in ericoid and some ectomycorrhizal fungi (Table 3). The view that these kinds of mycorrhiza are adapted for selective exploitation of organic sources of nitrogen and phosphorus is supported by the frequency with which extracellular acid proteinase, phosphatase, and phytase enzymes

Table 3. Extracellular enzymes of ericoid and ectomycorrhizal fungi which hydrolyse the principle sources of organic nitrogen and phosphorus in litter

Process	Substrate	Enzyme	1*	2*	Reference
Hydrolysis of macromolecular nitrogen	Protein	Proteinase	■	■■■	See Table 4
	Peptides	Peptidase	■	■?□	Bajwa and Read (1985), Abuzinadah and Read (1986a), Ramstedt and Söderhall (1983).
	Chitin	Chitinase	■	■?□	Dighton et al. (1987), Leake and Read (1990a) Mitchell et al. (1992), Kerley (1993), Hodge et al. (1993).
Mineralization of organic phosphorus	Organic P monoesters	Phosphomonoesterase (acid phosphatase)	■	■■■	See Table 4
	Phytin	(phytase)	■	■■?	See Table 4
		(alkaline phosphatase)	■	■?□	Straker and Mitchell (1986), Bae and Barton (1989).
	Organic P diesters Nucleic acid	Phosphodiesterase (Ribonuclease)	?	■?□	Griffiths and Caldwell (1992).
	Nucleic acid	(DNase)	■	■?□	Melin (1925), Mikola (1948), Laiho (1970), Leake and Miles (1996).

Key: 1* Ericoid mycorrhizal fungi, □ = enzyme recorded absent, ■ = enzyme present, ? = not known.
   2* Ectomycorrhizal fungi. □□□ = enzyme recorded absent, ■□□ = enzyme rarely detected, ■■□ = enzyme frequently recorded, ■■■ = enzyme common. ? = not known.

have been recorded in them (Table 4). Of the species examined, 80% have proteolytic capabilitities, and 100% have acid phosphatase activity, excepting a few isolates of *Cenococcum geophillum* which produce alkaline phosphatase instead. Phytase has been found in 19 out of 20 species of ectomycorrhizal fungi. Of the families examined, it is only in the Tricholomataceae that extracellular proteinase activity is not well developed: in over 70% of species of *Laccaria* and *Tricholoma* the ability to use protein as a nitrogen source is poorly developed or absent. However, in other members of the Agaricales like the Amanitaceae and Cortinariaceae, it is very well developed (Table 4). Some of the most proteolytically active species are found in the Boletales, particularly in species of *Suillus*. For example, using a single 4-mm diameter inoculum disc of mycelium of *Suillus variegatus*, Ryan and Alexander (1992) recorded hydrolysis of 32 mg of protein in less than 5 days.

Much more work still needs to be done in this area, since the samples of mycorrhizal fungi that have now been studied comprise less than 2% of the estimated total of more than 5000 species (Molina et al. 1992). Furthermore, the samples may not be particularly representative, since they are necessarily restricted to culturable isolates which have been derived mainly from sporocarps.

One of the most significant developments in studies of enzymatic capabilities of mycorrhizal fungi has been the increased attention given to the study of enzymes that are not directly involved in the release of N and P from organic matter (Table 5), but which may enhance or facilitate the activities of those which are. This second group of enzymes will be important where substrates are rendered recalcitrant due to reaction with phenolic acids, tannins and lignins, or protected from direct attack by physical barriers such as cell walls. Enzymes capable of degrading inhibitors and puncturing these barriers may be essential for effective nutrient mobilisation.

The possession, by ericoid and some ectomycorrhizas, of a suite of oxidative and hydrolytic enzymes can be expected to provide them with the biochemical potential to compete with saprophytic microorganisms for relatively complex nutrient forms such as the tannin-protein complexes typically found in the litter of ectomycorrhizal trees and in ericaceous dwarf shrubs. Ericoid endophytes and a few ectomycorrhizal mat-forming fungi have some ability to use tanned protein as a nitrogen source (Leake and Read 1989; Griffiths and Caldwell 1992; Bending 1994).

Some caution should be employed when making direct comparisons between catalytic capabilities of mycorrhizal and non-mycorrhizal fungi. While in pure culture the mycelium of symbiotic organisms may appear to show lower amounts of production or activity of a given enzyme than their saprotrophic counterpart on a unit weight basis, the relatively large biomass of ectomycorrhizal mycelium and its freedom from limitation imposed by carbon supply may offset such absolute differences. In addition, the morphological plasticity seen in the exploitation of microsites by mycelium is likely to be reflected in localised enzyme expression. Equally, the apparently weak expression of some enzyme systems such as cellulase does not necessarily imply that the enzyme is unimportant for the fungus. The ericoid endophyte *H. ericae* grows only weakly on cellulose as a carbon source, but readily penetrates the cellulose cell wall of the host root and may use the same localised capability to punch holes through cellulose walls in the litter, thus gaining access to complex insoluble nutrients contained within. This may be essential for the utilisation of intact, aseptically grown root material of *Vaccinium macrocarpon* as a sole nitrogen source by *H. ericae* which was demonstrated by Kerley (1993). The full significance of these enzyme systems in other mycorrhizal fungi will only be established after more rigorous investigation.

VI. Nutrient Uptake from Specific Organic Compounds, and Its Transfer to Host Plants

Within organic forest soils there are four major pools of nutrients, in order of increasing availability to plants: humified material, plant litter in varying stages of decomposition, live and dead microbial tissues and mineral nutrients (Fig. 1). Work with ericoid (Stribley and Read 1980) and ecto-mycorrhizas (Lundeberg 1970) provides no evidence of an ability to utilise nitrogen in humified materials. The recalcitrance of this type of substrate is borne out by its persistence and accumulation in forest and heathland soils.

In organic soils, the quantities of nutrients in microbial biomass are one to two orders of magni-

Table 4. The occurrence and activities of extracellular acid phosphatase, acid proteinase and phytase in ectomycorrhizal and ericoid mycorrhizal fungi. Proteinase activity is typically determined by the ability to use protein as the sole nitrogen source, to develop clear zones on protein-agar plates or to liquefy gelatine. Proteolytic capabilities are expressed as mycelium dry weight of fungus when grown on protein as sole N source as a percentage of the dry weight obtained on ammonium at the same N concentration. Acid phosphatase is determined by release of p-nitrophenyl (PNP) from p-nitrophenylphosphate. The activity of these enzymes is expressed as nM PNP released per mg mycelium dry weight per second. Phytase activity is typically determined qualitatively by growth and P uptake by mycelium provided with Ca, Fe or Al salt of phytin, and quantified by release of orthophosphate from these substrates. Phytase activity is expressed as nM P released per mg dry weight of mycelium per second

Order Family Species	Acid proteinase Enzyme production	Mycelium dry weight on protein as % of that on ammonium	Reference	Acid phosphatase Enzyme production	Activity: (nM PNP mg mycelium dry weight^{-1} s^{-1})	Reference	Phytase Enzyme production	Activity: (nM P mg mycelium dry weight^{-1} s^{-1})	Reference
Agaricales									
Tricholomataceae									
Laccaria amethystina	■		x						
Laccaria laccata	□	33	b, m	■	0.148, 0.093	6, 14, 15, 20, 34	■	0.084	6, 14, 15, 34
Laccaria bicolor	□	16	g	■	0.3–1.2	23(29)			
Laccaria proxima	■	19, 47	g; g						
Tricholoma albobrunneum	■		i; r	■		10			
Tricholoma atrosquamosum			x	■		7			
Tricholoma cingulatum	□		x						
Tricholoma flavovirens	□		r	■		7			
Tricholoma imbricatum	■		r	■		7			
Tricholoma matsutake	■		y						
Tricholoma pessundatum	□		r						
Tricholoma resplendens	■		i						
Tricholoma vaccinum	■		i						
Entolomataceae									
Entoloma sericeum	■			■	2.25* 4.30–5.08	1, 2	■	0.0167–0.0233	2, 6, 22
Amanitaceae									
Amanita brunnescens	■		i						
Amanita citrinum	■		i, r(2)						
Amanita flavoconia	■		i						
Amanita muscaria	□	173	b, i, m, q; r	■	0.04–0.175 0.074, 5.6–63.9	11, 20, 16	■		11
Amanita rubescens	■		f, i	■	0.31–0.50	2	■	0.010–0.015	2
Amanita spissa	■		x						
Amanita verna	■		r						

Table 4. *Continued*

Order Family Species	Acid proteinase Enzyme production	Mycelium dry weight on protein as % of that on ammonium	Reference	Acid phosphatase Enzyme production	Activity: (nM PNP mg mycelium dry weight^{-1} s^{-1})	Reference	Phytase Enzyme production	Activity: (nM P mg mycelium dry weight^{-1} s^{-1})	Reference
Cortinariaceae									
Cortinarius allutus	■		i						
Cortinarius amoenolens	■		i						
Cortinarius brunneus	■		i						
Cortinarius mucosus	■		i						
Hebeloma arenosa	■			■	0.0031	24			
Hebeloma crustulinforme	■	102–108	c, f, **g**$_{(2)}$, m, z	■	0.17, 0.161	2, 20	▨	0.0006–0.0008	2
Hebeloma cylindrosporum			i	■	0.0024–0.029	11			
Hebeloma edurum	■			■	2.95	6, 26$_{(81)}$; 8	■	0.031	6
Hebeloma longicaudum	■		i						
Hebeloma mesophaeum	■ ▨		r, x						
Hebeloma pusillum	■			■	2.11*	1			
Hebeloma subsaponaceum	■		x						
Hebeloma testaceum	■		x						
Hymenogastraceae									
Chondrogaster sp.	■		h						
Russulales									
Russulaceae									
Lactarius chelidonius	■		i						
Lactarius cimicarius	□		x						
Lactarius controversus	■		e						
Lactarius deliciosus	■		i$_{(6)}$, x						
Lactarius deterrimus	■		x$_{(2)}$						
Lactarius indigo	■		i						
Lactarius lanceolata	■		x						
Lactarius pubescens	■		i	■	0.008–0.013	11			
Lactarius quietus	□		r						
Lactarius rufus	■	58, 76; 23	**b,g**; **g**, w, x	■	0.022–0.068	11	■		11
Lactarius subdulcis	■		f						
Lactarius subpurpureus	■		i						
Lactarius thyinos	■		i$_{(2)}$						
Lactarius torminosus	■		i$_{(2)}$						
Lactarius sp.	■			■	1.75–2.28	2	■	0.0551–0.0736	2

Taxon		Col 1	Codes	Range 1	Ref 1	Value	Ref 2
Boletales							
Boletaceae							
Boletus edulis	■		i	39.5	16		
Boletus ornatipes							
Suillus bellinii	■□	102		7.73	6, 7	0.011	6
Suillus bovinus	■		b, q, m, v; r	3.86–4.54	6, 7, 10, 31	0.004–0.006	6(2)
Suillus brevipes				2.4–5.4	16		
Suillus caerulescens				49.0	16		
Suillus flavidus	□	107	r(2)	8.41–10.0	5, 6, 8, 12		
Suillus granulatus	■		a, r	0.085	25	0.009–0.056	6(2), 32
Suillus grevillei			p	10.8–19.6	16		25
Suillus lakeii				1.82–4.77			
Suillus luteus	□		p, v; r(2)	0.005–0.027	16, 35(43)	0.004–0.007	6(2), 11, 32
Suillus tomentosus	■	80–96, 60–75	g(2), v	19.4–19.6	7, 31		
Suillus variegatus	■	60	a				
Leccinum scabrum	□		x(2), x(2)				
Xerocomus badius							
Paxillaceae							
Paxillus involutus	■	171, 57–86	b, g(3), m, v	0.01–0.061; 24.5–25.9; 0.031–0.225; 0.113	2(8), 11, 14, 15		2, 11, 14, 15, 25
Hymenogastrales							
Rhizopogonaceae							
Rhizopogon colossus	■			70.8–71.7	19(2)		
Rhizopogon ellenae				2.78–40.8	19(2)		
Rhizopogon luteolus	■		r	91.7–142.2	19(2)		32, 33
Rhizopogon occidentalis		138	a, m	29.2–50.0	19()		
Rhizopogon roseolus				31.9–32.2	19()		
Rhizopogon subcaerulescens				0.0144	20		
Rhizopogon vinicolor				17.5–23.3	19()		
Rhizopogon vulgaris							
Gauteriaceae							
Gauteria monticola	■		h				
Phallales							
Hysterangiaceae							
Hysterangium garneri	■		h				
Hysterangium coricaceum	■		h				
Sclerodermatales							
Sclerodermataceae							
Scleroderma citrinum	■	99	b	1.03–1.31	2	0.0006–0.0206	2
Pisolithus tinctorius	■			2.05–2.95	4, 6(2), 8, 9, 13, 18, 27	0.009–0.042	1, 6(2), 27

Table 4. Continued

Order Family Species	Acid proteinase			Acid phosphatase			Phytase		
	Enzyme production	Mycelium dry weight on protein as % of that on ammonium	Reference	Enzyme production	Activity: (nM PNP mg mycelium dry weight^{-1} s^{-1})	Reference	Enzyme production	Activity: (nM P mg mycelium dry weight^{-1} s^{-1})	Reference
Aphyllophorales									
Corticiaceae									
Thelephora terrestris	■	66–104	g(2), i, m	■	0.051	20			
Tylospora fibrillosa	■ □		v						
Tylospora sp.	■	72–230	x(4), x(5) x(6)						
Piloderma croceum (bicolor)	■	92–108	g(2) i(3)	■	0.045	20			
Unidentified									
Picierhiza bicolorata	■		x						
Picierhiza punctata	□		x(2)						
Unnamed				■	1.172*	25	■	0.0197*	25
Unnamed				■	0.213*	25	■	0.0089*	25
Unnamed	■	62	g(2)						
ASCOMYCETES									
Cenococcum geophilum	□	172	a, f, s, x; r(2), x	■ (grey)	0.47–9.69* 0.028–0.19 0.675–0.947	1(4), 2, 10, 22(3)	■	0.0036–0.0064	2, 22(4), 32
E-STRAIN FUNGI	■	69–140	x(4), x(15)						
Tuberales									
Tuber albidum				■		30			
Tirmania nivea				■		29			
Tirmania pinoyi				■		29			
Terfizia claveryi				■		29			
Terfizia boudien				■		29			
ERICOID ENDOPHYTES									
Hymenoscyphus ericae	■	111	a, d, n, w	■	0.28	36, 37, 38, 41, 42, 43	■		43
Scytalidium vacinii	■		o						
Erica hipidula endophyte	■		p	■	0.002–0.028	44	■		43
Erica mauritanica endophyte	■		p						
Rhodothamnus endophyte	■		n						
Oidiodendron griseum	■		m, o						

Key:
□ Enzyme reported absent or of very low activity; ■ enzyme produced; ■ (grey) enzyme produced, but of low activity; ■ enzyme produced: the fungus may contribute significantly to nutrient mobilization. Where contrasted levels of enzyme activity have been reported for a species these are indicated and the references separated by a semicolon. References in bold typeface indicate the sources of the quantitative data on growth on protein versus ammonium (acid proteinase) and acid phosphatase and phytase activities. * V_{max} values. Numbers in brackets after the letter or number codes for a reference indicate the number of isolates or strains which were studied, where these are more than one.

Acid proteinase

a Abuzinadah (1986)
b Abuzinadah and Read (1986a)
c Abuzinadah and Read (1986b)
d Bajwa et al. (1985)
e Bending (1994)
f El-Badaoui and Botton (1989)
g Finlay et al. (1992)
h Griffiths and Caldwell (1992)
i Hutchison (1990a)
j Laiho (1970)
k Lundeberg G (1970)
l Lyr (1963)
m Leake (1988)
n Leake and Read (1990)
o Leake and Read (1991)
p Leake (unpubl.)
q Maijala et al. (1991)
r Melin (1925)
s Mikola (1948)
t Pachlewski and Chrusciak (1979)
u Ramstedt and Söderhall (1983)†
v Ryan and Alexander (1992)
w Spinner and Haselwandter (1985)
x Taylor A.F.S. (pers. comm.) and in Schulze (1995)
y Terashita et al. (1995)
z Zhu et al. (1990)

Ectomycorrhizal fungi

1 Antibus et al. (1986)
2 Antibus et al. (1992)
3 Bae and Barton (1989)
4 Berjaud and D'Auzac (1986)
7 Calleja and D'Auzac (1983)
9 Cao and Crawford (1993)†
10 Chrusciak (1984)
11 Dighton (1983)
12 Ducamp and Olivier (1989)
13 Gourp and Pargney (1991)
14 Hilger and Krause (1989)
15 Hilger et al. (1986)
16 Ho (198?)†
17 Ho and Tilak (1988)
18 Ho and Trappe (1985)
19 Ho and Trappe (1987)†
20 Ho and Zak (1979)
21 Kieliszewska-Rokicka (1992)
22 Kroehler et al. (1988)
23 Kropp (1990)
24 MacFall et al. (1991)

Acid phosphatase

25 McElhinney and Mitchell (1993)
26 Meysselle et al. (1991)
27 Mousin and Salsac (1986)
28 Mousin et al. (1988)
29 Naama-Al et al. (1993)
30 Pasqualini et al. (1992)
31 Sen (1990)†
32 Theodorou (1968)
33 Theodorou (1971)
34 Thomas (1985)
35 Zhu et al. (1988)†

Ericoid mycorrhizal fungi

36 Leake and Miles (1996)
37 Lemoine (1992)
38 Lemoine et al. (1992)
39 Mitchell and Read (1981)
40 Mitchell and Read (1985)
41 Pearson and Read (1975)
42 Shaw and Read (1989)
43 Straker and Mitchell (1986)
44 Straker (1986)

† Studies in which mycelium was homogenized or extracted in some way.

Only species giving positive results for proteinase have been noted from Hutchison (1990a) because his growth media contained ammonium and malt extract which are likely to repress proteinase production.

Table 5. Extracellular enzymes produced by ericoid and ectomycorrhizal fungi which hydrolyze or oxidize the structural components of litter which contain little or no nitrogen or phosphorus

Process	Substrate	Enzyme	1*	2*	Reference
Cuticle degradation [	Cutin, lipid, waxes	Fatty acid esterase	?	■■□	Hutchison (1990b), Caldwell et al. (1991)
Plant cell cell wall degradation	Pectin	Polygalacturonase	■	■■□	Perotto et al. (1993). Palmer and Hacskalo (1970), Lindeberg and Lindeberg (1977), Giltrap and Lewis (1982), Ramstedt and Söderhall (1983), Pachlewski and Chrusciak (1979)
	Cellulose	Cellulase	■	■■□	Norkrans (1950), Lyr (1963), Palmer and Hacskaylo (1970), Trojanowski et al. (1984), Maijala et al. (1991), Pachlewski and Chrusciak (1979)
	Hemicellulose	Xylanase	■	■?□	Cairney and Burke (1996). Lyr (1963), Palmer and Hacskaylo (1970), Terashita et al. (1995), Cao and Crawford (1993)
Oxidation of phenolic acids, tannins	Monophenols	Tyrosinase	■	■■□	Leake et al. (1989), Marr et al. (1986), Hutchison (1990a)
	Tannin	Peroxidase	■	■■□	Cairney and Burke (1994), Griffiths and Caldwell (1992), Bending (1994)
	Polyphenols	Polyphenol oxidase	■	■■□	Lindeberg (1948), Giltrap (1982), Ramstedt and Söderhall (1983), Leake and Read (1989), Griffiths and Caldwell (1992), Bending (1994)
	Polyphenols	Laccase	□	■■□	Bending (1994), Marr et al. (1986), Hutchison (1990a)
Lignin oxidation [	Lignin	Lignase	■	■■□	Haselwandter et al. (1990), Trojanowski et al. (1984), Griffiths and Caldwell (1992), Bending (1994)

Key: 1* Ericoid mycorrhizal fungi, □ = enzyme recorded absent, ■ = enzyme present, ? = not known.
  2* Ectomycorrhizal fungi, □□□ = enzyme recorded absent, ■□□ = enzyme rarely detected, ■■□ = enzyme frequently recorded, ■■■ = enzyme common; ? = not known.

tude higher than the amounts of mineral N and P (see Williams and Sparling 1984) and much of this nutrient is relatively labile. It has been estimated that 20% of nitrogen and between 7 and 18% of the organic phosphorus in a forest floor is contained in microbial tissues (Bååth and Söderström 1979), but the possibility that this nutrient pool may be exploited by mycorrhizal fungi has received little attention until recently. However, there is now increasing evidence that some mycorrhizal fungi are able to compete with saprophytes not only for the nutrients contained in plant litter but also for the nutrients within the saprophytic organisms themselves.

Chitin, the main structural polymer of insect exoskeletons and cell walls of higher fungi, is readily used as a nitrogen source by some mycorrhizal fungi (see Table 3). The use of some of the individual components of mycelia such as glucosamine, galactosamine, N-acetylglucosamine,

chitin and intact fungal biomass as nutrient sources by mycorrhizal and non-mycorrhizal plants of *Vaccinium macrocarpon* and *Pinus sylvestris* has been demonstrated by Kerley (1993). Amino acids (Bajwa and Read 1986; Abuzinadah and Read 1988), peptides (Bajwa and Read 1985; Abuzinadah and Read 1986a) and proteins (Bajwa et al. 1985; Spinner and Haselwandter 1985; Abuzinadah and Read 1986b) are known to be effective nitrogen sources for ericoid and many ectomycorrhizal fungi, which facilitate passage of nitrogen from these sources to host plants. This work indicates the potential in ericoid, and possibly some ectomycorrhizal fungi, to recycle nitrogen from soil microbial biomass as well as some components of plant litter.

Studies of the use of microbial P as nutrient sources by mycorrhizal fungi are in their infancy, although the major sources of P in fungi and bacteria are normally organic. One of the principal

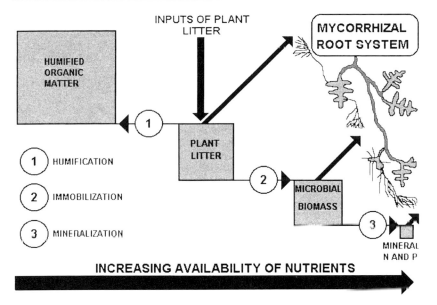

INPUTS OF PLANT LITTER

MYCORRHIZAL ROOT SYSTEM

HUMIFIED ORGANIC MATTER

1 HUMIFICATION

2 IMMOBILIZATION

3 MINERALIZATION

PLANT LITTER

MICROBIAL BIOMASS

MINERAL N AND P

INCREASING AVAILABILITY OF NUTRIENTS

Fig. 1. A model of the fate of organic nitrogen and phosphorus in plant litter, indicating the access to the specific pools: plant litter, microbial biomass and mineral nutrients provided by mycorrhizal roots. Mycorrhizal fungi have the enzymic capabilities to be directly involved in the three major nutrient-transforming processes: humification, immobilisation and mineralisation. Immobilisation is only temporary in mycorrhizal fungi before nutrients are transferred to host plants. Mineralisation is limited, and the pool of mineral nutrients is small, due to direct uptake and transfer of organic nutrients to plants by the mycorrhizas. Humification may be increased by the secretion of oxidase enzymes and selective exploitation and removal of N and P leading to increased C:N and C:P ratios and polymerisation of reactive phenolic and carbohydrate residues

phosphodiesters in soils in nucleic acid. Griffiths and Caldwell (1992) commented that:

"Of the many papers published on phosphatase production by mycorrhizal fungi, virtually all have used phosphomonoesterase substrates. However, most organic phosphorus in living tissues, and thus in fresh detritus, occurs as phosphodiesters. The abundance of phosphomonoesters, relative to phosphodiesters, in soil organic matter suggests that the phosphodiester pool is more important during litter decomposition."

As Jennings (1995) points out, the ease with which phosphomonesterases can be studied, and the fact that they occur frequently in fungi have resulted in other kinds of organic P-degrading enzymes being neglected. Soil analyses have revealed that phytates typically comprise the main identifiable fraction of organic P, often accounting for up to 50% of the total. Possible utilisation of phosphomonoester phytate-P by plant roots and mycorrhizas has attracted much attention, even though accumulation of these compounds in soil is a clear testimony to their recalcitrance in many environments.

Activity of the extracellular phosphodi-esterase, ribonuclease, has now been reported in some ectomycorrhizal fungi (Griffiths and Caldwell 1992). These studies have been extended with evidence that *Hymenoscyphus ericae* (Leake and Miles 1996) and *Suillus bovinus* (M.D. Myers, pers. comm.) can use DNA as a sole source of phosphorus, demonstrating their abilities to produce phosphodiesterase. The transfer nutrients from DNA to a host plant by *H. ericae* was confirmed by growing non-mycorrhizal and mycorrhizal *Vaccinium macrocarpon* axenically on nutrient agar containing orthophosphate or nuclei (nucleic acids) as sole phosphorus sources, and in a control treatment with no P added (Myers and Leake 1996). After 20 weeks, the plants were harvested. Non-mycorrhizal plants grown with nuclei had no greater dry weight or phosphorus content than uninfected plants grown without any phosphorus, and showed symptoms of severe phosphorus deficiency. In contrast, mycorrhizal plants grew as well on nuclei as on orthophosphate (Fig. 2) and obtained almost the same amount of P from the two sources.

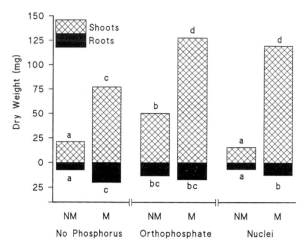

Fig. 2. Mean dry weight (mg) of roots and shoots of non-mycorrhizal (*NM*) and mycorrhizal (*M*) plants of *Vaccinium macrocarpon* after growth for 28 weeks on nutrient agar containing no added phosphorus or with phosphorus (0.2 mM) added as orthophosphate or nuclei. The means are back-transformed from log e values. Differences in shoot yields and also root yields, were determined separately using LSD on transformed data. For comparisons between shoot yields or root yields, means sharing the same letter are not significantly different ($P > 0.05$)

VII. Conclusions

The roots and mycelia of most mycorrhizal plants are ideally placed both spatially and temporally to play a direct role in recycling of organic nutrients. Colonisation of decomposing litter by these fungi is associated with increases in the activities of key nutrient-mobilising enzymes and with the mobilisation and export of nutrients from the litter. Studies of ericoid and ectomycorrhizal fungi in pure cultures have confirmed their abilities to degrade and obtain nutrients from the main N- and P-containing organic compounds in plant and microbial litter, and have revealed that these fungi produce a suite of nutrient-mobilising enzymes prominent amongst which are phosphatases and proteinases. When grown in association with host plants in axenic conditions, nutrient mobilisation by some mycorrhizal fungi from amino acids, peptides, proteins, amino sugars, chitin and nucleic acids has been demonstrated together with transfer of N and P into host plants. In contrast, non-mycorrhizal plants are unable to use the majority of these substrates as nutrient sources. The nutrient-mobilising potentials of mycorrhizal systems has thus clearly been established, but the extent to which these processes occur in nature is still unclear, since the studies to date have included only a rather restricted range of mycorrhizal fungi. The need to extend these studies to a wider range of representative species is currently being addressed.

Acknowledgments. We would like to record our grateful thanks to Dr. G. Bending, Dr. S. Kerley and Dr. A. Taylor for the use of their unpublished work in this chapter.

References

Abuarghub SM, Read DJ (1988) The biology of mycorrhiza in the Ericaceae XII. Quantitative analysis of individual "free" amoni acids in relation to time and depth in the soil profile. New Phytol 108:433–441

Abuzinadah RA (1986) The utilization of organic nitrogen sources by ectomycorrhizal fungi and their host plants. PhD Thesis, University of Sheffield, UK

Abuzinadah RA, Read DJ (1986a) The role of proteins in the nitrogen nutrition of ectomycorrhizal plants. I. Utilization of peptides and proteins by ectomycorrhizal fungi. New Phytol 103:481–493

Abuzinadah RA, Read DJ (1986b) The role of proteins in the nitrogen nutrition of ectomycorrhizal plants. III. Protein utilization by *Betula*, *Picea*, and *Pinus* in mycorrhizal association with *Hebeloma crustuliniforme*. New Phytol 103:507–514

Abuzinadah RA, Read DJ (1988) Amino acids as nitrogen sources for ectomycorrhizal fungi: Utilization of individual amino acids. Trans Br Mycol Soc 91:473–479

Adams JA (1986) Nitrification and ammonification in acid forest litter and humus as affected by peptone and ammonium N amendment. Soil Biol Biochem 18:45–51

Aguilera LM, Griffiths RP, Caldwell BA (1993) Nitrogen in ectomycorrhizal mat and non-mat soils of different-age Douglas-fir forests. Soil Biol Biochem 25:1015–1019

Alexander I (1989) Mycorrhizas in tropical forests. In: Proctor J (ed) Mineral nutrients in tropical forest and savanna ecosystems. Blackwell, Oxford, pp 169–188

Ali B (1969) Occurrence and characteristics of the vesicular-arbuscular endophyte of *Nardus stricta*. Nova Hedwigia 18:409–425

Ames RN, Reid CPP, Poryer LK, Cambardella C (1983) Hyphal uptake and transport of nitrogen from two 15N-labelled sources by *Glomus mosseae*, a vesicular-arbuscular mycorrhizal fungus. New Phytol 95:381–396

Antibus RK, Kroehler CJ, Linkins AE (1986) The effects of external pH, temperature, and substrate concentration on acid phosphatase activity of ectomycorrhizal fungi. Can J Bot 64:2383–2387

Antibus RK, Sinsabaugh RL, Linkins AE (1992) Phosphatase activities and phosphorus uptake from insositol phosphate by ectomycorrhizal fungi. Can J Bot 70:794–801

Bååth E, Söderström B (1979) Fungal biomass and fungal immobilization of plant nutrients in Swedish coniferous forest soils. Rev Ecol Biol Sols 16:477–489

Bae K, Barton LL (1989) Alkaline phosphatase and other hydrolases produced by *Cenococcum graniforme*, an ectomycorrhizal fungus. Appl Environ Microbiol 55: 2511–2516

Bajwa R, Read DJ (1985) The biology of mycorrhiza in the Ericaceae IX. Peptides as nitrogen sources for the ericoid endophyte and for mycorrhizal and non-mycorrhizal plants. New Phytol 101:459–467

Bajwa R, Read DJ (1986) Utilization of mineral and amino N sources by the ericoid endophyte *Hymenoscyphus ericae* and by mycorrhizal and non-mycorrhizal seedlings of *Vaccinium*. Trans Br Mycol Soc 87:269–277

Bajwa R, Abuarghub S, Read DJ (1985) The biology of Mycorrhiza in the Ericaceae. X. The utilization of proteins and the production of proteolytic enzymes by the mycorrhizal endophyte and by mycorrhizal plants. New Phytol 101:469–486

Bending GD (1994) The mobilization of organic nitrogen by mycorrhizal fungi with special reference to natural substrates. PhD Thesis, University of Sheffield, UK

Bending GD, Read DJ (1995a) The structure and function of the vegetative mycelium of ectomycorrhizal plants V. Foraging behaviour and translocation of nutrients from exploited litter. New Phytol 130:401–409

Bending GD, Read DJ (1995b) The structure and function of the vegetative mycelium of ectomycorrhizal plants VI. Activities of nutient mobilizing enzymes in birch litter colonized by *Paxillus involutus* (Fr.) Fr. New Phytol 130:411–417

Berjaud C, D'Auzac J (1986) Isolement et caractérisation des phosphatases d'un champignon ectomycorhizogène typique: *Pisolithus tinctorius*. Effects de la carence en phosphate. Physiol Veg 24:163–172

Blaschke H (1991) Distribution, mycorrhizal infection, and structure of roots of calcicole floral elements at treeline, Bavarian Alps, Germany. Arct Alp Res 23:444–450

Blaszkowski J (1994) Arbuscular fungi and mycorrhizae (Glomales) of the Hel Peninsula, Poland. Mycorrhiza 5:71–88

Boggie R, Hunter FR, Knight AH (1958) Studies on the root development of plants in the field using radioactive tracers. J Ecol 46:621–639

Bolan NS (1991) A critical review of the role of mycorrhizal fungi in the uptake of phosphorus by plants. Plant Soil 134:189–207

Bousquet N, Mousain D, Salsac L (1986a) Use of phytate by ectomycorrhizal fungi. In: Gianinazzi-Pearson V, Gianinazzi S (eds) Mycorrhizae: physiology and genetics. INRA, Paris, pp 363–368

Bousquet N, Mousain D, Salsac L (1986b) Influence de l'orthophosphate sur les activities phosphates de *Suillus granulatus* en culture in vitro. Physiol Veg 24:153–162

Brundrett MC, Kendrick B (1988) The mycorrhizal status, root anatomy, and phenology of plants in a sugar maple forest. Can J Bot 66:1153–1173

Brundrett M, Kendrick B (1990) The roots and mycorrhizas of herbaceous woodland plants I. Quantitative aspects of morphology. New Phytol 114:457–468

Cairney JWG, Burke RM (1994) Fungal enzymes degrading plant cell walls: their possible significance in the ectomycorrhizal symbiosis. Mycol Res 98:1345–1356

Cairney JWG, Burke RM (1996) Plant cell wall-degrading enzymes in ericoid and ectomycorrhizal fungi. In: Azcon-Aguilar C, Barea JM (eds) Mycorrhizas in integrated systems from genes to plant development. European Commission Rep EUR 16728 EN, pp 218–221

Caldwell BA, Castellano MA, Griffiths RP (1991) Fatty acid esterase production by ectomycorrhizal fungi. Mycologia 83:233–236

Calleja M, D'Auzac J (1983) Activités phosphatases et carance phosphatée chez des champignons supérieurs. Can J Bot 61:79–86

Calleja M, Mousain D, Lecouvreur B, D'Auzac J (1980) Influence de la carence phosphatée sur les activités phosphatases acids de trois champignons mycorrhiziens: *Hebeloma edurum* Metrod, *Suillus granulatus* (Fr. ex. L) O. Küntze et *Pisolithus tinctorius* (Pers) Coker et Couch. Physiol Veg 18:489–504

Cao W, Crawford DL (1993) Carbon nutrition and hydrolytic and cellulolytic activities in the ectomycorrhizal fungus *Pisolithus tinctorius*. Can J Microbiol 39:529–535

Carleton TJ, Read DJ (1991) Ectomycorrhizas and nutrient transfer in conifer-feather moss ecosystems. Can J Bot 69:778–785

Chrusciak E (1984) Effect of phosphorus and calcium on the formation of mycorrhiza of *Pinus sylvestris* in aseptic culture. Acta Mycol 20:213–218

Cromack K, Sollins P, Graustein WC, Speidel K, Todd AW, Spycher G, Li CY, Todd RL (1979) Calcium oxalate accumulation and soil weathering in mats of the hypogeous fungus *Hysterangium crassum*. Soil Biol Biochem 11:463–468

Crush JR (1973) Significance of endomycorrhizas in tussock grassland in Otago, New Zealand. N Z J Bot 11:645–660

Dighton J (1983) Phosphatase production by mycorrhizal fungi. Plant Soil 71:455–462

Dighton J (1991) Acquisition of nutrients from organic sources by mycorrhizal autotrophic plants. Experientia 47:362–369

Dighton J, Thomas ED, Latter PM (1987) Interactions between tree roots, mycorrhizas, a saprotrophic fungus and the decomposition of organic substrates in a microcosm. Biol Fertil Soils 4:145–150

Dodd JC, Burton CC, Burns RG, Jeffries P (1987) Phosphatase activity associated with the roots and the rhizosphere of plants infected with vesicular-arbuscular mycorrhizal fungi. New Phytol 107:163–172

Ducamp M, Olivier JM (1989) Comparison des activités phosphatases acides d'hétérocayons et d'homocayons de *Suillus granulatus*. Agronomie (Paris) 9:295–304

Durall DM, Todd AW, Trappe JM (1994) Decomposition of ^{14}C-labelled substrates by ectomycorrhizal fungi in association with Douglas fir. New Phytol 127:725–729

El-Badaoui K, Botton B (1989) Production and characterization of exocellular proteases in ectomycorrhizal fungi. Ann Sci For 46[Suppl]:728s–730s

Entry J, Rose CL, Cromack K (1991) Litter decomposition and nutrient release in ectomycorrhizal mat soils of a Douglas-fir ecosystems. Soil Biol Biochem 23:285–290

Finlay RD, Read DJ (1986) The structure and function of the vegetative mycelium of ectomycorrhizal plants. II The uptake and distribution of phosphorus by mycelial strands interconnecting host plants. New Phytol 103:157–165

Finlay RD, Frosegård Å, Sonnerfeldt AM (1992) Utilization of organic and inorganic nitrogen sources by ectomycorrhizal fungi in pure culture and in symbiosis with *Pinus contorta* Dougl. ex Loud. New Phytol 120:105–115

Fitter AH (1986) Spatial and temporal patterns of root activity in a species-rich alluvial grassland. Oecologia 69:594–599

Fogel R, Hunt G (1979) Fungal and arboreal biomass in a western Oregon Douglas-fir ecosystem: distribution patterns and turnover. Can J For res 9:245–256

Francis R (1985) The biology of vesicular-arbuscular mycorrhizas with special reference to their role in nutrient transfer between plants. PhD Thesis, University of Sheffield, Sheffield, UK

Frank AB (1894) Die Bedeutung der Mykorrhizapilze für die gemeine Kiefer. Forstwissenschaftl Centralbl 16:1852–1890

Garcia-Garrido JM, Garcia-Romera I, Ocampo JA (1992) Cellulase production by the vesicular-arbuscular mycorrhizal fungus Glomus mosseae (Nichol. and Gerd.) Gerd. and Trappe. New Phytol 121:221–226

Garcia-Romera I, Garcia-Garrido JM, Martinez-Molina E, Ocampo JA (1990) Possible influence of hydrolytic enzymes on vesicular arbuscular mycorrhizal infection of alfalfa. Soil Biol Biochem 22:149–152

Garcia-Romera I, Garcia-Garrido JM, and Ocampo JA (1991) Pectolytic enzymes in the vesicular-arbuscular fungus Glomus mosseae. FEMS Microbiol Letts 78:343–346

Giltrap NJ (1982) Production of polyphenol oxidases by ectomycorrhizal fungi with special reference to Lactarius spp. Trans Br Mycol Soc 78:75–81

Giltrap NJ, Lewis DH (1982) Catabolite repression of the synthesis of pectin-degrading enzymes by Suillus luteus (L. ex Fr.) S.F. Gray and Hebeloma oculatum Bruchet. New Phytol 90:485–497

Gimingham CH (1972) Ecology of heathlands. Chapman and Hall, London

Gourp P, Pargney JC (1991) Acid phosphatases immunocytolocalization of Pisolithus tinctorius L. during its confrontation with Pinus sylvestris (Pers.) Desv. root system. Cryptogamie Mycol 12:293–304

Griffiths RP, Caldwell BA (1992) Mycorrhizal mat communities in forest soils. In: Read DJ, Lewis DH, Fitter AH, Alexaner IJ (eds) Mycorrhizas in ecosystems. CAB International, Wallingford, pp 98–105

Hadley G (1969) Cellulose as a carbon source for orchid mycorrhiza. New Phytol 68:933–939

Harvey AE, Larsen MJ, Jurgensen MF (1976) Distribution of ectomycorrhizae in a mature Douglas-fir/larch forest in western Montana. Forest Sci 22:393–398

Haselwandter K, Bobleter O, Read DJ (1990) Degradation of ^{14}C-labelled lignin and dehydropolymer of coniferyl alcohol by ericoid and ectomycorrhizal fungi. Arch Microbiol 153:352–354

Hayman DS, Mosse B (1972) Plant growth responses to vesicular-arbuscular mycorrhiza III. Increased uptake of labile P from soil. New Phytol 71:41–47

Hepper CM, Warner A (1983) Role of organic matter in growth of a vesicular-arbuscular mycorrhizal fungus in soil. Trans Br Mycol Soc 81:155–156

Hilger AB, Krause HH (1989) Growth characteristics of Laccaria laccata and Paxillus involutus in liquid culture media with inorganic and organic phosphorus sources. Can J Bot 67:1782–1789

Hilger AB, Thomas KI, Krause HH (1986) The effects of several buffers on growth and phosphorus nutrition of selected ectomycorrhizal fungi. Soil Biol Biochem 18:61–68

Ho I (1989) Acid phosphatase, alkaline phosphatase and nitrate reductase activity of selected ectomycorrhizal fungi. Can J Bot 67:750–753

Ho I, Tilak KVBR (1988) A simple method for assessing acid phosphatase activity of ectomycorrhizal fungi. Trans Br Mycol Soc 91:346–347

Ho I, Trappe JM (1985) Phosphatase and nitrate reductase activity of Pisolithus tinctorius: intraspecific variation and ecological inferences. In: Molina R (ed) Proc 6th North American Conf on Mycorrhizae, 353pp

Ho I, Trappe JM (1987) Enzymes and growth substances of Rhizopogon sp. in relation to mycorrhizal hosts and infrageneric taxonomy. Mycologia 79:553–558

Ho I, Zak B (1979) Acid phosphatase activity of six ectomycorrhizal fungi. Can J Bot 57:1203–1205

Hodge A, Alexander IJ, Gooday GW (1993) Chitinolytic enzymes of pathogenic and ectomycorrhizal fungi. In: Peterson L, Schelkle M (eds) Abstracts of the 9th North American conference on mycorrhizae, Guelph, Ontario, Canada, 45pp

Hutchison L (1990a) Studies on the systematics of ectomycorrhizal fungi in axenic culture. II. The enzymatic degradation of selected carbon and nitrogen compounds. Can J Bot 68:1522–1530

Hutchison L (1990b) Studies on the systematics of ectomycorrhizal fungi in axenic culture. III. Patterns of polyphenol oxidase activity. Mycologia 82:424–435

Jayachandran K, Schwab AP, Hetrick BAD (1992) Mineralization of organic phosphorus by vesicular-arbuscular mycorrhizal fungi. Soil Biol Biochem 24:897–903

Jayaratne AHR (1992) Some aspects of the biology of vesicular-arbuscular mycorrhiza with special reference to the external mycelial phase. PhD Thesis, University of Sheffield, UK

Jennings DH (1995) The physiology of fungal nutrition. Cambridge University Press, Cambridge

Kerley SJ (1993) The role of mycorrhizal fungi in the mobilization of nitrogen from organic compounds with special reference to aseptically produced natural substrates. PhD Thesis, University of Sheffield, UK

Kieliszewska-Rokicka B (1992) effect of nitrogen level on acid phosphatase activity of eight isolates of ectomycorrhizal fungus Paxillus involutus cultured in vitro. Plant Soil 139:229–238

Kroehler CJ, Antibus RK, Linkins AE (1988) The effects of organic and inorganic phosphorus concentration on the acid phosphatase activity of ectomycorrhizal fungi. Can J Bot 66:750–756

Kropp BR (1990) Variation in acid phosphatase activity among progeny from controlled crosses in the ectomycorrhizal fungus Laccaria bicolor. Can J Bot 68:864–866

Laiho O (1970) Paxillus involutus as a mycorrhizal symbiont of forest trees. Acta Forest Fenn 106:1–72

Leake JR (1988) Causes and effects of soil acidification by Calluna vulgaris (L.) Hull, with special reference to the role of mycorrhizas. PhD Thesis, University of Sheffield, UK

Leake JR, Miles W (1996) Phosphodiesters as mycorrhizal P sources I. Phosphodiesterase production and the utilization of DNA as a phosphorus source by the ericoid mycorrhizal fungus Hymenoscyphus ericae. New Phytol 132:435–443

Leake JR, Read DJ (1989) The effect of phenolic compounds on nitrogen mobilization by ericoid mycorrhizal systems. Agric Ecosyst Environ 29:225–236

Leake JR, Read DJ (1990a) Chitin as a nitrogen source for mycorrhizal fungi. Mycol Res 94:993–1008

Leake JR, Read DJ (1990b) Proteinase activity in mycorrhizal fungi I. The effect of extracellular pH on the production and activity of proteinase by ericoid endophytes from soils of contrasted pH. New Phytol 115:243–250

Leake JR, Read DJ (1991) Experiments with ericoid mycorrhiza. In: Norris JR, Read DJ, Varma AK (eds) Methods in microbiology, vol 23. Academic Press, London, pp 435–459

Leake JR, Shaw G, Read DJ (1989) The role of ericoid mycorrhizas in the ecology of ericaceous plants. Agric Ecosyst Environ 29:237–250

Lemoine MC (1992) Etudes des interactions cellulaires entre symbiontes endomycorrhiziens chez les ericacees. Role des phosphatases acides fongiques: approaches physiologiques, biochemiques, immunologiques et cytologiques. Doctoral thesis, University of Nancy, France

Lemoine MC, Gianinazzi-Pearson V, Gianinazzi S, Straker CJ (1992) Occurence and expression of acid phosphatase of Hymenoscyphus ericae (Read) Korf and Kernan, in isolation or associated with plant roots. Mycorrhiza 1:137–146

Lindeberg G (1948) On the occurrence of polyphenoloxidase in soil-inhabiting Basidiomycetes. Physiol Plant 1:196–205

Lindeberg G, Lindeberg M (1977) Pectinolytic ability of some mycorrhizal and saprophytic Hymenomycetes. Arch Mikrobiol 43:438–447

Lundeberg G (1970) Utilization of various nitrogen sources, in particular bound soil nitrogen, by mycorrhizal fungi. Stud Forest Suec 79:1–95

Lyr H (1963) Zur Frage des Streuabbaus durch ectotrophi Mykorrhizapilze. In: Rawald W, Lyr H (eds) Mykorrhiza. International Mykorrhiza Symposium, Wiemar, pp 123–145

MacDonald RM, Lewis M (1978) The occurrence of some acid phosphatases and dehydrogenases in the vesicular-arbuscular mycorrhizal fungus Glomus mosseae. New Phytol 80:135–141

MacFall J, Slack SA, Iyer J (1991) Effects of Hebeloma arenosa and phosphorus fertility on root acid phosphatase activity of red pine (Pinus resinosa) seedlings. Can J Bot 69:380–383

Maijala P, Fagerstedt KV, Raudaskoski M (1991) Detection of extracellular cellulolytic and proteolytic activity in ectomycorrhizal fungi and Heterobasidium annosum (Fr.) Bref. New Phytol 117:643–648

Mamolos AP, Elissou GK, Veresoglou DS (1995) Depth of root activity of coexisting grassland species in relation to N and P additions, measured using non-radioactive tracers. J Ecol 83:643–652

Marr CD, Grund DW, Harrison KA (1986) The taxonomic potential of laccase and tyrosinase spot tests. Mycologia 78:169–184

McElhinney C, Mitchell DT (1993) Phosphatase activity of four ectomycorrhizal fungi found in a Sitka spruce-Japanese larch plantation in Ireland. Mycol Res 97:725–732

McGonigle TP, Fitter AH (1990) Ecological specificity of vesicular-arbuscular mycorrhizal associations. Mycol Res 94:120–122

Melin E (1925) Untersunchungen über die Bedeutung der Bedeutung der Baummykorrhiza. Eine ökologisch physiologische Studie. Fischer, Jena, 151pp

Meyer FH (1973) Distribution of ectomycorrhizae in native and man-made forests. In: Marks GC, Kozlowski TT (eds) Ectomycorrhiae: their ecology and physiology. Academic Press, New York, pp 79–105

Meyseselle JP, Gay G, Debaud JC (1991) Intraspecific genetic variation of acid phosphatase activity in monokaryotic and dikaryotic populations of the ectomycorrhizal fungus Hebeloma cylindrosporum. Can J Bot 69:808–813

Mikola P (1948) On the physiology and ecology of Cenococcum graniforme especially as a mycorrhizal fungus of birch. Commun Inst Forest Fenn 36:3

Mikola P, Laiho O (1962) Mycorrhizal relations in the raw humus layer of northern spruce forests. Commun Inst Forest Fenn 55:1–13

Mitchell DT, Read DJ (1981) Utilization of inorganic and organic phosphates by the mycorrhizal endophytes of Vaccinium macrocarpon and Rhododendron ponticum. Trans Br Mycol Soc 76:255–260

Mitchell DT, Read DJ (1985) Growth of African and European ericoid mycorrhizal endophytes on a range of substrates. Trans Br Mycol Soc 84:355–357

Mitchell DT, Sweeney M, Kennedy A (1992) Chitin degradation by Hymenoscyphus ericae and the influence of H. ericae on the growth of ectomycorrhizal fungi. In: Read DJ, Lewis DH, Fitter AH, Alexander IJ (eds) Mycorrhizas in ecosystems. CAB International, Wallingford, UK pp 246–251

Mohandas S (1992) Effect of VAM inoculations on plant growth, nutrient level and root phosphatase activity in papaya (Carica papaya cv. Coorg Honey Dew). Fertil Res 31:263–267

Molina R, Massicotte H, Trappe JM (1992) Specificity phenomena in mycorrhizal symbiosis: community ecological consequences and practical applications. In: Allen MF (ed) Mycorrhizal functioning. Champman and Hall, New York, pp 357–423

Mosse B (1959) Observations on the extra-matrical mycelium of a vesicular-arbuscular endophyte. Trans. Br Mycol Soc 42:439–448

Mosse B, Hayman DS, Arnold DJ (1973) Plant growth responses to vesicular-arbuscular mycorrhiza V. Phosphate uptake by three plant species from P-deficient soils labelled with ^{32}P. New Phytol 72:809–815

Mousin D, Salsac L (1986) Utilization du phytate et activités phosphatases acides chez Pisolithus tinctorius, basidiomycète mycorhizien. Physiol Veg 24:193–200

Mousin D, Bousquet N, Polard C (1988) Comparison des activités phosphatases d'homobasidiomycetes ectomycorrhiziens en culture in vitro. Eur J For Pathol 18:299–309

Myers MD, Leake JR (1996) Phosphodiesters as mycorrhizal P sources II. Ericoid mycorrhiza and the utilization of nuclei as a phosphorus source by Vaccinium macrocarpon. New Phytol 132:445–451

Naama-Al MM, Ewaze JO, Nema JH (1993) Acid phosphatase activity of four mycorrhizal fungi. Cryptogamic Bot 4:19–22

Nicolson TH (1959) Mycorrhiza in the Gramineae I. Vesicular-arbuscular endophytes, with special reference to the external phase. Trans Br Mycol Soc 42:421–438

Norkrans B (1950) Studies in growth and cellulytic enzymes of Tricholoma. Symb Botan Upsal 11:1–126

Ogawa M (1977) Ecology of higher fungi in Tseuga diversifolia and Betula ermani-Abies mariesii forests of the subalpine zone. Trans Mycol Soc Jpn 18:1–19

Pachlewski R, Chrusciak E (1979) Aktywnosc enzymatyczna grzybow mikoryzowych. Acta Mycol 15:3–9

Pacovsky RS, Silva PD, Carvalho MT, Tsai SM (1991) Growth and nutrient allocation in Phaseolus vulgaris L. colonized with endomycorrhizae or rhizobium. Plant Soil 132:127–138

Palmer JG, Hacskaylo E (1970) Ectomycorrhizal fungi in pure culture I. Growth on single carbon sources. Physiol Pl 23:1187–1197

Pasqualini S, Panara F, Antonielli M (1992) Acid phosphatase activity in *Pinus pinea-Tuber albidum* ectomycorrhizal association. Can J Bot 70:1377–1383

Paul EA, Clark FE (1989) Soil biology and biochemistry. Academic Press, New York

Pearson V, Read DJ (1975) The physiology of the mycorrhizal endophyte of *Calluna vulgaris*. Trans Br Mycol Soc 64:1–7

Perotto R, Bettini V, Bonfante P (1993) Production of polygalacturanases by ericoid fungi. Poster abstract, 9th North American conference on mycorrhizae. Guelph, Ontario, Canada, 154 pp

Persson H (1978) Root dynamics in a young Scots pine stand in central Sweden. Oikos 30:508–519

Persson HA (1983) The distribution and productivity of fine roots in boreal forests. Plant Soil 71:87–101

Ponge JF (1990) Ecological study of a forest humus by observing a small volume I. Penetration of pine litter by mycorrhizal fungi. Eur J For Pathol 20:290–303

Powell CLL (1975) Plant growth responses to vesicular-arbuscular mycorrhiza VIII. Uptake of P by onion and clover infected with different *Endogone* spore types in ^{32}P labelled soils. New Phytol 75:563–566

Rabatin SC (1979) Seasonal and edaphic variation in vesicular-arbuscular mycorrhizal infection of grasses by *Glomus tenuis*. New Phytol 83:95–102

Rabatin SC (1980) The occurrence of the vesicular-arbuscular mycorrhizal fungus *Glomus tenuis* with moss. Mycologia 72:191–195

Rabatin SC, Stinner BR, Paoletti MG (1993) Vesicular-arbuscular mycorrhizal fungi, particularly *Glomus tenue*, in Venezuelan bromeliad epiphytes. Mycorrhiza 4:17–20

Ramstedt M, Söderhall K (1983) Proteinase, phenoloxidase and pectinase activities in mycorrhiza fungi. Trans. Br Mycol Soc 81:157–161

Rayner ADM, Boddy L (1988) Fungal decomposition of wood: its biology and ecology. Wiley International, Chichester

Read DJ (1991) Mycorrhizas in ecosystems. Experientia 47:376–391

Read DJ, Haselwandter K (1981) Observations on the mycorrhizal status of some alpine plant communities. New Phytol 88:341–352

Reiners WA (1965) Ecology of a heath-shrub synusia in the pine barrens of Long Island, New York. Bull Torrey Bot Club 92:448–464

Richter DL, Bruhn JN (1989) *Pinus resinosa* ectomycorrhizae: seven host-fungus combinations synthesized in pure culture. Symbiosis 7:211–228

Ryan EA, Alexander IJ (1992) Mycorrhizal aspects of improved growth of spruce when grown in mixed stands on heathlands. In: Read DJ, Lewis DH, Fitter AH, Alexander IJ (eds) Mycorrhizas in ecosystems. CAB International, Wallingford, pp 237–245

Sanders FE, Tinker PB (1971) Mechanism of absorption of phosphate from soil by *Endogone* mycorrhizas. Nature 233:278–279

Schulze ED (1995) NIPHYS – nitrogen physiology of forest plants and soils. EEC contract No EV5V-CT92-0143. Final Rep, Bayreuth, pp 25–34

Sen R (1990) Intraspecific variation in two species of *Suillus* from Scots pine (*Pinus sylvestris* L.) forests based on somatic incompatibility and isozyme analyses. New Phytol 114:617–626

Shaw G, Read DJ (1989) The biology of mycorrhiza in the Ericaceae. XIV. Effects of iron and aluminium on the activity of acid phosphatase in the ericoid endophyte *Hymenoscyphus ericae* (Read) Korf and Kernan. New Phytol 113:529–533

Singer R (1984) The role of fungi in Amazonian forests and in reforestation. In: Sioli H (ed) The Amazon. Limnology and landscape ecology of a mighty tropical river and its basin. Junk, Dortrecht, pp 603–614

Singer R, Araújo IJS (1979) Litter decomposition and ectomycorrhiza in Amazonian forests. I. A comparison of litter decomposition and ectomycorrhizal basidiomycetes in latosol-terra firme rain forest and white podzol campinarana. Acta Amazon 9:25–41

Siqueira JO, Hubbell DH (1986) Effect of organic substrates on germination and germ tube growth of vesicular arbuscular mycorrhizal fungus spores in vitro. Pesquisa Agrop Brasil 21:523–528

Soma K, Saito T (1979) Ecological studies of soil organisms with reference to the decomposition of pine needles I. Soil macrofaunal and mycofloral surveys in coastal pine plantations. Rev Ecol Biol Sol 16:337–354

Spinner S, Haselwandter K (1985) Proteins as nitrogen sources for *Hymenoscyphus* (= *Pezizella*) *ericae*. In: Molina R (ed) Proceedings of the 6th North American conference on mycorrhizae. Forest Research Laboratory, Oregon State University, Corvallis, Oregon, 422pp

St. John TV, Coleman DC, Reid CPP (1983a) Growth and spatial distibution of nutrient-absorbing organs; selective exploitation of soil heterogeneity. Plant Soil 71:487–493

St. John TV, Coleman DC, Reid CPP (1983b) Association of vesicular-arbuscular mycorrhizal hyphae with soil organic particles. Ecology 64:957–959

Straker CJ (1986) Aspects of phosphorus nutrition in endomycorrhizal fungi of the Ericaceae. PhD Thesis, University of Cape Town, South Africa

Straker CJ, Mitchell DT (1986) The activity and characterization of acid phosphatases in endomycorrhizal fungi of the Ericaceae. New Phytol 104:243–256

Stribley DP, Read DJ (1980) The biology of mycorrhiza in the Ericaceae. VIII. The relationship between mycorrhizal infection and the capacity to utilize simple and complex organic nitrogen sources. New Phytol 86:365–371

Tanner CC, Clayton JS (1985) Vesicular arbuscular mycorrhiza studies with a submerged aquatic plant. Trans Br Mycol Soc 85:683–688

Tarafdar JC, Marschner H (1994) Phosphatase activity in the rhizosphere and hyphosphere of VA mycorrhizal wheat supplied with inorganic and organic phosphorus. Soil Biol Biochem 26:387–395

Taylor CMA, Tabbush PM (1990) Nitrogen deficiency in Sitka spruce plantations. HMSO, London (Forestry Commission Bulletin 89)

Terashita T, Kono M, Yoshikawa K, Shishiyama J (1995) Productivity of hydrolytic enzymes by mycorrhizal mushrooms. Mycoscience 36:221–225

Theodorou C (1968) Inositol Phosphates in needles of *Pinus radiata* D. Don and the phytase activity of mycorrhizal fungi. Proc 9th Int Congr on Soil science Adelaide, vol 3, pp 480–490

Theodorou C (1971) The phytase activity of the mycorrhizal fungus *Rhizopogon roseolus*. Soil Biol Biochem 3:89–90

Thomas KI (1985) The utilization of inositol hexaphosphate by selected ectomycorrhizal symbionts grown in pure culture. For Abstr 46:649

Tommerup IC, Abbott LK (1981) Prolonged survival and viability of VA mycorrhizal hyphae after root death. Soil Biol Biochem 12:431–433

Trojanowski J, Haider H, Hüttermann A (1984) Decomposition of ^{14}C-labelled lignin, holocellulose and lignocellulose by mycorrhizal fungi. Arch Microbiol 139:202–206

Unestam T (1991) Water repellency, mat formation, and leaf-stimulated growth of some ectomycorrhizal fungi. Mycorrhiza 1:13–20

Wallén B (1983) Translocation of ^{14}C in adventitiously rooting Calluna vulgaris on peat. Oikos 40:241–248

Warner A (1984) Colonization of organic matter by vesicular-arbuscular mycorrhizal fungi. Trans Br Mycol Soc 82:352–354

Warner A, Mosse B (1980) Independent spread of vesicular-arbuscular mycorrhizal fungi in soil. Trans Br Mycol Soc 74:407–410

Williams RL, Sparling GP (1984) Extractable N and P in relation to microbial biomass in UK acid organic soils. Plant Soil 76:139–148

Zhu H, Higginbotham KO, Dancik BP, Navratil S (1988) Intraspecific genetic variability of isozymes in the ectomycorrhizal fungus Suillus tomentosus. Can J Bot 66:588–594

Zhu H, Guo D, Dancik BP (1990) Purification and characterization of an extracellular acid proteinase from the ectomycorrhizal fungus Hebeloma crustuliniforme. App Environ Microbiol 56:837–743

19 Decomposition of Plant Litter by Fungi in Marine and Freshwater Ecosystems

M.O. Gessner[1], K. Suberkropp[2] and E. Chauvet[3]

CONTENTS

I. Introduction

A. Fungi and Plant Litter Decomposition

One of the most basic functions of fungi in ecosystems lies in the decomposition of plant remains such as leaf litter and wood (Harley 1971). The complementary perception that fungi contribute significantly to the decomposition process is a classic paradigm in terrestrial ecology (see Dighton, Chap. 17, this Vol.) but has not gained general acceptance for aquatic habitats. Although there is evidence that fungi can also be important, if not the predominant decomposers in many aquatic environments (Suberkropp and Klug 1981; Maltby 1992a; Newell 1993a; Bärlocher and Biddiscombe 1996), there is a continuing debate about their contribution relative to that of other organisms, and sometimes fungi are discounted as significant decomposers of plant material in both freshwater

and marine systems (Iversen 1973; Benner et al. 1986; Blum et al. 1988; Moran and Hodson 1989).

We feel that apparent misconceptions concerning the role of fungi in decomposition are due to a number of problems associated with the current understanding of this process in aquatic habitats. One of the prime reasons is clearly the lack of definitive data concerning fungi (and other microorganisms) for many aquatic decomposition systems. Although a considerable body of literature on decomposition in aquatic environments has accumulated (see reviews by Anderson and Sedell 1979; Bird and Kaushik 1981; Brinson et al. 1981; Mellilo et al. 1984; Polunin 1984; Webster and Benfield 1986; Harrison 1989; Adam 1990; Boulton and Boon 1991; Maltby 1992b; Robertson et al. 1992; Enríquez et al. 1993; Newell 1993a), only a small number of studies have explicitly addressed the role of saprotrophic fungi in the process. Conversely, many of the studies that are primarily concerned with aquatic fungi typically provide little information on their role in the decomposition process (see Kohlmeyer and Kohlmeyer 1979; Webster and Descals 1981; Ulken 1984; Schaumann 1993; Hyde and Lee 1995).

The methodological difficulties in analyzing natural microbial assemblages in a quantitative fashion (Newell 1992a; Hobbie and Ford 1993) add further to the insufficient and biased data base. For the filamentous saprotrophic fungi, this problem is due, in large part, to the intimate association between the fungus and the decomposing plant tissue (Newell et al. 1996), i.e., the fact that hyphae penetrate the substrate rather than simply adhere to its outer surface (Fig. 1; Newell 1994a). However, in recent years, there have been a number of approaches to overcome this problem; these methodological aspects have been reviewed by Newell (1992a, 1994a) and Gessner and Newell (1996).

Another major problem relates to the artifacts introduced with the collection and deploy-

[1] EAWAG, Forschungszentrum für Limnologie, 6047 Kastanienbaum, Switzerland
[2] Department of Biological Sciences, University of Alabama, Tuscaloosa, Alabama 35487-0344, USA
[3] Centre d'Ecologie des Systèmes Aquatiques Continentaux, CNRS-UPS, 29 rue Jeanne Marvig, 31055 Toulouse Cedex, France

The Mycota IV
Environmental and Microbial Relationships
Wicklow/Söderström (Eds.)
© Springer-Verlag Berlin Heidelberg 1997

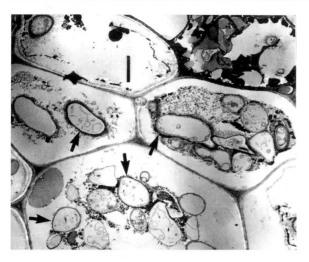

Fig. 1. Transmission electron micrograph of a cross section of a yellow poplar leaf (*Liriodendron tulipifera* L.) that had been decomposing in a hardwater stream for 3 weeks. The leaf was fixed in 2% glutaraldehyde in 10 mM Na cacodylate, pH 7, immediately after it was removed from the stream. It was then postfixed in 2% osmium tetroxide, dehydrated, embedded, and sectioned. Note the masses of fungal hyphae within leaf cells. Four hyphae are indicated by *arrows. Bar* 2 μm

ment of plant litter in an unnatural manner. Although this problem can be circumvented by carefully analyzing and simulating the specific situations in which litter decomposes in nature, studies of plant litter decomposition have often been carried out under experimental conditions that do not resemble natural conditions and in some cases clearly select against the colonization by fungi (Boulton and Boon 1991; Newell 1993a; Bärlocher and Biddiscombe 1996). In extreme cases, living, rather than senescent, plant material may be harvested, oven-dried, ground to fine powder, preleached, or incubated in artificial tanks in order to describe the microbial dynamics during decomposition. Clearly, results from such studies add little to our understanding of naturally occurring processes. Conversely, as the conditions of the natural milieu are taken into account and if reliable quantitative methods are further developed and implemented, we will certainly come closer to a true understanding of fungi and their role in decomposition.

We propose, in view of the preceding shortcomings, that the ongoing discussion about the importance of the various organisms participating in decomposition be dealt with against a background of clearly defined criteria. To ascertain whether fungi are important in a given decomposition system, the following conditions should be met: (1) Fungi must be present in the system. Ideally, they should be detected by either direct techniques or indirect methods involving chemical markers such as ergosterol, phospholipid fatty acid (PLFA) profiles, cell components detectable by immunological techniques, or specific DNA sequences (Newell 1992a; Kemp et al. 1993; Gessner and Newell 1996). Culturing techniques may be valuable but have the danger of introducing substantial bias. (2) The identified species must be able to grow under natural conditions, i.e., on the available substrate and under the set of prevailing environmental factors such as salinity, oxygen concentration, temperature, humidity, and others. (3) The fungi should have the potential to elaborate the enzymes necessary to degrade plant polymers and to produce them in amounts sufficient to bring about significant plant degradation in nature. (4) They must be capable of causing mass loss of organic matter or at least accelerating mass loss when growing in mixed assemblages. (5) Finally, they should be competitive in nature, i.e., they should colonize and exploit their resource and grow at a competitively rapid rate or else be able to oust established species. A good indication of how fungi perform is the accumulation of fungal biomass. This process is best measured dynamically, i.e., by determining instantaneous fungal growth rates, but as yet results of only a few such studies have been published (Newell and Fallon 1991; Newell et al. 1995; Suberkropp 1995; Weyers and Suberkropp 1996; Gessner and Chauvet 1997).

The ultimate proof of fungal participation in decomposition consists in demonstrating fungus-specific degradative activity. This may be indicated when activities of a fungus grown on leaf litter in microcosms (Hicks and Newell 1984; Suberkropp 1991) are similar to the activities observed in the ecosystem. Alternatively, careful application of antibiotics and fungal inhibitors (Padgett 1993), following or coupled with the kind of investigations described above, has been proposed as a way towards this end. Unfortunately, this technique is loaded with potential pitfalls (Oremland and Capone 1988), and to date no study has fully satisfactorily undertaken this endeavor. Often, concepts are derived from observations of fungal presence, isolation and subsequent identification of strains, and sometimes measurements of enzymatic potentials of selected isolates. Clearly, considerably more process-oriented quantitative work is necessary before the current conclusions about the role of fungi in aquatic decomposition processes can be placed on a firm basis. However, as we will attempt to show below, fungi are definitely important in some of the aquatic systems that have been studied so far, and we would predict that their critical role will emerge in various other systems when more information on fungal productivity and activity becomes available.

B. A Conceptual Model of Fungus-Mediated Decomposition in Aquatic Habitats

The decomposition of plant remains, or more generally speaking, coarse particulate organic matter (CPOM), in aquatic environments can be viewed as the sum of biotic and abiotic transformations resulting in the formation of biomass (fungal, bacterial or animal), carbon dioxide and other mineral substances, dissolved organic matter (DOM), and fine particulate organic matter (FPOM). Decomposition is therefore a generic term that encompasses both mineralization processes and other transformations which affect the physical state and the chemical composition of organic matter. From a fungal perspective, the interplay between the biota and the abiotic factors regulating the various litter transformations or subprocesses occurring during decomposition and leading to the mentioned primary products of decomposition can be conceptualized as shown in Fig. 2. The basic feature of this concept is that internal (litter quality) and external (environmental conditions) variables control the activity of fungi which in turn determines various outcomes of the decomposition process. Biotic factors such as fungal community structure and bacterial and detritivore activity set the stage or modify the fungal activity patterns. The outcomes of fungal activity can be divided into those affecting the decomposition process (on the left in Fig. 2) and those affecting fungal performance (on the right in Fig. 2). As for the process-related outcomes, most emphasis in the past has been placed on the rates of litter mass loss, with nutrient dynamics, especially nitrogen immobilization, having also been studied to some extent. Little attention has been devoted to what we call the pattern of mass loss, which is to say the proportions to which the pri-

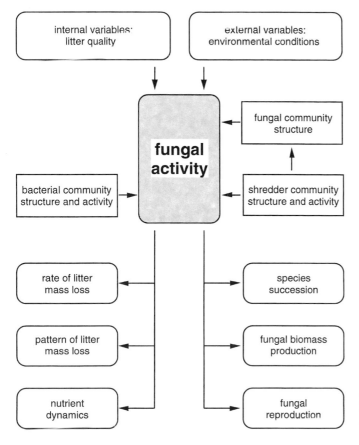

Fig. 2. Simplified schematic representation of a fungus-dominated litter decomposition system as viewed from a fungal and a process perspective

mary decomposition products are generated at different times during litter decomposition and fungal life history. When considering these patterns, the overall resolution of the process is enhanced, while the integration of the various subprocesses theoretically yields the decomposition rate. The outcomes from the fungal point of view concern life-history patterns observable as shifts in fungal community structure, somatic growth, or mycelial biomass production, and allocation of resources to reproduction.

C. The Current Data Base: an Overview

Apart from free-floating macroalgae and driftwood in the open ocean, the production and decomposition of plant remains in freshwater and marine systems is confined to narrow interfaces, or broader transition zones, between land and water. In spite of this geographical restriction, the area covered by such ecosystems is significant on a global scale (Day et al. 1989; Mitsch and Gosselink 1993). Systems include coastal marine areas, in-

Table 1. Examples of aquatic ecosystems receiving substantial amounts of plant remains and fungi potentially involved in the decomposition of this material. Zygomycota and Basidiomycota are notably scarce in all habitats

Habitat or ecosystem	Type of organic matter	Associated fungi[a]	Example of species or genus	Selected references[b]
Marine systems				
General	Wood	Ascomycota	*Savoryella lignicola*	15, 16, 18
		Deuteromycota	*Zalerion maritimum*	15, 16, 18
		Basidiomycota	*Nia vibrissa*	15, 16, 18
Salt marshes	Emergent macrophytes	Ascomycota	*Phaeosphaeria spartinicola*	22, 24, 25, 33
		Oomycota	*Pythium grandisporangium*	21, 23
Seaweed beds	Macroalgae	Labyrinthulomycota	*Thraustochytrium motivum*	26
		Oomycota?	*Haliphthoros milfordensis*	28
		Ascomycota	*Corollospora*	15
		(yeasts)	*Metschnikowia*	15
		Deuteromycota	*Dendryphiella salina*	15, 26
Seagrass beds	Submerged leaves	Fungi unimportant?		5, 27
Mangroves	Leaves	Labyrinthulomycota	*Schizochytrium aggregatum*	6, 9, 34
		Oomycota	*Halophytophthora vesicula*	8, 23
		Ascomycota	*Lulworthia*	8
		Deuteromycota	*Trichoderma*	8, 31
	Wood	Ascomycota	*Lulworthia grandispora*	13, 14, 15, 20
		Deuteromycota	*Cirrenalia pygmea*	13, 14, 15, 20
		Basidiomycota	*Halocyphina villosa*	13, 20
Freshwater systems				
Lake littoral and wetlands	Emergent macrophytes	Ascomycota	*Leptosphaeria*	1, 3, 30
		Deuteromycota	*Cladosporium*	1, 3
	Floating-leaved macrophytes	Deuteromycota	*Colletotrichum nymphaeae*	11, 17
		Oomycota	*Pythium*	17
	Terrestrial leaf litter	Deuteromycota "aero-aquatic hyphomycetes"	*Helicodendron*	4, 10, 35
Streams and rivers	Terrestrial leaf litter	Oomycota	*Pythium*	2, 19
		Deuteromycota	*Epicoccum purpurescens*	2, 19
		"aquatic hyphomycetes"	*Tetracladium marchalianum*	2, 7, 12, 19, 32, 35, 36
	Wood	Ascomycota	*Nectria*	19, 30
		Deuteromycota "aquatic hyphomycetes"	*Heliscus lugdunensis*	19, 29

[a] Including true fungi and fungus-like protists.
[b] 1, Apinis and Taligoola (1974); 2, Bärlocher (1992); 3, Bärlocher and Biddiscombe (1996); 4, Bergbauer et al. (1992); 5, Blum et al. (1988); 6, Bremer (1995); 7, Chamier (1985); 8, Fell and Master (1980); 9, Findlay et al. (1986); 10, Fisher and Webster (1981); 11, Gaur et al. (1992); 12, Hasija and Singhal (1991); 13, Hyde and Jones (1988); 14, Hyde and Lee (1995); 15, Kohlmeyer and Kohlmeyer (1979); 16, Kohlmeyer and Volkman-Kohlmeyer (1991); 17, Kok et al. (1992); 18, Leightley (1980); 19, Maltby (1992a); 20, Mouzouras (1989); 21, Newell (1992b); 22, Newell (1993a); 23, Newell and Fell (1995); 24, Newell and Wasowski (1995); 25, Newell et al. (1996); 26, Raghukumar et al. (1992); 27, Sathe and Raghukumar (1991); 28, Schaumann (1993); 29, Shearer (1992); 30, Shearer (1993); 31, Singh and Steinke (1992); 32, Suberkropp (1992b); 33, Torzilli and Andrykovitch (1986); 34, Ulken (1984); 35, Webster and Descals (1981); 36, Wicklow and Carroll (1981).

land wetlands such as freshwater swamps and the littoral zones of lakes, and running waters which often receive particulate organic matter from adjacent riparian vegetation. Decomposition of plant material in these aquatic ecosystems encompasses a multitude of situations (Table 1). Habitats may be constantly submerged, periodically or occasionally flooded, or permanently exposed to air as is the case for standing-dead emergent macrophytes. A range of environmental constraints, and opportunities, accompany the differences in the hydrologic regime and the general physical setting. The organic matter produced in, or imported to the aquatic environment likewise differs greatly in terms of particle size, physical structure, chemical makeup, and the timing when it becomes available. Associated with these differences in substrate properties and habitat structure is a high diversity in the affiliated fungi, both among and within systems (Table 1). This chapter focuses on three particular habitats whose general metabolism depends largely on plant litter and in which the potential role of fungi in decomposition has been examined to some extent: *Spartina* salt marshes, mangrove swamps or mangals, and freshwater streams.

II. Model Systems

A. Standing-Dead *Spartina* Leaves

Salt marshes cover much of the world's seashores outside the tropics. Where currents provide sufficient fine materials for the formation of soft-bottom sediments, they develop in the land-water transition zone extending from the low to the high tidal mark. The vegetation of salt marshes is commonly characterized by perennial grasses, but low shrubs may also occur. Along the Atlantic coast of North America, the smooth cord-grass, *Spartina alterniflora* Loisel., is the dominant species, often occurring in nearly monospecific stands, and most studies on the role of fungi in the decomposition process have been carried out here. In other parts of the world, however, the salt-marsh flora may be much more diversified (Adam 1990). As with many wetlands, the productivity of salt marshes is generally considered to be high with annual aboveground production averaging 600 g dry mass m^{-2} (Mitsch and Gosselink 1993). A significant portion of this production is channeled into

detritus-based food webs where invertebrate detritivores, e.g., saltmarsh periwinkle (Newell 1993a; Bärlocher and Newell 1994) may consume significant amounts of both leaf material and fungal biomass.

Although there have been a number of studies on the decomposition of *Spartina* litter, conclusions concerning the role of fungi in these ecosystems have varied greatly from study to study. Much of the disagreement can be traced to differences in procedures for collecting, preparing, and placing the litter that is used to examine decomposition dynamics (Newell 1993a, 1996; see Introduction). For example, if green leaves are harvested and oven-dried before being placed on sediments, fungi have typically been discounted. When decomposition studies have simulated natural conditions and used naturally senesced leaves that have been left attached to plants, then fungi are found to be the major decomposers of this litter (Newell et al. 1989; Newell 1993a, 1996). Newell (1993a) provides a more detailed summary of the different procedures used in preparing *Spartina* litter for decomposition studies and the relationship of these procedures to our understanding of the decomposition subsystem in salt marshes. This section will focus on results of studies that have examined the decomposition of *Spartina* leaf litter as it typically occurs, i.e., as standing-dead (Buth and Voesenek 1988; Newell 1993a).

A variety of both marine and terrestrial fungi have been found associated with standing-dead *Spartina* litter. Members of the Ascomycota, including species of *Phaeosphaeria*, *Pleospora*, *Leptosphaeria*, and *Buergenerula*, dominate these systems (Gessner 1977; Newell 1993a; Newell and Wasowski 1995). Depending on characteristics of the habitat, i.e., degree and regularity of inundation by tides, there may be a vertical distribution of fungi on standing-dead leaves and culms of *S. alterniflora*, with marine fungi occurring on lower portions that are regularly submerged by tides and terrestrial fungi occurring on upper parts of the plant that are not generally submerged (Gessner 1977). The fungal communities associated with *Spartina* litter do not appear to be particularly complex. In certain systems, a single species may account for the majority of the fungal biomass (Newell et al. 1989) and reproductive output (Newell and Wasowski 1995). This may be a result of the harsh and highly variable environmental conditions associated with these substrates, such as periodic drying and submergence in seawater (see below).

The fungi isolated from decomposing *Spartina* litter produce a variety of extracellular enzymes necessary for the degradation of plant cell walls. In laboratory studies, fungal isolates typically produce the necessary enzyme complexes to degrade cellulose and hemicelluloses, including those containing xylose and arabinose (Gessner 1980; Torzilli 1982). Mixed results concerning the ability of isolates to degrade pectin have been obtained. Gessner (1980) found that only 5 of 20 isolates tested produced polygalacturonases. Torzilli (1982) detected pectinolytic activity when assays were carried out at pH 8 (pectin lyase) for all three species tested, but only one species produced pectinolytic activity when assayed at pH 5 (polygalacturonase). When four fungal species were provided with isolated cell walls from *S. alterniflora*, all of them grew, and filtrates from these cultures caused release of reducing sugars, indicating the ability of these fungi to use the polysaccharides in native cell wall material (Torzilli 1982).

Three species grown in culture also caused losses in both the cellulose and hemicellulose fractions but not the lignin fraction of *Spartina* tissue (Torzilli and Andrykovitch 1986). When provided with *Spartina* lignocellulose in which the lignin had been specifically radiolabeled, *Phaeosphaeria spartinicola* Leuchtmann caused a loss of 6% (3.3% mineralized and 2.7% solubilized) in the lignin fraction in 45 days (Bergbauer and Newell 1992). In the same time interval, *P. spartinicola* caused 26% loss in total lignocellulose (22% mineralized and 4% solubilized). In addition, transmission electron micrographs of decaying *Spartina* leaves collected in the field revealed symptoms of soft rot from each of four ascomycete species examined, further demonstrating the enzymatic ability of fungi to degrade cell wall material in its native state (Newell et al. 1996). Thus, the fungi found colonizing standing-dead litter of *S. alterniflora* have the necessary enzymatic complement to degrade the lignocellulosic tissues of this plant and elaborate these enzymes when growing in their natural substrate.

Decomposition rates of standing-dead *Spartina* litter are moderate in comparison to other types of litter decomposing in freshwaters and faster than decomposition rates of grasses in terrestrial environments (Webster and Benfield 1986; Newell 1993a). In part, this may reflect the periodic wetting that standing-dead litter experiences from dew formation, the movement of tides, and rainfall events. Microbial respiration associated with standing-dead leaves of *Spartina* is affected by both temperature and moisture (Newell et al. 1985; see also Gallagher et al. 1984). Rates of respiration increase dramatically shortly after leaf wetting provided that the temperature is favorable. Respiration rates range from 50–200 µg CO_2-$C g^{-1}$ litter dry mass h^{-1} which, if extrapolated to a square meter of marsh sediment surface, are similar to rates of carbon loss as a result of the benthic oxygen demand (Newell et al. 1985). When *Spartina* leaves are wet, the associated microbiota exhibit a Q_{10} of 1.4–2.3 (Gallagher and Pfeiffer 1977; Newell et al. 1985). Conversely, if standing-dead portions of salt marsh plants are dry, e.g., less than –6 MPa water potential (Newell et al. 1991), then rates of carbon dioxide evolution become very low irrespective of the temperature (1–10 µg CO_2-$C g^{-1} h^{-1}$). If fungi are the major decomposers of standing-dead *Spartina* leaves, these respiration data would suggest that fungi become active during periods when litter is wetted as a result of dew, rain, or tides, and cause significant decomposition of *Spartina* leaves before they are consumed by detritivores, or are fragmented into small pieces and fall to the sediment.

Newell and coworkers have estimated the fungal biomass associated with decomposing *Spartina* leaves with a number of techniques, including the determination of total mycelial volume, glucosamine content, ergosterol content, and an immunoassay (Newell 1992a). In a direct comparative study, Newell et al. (1989) found that hyphal biovolumes provided the lowest estimates (1.8% of leaf organic matter) with the immunosorbent method (ELISA) giving the highest (20% of leaf organic matter) estimates of peak fungal biomass. Estimates of biomass derived from ergosterol concentrations were intermediate (5% of leaf organic matter), but the latter value would be approximately double (Fig. 3) if it were based on a more realistic conversion factor (see Newell 1994b). Although problems in interpreting ergosterol concentrations remain, in particular the potential variability of conversion factors with respect to different growth conditions (Gessner and Chauvet 1993; Newell et al. 1987; Newell 1994b) this method currently appears to be the most promising for measuring living fungal biomass associated with decomposing leaves (Newell 1992a, 1994a; Gessner and Newell 1996). Estimates using

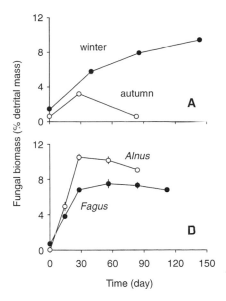

Fig. 3A,B. Dynamics of fungal biomass associated with autumn and winter-spring cohorts of standing-dead smooth cordgrass (*Spartina alterniflora* Loisel.) leaves in a salt marsh (**A** after Newell et al. 1989; Newell 1994b) and decomposing alder [*Alnus glutinosa* (L.) Gaertn.] and beech (*Fagus sylvatica* L.) leaf litter in a stream (**B** after Gessner and Chauvet 1994). *Bars* in **B** standard deviations

ergosterol determinations for fungal biomass and biovolume for bacterial biomass indicate that at maximum biomass, fungi account for over 99% of the total microbial biomass (Newell 1993a). A considerable portion of the fungal biomass associated with *Spartina* litter can be in the form of ascomata of the dominant fungi on leaves. At some points in time, these reproductive structures may account for 31% of the living fungal biomass and about 9% of total fungal production (Newell and Wasowski 1995).

During decomposition of standing-dead *Spartina* leaves, nitrogen concentration first declines and then increases (Newell et al. 1989). Initial declines occurring in senescing leaves that remain attached to the plant have been attributed to translocation of nitrogen back into the rhizome (Newell et al. 1989). Since increases in fungal biomass are greatest during the initial stages of decomposition when nitrogen concentrations in the leaf material are low (<1%), nitrogen may become the limiting nutrient in this system. This idea is supported by the finding that fungal biomass associated with standing-dead *Spartina* leaves attains higher values in plots which received additions of nitrogen fertilizer than in untreated control plots (Newell 1993a). If N is limiting, aquisition from

external sources would appear to be a sensible strategy, even if it is metabolically costly. In this light, speculation that a fungal-bacterial consortium associated with decaying *Spartina* leaves fixes atmospheric nitrogen would warrant closer examination (Newell 1993a).

B. Submerged Mangrove Leaves

Mangrove swamps replace salt marshes at low latitudes. They differ from salt marshes in their physical structure due to the predominance of trees and shrubs, but otherwise share many of the general ecological features with their counterparts at higher latitudes (Day et al. 1989; Mitsch and Gosselink 1993). Mangroves extend from saline habitats to tidal freshwaters with an abundant development in brackish estuarine waters. Based on the geomorphological and hydrological template, several types of mangroves including fringe, riverine, and basin forests can be distinguished (Day et al. 1989; Mitsch and Gosselink 1993). As in salt marshes, plant species diversity of mangroves is higher in the Indo-West Pacific than in the New World, and this affects the quality and diversity of both leafy and woody litter available for decomposition. Estimates of annual litter production range from 110 to 2370 g m^{-2}, with the average exceeding average above-ground salt marsh productivity (Bunt 1995; Saenger and Snedaker 1993).

Decomposition of leaf litter in mangrove ecosystems has been studied to a considerably lesser extent than in salt marshes. Most of the early studies were carried out in subtropical areas at the upper distributional limit of the mangal (e.g., Fell and Master 1980; Steinke and Ward 1987; Blum et al. 1988; Steinke et al. 1990) rather than in the tropics, where mangroves show their best development. Consequently, caution is needed when generalizing from the currently existing data base. However, even with the limited information available, it is clear that considerable variation exists in the decomposition of leaf litter in mangrove forests. Within a given stand, differences may result from the relative position of litter along the flooding-duration gradient. Continuously submerged leaves tend to decompose more rapidly than leaves deposited on the intertidal forest floor (Steinke and Ward 1987; Robertson el al. 1992). If, however, as in tropical Australia, sesarmid crabs are present, their activity may lead to an extremely rapid disappearance of leaves in the high intertidal zone (Robertson el al. 1992). Further differences in the rates and patterns of decomposition may result from varying environmental conditions and leaf quality (Robertson et al. 1992; Steinke et al. 1990).

Directly linked to the general paucity of data on the decomposition of mangrove leaves is the scarce information on the role of fungi in the process (Hyde and Lee 1995). Pioneering work on mangrove leaf decomposition, however, was carried out from a mycological perspective (Fell and Master 1980), and this has resulted in a continued, albeit restricted, interest and some basic information for this decomposition system. The eumycotic fungi typical of the salt-marsh system also occur in decomposing mangrove leaves, but Deuteromycota rather than Ascomycota have most frequently been recorded (Fell and Master 1980; Singh and Steinke 1992). Measurements of leaf-associated eumycotic fungal mass with the ergosterol assay indicate, however, that the higher fungi associated with submerged mangrove leaves are much less productive than was found in salt-marsh grasses (Newell and Fell 1992a). If this finding turns out to be a general feature of mangrove leaves, it would indicate that the eumycotic fungi, although regularly present, assume minor importance in the decomposition process. Conceivably, Deuteromycota and Ascomycota are better represented in leaf litter deposited in the drier upper intertidal or supratidal zones of mangrove swamps (Newell and Fell 1992a).

Increasing evidence suggests that members of the Oomycota, which lack ergosterol in their membranes (Newell 1992a), assume significant importance in the decomposition of submerged mangrove leaves (Newell 1996). In a series of experiments it was demonstrated that zoospores of *Halophytophthora* not only colonize freshly submerged mangrove leaves at a rapid rate but also show a preference for this substrate (Newell 1992b; Newell and Fell 1992b; Newell et al. 1987), although the mechanism of this rapid leaf occupation is as yet unclear (Newell and Fell 1995). Earlier work had shown that *Halophytophthora vesicula* (Anast. & Churchland) Ho & Jong and *H. spinosa* var. *spinosa* (Fell & Master) Ho & Jong are regular members of the fungal assemblages associated with mangrove leaves in southern Florida, and cause leaf mass loss when growing in seminatural culture systems (Fell and Master 1980; Fell et al. 1980). Taken together, this evidence points to members of the Oomycota as potentially important decomposers of mangrove leaf litter although specific data on their productivity in, and degradation of, mangrove leaves in the natural environment is, as yet, lacking.

If the filamentous lower rather than the higher fungi dominate the fungal assemblage of decomposing mangrove leaves, it is not surprising that concentrations of ergosterol (a membrane constituent largely restricted to the eumycotic fungi and lacking in the Oomycota and Chytridiomycota) were found to be exceeding low in this material (Newell and Fell 1992a). Mycelia within the leaf matrix should nevertheless be detectable, and histological examination by light and electron microscopy has indeed revealed hyphal growth within the decomposing leaf tissue of a variety of mangrove species (Fell and Master 1980; Steinke et al. 1990). These observations have led several authors to conclude that filamentous fungi are important agents in the degradation of mangrove leaves, but unfortunately none of these studies provided quantitative evidence, such as estimates of fungal mass.

Other studies have questioned the importance of filamentous fungi (including Oomycota) in the decomposition of mangrove leaves. Blum et al. (1988) came to the conclusion that fungi are of minor importance after comparing standing stocks of bacteria and fungi in decomposing leaves of *Rhizophora mangle* L. using a direct microscopic technique. In this study, peak biomass of filamentous fungi was estimated to be as low as 0.08% of the detrital mass. Measurements of hyphal lengths tend to result in substantial underestimates of fungal biomass in leaf litter (Newell 1992a); however, whether the exceedingly low values of <0.1% can be entirely accounted for by the unavoidable methodological inefficiencies (see Newell 1992a; cf. Table 2) or whether the quasi-absence of mycelia was an inherent feature of the examined litter (Blum et al. 1988) remains currently unknown. Benner et al. (1986) also concluded that fungi were unimportant in the mineralization of differentially radiolabeled lignocelluloses which were prepared by grinding labeled whole leaves to particles <425 μm and subsequent extraction of nonstructural leaf constituents. Bacterial and fungal respiration were distinguished by size fractionation. With this experimental setup, bacteria were found to account for 100% of the evolved CO_2, although plating on corn-meal agar showed that fungi were present in the

Table 2. Estimates of maximum fungal biomass associated with decomposing leaf litter in streams. Measured hyphal lengths were converted to mycelial mass by assuming an average hyphal diameter of 3 μm and a density of 500 fg μm^{-3} (cf. Findlay and Arsuffi 1989; Newell 1992a). ATP was converted to fungal biomass assuming that fungal ATP accounted for 90% of the total ATP (cf. Baldy et al. 1995; Findlay and Arsuffi 1989) and an average ATP concentration of 1.75 mg g^{-1} mycelial dry mass (Suberkropp 1991; Suberkropp et al. 1993). Ergosterol was converted to fungal biomass assuming an average mycelial concentration of 5.5 mg g^{-1} dry mass (Gessner and Chauvet 1993) unless more specific data was available

Fungal biomass (mg g^{-1} detrital mass)	No. of streams	No. of leaf types	Method	Reference[a]
0.12	1	1	Biovolume[b]	10
0.7	1	1	Biovolume[b]	2
13–21	1	3	Biovolume[c]	3
8–49	1	3	Biovolume[c]	4
23	1	1	ATP	4
37–147	1[d]	2	ATP	11, 17
20–130	4	3	ATP	12
8–32	2	1	ATP	13
15–111	8[d]	1	ATP	15
2.6–13	3	3	ATP	9
47–83	2	2	ATP	16
127–158	2	2	Ergosterol	16
1.1–8	4	2	Ergosterol	8
49–99	1	3	Ergosterol	1
61–155	1[d]	7	Ergosterol	5, 6, 7, 16
45–177	3	1	Ergosterol	14

[a] 1, Baldy et al. (1995); 2, Bärlocher and Kendrick (1974); 3, Buttimore et al. (1984); 4, Findlay and Arsuffi (1989); 5, Gessner and Chauvet (1994); 6, Gessner and Schwoerbel (1991); 7, Gessner et al. (1993); 8, Griffith and Perry (1994); 9, Griffith et al. (1995); 10, Iversen (1973); 11, Lawson et al. (1984); 12, Rosset et al. (1982); 13, Suberkropp (1991); 14, Suberkropp (1995); 15, Suberkropp and Chauvet (1995); 16, Suberkropp et al. (1993); 17, Suberkropp and Klug (1976)
[b] Hyphal length determined after clearing of whole leaf material.
[c] Hyphal length determined after grinding and collecting leaf pieces on membrane filters.
[d] Different sites or years or both in the same stream.

unfractionated sample. In favor of the fungi we think, however, that the artificial structural changes of the leaf material and the marked reduction in particle size rule out a meaningful interpretation of this result in terms of the naturally occurring decomposition of coarse particulate plant residues. This does not, of course, preclude that the experimental approach chosen by Benner et al. (1986) might be useful in investigating the fate of fine particulate organic matter (FPOM) from a biogeochemical perspective with the prepared lignocellulose representing a convenient seminatural model substrate.

Recent evidence suggests that, in addition to filamentous eukaryotes, the unicellular Labyrinthulomycota (Labyrinthulales and Thraustochytriales), another group of heterotrophic protists, are actively involved in the decomposition of submerged mangrove leaves (Bremer 1995). Both labyrinthulids and thraustochytrids were consistently isolated from mangrove leaves at all stages of decomposition but not from those attached to the tree. Scanning electron microscopy after freeze fracture of leaves inoculated

with selected strains in the laboratory revealed leaf penetration and, for one species (*Schizochytrium aggregatum* Goldstein and Belsky), localized degradation of internal leaf tissue. Finally, several isolates caused substantial leaf mass loss when grown on *Sonneratia* leaf litter in a microcosm, with the one isolate tested also being able to degrade carboxymethylcellulose at a particular life stage (4–5-day-old cultures). Taken together, these findings thus point to a notable potential of Labyrinthulomycota to decompose mangrove leaf litter in their natural habitat.

Mangrove leaves experience increases in nitrogen concentrations and concomitant decreases in C/N ratios during decomposition (Robertson et al. 1992; Steinke and Ward 1987). When breakdown rates are sufficiently low, absolute amounts of nitrogen may sometimes exceed initial values (e.g., Benner et al. 1990), indicating nitrogen uptake from external sources. Various hypotheses have been proposed to explain this immobilization of nitrogen by decomposing leaf litter, but as yet the phenomenon is not fully understood (Melillo et al. 1984; Buchsbaum et al.

1991). Whatever the mechanism, fungi appear to be involved in the nitrogen enrichment of decomposing mangrove leaves. Using a simple laboratory microcosm, Fell et al. (1980) incubated senescent *Rhizophora* leaves in the presence and absence of *Halophytophthora* species; the surrounding seawater was amended with NH_4No_3 to give various final concentrations of dissolved N. Analysis of the leaf litter after 10 days for mass loss and nitrogen concentration demonstrated that the presence of fungi was necessary to induce nitrogen immobilization at this early decomposition stage, and this effect was stronger with increasing amounts of N provided in the seawater.

C. Terrestrial Leaves in Streams

The most striking feature of lotic ecosystems is perhaps the unidirectional flow of water. Streams and rivers have therefore long been considered as mere transport systems, for water, and also for solutes and particulate matter. Current concepts, however, emphasize that running waters are actively metabolizing ecosystems (Calow and Petts 1992; Stanford and Covich 1988). In headwater streams with closed canopies, much of the energy available to the heterotrophic aquatic biota enters the wetted channel in the form of allochthonous particulate organic matter such as leaf litter and woody debris which is derived from the riparian vegetation (Bird and Kaushik 1981; Maltby 1992a). In this respect, running waters are similar to mangrove swamps. Total CPOM inputs to such streams are typically on the order of $500\,g\,dry$ $mass\,m^{-2}$ but may exceed $1000\,g\,m^{-2}$ (Weigelhofer and Waringer 1994 and references therein). The diversity of leaves entering streams is generally high, and rates of breakdown vary considerably among leaf species (Webster and Benfield 1986; Gessner and Chauvet 1994). This and other aspects of leaf decomposition in streams (Webster and Benfield 1986; Boulton and Boon 1991; Maltby 1992b) and lotic food webs, which to a large extent are based on allochthonous leaf material, have received a great deal of attention, with many investigations focusing on the macroinvertebrate detritivores known as shredders, which consume leaf litter and its associated microbiota (Bärlocher 1985; Suberkropp 1992a).

The fungi associated with decomposing leaves in streams have received more attention than the fungi in the two marine ecosytems discussed above. Representatives from both Oomycota and Chytridiomycota can be detected in streams, but so far studies have not indicated that these organisms play a major role in leaf breakdown (Bärlocher 1992). Likewise, although a diverse assemblage of Ascomycota is associated with wood in streams, such teleomorphic stages are uncommon on decomposing leaves (Shearer 1992, 1993). Terrestrial hyphomycetes are part of the phylloplane microbiota and thus colonize leaves before they enter streams (Bärlocher 1992; Maltby 1992b); their role in decomposition once leaves fall into streams is not certain, but their activity appears to be limited at the low wintertime temperatures which prevail in many temperate regions after leaf fall (Bärlocher 1992; Maltby 1992a).

The fungi that are most actively involved in the decomposition of leaves in streams appear to be the aquatic hyphomycetes, which sometimes are also called Ingoldian fungi or amphibious hyphomycetes (Bärlocher 1992; Webster and Descals 1981; Wicklow and Carroll 1981). The recognition of the fundamental importance of aquatic hyphomycetes has come from observations concerning their regular occurrence and sporulation on decomposing leaves in streams as well as from studies of their activities on leaf litter, including measurements of high enzymatic potential in the laboratory (Bärlocher 1992; Suberkropp 1992b and references therein). A number of studies have also emphasized the correlations between leaf litter inputs and increases in concentrations of aquatic hyphomycete conidia transported by stream water or trapped in foam floating at the water surface: increases in conidial concentrations of several orders of magnitude commonly occur during periods of bulk leaf inputs (Suberkropp 1992b). Aquatic hyphomycetes are well adapted to the stream environment (Bärlocher 1992; Suberkropp 1992b), with their tetraradiate and sigmoid conidia apparently representing morphological adaptations for attachment in flowing waters (Webster and Descals 1981; Webster 1987). In addition, these fungi have the ability to grow at the low water temperatures prevailing in the season following leaf fall in temperate regions.

Aquatic hyphomycetes produce a variety of extracellular enzymes that degrade the structural polysaccharides of leaves (Suberkropp and Klug 1981; Chamier 1985; Hasija and Singhal 1991; Suberkropp 1992b and references therein). Enzymes that hydrolyze cellulose (endoglucanases, exoglucanases, and exoglucosidase) and hemicelluloses (xylanases, xylosidase, and arabinosidase) are produced by a number of species growing in culture on pure substrates or on leaf material. Aquatic hyphomycetes also typically produce several enzymes that degrade pectin (Suberkropp and Klug 1980; Chamier and Dixon 1982). Pectin degradation leads to the softening and maceration of plant tissue, resulting in the release of mesophyll cells as FPOM and the exposure of other cell-wall polymers to the activity of polysaccharidases (Chamier 1985; Suberkropp 1992b). Both polygalacturonase and pectin lyase depolymerize pectin, with the latter enzyme thought to play a greater role in leaf maceration (Suberkropp and Klug 1981; Jenkins and Suberkropp 1995). In addition, there is evidence from laboratory studies that aquatic hyphomycetes are capable of degrading lignin-like substrates (Fisher et al. 1983; Zemek et al. 1985; Zare-Maivan and Shearer 1988; Abdullah and Taj-Aldeen 1989). However, the difficulties in assessing lignin degradation using culture methods and the use of different methodologies limit our knowledge of the ligninolytic capabilities of these fungi in natural circumstances (Chamier 1985). Other major plant compounds such as proteins and lipids (Zemek et al. 1985; Zare-Maivan and Shearer 1988; Abdullah and Taj-Aldeen 1989) also can be degraded by the enzymatic activity of aquatic hyphomycetes. Thus, it appears that most plant compounds, except perhaps lignin, can be metabolized by the majority of aquatic hyphomycete species, and, in accordance with this conclusion, pure cultures of aquatic hyphomycetes have been shown to degrade leaf material in the laboratory (Suberkropp and Klug 1980; Chauvet and Mercé 1988; Chergui and Pattee 1991; Suberkropp 1991). The apparent lack of broad-scale specialization among species indicates that the aquatic hyphomycetes are a relatively homogenous and generalist group with respect to nutritional niche breadth and functional role in the ecosystem (Suberkropp 1992b).

Little information is available concerning the enzyme activities associated with decomposing leaf litter in streams. Since fungi are not the only microbial group colonizing leaf litter, information derived from such studies in natural conditions may be limited unless accompanied by a simultaneous estimation of the fungal occurrence. Fungal biomass, determined as ergosterol content, was closely correlated with exocellulase activity on maple leaves decomposing in a mid-size river, suggesting that this hydrolytic activity was due to fungi (Golladay and Sinsabaugh 1991). Conversely, when studying leaf decomposition in two streams differing in alkalinity, Jenkins and Suberkropp (1995) found that the leaf-associated activity of three hydrolytic enzymes (xylanase, endocellulase, and galacturonase) was lower in the hardwater stream. However, softening and breakdown of leaf litter proceeded faster, higher amounts of fungal biomass built up (as ATP content, see below), and sporulation of aquatic hyphomycetes was more abundant in the hardwater stream than in the softwater stream. Consequently, Jenkins and Suberkropp (1995) concluded that the hydrolytic enzymes they examined were poor indicators of leaf breakdown in streams. Pectin lyase activity, in contrast, was higher in the hardwater stream, concomitant with faster leaf breakdown and greater fungal activity (Jenkins and Suberkropp 1995). Similar results were obtained by Griffith et al. (1995), indicating that pectin degradation is a key process during early stages of leaf decomposition in streams (Suberkropp and Klug 1981).

Colonization of leaf litter by aquatic hyphomycetes is initiated by the impacting and trapping of conidia on leaf surfaces after leaves enter a stream. Subsequent germination is rapid, within 2–6 h in most species (Read et al. 1992). Once established, the fungal hyphae extend rapidly inside the leaf matrix so that significant quantities of mycelial mass are built up (Fig. 1) within a few weeks after leaf colonization (Fig. 3). Maximum fungal biomass can reach more than 15% of the total detrital mass but some early estimates of fungal biomass which are based on measurements of hyphal length are much lower than 1% (Table 2). Evidence from comparative studies in other decomposition systems suggests that the exceedingly low estimates obtained via determinations of hyphal length, in particular after clearing of leaves, are artifacts (Newell 1992a). Measurements of ATP and the fungal membrane-component ergosterol probably result in more realistic estimations (Newell 1992a; Suberkropp et al. 1993; Gessner and Newell 1996). However, it is clear from Table 2 that, in addition to differences due to methodology, considerable variation in fungal biomass production exists among systems, whether these be streams, leaf species, or both. Although ATP is not specific for fungi and hence reflects the total living biomass associated with leaves, it can probably be used as an index of fungal mass associated with decomposing leaves in streams in most cases. Two lines of evidence support this argument. First, independent estimates of fungal and bacterial biomass indicate that fungi typically account for greater than 90% of the microbial biomass developing on decomposing leaves in streams (Findlay and Arsuffi 1989; Baldy et al. 1995; Weyers and Suberkropp 1996), and major sources of ATP other than bacteria are generally absent (Golladay and Sinsabaugh 1991; Suberkropp et al. 1993). Secondly, while ATP tends to give lower estimates of fungal biomass than ergosterol (Table 2), both indices follow similar patterns of change during leaf decomposition, further supporting the idea that fungi account for most of the microbial biomass and hence the ATP associated with leaves in these systems (Suberkropp et al. 1993).

The growth of fungal hyphae within the leaf matrix is closely followed by the production of conidiophores, which start to release conidia in as little time as 6–10 days after leaves enter a stream. Sporulation rates rapidly increase to maxima and then decrease. Maximum sporulation rates can reach seven conidia per μg of detrital matter per day (Suberkropp et al. 1993; Gessner and Chauvet 1994; Bärlocher et al. 1995). However, in streams containing low concentrations of nutrients or when fungi are growing in leaves of poor quality, maximum sporulation rates may reach only a few percent of this value (Bärlocher 1982; Suberkropp 1991; Suberkropp and Chauvet 1995). A striking feature of aquatic hyphomycetes colonizing leaf litter is that sporulation begins early during the growth phase. This has been demonstrated not only from laboratory microcosms containing single species (Suberkropp 1991), but also from field experiments where sporulation rates of natural communities were measured simultaneously with fungal biomass (Bärlocher 1982 in combination with Rosset et al. 1982; Suberkropp 1991; Suberkropp et al. 1993; Gessner and Chauvet 1994; Baldy et al. 1995; Maharning and Bärlocher 1996). The maximum rates of sporulation are controlled both by internal factors such as leaf litter quality (Gessner and Chauvet 1994) and external factors such as temperature and nutrient availability in stream water (Suberkropp and Chauvet 1995, see below).

In lotic ecosystems, as in mangrove swamps, increases in the nitrogen and phosphorus concentrations of decomposing leaf material have frequently been reported (Webster and Benfield 1986). Absolute increases in nitrogen were found in several studies (e.g., Chauvet 1987), indicating that nitrogen was obtained from the water flowing across the leaves. The idea that microorganisms associated with leaf litter decomposing in streams immobilize nutrients is consistent with results of laboratory experiments performed with fungal cultures (e.g., Suberkropp et al. 1983; Chergui and Pattee 1991). Most aquatic hyphomycetes can assimilate amino acids (Thornton 1965) and can produce enzymes allowing them to obtain phosphorus from organic P compounds present in dead leaves (Suberkropp and Jones 1991). However, addition of inorganic phosphorus can stimulate the rates of microbial respiration and leaf decomposition (Elwood et al. 1981) implying that P can limit microorganisms developing on leaves and is obtained, at least in part, from stream water. Correlations between patterns of phosphorus uptake and changes in respiration associated with leaf detritus also occur (Mulholland et al. 1984).

III. Emerging Principles

A. Organic Budgets

Our understanding of the decomposition of plant litter and the roles of different groups of microorganisms improves as we obtain more quantitative data concerning the rates of transformation and partitioning of carbon and nutrients during decomposition. In two of the systems previously discussed, i.e., standing-dead *Spartina* leaves in salt marshes and deciduous leaf litter in streams, attempts have been made to develop organic matter budgets and estimate carbon flow for leaves colonized in the field and in the laboratory. In order to summarize these tentative budgets here, the values presented in Fig. 4 were calculated for a particular decomposition stage. As a consequence, they do not reflect the dynamic changes that can occur during the decomposition sequence and hence are intended to be illustrative rather than comprehensive and conclusive. For example, the mycelial and bacterial components of cordgrass are presented for leaves at the point in time at which 28.8% of the leaf mass is remaining. During decomposition, the carbon flow to fungal mass was estimated to be much greater, namely 6.7%

field budgets

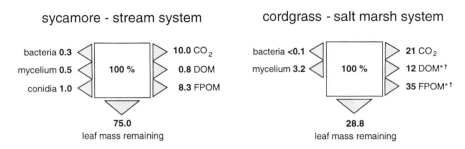

sycamore - stream system

bacteria 0.3
mycelium 0.5   100 %
conidia 1.0

10.0 CO_2
0.8 DOM
8.3 FPOM

75.0
leaf mass remaining

cordgrass - salt marsh system

bacteria <0.1
mycelium 3.2   100 %

21 CO_2
12 DOM*†
35 FPOM*†

28.8
leaf mass remaining

microcosm budgets

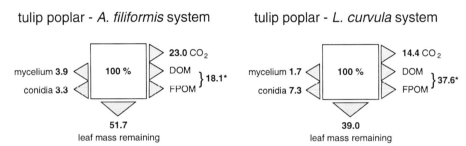

tulip poplar - *A. filiformis* system

mycelium 3.9   100 %
conidia 3.3

23.0 CO_2
DOM } 18.1*
FPOM

51.7
leaf mass remaining

tulip poplar - *L. curvula* system

mycelium 1.7   100 %
conidia 7.3

14.4 CO_2
DOM } 37.6*
FPOM

39.0
leaf mass remaining

Fig. 4. Organic matter budgets for sycamore (*Platanus occidentalis* L.) leaf litter in a stream, standing-dead smooth cordgrass (*Spartina alterniflora* Loisel.) leaves in a salt marsh (winter-spring cohort), and tulip poplar (*Liriodendron tulipifera* L.) leaves inoculated with pure cultures of aquatic hyphomycetes in a microcosm. Data modified after Findlay and Arsuffi (1989), Newell et al. (1989) together with Newell (1994b), and Suberkropp (1991). An *asterisk* (*) refers to compartments determined by difference; the *cross* (†) indicates that material translo-cated to rhizomes is included. Percentages of conidia represent cumulative values. Percentages of bacteria and mycelium represent the amounts calculated for the last date of the period investigated. Note, therefore, that the total carbon flow to the latter compartments was much higher. For example, an estimate for the cordgrass system indicates that over the course of decomposition 15% of the inital leaf carbon were converted to mycelial carbon (Newell et al. 1989)

of the initial mass based on ergosterol concentrations and 15% based on an enzyme-linked immunosorbent assay, than the final value of 3.2% appearing in Fig. 4 (Newell et al. 1989; Newell 1994b). However, because much of the fungal organic matter produced during leaf decomposition is eventually lost as CO_2 or as another form, final rather than maximum fungal biomass values are included in the budgets. We anticipate that the construction of such budgets **at various decomposition stages** will increase our understanding of the decomposition process by allowing examination of the pathways of carbon and nutrient flow as well as their temporal dynamics, and thus augment resolution of the process analysis. In addition, information about the allocation of leaf material to various compartments allows integration of the decomposition process into ecosystem models and can point to funda-mental questions that can guide future research.

One of the fates of plant litter is its conversion into other forms of organic matter such as DOM and FPOM. Only one of the budgets in Fig. 4 determined losses of DOM and FPOM from detritus directly, and our calculations for the remaining budgets have determined losses in both these pools by difference. The amount of organic material released as FPOM was about ten times that released as DOM from sycamore leaves decomposing in a stream (Fig. 4; Findlay and Arsuffi 1989). For the two other types of leaves examined in that study, 1.3 (oak) and 20 (elm) times as much FPOM was released as DOM, indicating that the ratio of fine particulate to dissolved matter can be quite variable with different litter types but is generally greater than 1. Newell et al. (1989) also indicated that losses of FPOM were about three times greater than those of DOM during decomposition of *Spartina* leaves (Fig. 4), suggesting that this might be a general phenomenon occurring during fungus-mediated decomposition of plant residues in aquatic environments. In streams, such macerating capabilities have been shown to be related to the production of pectin-degrading enzymes (Suberkropp and Klug 1980, 1981; see above).

From the studies of organic matter budgets constructed for aquatic hyphomycetes growing on leaves in a stream and in the laboratory (Fig. 4), it is evident that a significant proportion of production by these fungi is used for the formation of conidia (Findlay and Arsuffi 1989; Suberkroop 1991). Therefore, evaluation of total fungal production on leaves decomposing in streams must include both compartments, i.e., mycelium and conidia. For *Anguillospora filiformis* Greathead and *Lunulospora curvula* Ingold grown on leaves in stream microcosms, conidium production represents 46 and 81%, respectively, of total fungal production and accounts for 7 and 12%, respectively, of the loss in leaf mass (Suberkropp 1991). In natural situations, competition with other fungi and bacteria, consumption by invertebrates, and other factors may reduce fungal reserves so that conidium production may be lower. However, rough estimates of conidium production by the fungal communities colonizing leaves in streams indicate that conidia may represent 1–4% of the mass loss of leaves, depending on the types of litter, stream characteristics, and the duration of the study (Findlay and Arsuffi 1989; Suberkropp 1991; Baldy et al. 1995). An estimate of fungal reproductive output is also available for fungi growing on *Spartina* leaves. Substantial amounts of fungal biomass are allocated to ascomata by these fungi, and liberation of ascospores may represent a significant loss of carbon from this type of cordgrass litter. Newell and Wasowski (1995) estimated that, during periods of leaf wetness, an average of 16 ascospores are released per cm^2 of *Spartina* leaf area per hour (which we calculate to be about 0.5 μg dry mass per day), and stress that this figure is probably an underestimate. Since fungal spores typically contain higher concentrations of nutrients (e.g., N and P) than either the mycelium or the surrounding environment (Dowding 1976), spore release may represent a more significant pathways for losses of nitrogen and phosphorus from decomposing leaves than of carbon.

The high variability in spore output observed by Newell and Wasowski (1995) and the variation in conidium production between the two fungal species examined in microcosms and the community occurring on leaves colonized in a stream (Fig. 4) suggest that the allocation of resources from fungal biomass to reproductive structures may vary to a great extent, depending on environmental conditions as well as on the species composition of the fungal community. Further research concerning allocation of resources to reproduction by fungi, and the factors that affect this allocation, is needed before we understand this important aspect of fungal ecology in aquatic habitats.

Two other general trends that emerge from the analysis of organic matter flow (Fig. 4) are that fungal biomass increases at an early stage (Fig. 3) and that it reaches a significant proportion of the total detrital mass (up to 15%, Table 2). Given the dynamics of fungal growth in the initial stages of leaf decomposition, measurements of biomass are insufficient to fully appreciate the productivity of fungi in decomposing litter. Consequently, Newell and Fallon (1991) designed a method for estimating instantaneous growth rates of fungi in decomposing litter by determining rates of incorporation of [^{14}C]acetate into ergosterol (Newell 1993b; Gessner and Newell 1996). In a preliminary field test of the method, fungal growth rates varied from 0.04 to 0.05 day^{-1} (see also Newell 1993a; Newell et al. 1995) and total fungal organic production was 64% of the loss in organic matter from cordgrass and 36% from the sedge *Carex walteriana* Bailey (Newell and Fallon 1991). Using the same method to estimate fungal production associated with yellow poplar leaves decomposing in streams, Suberkropp (1995) found growth rates ranging from 0.01 to 0.20 day^{-1}. Highest rates occurred in the early stages of the decomposition process, consistent with the greatest increases in standing stocks during this time. Depending on the characteristics of the stream in which the leaves were decomposing, total fungal organic production ranged from 10 to 15% of the organic loss from leaves. As additional data of this kind become available, more accurate budgets of fungal participation and relative flow of organic matter through fungal populations should be possible, eventually permitting a more adequate appreciation of fungal importance in the decomposition process and their status in aquatic ecosystems.

B. Fungal Regulation of Decomposition

A variety of factors may influence the decomposition of coarse particulate plant residues in aquatic environments. Among those most frequently noted are temperature, nutrient concentrations of the decomposing plant material, in particular nitrogen, the proportion of refractory plant constituents, nutrient concentrations in and pH of the surrounding water, oxygen availability, and the presence of detritivores (see Melillo et al. 1984; Webster and Benfield 1986). A number of studies have also revealed "site effects", which integrate a range of environmental variables, on rates of organic matter breakdown (Robertson et al. 1992; Webster and Benfield 1986).

Little is known as to the role of fungi (and other microorganisms) in the regulation of this

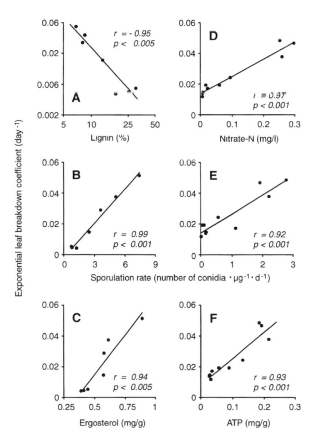

process, but recent evidence from stream systems points to their critical importance. In a study of seven leaf types, the initial content of refractory leaf constituents (lignin) was strongly correlated with the exponential rate of leaf breakdown (Fig. 5a). Significantly, the sporulation rates of aquatic hyphomycetes and the amount of fungal biomass developing in the decomposing leaves were also correlated with both breakdown rate (Fig. 5b,c) and lignin content. This, together with the high fungal mass that built up in the decomposing leaf material, suggests that leaf breakdown rate was controlled through a kinetic carbon limitation of the leaf-associated fungi (Gessner and Chauvet 1994). Suberkropp and Chauvet (1995), likewise, found strong correlations between the average concentration of nitrate in the water of various streams, the activity of leaf-associated fungi in these streams and the exponential breakdown rates of leaves (Fig. 5d-f), indicating that exogenous nitrogen availability was a prime factor accounting for differences in the degradative activity of fungi and hence breakdown rate in that study.

These two stream studies represent specific cases of, and thus lend support to, the general conceptual model of organic matter breakdown outlined in Fig. 2. The same or similar factors are probably operative in other aquatic systems. For example, phosphorus rather than nitrogen may be the external nutrient that is limiting, and lignin contents may be less important than concentrations of other phenolic compounds such as tannins. Whether one or the other of the various controlling factors will be important in a given situation will depend on how they are balanced against each other (Gessner and Chauvet 1994; cf. Sinsabaugh et al. 1993). The effect of exogenous nitrogen (or another nutrient), for example, would appear to be less pronounced when fungal demands can be met by nitrogen sources within the decomposing plant material, provided the fungi in question have the enzymatic potential to make use of the nitrogen locked up in the plant tissue (Newell 1993a). This potential, however, seems to be the rule rather than the exception in aquatic fungi (see above, Thornton 1965; Jones and Byrne 1976; Suberkropp and Jones 1991). Likewise, nutrient availability, whether external or internal, would be less critical when labile carbon is in limited supply, as may often be the case in leaf species with high concentrations of refractory carbon compounds (Gessner and Chauvet 1994) such as

Fig. 5A–F. Correlations between leaf breakdown coefficients and concentrations of refractory leaf constituents (**A**), nitrate-N concentrations in stream water (**D**), maximum sporulation rates of aquatic hyphomycetes (**B** and **E**), maximum ergosterol fungal biomass (**C**), and maximun ATP microbial biomass (**F**). *Panels on left* Gessner and Chauvet (1994), *panels on right* Suberkropp and Chauvet (1995)

lignin and cutin. It is even conceivable that the apparent balance between the various resources necessary for fungal growth is altered by modifiers such as toxins or inhibitors, when these have specific effects on metabolic processes of fungi or the products thereof. If a high concentration of tannins in plant residues, for example, leads to a complexation of fungal exoenzymes, a relatively higher fungal nitrogen demand may result compared to a situation where tannins are absent. In such a case, nitrogen may become limiting even if, for example, N:P ratios indicate that phosphorus should be. Thus, the regulation of decomposition processes in aquatic environments may vary broadly according to the relative impact and interactions of a range of controlling factors, including both external ones related to the environment and factors intrinsic to the decomposing plant material.

IV. Conclusions

A diversity of aquatic habitats occur at land-water interfaces where the productivity of plants is generally high and large amounts of plant litter enter the detrital pool. A wide range of environmental conditions (e.g., temperature, salinity, nutrient availability) exists in these habitats where different types of plant litter from both aquatic and terrestrial sources are decomposing, and the diversity of fungi that are found and potentially active is high. Due to the paucity of data for most systems, the overall role of fungi in the decomposition of plant litter in aquatic environments is difficult to assess at present. Quantitative data on the occurrence, production, and activity of fungi in aquatic habitats are needed in order to evaluate their significance as agents of decomposition and nutrient cycling, and their role in aquatic food webs. However, there have been enough studies in a few habitats to demonstrate that the role of fungi can be substantial in both marine and freshwater ecosystems. For example, fungi have been shown to be important decomposers of both standing-dead *Spartina* leaves in salt marshes and terrestrial leaf litter in streams. In these habitats, fungal biomass can exceed 10% of total detrital mass and may account for a significant proportion of leaf mass loss. In addition, fungi can immobilize nutrients such as nitrogen and phosphorus. The dominant species of fungi in these habitats typically possess the enzymatic potential necessary to degrade the structural components of leaf litter. In addition, there is some evidence to implicate fungi as agents of leaf decomposition in mangrove swamps. In other habitats, such as freshwater wetlands, fungi are probably important as well but these systems have not received enough attention to assess the role of fungi in them with any confidence. In the systems which have been studied so far, fungus-mediated decomposition leads to the mineralization of leaf carbon to CO_2 as well as to the production of other forms of organic matter such as DOM and FPOM. Fungal activity and, consequently, leaf breakdown rates are regulated both by internal (litter quality) and external (temperature, nutrient concentrations) factors. As fungi grow in leaf litter, their production is partitioned between mycelium and reproductive structures. How this allocation of resources is controlled is, however, not yet understood.

Acknowledgments. We are grateful to Steven Y. Newell for providing us with unpublished data and commenting critically on the manuscript. We also thank Jolanta Nunley for help in obtaining the electron micrograph, and Frank Brodrecht, Geneviève Guiraud, Marie-Hélène O'Donoghue, and Anne Schürer for help in organizing the literature. During the preparation of this chapter, M.O. Gessner was with the Center for Ecosystem Research, University of Kiel, Germany, which was supported by BMBF grant no. 0339077E.

References

Abdullah SK, Taj-Aldeen SJ (1989) Extracellular enzymatic activity of aquatic and aero-aquatic conidial fungi. Hydrobiologia 174:217–223

Adam P (1990) Salt marsh ecology. Cambridge University Press, Cambridge, pp 344–348

Anderson NH, Sedell JR (1979) Detritus processing by macroinvertebrates in stream ecosystems. Annu Rev Entomol 24:351–377

Apinis AE, Taligoola HK (1974) Biodegradation of *Phragmites communis* Trin. by fungi. In: Kilbertus G, Reisinger O, Concela Da, Fonseca JA (eds) Biodegradation et humification. Pierron, pp 24–32

Baldy V, Gessner MO, Chauvet E (1995) Bacteria, fungi and the breakdown of leaf litter in a large river. Oikos 74:93–102

Bärlocher F (1982) Conidium production from leaves and needles in four streams. Can J Bot 60:1487–1494

Bärlocher F (1985) The role of fungi in the nutrition of stream invertebrates. Bot J Linn Soc 91:83–94

Bärlocher F (1992) The ecology of aquatic hyphomycetes. Springer, Berlin Heidelberg New York (Ecological Studies, vol 94)

Bärlocher F, Biddiscombe NR (1996) Geratology and decomposition of *Typha latifolia* and *Lythrum salicaria* in a freshwater marsh. Arch Hydrobiol 136:309–325

Bärlocher F, Kendrick B (1974) Dynamics of the fungal populations on leaves in a stream. J Ecol 62:761–791

Bärlocher F, Newell SY (1994) Growth of the saltmarsh periwinkle *Littoraria irrorata* on fungal and cordgrass diets. Mar Biol 118:109–114

Bärlocher F, Canhoto C, Graça MAS (1995) Fungal colonization of alder and eucalypt leaves in two streams in Central Portugal. Arch Hydrobiol 133:457–470

Benner R, Moran MA, Hodson RE (1986) Biogeochemical cycling of lignocellulosic carbon in marine and freshwater ecosystems: relative contributions of procaryotes and eucaryotes. Limnol Oceanogr 31:89–100

Benner R, Hatcher PG, Hedges JI (1990) Early diagenesis of mangrove leaves in a tropical estuary: bulk chemical characterization using solid-state ^{13}C NMR and elemental analyses. Geochim Cosmochim Acta 54:2003–2013

Bergbauer M, Newell SY (1992) Contribution to lignocellulose degradation and DOC formation from a salt marsh macrophyte by the ascomycete *Phaeosphaeria spartinicola*. FEMS Microbiol Ecol 86:341–348

Bergbauer M, Moran MA, Hodson RE (1992) Decomposition of lignocellulose from a freshwater macrophyte by aero-aquatic fungi. Microb Ecol 23:159–167

Bird GA, Kaushik NK (1981) Coarse particulate organic matter in streams. In: Lock MA, Williams DD (eds) Perspectives in running water ecology. Plenum Press, New York, pp 41–68

Blum LK, Mills AL, Zieman JC, Zieman RT (1988) Abundance of bacteria and fungi in seagrass and mangrove detritus. Mar Ecol Prog Ser 42:73–78

Boulton AJ, Boon PI (1991) A review of methodology used to measure leaf litter decomposition in lotic environments: time to turn over an old leaf? Aust J Mar Freshw Res 42:1–43

Bremer GB (1995) Lower marine fungi (Labyrinthulomycetes) and the decay of mangrove leaf litter. Hydrobiologia 295:89–95

Brinson MM, Lugo AE, Brown S (1981) Primary productivity, decomposition and consumer activity in freshwater wetlands. Annu Rev Ecol Syst 12:123–161

Buchsbaum R, Valiela I, Swain T, Dzierzeski M, Allen S (1991) Available and refractory nitrogen in detritus of coastal vascular plants and macroalgae. Mar Ecol Prog Ser 72:131–143

Bunt JS (1995) Continental scale patterns in mangrove litter fall. Hydrobiologia 259:135–140

Buth GJC, Voesenek LACJ (1988) Respiration of standing and fallen plant litter in a Dutch salt marsh. In: Verhoeven JTA, Heil GW, Werger MJA (eds) Vegetation structure in relation to carbon and nutrient economy. SPB Academic, The Hague, pp 51–60

Buttimore CA, Flanagan PW, Cowan CA, Oswood MW (1984) Microbial activity during leaf decomposition in an Alaskan subarctic stream. Holarct Ecol 7:104–110

Calow P, Petts GE (1992) The rivers handbook. Hydrological and ecological principles, vol 1. Blackwell Scientific, Oxford

Chamier A-C (1985) Cell-wall-degrading enzymes of aquatic hyphomycetes: a review. Bot J Linn Soc 91:67–81

Chamier A-C, Dixon PA (1982) Pectinases in leaf degradation by aquatic hyphomycetes: the enzymes and leaf maceration. J Gen Microbiol 128:2469–2483

Chauvet E (1987) Changes in the chemical composition of alder, poplar, and willow leaves during decomposition in a river. Hydrobiologia 148:35–44

Chauvet E, Mercé J (1988) Hyphomycètes aquatiques: importance dans la décomposition des litières. Rev Sci Eau 1:203–216

Chergui H, Pattee E (1991) An experimental study of the breakdown of submerged leaves by hyphomycetes and invertebrates in Morocco. Freshw Biol 26:97–110

Day JW Jr, Hall CAS, Kemp WM, Yáñez-Arancibia A (1989) Estuarine ecology. Wiley, New York

Dowding P (1976) Allocation of resources, nutrient uptake and release by decomposer organisms. In: Anderson JM, Macfadyen A (eds) The role of terrestrial and aquatic organisms in decomposition processes. Blackwell Scientific, Oxford, pp 169–183

Elwood JW, Newbold JD, Trimble AF, Stark RW (1981) The limiting role of phosphorus in a woodland stream ecosystem: effects of P enrichment on leaf decomposition and primary producers. Ecology 62:146–158

Enríquez S, Duarte CM, Sand-Jensen K (1993) Patterns in decomposition rates among photosynthetic organisms: the importance of detritus C:N:P content. Oecologia 94:457–471

Fell JW, Master IM (1980) The association and potential role of fungi in mangrove detrital systems. Bot Mar 23:257–263

Fell JW, Master IM, Newell SY (1980) Laboratory model of the potential role of fungi in the decomposition of red mangrove (*Rhizophora mangle*) leaf litter. In: Tenore KR, Coull BC (eds) Marine benthic dynamics. University of South Carolina Press, Columbia, pp 359–372

Findlay RH, Fell JW, Coleman NK, Vestal JR (1986) Biochemical indicators of the role of fungi and thraustochytrids in mangrove detrital systems. In: Moss ST (ed) The biology of marine fungi. Cambridge University Press, Cambridge, pp 91–103

Findlay SEG, Arsuffi TL (1989) Microbial growth and detritus transformations during decomposition of leaf litter in a stream. Freshw Biol 21:261–269

Fisher PJ, Webster J (1981) Ecological studies on aero-aquatic hyphomycetes. In: Wicklow DT, Carroll GC (eds) The fungal community: its organization and role in the ecosystem. Dekker, New York, pp 709–730

Fisher PJ, Davey RA, Webster J (1983) Degradation of lignin by aquatic and aero-aquatic hyphomycetes. Trans Br Mycol Soc 80:166–168

Gallagher JL, Pfeiffer WJ (1977) Aquatic metabolism of the communities associated with attached dead shoots of salt marsh plants. Limnol Oceanogr 22:562–564

Gallagher JL, Kibby HV, Skirvin KW (1984) Community respiration of decomposing plants in Oregon estuarine marshes. Estuarine Coastal Shelf Sci 18:421–431

Gaur S, Singhal PK, Hasija SK (1992) Relative contribution of bacteria and fungi to water hyacinth decomposition. Aquat Bot 43:1–15

Gessner MO, Chauvet E (1993) Ergosterol-to-biomass conversion factors for aquatic hyphomycetes. Appl Environ Microbiol 59:502–507

Gessner MO, Chauvet E (1994) Importance of stream microfungi in controlling breakdown rates of leaf litter. Ecology 75:1807–1817

Gessner MO, Chauvet E (1997) Growth and production of aquatic hyphomycetes in decomposing leaf litter. Limnol Oceanogr (in press)

Gessner MO, Newell SY (1996) Bulk quantitative methods for the examination of eukaryotic organoosmotrophs in plant litter. In: Hurst CJ, Newell SY, Christian RR (eds) Manual of environmental microbiology, section IV. Aquatic environments. American Society for Microbiology, Washington, DC, pp 295–308

Gessner MO, Schwoerbel J (1991) Fungal biomass associated with decaying leaf litter in a stream. Oecologia 87:602–603

Gessner MO, Thomas M, Jean-Louis A-M, Chauvet E (1993) Stable successional patterns of aquatic hyphomycetes on leaves decaying in a summer cool stream. Mycol Res 97:163–172

Gessner RV (1977) Seasonal occurrence and distribution of fungi associated with *Spartina alterniflora* from a Rhode Island estuary. Mycologia 69:477–491

Gessner RV (1980) Degradative enzyme production by salt-marsh fungi. Bot Mar 23:133–139

Golladay SW, Sinsabaugh RL (1991) Biofilm development on leaf and wood surfaces in a boreal river. Freshw Biol 25:437–450

Griffith MB, Perry SA (1994) Fungal biomass and leaf litter processing in streams of different water chemistry. Hydrobiologia 294:51–61

Griffith MB, Perry SA, Perry WB (1995) Leaf litter processing and exoenzyme production on leaves in streams of different pH. Oecologia 102:460–466

Harley JL (1971) Fungi in ecosystems. J Appl Ecol 8:627–642

Harrison PG (1989) Detrital processing in seagrass systems: a review of factors affecting decay rates, remineralization and detritivory. Aquat Bot 23:263–288

Hasija SK, Singhal PK (1991) Degradation of plant litter by aquatic hyphomycetes. In: Arora DK, Rai B, Mukerji KG, Knudsen GR (eds) Handbook of applied mycology: soil and plants, vol 1. Dekker, New York, pp 481–505

Hicks RE, Newell SY (1984) The growth of bacteria and the fungus *Phaeosphaeria typharum* (Desm.) Holm (Eumycota: Ascomycotina) in salt-marsh microcosms in the presence and absence of mercury. J Exp Mar Biol Ecol 78:143–155

Hobbie JE, Ford TE (1993) A perspective on the ecology of aquatic microbes. In: Ford TE (ed) Aquatic microbiology: an ecological approach. Blackwell Scientific, Oxford, pp 1–14

Hyde KD, Jones EBG (1988) Marine mangrove fungi. P S Z N I: Mar Ecol 9:15–33

Hyde KD, Lee SY (1995) Ecology of mangrove fungi and their role in nutrient cycling: what gaps occur in our knowledge? Hydrobiologia 295:107–118

Iversen TM (1973) Decomposition of autumn-shed beech leaves in a spring brook and its significance for the fauna. Arch Hydrobiol 72:305–312

Jenkins CC, Suberkropp K (1995) The influence of water chemistry on the enzymatic degradation of leaves in streams. Freshw Biol 33:245–253

Jones EBG, Byrne PJ (1976) Physiology of the higher marine fungi. In: Jones EBG (ed) Recent advances in aquatic mycology. Wiley, New York, pp 135–176

Kemp PF, Sherr BF, Sherr EB, Cole JJ (1993) Handbood of methods in aquatic microbial ecology. Lewis, Boca Raton

Kohlmeyer J, Kohlmeyer E (1979) Marine mycology. Academic Press, New York

Kohlmeyer J, Volkmann-Kohlmeyer B (1991) An illustrated key to the filamentous higher marine fungi. Bot Mar 34:1–61

Kok CJ, Haverkamp W, van der Aa HA (1992) Influence of pH on the growth and leaf-maceration ability of fungi

involved in the decomposition of floating leaves of *Nymphaea alba* in an acid water. J Gen Microbiol 138:103–108

Lawson DL, Klug MJ, Merritt RW (1984) The influence of the physical, chemical, and microbiological characteristics of decomposing leaves on the growth of the detritivore *Tipula abdominalis* (Diptera: Tipulidae). Can J Zool 62:2339–2343

Leightley LE (1980) Wood decay activities of marine fungi. Bot Mar 23:387–395

Maharning AR, Bärlocher F (1996) Growth and reproduction in aquatic hyphomycetes. Mycologia 88:80–88

Maltby L (1992a) Heterotrophic microbes. In: Calow P, Petts GE (eds) The river handbook: hydrological and ecological principles, vol 1. Blackwell, Oxford, pp 165–194

Maltby L (1992b) Detritus processing. In: Calow P, Petts GE (eds) The river handbook: hydrological and ecological principles, vol 1. Blackwell, Oxford, pp 331–353

Melillo JM, Naiman RJ, Aber JD, Linkins AE (1984) Factors controlling mass loss and nitrogen dynamics of plant litter decaying in northern streams. Bull Mar Sci 35:341–356

Mitsch W, Gosselink J (1993) Wetlands, 2nd edn. Van Nostrand Reinhold, New York

Moran MA, Hodson RE (1989) Bacterial secondary production on vascular plant detritus: relationship to detritus composition and degradation rate. Appl Environ Microbiol 55:2178–2189

Mouzouras R (1989) Decay of mangrove wood by marine fungi. Bot Mar 32:65–69

Mulholland PJ, Elwood AD. Newbold JD, Webster JR, Ferren LA, Perkins RE (1984) Phosphorus uptake by decomposing leaf detritus: effect of microbial biomass and activity. Verh Int Ver Limnol 22:1899–1905

Newell SY (1992a) Estimating fungal biomass and productivity in decomposing litter. In: Carroll GC, Wicklow DT (eds) The fungal community: its organization and role in the cosystem, 2nd edn. Dekker, New York, pp 521–561

Newell SY (1992b) Autumn distribution of marine Pythiaceae across a mangrove-saltmarsh boundary. Can J Bot 70:1912–1916

Newell SY (1993a) Decomposition of shoots of a salt-marsh grass: methodology and dynamics of microbial assemblages. In: Jones JG (ed) Advances in microbial ecology, vol 13. Plenum Press, New York, pp 301–326

Newell SY (1993b) Membrane-containing fungal mass and fungal specific growth rate in natural samples. In: Kemp PF, Sherr BF, Sherr EB, Cole JJ (eds) Handbook of methods in aquatic microbial ecology. Lewis, Boca Raton, pp 579–584

Newell SY (1994a) Ecomethodology for organoosmotrophs: prokaryotic unicellular versus eukaryotic mycelial. Microb Ecol 28:151–157

Newell SY (1994b) Total and free ergosterol in mycelia of saltmarsh ascomycetes with access to whole leaves or aqueous extracts of leaves. Appl Environ Microbiol 60:3479–3482

Newell SY (1996) Established and potential impacts of eukaryotic mycelial decomposers in marine/terrestrial ecotones. J Exp Mar Biol Ecol 200:187–206

Newell SY, Fallon RD (1991) Toward a method for measuring fungal instantaneous growth rates in field samples. Ecology 72:1547–1559

Newell SY, Fell JW (1992a) Ergosterol content of living and submerged, decaying leaves and twigs of red mangrove. Can J Microbiol 38:979–982

Newell SY, Fell JW (1992b) Distribution and experimental responses to substate of marine oomycetes (*Halophytophthora spp.*) in mangrove ecosystems. Mycol Res 96:851–856

Newell SY, Fell JW (1995) Do halophytophtoras (marine Pythiaceae) rapidly occupy fallen leaves by intraleaf mycelial growth? Can J Bot 73:761–765

Newell SY, Wasowski J (1995) Sexual productivity and spring intramarsh distribution of a key saltmarsh microbial secondary producer. Estuaries 18:241–249

Newell SY, Fallon RD, Cal Rodriguez RM, Groene LC (1985) Influence of rain, tidal wetting and relative humidity on release of carbon dioxide by standing-dead salt-marsh plants. Oecologia 68:73–79

Newell SY, Miller JD, Fallon RD (1987) Ergosterol content of salt-marsh fungi: effect of growth conditions and mycelial age. Mycologia 79:688–695

Newell SY, Fallon RD, Miller JD (1989) Decomposition and microbial dynamics for standing, naturally positioned leaves of the salt-marsh grass *Spartina alterniflora*. Mar Biol 101:471–481

Newell SY, Arsuffi TL, Kemp PF, Scott LA (1991) Water potential of standing-dead shoots of an intertidal grass. Oecologia 85:321–326

Newell SY, Moran MA, Wicks R, Hodson RE (1995) Productivities of microbial decomposers during early stages of decomposition of leaves of a freshwater sedge. Freshw Biol 34:135–148

Newell SY, Porter D, Lingle WL (1996) Lignocellulolysis by ascomycetes (Fungi) of a saltmarsh grass (smooth cordgrass). Microsc Res Techn 33:32–46

Oremland RS, Capone DG (1988) Use of "specific" inhibitors in biogeochemistry and microbial ecology. In: Marshall KC (ed) Advances in microbial ecology, vol 10. Plenum Press, New York, pp 285–383

Padgett DE (1993) Distinguishing bacterial from nonbacterial decomposition of *Spartina alterniflora* by respirometry. In: Kemp PF, Sherr BF, Sherr EB, Cole JJ (eds) Handbook of methods in aquatic microbial ecology. Lewis, Boca Raton, pp 465–469

Polunin NVC (1984) The decomposition of emergent macrophytes in fresh water. Adv Ecol Res 14:115–166

Raghukumar C, Nagarkar S, Raghukumar S (1992) Association of thraustochytrids and fungi with living marine algae. Mycol Res 96:542–546

Read SJ, Moss ST, Jones EBG (1992) Attachment and germination of conidia. In: Bärlocher F (ed) The ecology of aquatic hyphomycetes. Ecological studies, vol 94. Springer, Berlin Heidelberg New York, pp 135–151

Robertson AI, Alongi DM, Boto KG (1992) Food chains and carbon fluxes. In: Robertson AI, Alongi DM (eds), Tropical mangrove ecosystems. Coastal and estuarine studies, vol 41. American Geophysical Union, Washington, DC, pp 293–325

Rosset J, Bärlocher F, Oertli JJ (1982) Decomposition of conifer needles and deciduous leaves in two Black Forest and two Swiss Jura streams. Int Revue Ges Hydrobiol 67:695–711

Saenger P, Snedaker SC (1993) Pantropical trends in mangrove above-ground biomass and annual litterfall. Oecologia 96:293–299

Sathe V, Raghukumar S (1991) Fungi and their biomass in detritus of the seagrass *Thalassia hemprichii* (Ehernberg) Ascherson. Bot Mar 34:271–277

Schaumann K (1993) Marine Pilze. In: Meyer-Reil L-A, Köster M (eds) Mikrobiologie des Meeresbodens. Fischer, Jena, pp 144–195

Shearer CA (1992) The role of woody debris. In: Bärlocher F (ed) The ecology of aquatic hyphomycetes. Ecological studies, vol 94. Springer, Berlin Heidelberg New York, pp 77–98

Shearer CA (1993) The freshwater Ascomycetes. Nova Hedwigia 56:1–33

Singh N, Steinke TD (1992) Colonization of decomposing leaves of *Bruguiera gymnorrhiza* (Rhizophoraceae) by fungi, and in vitro cellulolytic activity of the isolates. S Afr J Bot 58:525–529

Sinsabaugh RL, Antibus RK, Linkins AE, McClaugherty CA, Rayburn L, Repert D, Weiland T (1993) Wood decomposition: nitrogen and phosphorus dynamics in relation to extracellular enzyme activity. Ecology 74:1586–1593

Stanford JA, Covich AP (1988) Community structure and function in temperate and tropical streams. J N Am Benthol Soc 7:261–529

Steinke TD, Ward CJ (1987) Degradation of mangrove leaf litter in the St Lucia Estuary as influenced by season and exposure. S Afr J Bot 53:323–328

Steinke TD, Barnabas AD, Somaru R (1990) Structural changes and associated microbial activity accompanying decomposition of mangrove leaves in Mgheni Estuary. S Afr J Bot 56:39–48

Suberkropp K (1991) Relationships between growth and sporulation of aquatic hyphomycetes on decomposing leaf litter. Mycol Res 95:843–850

Suberkropp K (1992a) Interactions with invertebrates. In: Bärlocher F (ed) The ecology of aquatic hyphomycetes. Ecological studies, vol 94. Springer, Berlin Heidelberg New York, pp 118–133

Suberkropp K (1992b) Aquatic hyphomycete communities. In: Carroll GC, Wicklow DT (eds) The fungal community: its organization and role in the ecosystem, 2nd edn. Dekker, New York, pp 729–747

Suberkropp K (1995) The influence of nutrients on fungal growth, productivity and sporulation during leaf breakdown in streams. Can J Bot 73(Suppl 1):S1361–S1369

Suberkropp K, Chauvet E (1995) Regulation of leaf breakdown by fungi in streams: influences of water chemistry. Ecology 76:1433–1445

Suberkropp K, Jones EO (1991) Organic phosphorus nutrition of some aquatic hyphomycetes. Mycologia 83:665–668

Suberkropp K, Klug MJ (1976) Fungi and bacteria associated with leaves during processing in a woodland stream. Ecology 57:707–719

Suberkropp K, Klug MJ (1980) The maceration of deciduous leaf litter by aquatic hyphomycetes. Can J Bot 58:1025–1031

Suberkropp K, Klug MJ (1981) Degradation of leaf litter by aquatic hyphomycetes. In: Wicklow DT, Carroll GC (eds) The fungal community: its organization and role in the ecosystem. Dekker, New York, pp 761–775

Suberkropp K, Arsuffi TL, Anderson JP (1983) Comparison of degradative ability, enzymatic activity, and palatability of aquatic hyphomycetes grown on leaf litter. Appl Environ Microbiol 46:237–244

Suberkropp K, Gessner MO, Chauvet E (1993) Comparison of ATP and ergosterol as indicators of fungal biomass associated with decomposing leaves in streams. Appl Environ Microbiol 59:3367–3372

Thornton DR (1965) Amino acid analysis of fresh leaf litter and the nitrogen nutrition of some aquatic hyphomycetes. Can J Microbiol 11:657–662

Trozilli AP (1982) Polysaccharidase production and cell wall degradation by several salt marsh fungi. Mycologia 74:297–302

Torzilli AP, Andrykovitch G (1986) Degradation of *Spartina* lignocellulose by individual and mixed cultures of salt-marsh fungi. Can J Bot 64:2211–2215

Ulken A (1984) The fungi of the mangal ecosystem. In: Por FD, Dor I (eds) Hydrobiology of the mangal. The ecosystem of the mangrove forests. Developments in hydrobiology, vol 20. Junk, The Hague, pp 27–33

Webster J (1987) Convergent evolution and the functional significance of spore shape in aquatic and semi-aquatic fungi. In: Rayner ADM, Brasier CM, Moore D (eds) Evolutionary biology of the fungi. Cambridge University Press, Cambridge, pp 191–200

Webster J, Descals E (1981) Morphology, distribution,and ecology of conidial fungi in freshwater habitats. In: Cole GT, Kendrick B (eds) Biology of conidial fungi, vol 1. Academic Press, New York, pp 295–355

Webster JR, Benfield EF (1986) Vascular plant breakdown in freshwater ecosystems. Annu Rev Ecol Syst 17:567–594

Weigelhofer G, Waringer JA (1994) Allochthonous input of coarse particulate organic matter (CPOM) in a first to fourth order Austrian forest stream. Int Revue Ges Hydrobiol 79:461–471

Weyers HS, Suberkropp K (1996) Fungal and bacterial production during the breakdown of yellow poplar leaves in two streams. J N Am Benthol Soc 15:408–420

Wicklow DT, Carroll GC (1981) (eds) The fungal community: its organization and role in the ecosystem. Dekker, New York, pp 679–776

Zare-Maivan H, Shearer CA (1988) Extracellular enzyme production and cell wall degradation by freshwater lignicolous fungi. Mycologia 80:365–375

Zemek J, Marvanová L, Kuniak L, Kadlecíková B (1985) Hydrolytic enzymes in aquatic Hyphomycetes. Folia Microbiol 30:363–372

20 Biomass and Productivity Estimates in Solid Substrate Fermentations

M.J.R. Nout[1], A. Rinzema[1], and J.P. Smits[2]

CONTENTS

I. Introduction

There is an overlap of interests amongst microbial ecologists studying fungal decay of litter and wood, or hyphal growth in soil, and fermentation microbiologists involved in solid substrate fermentations (SSF) in the sense that they all need to quantify fungal biomass in materials from which it cannot be easily extracted. In addition, there is a common interest in biochemical and metabolic activities of the fungal mass.

[1] Department of Food Science, Agricultural University, Bomenweg 2, 6703 HD Wageningen, The Netherlands
[2] Division Agrotechnology and Microbiology, TNO Nutrition and Food Research Institute, P.O. Box 36, 3700 AJ Zeist, The Netherlands

Fermentations are the result of substrate modification by microbial activity, and aim to obtain marketable products ranging from foods and beverages, animal feeds, enzymes, to antibiotics, etc. The physical state of the substrate to be modified varies, and is closely associated with the type of microorganism to be used, as well as the reactor in which the fermentation takes place. In principle, two techniques are distinguished, namely SmF (submerged fermentation, or liquid fermentation) and SSF (solid substrate fermentation).

In liquid fermentations, water-soluble substrate is present in an aqueous continuous phase, and stirred tank reactors are most commonly used in such fermentations in which bacteria, yeasts, as well as many molds, can be grown.

This chapter deals with solid substrate fermentation (SSF), characterized by a porous substrate with a gaseous continuous phase. Although the porous solid substrate contains water with dissolved water-soluble substrate components, free flowing water might be absent (Lonsane et al. 1985). The latter type of fermentation is also referred to as solid-state fermentation. Here, the physical conditions are particularly suitable for filamentous fungi. SSF is of economic importance in the production of fermented foods (Nout 1995) and feeds, as well as microbial metabolites and biomass, e.g., spores, Table 1 summarizes major SSF applications.

Compared with liquid fermentations, the major attractive aspects of SSF are (1) better product quality (e.g., spores, enzyme mixtures); (2) higher product yield and concentration (e.g., enzymes, flavors, gibberellic acid; Kumar and Lonsane 1987) which may reduce downstream processing costs; (3) cheaper processing of many foods and feeds which are solid in nature; (4) potential for less wastewater and off-gas production; (5) furthermore, it has been claimed that SSF enables smaller reactor volumes to be used; this still needs to be demonstrated at a large-scale production level. On the other hand, SSF has drawbacks com-

The Mycota IV
Environmental and Microbial Relationships
Wicklow/Söderström (Eds.)
© Springer-Verlag Berlin Heidelberg 1997

Table 1. Some representative solid substrate fermentations

Microorganism	Substrate	Product	Fermentation system	Reference
Rhizopus, Neurospora spp.	Cassava roots	Detoxified flour	Heap	Essers et al. (1994)
Lactic acid bacteria	Chopped maize	Silage (animal feed)	Pit	Cannel and Moo-Young (1980)
Rhizopus oligosporus *Rhizopus oryzae*	Soybean	Tempe	Tray	Nout and Rombouts (1990)
Trichoderma viride	Sugar beet pulp	Edible protein	Packed bed	Durand et al. (1988)
Aspergillus awamori	Wheat bran	Enzymes	Rotating drum	Silman (1980)
Saccharomyces cerevisiae	Potatoes	Proteins	Fluidized bed	Hong et al. (1989)
Aspergillus oryzae	Rice	Koji (fermentation starter)	Agitated tank	Cook and Campbell-Platt (1994)
Aspergillus phoenicis	Sugar beet pulp	β-Glucosidase	Agitated tank	Deschamps and Huet (1984)
Lactobacilli and yeasts	Animal waste and maize	Animal feed	Continuous	Hrubant et al. (1989)
Beauveria bassiana	Clay granules	Spores for biological pest control	Packed bed	Desgranges et al. (1993)

pared to liquid fermentations, notably its heterogeneity and restricted transport of mass and energy, which tend to result in overheating, local dehydration, depletion of oxygen, and accumulation of carbon dioxide if no process control measures are applied.

It is of interest to compare the rationale of biomass estimation and analytical approach of microbial ecologists studying fungal communities in, e.g., soil and litter, with those of biotechnologists involved in SSF. As formulated by Newell (1992), major aims of estimating biomass in microbial ecology of litter and soil are (1) to monitor fungal colonization patterns and decomposition rates; (2) to study dynamics of fungal mycelial growth throughout the season; and (3) to arrive at an understanding of, and describe nutrient flows and environmental controls on decay rates of litter. In fermentation technology, the scenario can be quite different from that in nature. The fungi used as inocula are (in most cases) of known origin and have been studied and selected as pure cultures in the laboratory; fermentation conditions are kept under close control and are optimized as well as possible; and consequently the incubation times during fermentation are as short as possible.

The major reason for biotechnologists to estimate biomass or related properties of fungi is that processes must be designed, monitored, and controlled. Ideally, predictive statements on performance must be made, often with the help of computer models. The amount of biomass in food

fermentation is usually significantly higher than in soil and, to an extent, in litter. Fermentations usually take a few days to complete, during which period a rather synchronous growth takes place from germinating spores to adult sporulated mycelium. Consequently, age effects can be very pronounced compared with the situation in nature. Because of the high biomass density in SSF, overheating and disturbance of the gas atmosphere becomes a problem. Metabolic activity consequently plays a prominent role in SSF research.

Perhaps a special, and challenging aspect of SSF ecology is the study of mixed fungal populations. The monitoring of individual fungal species as part of a mixed flora requires species-specific detection techniques. Although conditions in soil and litter microbiology may differ considerably from those in fermentations, there is sufficient overlap of motives and interest of microbiologists to monitor biomass; consequently, there is ample opportunity for exchange and sharing of expertise.

In the context of this chapter, a definition of biomass is very much inspired by the purpose of the fermentation process. In principle, total bio- and necromass (TBNM) would include all mycelial matter, together with asexual as well as sexual bodies. In addition, some extracellular material (e.g., polysaccharides) adhering to the hyphae could be considered to belong to TBNM. More specific types of biomass could be termed: growing biomass (GBM), biologically active biomass (ABM), and reproductive biomass (RBM). From a functional point of view, these types of biomass would serve distinct purposes.

Table 1 shows that SSF is used for the production of edible microbial matter (protein, feed, tempe) In such cases, the total bio- and necromass (TBNM) is of relevance. No distinction will be made between bio- and necromass for feeding purposes. During the fermentation process, all that is important is to monitor TBNM and achieve a maximum yield of dry matter.

The situation may be different where the production of metabolites is the aim of the fermentation process. In this situation, the biomass of relevance is the GBM and ABM part, notably the hyphal tips (peripheral zone) and the hyphae directly behind them (productive zone) (Bull and Trinci 1977) Much of the older mycelium and spore-bearing and fruiting bodies are metabolically less interesting, except in cases where desirable substances such as antiinsectan compounds need to be extracted from, e.g., fungal sclerotia (Wicklow et al. 1994). Optimization of such SSF would require an understanding and exploitation of the cause-effect relationship between active metabolism and metabolite production (Huang et al. 1984). Hence, in such situations, it is more relevant to monitor and, if possible, predict the fungal growth and production of biologically active substances.

In fermentation technology, the production of starter material (inoculum) of maximum activity is essential. In fungal SSF, most inocula consist of conidiospores or sporangiospores. Mycelium fragments usually have relatively poor reproductive ability. It is thus obvious that during inoculum production the yield of RBM biomass is the property to be monitored.

The above examples serve to indicate that the purpose of the fermentation is an important factor determining the fungal properties of relevance for process monitoring and control. Another important aspect affecting the choice of methodology is the substrate used in the fermentation: can sensors be introduced into it, or may the substrate interfere with measuring techniques? In addition, it is important to realize that there is a variety of fermentation systems (e.g., with stationary, or with agitated substrate; using batch or continuous flow approaches) which may limit the choice of sampling and sensing devices to be used for process monitoring and control.

Probably one of the most challenging aspects of SSF is the penetration and attachment of fungal mycelium into the substrate, making it impracticable to physically separate it for the purpose of quantification. Thus, monitoring of the progression of SSF often has to rely on in-situ (in the fermenting product) measurements. This may be done by direct methods (measuring biomass, either TBNM, GBM, ABM, or RBM) or indirectly (measuring chemical constituents or metabolites). A distinction will be made between off-line (requiring samples to be taken which are to be analyzed in the laboratory) and on-line approaches (using sensors placed in the fermentation reactor, resulting in direct signals).

Special attention will also be paid to the use of simulation media in the fundamental study of fungal growth and metabolism. Simulation media may enable the researcher to overcome the problem of fungal invasion of natural substrate, because they can either be degraded, or they are inert and enable a physical separation of fungal mass from the substrate.

The estimation and prediction of biomass productivity is of essential importance in industrial fermentation systems. A critical overview will be given of the various approaches to describe fungal productivity by way of mathematical models.

Several excellent reviews were published on the subject of fungal biomass determination from the food science (Jarvis et al. 1983) as well as the natural ecology point of view (Dighton and Kooistra 1993; Newell 1992).

In the following parts of this chapter, an overview will be presented of methods used for biomass monitoring in SSF research. This overview cannot pretend to be exhaustive, and the reader is encouraged to consult, e.g., the extensive review of Newell (1992) in addition.

II. Methodology

A distinction will be made between off-line and on-line approaches to monitor biomass. In the first case, a sample of fermenting material is drawn and analyzed, typically taking as little as 30 min to as long as a few days before results are available. In fermentation control, feedback actions (i.e., adjustments and fine-tuning of reactor settings determining environmental conditions, or determination of the right moment for harvesting) require a minimum of delay caused by analysis. On-line approaches aim to provide real-time data

Table 2. Monitoring fungal matter in solid substrate fermentations

Method	Principle	Scientific origin	Application	Limitations	Reference
Simulation models					
1. Colony growth	Hyphal extension	Ecology	Growth rate on agar media	Colony density variable	De Reu et al. (1995); Desgranges and Durand (1990); Gervais et al. (1988); Wiegant et al. (1992)
2. Gravimetry	Separation from substrate and weighing	?	Total bio- and necromass	Not always representative of SSF	De Ruiter et al. (1992b); Gutierrez-Rojas et al. (1995); Larroche et al. (1986); Lhomme and Roux (1992); Matcham et al. (1985); Nout et al. (1987); Zhu et al. (1994)
Off-Line					
1. Microscopy	Visualization in agar film or membrane filter; vital staining optional	Soil and food microbiology	Total biovolume and bio- and/or necromass; morphology	Recovery, extraction, homogenization, detection	Cook et al. (1991); Davey et al. (1991); Hyeong-Cheol et al. (1995); Mitchell et al. (1990); Newell (1992); Packer and Thomas (1990); Scheu and Parkinson (1994); Schnürer (1993)
2. Biochemistry					
Chitin/ glucosamine	Cell wall component	SSF	Total bio- + necromass	Nonfungal chitin, mycelial age	Arima and Uozumi (1967); Boyle and Kropp (1992); Ito (1993); Matcham et al. (1985); Narahara et al. (1982); Wissler et al. (1983)
Ergosterol	Sterol of fungal membrane	SSF, food	Total (bio?) mass	Substrate composition, oxygen level, method of analysis	Newell et al. (1988); Nout et al. (1987); Schnürer (1993); Seitz et al. (1979)
Protein	Nitrogen *6.25* conversion factor	SSF, food	Protein enrichment; less suitable for biomass estimation	Conversion factors, substrate nitrogen level, age	Auria et al. (1993); Boyle and Kropp (1992); Castillo et al. (1994); Gordon et al. (1990); Grajek and Gervais (1987); Sargantanis et al. (1993); Saucedo-Castenada et al. (1990); Xue et al. (1992)
Enzyme activities	Laccase, reductase, esterase	Mushrooms, SSF	Metabolic activity	Interpretation needs species-specific knowledge	Boyle and Kropp (1992); Le Page et al. (1988); Matcham et al. (1985); Wiegant et al. (1992)
Immuno-assay	Specific reaction with extracellular polysaccharides	Food, biotechnology	Total bio- and necromass	Cross-reactivity	De Ruiter et al. (1992a,b); Newell (1992)
ATP	Intracellular metabolism	Food	Metabolic activity	ATP content of substrate	Cochet et al. (1984)
DNA	Protein synthesis	?	Growth	Limited calibration; specificity?	Bajracharya and Mudgett (1980)

Table 2. *Continued*

Method	Principle	Scientific origin	Application	Limitations	Reference
3. Microbiology Viable count	Living propagules	?	Starter production; differentiation of mixed cultures	Uneven distribution; variability; emphasis on sporulation	Cook et al. (1991); Jarvis et al. (1983); Molimard et al. (1995); Nout (1984); Schnürer (1993), Silman et al. (1983);
On-Line **1. Respiration** Substrate-induced respiration	Metabolism of added C-source during short incubation period	Soil	Biomass in substrate-poor medium	Not practicable in most foods	Newell (1992); Wiegant (1991)
CO_2 in effluent gas	CO_2 from growth and maintenance	SSF	Control and modeling	Ratio growth: maintenance varies with age	Ito (1993); Larroche et al. (1986); Narahara et al. (1982); Saucedo-Castenada et al. (1990)
2. Heat production Thermocouples in product, or non-contact infrared or heatflux sensors	Metabolic heat	SSF	Process control		De Reu et al. (1993); Saucedo-Castenada et al. (1990); Silman (1988)
3. Electrical properties Conductance and Impedance	Electrical changes in substrate due to metabolites	Food	Monitoring of first phase of growth	Sensor placement in SSF critical	Jarvis et al. (1993); Owens et al. (1992)
Capacitance	Electrical charges of microbial membranes	Food, SSF	Signal of biomass	Sensor placement in SSF critical	Davey et al. (1991); Penaloza et al. (1992)
4. Volatiles	Headspace GLC of metabolites	Food	Metabolism and sporulation	Needs further development	Borjesson et al. (1993)
5. Electromagnetic sensing FTIR-PAS	Protein spectrum	SSF	Mycelial (?) protein	Requires substrate-specific calibration	Gordon et al. (1990)
RIR	Spectrum of water, lipid, starch, protein	SSF	Biochemical activity and substrate modification	Requires substrate-specific calibration	Silman et al. (1983)
NIR	Spectra of glucosamine and ergosterol	SSF	Total bio- and necromass	?	Desgranges et al. (1991b)
Reflected light	Color changes due to biomass and biochemical substrate modification	SSF	Total bio- and necromass, and substrate modification	Needs substrate-specific calibration	Ramana Murthy et al. (1993)
6. Consistency Compression, rupture, penetration tests	Firmness by mycelial binding	SSF	Biomass and substrate modification	Needs substrate-specific calibration	Ariffin et al. (1994); Auria et al. (1993); Kronenberg and Hang (1985); Nout et al. (1985)
Manometry	Pressure drop in reactor	SSF	Biomass and substrate modification	Needs substrate- and reactor-specific calibrations	Auria et al. (1993)

using sensors in the reactor vessel, thus omitting sampling.

For the purpose of calibration of indirect methods, and for laboratory-scale fundamental investigations, several simulations of SSF conditions are created, enabling a direct, often gravimetric, quantification of biomass. As many of the in-situ methods have been compared with direct biomass determinations under simulated conditions, we will first consider these. As a general guide, Table 2 summarizes the major approaches and their relevant properties.

A. Simulation Media

1. Inert Carriers and Gravimetry

Several studies on fungal biomass have been related to total mass produced in liquid shaking cultures (Matcham et al. 1985; De Ruiter et al. 1992b). This method has its roots in classical microbiology; it gives a direct measure of TBNM without distinction between biomass and necromass (weight $TBNM g^{-1}$ substrate). Obviously, this is the simplest way to produce easily separable fungal mass, i.e., by filtration followed by washing and drying. However, conditions in liquid cultures may not be representative of SSF situations. In the area of SSF research a number of methods employing inert carriers were developed. We found that the porous structure of solid substrate could be simulated in agar-solidified filterable growth media (Nout et al. 1987). The agar medium may be used in the shape of particles, and can be separated from the biomass by melting and diluting in water in a microwave oven. A different approach is to use inert carriers such as pumice stones, vermiculite (Lhomme and Roux 1992), amberlite resin (Gutierrez-Rojas et al. 1995), or pozzolano (Larroche et al. 1986), which can be imbibed with nutrient solutions. Total fungal mass can be weighed after extracting the medium by washing, followed by drying. Compared to these rigid materials, the use of fibrous or elastic inert carriers such as polyester, glass wool (Lhomme and Roux 1992), or polyurethane foam (Ozawa et al. 1993; Zhu et al. 1994) has even more advantages, as the fermentation liquid can be squeezed out and can be replaced with fresh medium for semicontinuous cultivation. In many cases, no attempt is made to physically separate the fungal mass from its carrier; a weight difference after drying is assumed to be the bio- plus necromass. It is clear that the weight ratio of fungal mass: carrier will influence the accuracy and significance of such mass data.

2. Estimation of Specific Growth Rate from Colony Diameter Increase

Perhaps the simplest approach to quantifying fungal growth is to measure, as a function of incubation time, the diameter of colonies emerging from a single-point inoculation on solid agar media in petri dishes. This approach, which originates from fungal ecology, is a direct method with mass expressed as (area of $TBNM g^{-1}$ substrate); no distinction can be made between biomass and necromass. Assuming that the fungal colony is homogeneous, its extension could be used to estimate the specific growth rate μ as a function of, e.g., water activity (Gervais et al. 1988; Gibson et al. 1994), CO_2 concentration in the gas phase (Desgranges and Durand 1990), or of interspecies interactions (Wiegant et al. 1992). Trinci (1969, 1971) demonstrated that the effect of temperature on the radial growth rate of fungal colonies reflects the effect on the specific growth rate. A constant colony density is not a prerequisite for this type of measurement, but a constant length of the peripheral growth zone and a constant ratio between colony radial growth rate and linear hyphal extension rate are required. Nevertheless, this approach has only limited value, as Molin et al. (1992) demonstrated that the latter requirement is not fulfilled when the effect of the water activity is studied, because the distribution of the branching angle changes with water activity.

In this approach, the use of pregrown microcolonies (De Reu et al. 1995) achieved highly reproducible growth data. However, these authors also observed that the colony diameter is not a reliable index for TBNM as the density of fungal colonies was found to be strongly influenced by, e.g., nutrients and O_2 levels (De Reu et al. 1995).

B. Off-Line Approaches

1. Microscopy

Direct microscopic detection of fungal material has been in use in mycology and in food quality control since as early as 1911. The Howard mold count as developed for tomato products lacks precision (Jarvis et al. 1983) as the food matrix masks an unknown number of fungal parts and the method lacks the crucial homogenization step. In principle, microscopic examination of fungal fermented materials offers unique opportunities, e.g., to study the penetration of hyphae into solid media (Mitchell et al. 1990), to examine the mycelial morphology (i.e., total hyphal length, mean hyphal length, number of growth tips, branching frequency) as an indicator of metabolic activity (Packer and Thomas 1990), or to distinguish different organisms growing in mixed culture (Cook et al. 1991). Again, this approach is a direct method (length or area of BNM g^{-1} substrate) with a possibility of distinguishing NM from BM.

For quantitative determination of mycelial mass in solid fermented materials, the first step required is an extraction/homogenization. This may be relatively easy from a laboratory agar medium (Schnürer 1993) or from soil (Scheu and Parkinson 1994), but sticky or fatty fermented foods tend to smear, making separation of the biomass more difficult. A compromise needs to be found between adequate homogenization of the sample, and prevention of excessive damage to the biomass (Newell 1992).

Next, the homogenate is either diluted in agar and the mycelium is examined in agar films (Cook et al. 1991; Davey et al. 1991), or the mycelium is separated on a membrane filter for epifluorescence microscopy. The Jones-Mollison technique for agar film preparation was described in 1948 for the examination of soil samples and is still widely used. Epifluorescence (Scheu and Parkinson 1994) is suitable only when the fermented material can be adequately separated by prefiltration or sedimentation, failing which an extent of underestimation of biomass will occur (Newell 1992).

A seemingly simple approach to quantify fungal mass is by estimating its total hyphal length, in fermented products expressed as km g^{-1} dry substrate (Cook et al. 1991; Davey et al. 1991). However, the diameter of the mycelium, usually an average of measurements taken at a number of randomly chosen locations, is required to calculate the biovolume (Schnürer 1993). Finally, a conversion of biovolume to biomass is possible when the specific density and dry matter content (Schnürer 1993) of the particular mold are known. Conversion factors from volume to dry matter or biomass-carbon are rather variable (Newell 1992). Young mycelium had a higher specific density than old (empty?) mycelium (Newell and Statzell-Tallman 1982). This variability may result in considerable inaccuracy in materials containing a mixture of (unknown) molds of uncertain history. In order to achieve realistic mass data, the mold inocula should be calibrated for their density and dry matter content as a function of physiological conditions; this limits the biovolume approach to situations where the physiological conditions do not vary significantly during the fermentation. Thus, although the approach of estimating hyphal length may seem simple, in practice it has severe limitations.

For qualitative purposes, visualization of biomass can be greatly enhanced by contrast stains such as fluorescent brighteners (Jarvis et al. 1983), phenolic aniline blue (Davey et al. 1991), and calcofluor white M2R (Scheu and Parkinson 1994). Given the fact that polyclonal and monoclonal antibodies have been raised to the extracellular immunogenic polysaccharides of *Aspergillus*, *Penicillium*, and various mucoralean molds (De Ruiter et al. 1992a), the immunofluorescent marker technique should be useful for detection of their biomass. This has not yet been reported, however, and it may well be that confounding factors, e.g., autofluorescence of the food matrix, or variable or cross-reactivity create problems. In addition, image analysis technology can significantly speed up the process of examining cultures (Packer and Thomas 1990).

An important aspect is whether one should be interested in the total mass (TBNM) which has been produced, or only in the metabolically active part. A semiquantitative approach to metabolically active biomass employs fluorogenic substrates, notably FDA (Fluorescein Di Acetate) to visualize living spores and hyphae. In this method, which was described earlier for soil analysis (Hyeong-Cheol Yang et al. 1995; Newell 1992), FDA diffuses into the living cell and is split by the prevailing esterase activity. Fluorescein cannot leave the cell because of its polarity; using imaging technology, the intensity of the fluorescence can be quantified as an index of vitality. As one needs to distinguish the intracellular esterase from the extracellular environment, the extraction and homogenization treatment must be adequate but not destructive. Whereas the use of high-speed homogenizers is reported in many publications, we suggest the use of a stomacher-type homogenizer such as used in food microbiology (Nout et al. 1987).

2. (Bio)chemistry

As an alternative of direct detection and quantification of fungal mass, characteristic (bio)chemical compounds or activities could be measured as an index of mass and/or metabolic activity. Generally, such index compounds represent only a small fraction of the mycelial weight (i.e., glucosamine at about 2% of TBNM). Consequently, a weak point of such indirect approaches is that small deviations of detected levels of index compounds would translate to considerable variations in extrapolated mass values.

Glucosamine / Chitin. Most fungi contain chitin in their cell walls. Chitin can be chemically determined as its monomer, N-acetyl-glucosamine, or simply glucosamine. One of the first reports of this technique originates from SSF research and was aimed to monitor the growth of *Aspergillus oryzae* in rice koji (Arima and Uozumi 1967); the mold was reported to contain 114 mg glucosamine per gram of mycelium dry matter. The method is used widely in wood and litter microbiology (Jarvis et al. 1983; Boyle and Kropp 1992), as well as in some solid substrate fermentations (Narahara et al. 1982; Wissler et al. 1983; Matcham et al. 1985; Ito 1993). The basic assumption is that fungi possess equal amounts of glucosamine per unit mycelial mass in liquid as well as in solid cultures. Surprisingly, this has never been validated using model solid substrates. The glucosamine method (mg glucosamine g^{-1} substrate) is an indirect approach representing TBNM without distinguishing NM and BM.

According to Jarvis et al. (1983) and Newell (1992) there are several sources of variability associated with the glucosamine method. The most important relate to the complexity and variability of the chemical method, and to the significant increase of fungal glucosamine content with progressing mycelial age (Wissler et al. 1983; Boyle and Kropp 1992). A number of contradicting data have been published on this issue. Desgranges et al. (1991a) found that *Beauveria bassiana* cultivated on agar media for 120 h had a constant glucosamine:dry matter ratio, despite the fact that extensive sporulation occurred, whereas other biomass indicators had a varying ratio. However, the glucosamine content of the dry matter

varied considerably with medium composition. Ride and Drysdale (1971) obtained the same result for *Fusarium oxysporum* during 2 weeks. On the other hand, Whipps and Lewis (1980) have shown that the composition of *Fusarium oxysporum* changes with age after 5 weeks. In experiments with pregrown microcolonies of *Rhizopus oligosporus* on defined agar media, we observed a continuous increase of the glucosamine/dry matter ratio with age (Rinzema et al., unpublished, cf. Fig. 1). It has been assumed that age effects do not play a disturbing role in the fermentation of rice with *A. oryzae* (Narahara et al. 1982; Ito 1993), or of rye with *Agaricus bisporus* (Matcham et al. 1985). However, this assumption has not always been verified. For instance, Narahara et al. (1982) measured only glucosamine, CO_2, and O_2 without measuring fungal dry matter. Obviously, it is not possible to draw conclusions on age effects on the basis of such data. Clearly, age effects must be checked before glucosamine can be taken as an indicator for fungal dry matter, taking into consideration the relevant "time window" for the process under study.

The fact that nonfungal (i.e., invertebrate) biomass may contribute to glucosamine levels does not need to influence the results, since these levels will, in general, not change during fungal growth.

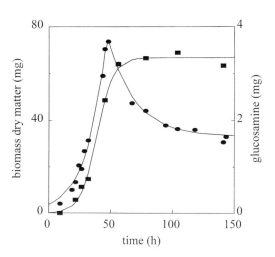

Fig. 1. Biomass dry matter (●) and glucosamine (■) development during cultivation of *R. oligosporus* colonies on agar with mineral glucose medium. *Symbols* indicate measurements, *lines* indicate fitted models. A combination of exponential growth and decay and the logistic law were used to fit the dry matter and glucosamine date, respectively. (A. Rinzema et al., unpubl.)

Ergosterol. Ergosterol is a membrane-associated sterol which occurs almost exclusively in fungi. This makes it more characteristic for fungal biomass than glucosamine. One of the first reported applications of the ergosterol method stem from the quality control of cereals and seeds (Seitz et al. 1979). This method is an indirect method (mg ergosterol g^{-1} BM) representing BM, including fungal spores (Desgranges et al. 1991a).

The analysis involves extraction in methanol and saponification of ergosteryl-exters, followed by HPLC with UV detection at 281 mm (Seitz et al. 1979, Newell et al. 1988; Schnürer 1993; Scheu and Parkinson 1994). There is a need for standardization, as sample pretreatment (drying, freezing, lyophilization) may lead to variable reductions of recovery (Newell et al. 1988). Of *Rhizopus oligosporus* biomass, 86% of ergosterol was extractable as such, whereas about 10% occurred as saponifiable esters (Nout et al. 1987). It would appear that some ergosterol could be extracted only after previous saponification of the sample (Nout et al. 1987), but this procedure had a lower overall recovery (Newell et al. 1988).

Reported ergosterol contents are usually of the order of 5 $\mu g\,mg^{-1}$ biomass dry weight, but significant variations may be expected from sources including species (Newell 1992) and growth conditions. It was suggested (Nout et al. 1987) and demonstrated (Desgranges et al. 1991a) that the substrate composition, e.g., lipid content, has a strong influence on ergosterol content of biomass. It was shown that reduced oxygen levels reduced the ergosterol content of *Rhizopus oligosporus* (Nout et al. 1987). On the other hand, mycelial age and growth temperature do not appear to have a great impact on the ergosterol content (Nout et al. 1987; Newell 1992). On the whole, this shows that the usefulness of ergosterol as a fermentation parameter depends very much on the extent of calibration of the fermentation organism.

An interesting aspect of ergosterol is that it is degraded in dying mycelium (Newell 1992), and correlates well with markers for metabolic activity (Scheu and Parkinson 1994). It was suggested that the combination of glucosamine and ergosterol determinations may enable a differentiation of NM vs. BM (Newell 1992). However, both chemical analyses are time-consuming and thus less suitable for monitoring relatively rapid SSF processes.

Protein. Fungal matter contains various proteins present in the cytoplasm, membrane as well as cell wall. In principle, fungal protein could thus be used as a chemical index for mass. Protein determinations such as commonly used in water and soil mycology constitute an indirect approach (g protein g^{-1} substrate) representing TBNM. Whereas this could be a feasible approach in environments with a low endogenous protein content such as water (Newell and Statzell-Tallman 1982), soil (Scheu and Parkinson 1994), or wood (Boyle and Kropp 1992), most food substrates contain proteins which cannot be easily distinguished from fungal cellular protein. The complexity of applying protein measurement for estimation of fungal mass increases even more when the microorganism is able to produce extracellular protein.

Nevertheless, protein determinations are commonly carried out to monitor SSF. Several processes were developed with the purpose of increasing the nutritive value of low-protein material by increasing the protein content, e.g. by growing *Aspergillus tamarii* (Xue et al. 1992) or *Trichoderma viride* (Grajek and Gervais 1987; Durand and Chereau 1988) on sugar beet pulp, or *Aspergillus niger* on cassava (Saucedo-Castenada et al. 1990), wheat bran (Auria et al. 1993) or sweet potato residue (Yang et al. 1993). In such instances, it makes sense to measure true protein, e.g., by the Lowry method, as this is the targeted compound of the process. It would not be essential to convert protein into fungal mass in such cases.

Several attempts were made to convert protein data, obtained from either Lowry or Kjeldahl nitrogen measurements, to biomass. Conversion factors (protein % of biomass dry weight) used include 27.6 for *Rhizopus oligosporus* (Sargantanis et al. 1993), from 21 (Auria et al. 1993) to as high as 47 (Saucedo-Castenada et al. 1990) for *Aspergillus niger*, and 37 for a mixture of *Trichoderma reesei* and *Aspergillus niger* (Castillo et al. 1994). In addition, protein content of liquid-grown mycelium of *Aspergillus niger* has been used to calibrate protein data of SSF in inert media (Gutierrez-Rojas et al. 1995). Studies on growth media with high and low nitrogen levels indicated that the protein content of fungal biomass may correlate with the nitrogen content of the environment, whereas older mycelium tended to have significantly lower protein contents (Boyle and Kropp 1992), possibly because of cytoplasmic depletion.

In addition to chemical protein analysis, it would be possible to estimate true protein by

FTIR-PAS (Fourier transform-photoacoustic spectroscopy; Gordon et al. 1990). Calibrated on Lowry protein determinations, it appears that this method requires testing of serially diluted samples, as the linearity is rather limited. Investigations have also been made into the incorporation of mineral nitrogen into biomass protein using ^{15}N in $(^{15}NH_4)_2SO_4$ (Wissler et al. 1983). Comparison of the percentage ^{15}N in available nitrogen before fermentation and in the protein fraction after fermentation showed that this method gave results comparable to the Kjeldahl determination of TCA-precipitated protein.

Enzyme Activities. Perhaps the most important aspect of fermentation is the (extracellular) enzyme activity produced by the fermenting microorganisms. Substrates for SSF are usually preheated for pasteurization, hence they will have minimum endogenous enzyme activity. In such cases, any easily detectable enzyme activity could be measured to monitor the progress of fungal metabolism. In native substrates with still active endogenous enzymes, a more fungal-specific enzyme activity will be more appropriate. The measurement of enzyme activities originates from composting and fermentation microbiology, and it could be referred to as an indirect method (units g^{-1} substrate) related to metabolically active biomass. However, as enzyme activities and productivities tend to be influenced by many environmental factors, it is not always practicable to translate enzyme activities into biomass. To the present, laccase has been used to monitor *Agaricus bisporus* in compost, measured either spectrophotometrically after reacting with syringaldazin (Wiegant et al. 1992) or by oxygen consumption when reacting with p-phenylenediamine (Matcham et al. 1985). Esterase activity can be conveniently measured by fluorometry measuring the release of fluorescein from FDA (fluorescein diacetate) (Schnürer and Rosswall 1982; Boyle and Kropp 1992). Another option is to measure reductase activity by spectrophotometric detection of formazan formed by reduction of INT [2-(p-iodophenyl)-3-(p-nitrophenyl)-5-phenyltetrazoliumchloride] (Le Page et al. 1988). Particularly during the initial active growth phase, accumulation of enzyme activity tends to correlate with increased biomass. From a pragmatic and functional point of view, this may yield valuable information during SSF.

Immuno-Assay. Polyclonal or monoclonal antibodies raised against immunogenic extracellular polysaccharides (EPS) in the fungal cell wall can be used in immunoassays for fungal cell walls. This is an indirect technique (units g^{-1} substrate) which appears to originate in food microbiology. The level of EPS relates to the surface area and thus the volume of biomass. Newell (1992) discusses in detail the principle of immunoassays, and the detection systems utilized. Immunoassays for fungal mass constitute a powerful and highly sensitive detection tool for food quality assurance (De Ruiter et al. 1992a,b). The thermal stability of the EPS enables detection of fungal contamination, even in heatprocessed foods, e.g., fruit juices and jams. The reactivity of antibodies still needs considerable investigation. This includes aspects such as cross-reactivity with (food) substrate components or with nontarget fungal genera, and occasional lack of reactivity in the presence of certain substrates, or with certain species within a target genus.

Immunoassay for the purpose of general fungal biomass determination is at present not in practical use; but the technique could prove extremely valuable in differentiation of fungal growth in mixed cultures.

ATP. Determination of microbial ATP using the luciferin-luciferase reaction is commonly used in food microbiology as an indirect approach (nmol ATP g^{-1} substrate) to detect bacteria and yeasts in foods. Extraction of microbial cells from the food is required in order to eliminate interference by food intrinsic ATP (Jarvis et al. 1983). The detection is sensitive and rapid. Intracellular ATP of *Trichoderma reesei* has been measured in cellulosic media (Cochet et al. 1984). The ATP content increased rapidly during the initial active growth phase and declined thereafter. ATP content proved to be an interesting indicator to mark the end of the active growth phase, which marks the start of cellulase excretion by the mold.

DNA. The accumulation of intracellular DNA could serve as an indicator of fungal growth. This could also be termed an indirect method (mg DNA g^{-1} substrate) of which it is not yet fully investigated whether it relates to biomass only, or to the total of biomass and necromass. Indeed, it was shown for *Aspergillus oryzae* growing on rice (Bajracharya and Mudgett 1980), that the DNA content increased rapidly during the initial growth phase. Calibrated against liquid-cultivated mycelium the biomass contained 0.3% DNA which decreased to 0.2% at the end of the active growth period. Further data were not given, but it appears that the DNA measurement could serve a purpose similar to the ATP determination. It remains to be investigated what background inaccuracy is caused by endogenous DNA from the fermented substrate. Possibly the level of RNA would be an even better indicator of fungal ABM.

Volatiles. Cereal grain-associated strains of *Penicillium* and *Aspergillus* spp. were found to generally produce volatiles including 2-methyl-1-propanol, 3-methyl-1-butanol and 3-methyl-furan (Borjesson et al. 1993). Quantification of these substances using headspace gas liquid chromatography (GLC) showed that their levels increased with incubation time, and especially with sporulation. The monitoring of headspace volatiles (μg g^{-1} substrate) could thus be used as an indirect approach relating to physiological activity of biomass. Whereas the approach of GLC headspace analysis will be of primary interest to detect spoilage in a very early stage, it might conceivably be applied to SSF, e.g., to serve as a functional signal in the production of aroma compounds.

3. Viable Count Methods

Whereas the cultivation of viable cells (colony-forming units g^{-1} substrate) is a common and valid technique when dealing with unicellular microorganisms including bacteria and most yeasts, filamentous fungi present several problems. The major problem is that the number of viable mold propagules (colony-forming units) represent an extent of sporulation, rather than mycelial biomass, as was shown, e.g., with *Aspergillus awamori* (Silman et al. 1983). In addition, uneven distribution of fungal mass through the substrate matrix results in large standard deviations. Factors influencing the variability of viable mold counting techniques are discussed by Jarvis et al. (1983) and Nout (1984).

Despite these drawbacks, viable count methods have a specific role to play as a direct reproductive biomass-related approach. When mass-producing mold inoculum, spore count relates directly to potential yield of inoculum. In addition, viable counting was shown to correlate

well with hyphal length when non-sporulating molds were studied, e.g., *Fusarium culmorum* (Schnürer 1993). Another functional application is the study of the growth of individual species in fungal mixed cultures, e.g., with *Penicillium camembertii* and *Geotrichum candidum* in camembert cheese (Molimard et al. 1995) and to distinguish the biocontrol agent *Trichoderma harzianum* from *Fusarium culmorum* in stored wheat (Cheetham et al. 1995). The use of viable counting in combination with direct microscopic determination of hyphal length was successfully employed to demonstrate that in tapé (fermenting rice) higher counts of yeasts comparable to molds overestimate their importance in the fermentation, as the molds represent 85–95% of the total biomass (Cook et al. 1991).

C. On-Line (In-Line) Approaches

In principle, this section deals with on-line methods using sensors within the reactor system giving real-time signals which can be interpreted instantaneously, and which can thus be directly used for optimization of the fermentation process by fine-tuning of environmental conditions. Also mentioned are some approaches requiring sampling and analysis; the characteristic of the analysis being that it is rapid, i.e., yielding data within minutes.

1. Respiratory CO$_2$ Production

Two distinct approaches are used to relate respiratory activity to biomass. The first, substrate-induced respiration (SIR) has been developed to estimate living biomass in soil samples. To a sample of soil, a solution is added containing minerals, glucose, and a suitable antibiotic, e.g., streptomycin, to inhibit bacterial metabolism (Wiegant 1991; Newell 1992). The suspension is incubated during a few hours in closed serum bottles and the released CO$_2$ (μmol h^{-1}g^{-1} substrate) is measured in the headspace by, e.g., GLC. The method can be calibrated with known quantities of mycelium. Alternatively, Wiegant (1991) determined a specific CO$_2$ production rate of 35 ml h^{-1}g^{-1} of *Scytalidium thermophilum* mycelium when incubated in liquid medium at 45 °C. Assuming that during the test no growth takes place, and that dry biomass contains 50% carbon, the biomass was estimated to be 14 mg biomass C per ml h^{-1} of produced CO_2. As most fermenting materials contain significant quantities of substrate, the controls would give too strong a background. This makes the method less attractive for use in SSF.

The second approach is to measure on-line CO_2 levels in the effluent gas of bioreactors. In its simplest form, CO_2 production is monitored as a function of time, as was done with *Aspergillus awamori* growing on wheat bran (Silman et al. 1983). Although this does not yield biomass data, valuable information is obtained with relative ease, about the onset of the fermentation, and the metabolic activity of the mold. Studies with *Aspergillus oryzae* growing in rice koji, in which complementary biomass estimations were made, indicate that CO_2 development is detectable long before a significant increase of biomass can be measured, and that the cumulative CO_2 production does not really reflect the total accumulated biomass, but gives a better representation of the assimilation of the carbon source (Larroche et al. 1986). It was shown that, initially, CO_2 production was mainly due to growth of *A. oryzae*, and that, gradually, the growth-related CO_2 strongly declined but that maintenance-associated CO_2 production became dominant (Ito 1993). Using mathematical models, a prediction may be made of the CO_2 evolution during the fermentation. This was attempted for *Aspergillus niger* growing on cassava (Saucedo-Castenada et al. 1990) and for *Rhizopus oligosporus* growing on soybean (De Reu 1995).

2. Respiratory O_2 Consumption

Several authors (Narahara et al. 1982; Kim et al. 1985; Rodriguez Leon et al. 1988; Soccol et al. 1993) used Pirt's linear growth law to fit the observed relation between specific respiration rate and specific growth rate, i.e.:

$$Q = 1/Y^*\mu + m,$$

where Q is the specific respiration rate (O_2 consumption or CO_2 production), μ is the specific growth rate, Y is a stoichiometric coefficient, and m is the maintenance coefficient. However, Pirt's law is valid only for growing biomass (Beeftink et al. 1990). At a certain point, the fungal biomass will start to differentiate or decay, and deviations from Pirt's law are likely to occur. These deviations may depend upon the biomass indicator used, because the ratio between the indicator and biomass dry weight can vary with the age of the biomass, as was demonstrated for glucosamine and nitrogen.

3. Heat Production

During SSF, considerable amounts of heat are generated. The heat production (J g^{-1} substrate) in a mass of fermenting solid substrate can be a very effective index of metabolic activity, relating to growing biomass. The use of heat-sensing technology has developed particularly in food and biotechnology SSF. Local temperatures as well as temperature gradients can be monitored using thermocouples. This was attempted for *Aspergillus niger* growing on cassava (Saucedo-Castenada et al. 1990), and with better results during the heating phase of a packed bed reactor with *Rhizopus oligosporus* growing on soybeans (De Reu 1995). Whereas thermocouples need to be placed in the product, noninvasive devices, e.g., a heatflux sensor attached to the exterior reactor wall, or an infrared sensor placed about 5 cm from the reactor wall (Silman 1988) have been used in *Aspergillus awamori* fermentations. The latter techniques have as an advantage that they do not compromise the sterility of the product. On the other hand, they would be less suitable for measuring gradients.

Another aspect of heat accumulation is that control mechanisms are required in order to avoid the substrate temperature exceeding T_{max} for growth. Using thermocouple technology in a rotating drum reactor, it is possible to activate mixing and cooling on the basis of temperature signals (De Reu et al. 1993). This is an example of how on-line monitoring can function as a fermentation control leading to optimization of the process. It remains to be demonstrated that this simple control method can be applied to industrial solid-state fermentors with lower area-to-volume ratio.

4. Electrical Properties

During microbial growth, the conductance (S g^{-1} substrate) of the surrounding medium decreases due to uptake of NH_4^+ and reactions of thus released H^+ with buffers, while it is increased by release of NH_4^+, organic acids and, e.g., $H_2CO_3^-$ (Owens et al. 1992). Measurement of conductance and impedance have become common in food microbiology to detect levels of approx. 10^6 fungal conidia in food, within about 4 h (Jarvis et al.

1983). More interesting for use in SSF, the conductance as measured with an electrode permanently located in the fermenting mass gives a useful on-line signal of the onset of fungal metabolism such as demonstrated for *Rhizopus oligosporus* in quinoa (Penaloza et al. 1992) and soybean (Davey et al. 1991).

Another aspect of electrical properties is that the membranes of intact cells are electrically charged. It is thus possible to concentrate, separate, and even identify unicellular organisms in a liquid medium by dielectrophoresis (Betts 1995). Whereas this technique is not (yet) suitable for on-line use in SSF, monitoring of the capacitance (F g^{-1} substrate) of the SSF mass at 0.30 MHz has been shown (Davey et al. 1991; Penaloza et al. 1992) to give on-line signals representing living biomass. After an initial steep increase of the signal, the capacitance decreases when the mycelium starts dying. The technique allows only permanent placement of a sensor and is rather sensitive towards reactor geometry and packing density of the substrate. From this point of view, it would therefore be best suited to applications in nonagitated fermentation systems. However, the gradients occurring in nonagitated systems render the use of a fixed sensor less representative.

5. Electromagnetic Sensing

Reflected infrared analysis (RIR) is applied in food analysis to estimate levels of, e.g., moisture, protein, fat, and carbohydrates. It is a fast method and is suitable for automatization. RIR has been applied in SSF (Silman et al. 1983) and was found suitable to monitor changes of water, oil, starch, and protein content in the substrate. It could thus be used to indicate biochemical substrate modification. As mentioned earlier, FTIR-PAS (Fourier transform infra-red-photo acoustic spectroscopy) has been used for noninvasive protein sensing. RIR methods for direct measurement of glucosamine and ergosterol, and residual sucrose and nitrogen levels on solid fermentation media have been compared to manual methods (Desgranges et al. 1991a,b). The RIR methods were as accurate as the manual methods. Generally, infra red sensing requires extensive calibration of the substrate material to be used in order to provide a reliable baseline for the spectra. This requirement cannot always be met adequately, especially not when the substrate is significantly modified due to microbial activity during fermentation.

As a result of biomass formation, sporulation as well as biochemical reactions taking place in the substrate, the color of the fermenting mass changes. Tristimulus analysis of reflected light was attempted (Ramana Murthy et al. 1993) to monitor *Aspergillus niger* fermentation of wheat bran. As the calibration did not take enzymatic browning into consideration, and only limited data were given of the fermentation curve, it is not possible to evaluate the usefulness of this technique.

6. Consistency

Particularly during static SSF (as opposed to agitated SSF), the substrate particles become strongly bound into a solid mass by the mycelium. Consequently, its resistance to penetration will increase. Several methods were developed to monitor the progress of the SSF by measuring force required to break SSF material (Nout et al. 1985), to bend a solid slab of tempe (Ariffin et al. 1994), to penetrate fermented mass (Kronenberg and Hang 1985), or to measure the consistency of the substrate freed from mycelium (De Reu et al. 1996). These approaches appear primitive, but they can yield valuable information on functional properties, e.g., firmness and sliceability of tempe, or loss of firmness, and thus attractiveness, by extensive degradation of nonstarch polysaccharides (De Reu et al. 1996). Most of these methods would be in-line rapid assessments; it is conceivable, though, to install consistency sensors on-line.

Changes in porosity due to occupation of the interstitial gas spaces result in an increased pressure drop across aerated packed-bed reactors. An increase in pressure drop was correlated with respective phases of *Aspergillus niger* growth in SSF (Auria et al. 1993). Although this is a parameter which is easy to measure, extreme care should be taken because of the shrinkage which often takes place in static SSF due to evaporation. When the mass in a packed-bed reactor shrinks, false air channels may be formed which obscure pressure drop measurements.

D. Comparison of Estimates

In several of the studies mentioned above, a combination of methods has been used to monitor fungal development in SSF. In the majority of cases, indirect measurements were linked to gravimetric (Narahara et al. 1982; Larroche et al. 1986;

Nout et al. 1987; Auria et al. 1993) or microscopic (Nout et al. 1987; Davey et al. 1991; Schnürer 1993) biomass data. Of the indirect measurements, chitin (Wissler et al. 1983; Matcham et al. 1985; Desgranges et al. 1991a,b; Boyle and Kropp 1992), ergosterol (Matcham et al. 1985; Nout et al. 1987; Desgranges et al. 1991a,b; Schnürer 1993), protein (Wissler et al. 1983; Boyle and Kropp 1992; Rinzema et al. 1996), and laccase (Matcham et al. 1985; Wiegant et al. 1992) are predominant. In addition, CO_2 in effluent gas (Narahara et al. 1982; Larroche et al. 1986; Saucedo-Castenada et al. 1990; Desgranges et al. 1991a,b; Auria et al. 1993; De Reu 1995) has been used in several multimethod studies. We made an attempt to chart in Table 3 the order of magnitude of the various parameters regardless of fungal species or substrate used, at two phases of development, i.e., half-way in the active growth phase, and during the stationary phase.

Most total fungal mass data estimated from gravimetric or microscopic data are of the same order of magnitude, i.e., up to 50–100 mg TBNM per gram of fermented substrate in the stationary phase (all on a dry weight basis). An exception are the estimates based on hyphal lengths reported by Davey et al. (1991), which we converted using the factor used by Schnürer (1993); obviously, the resulting biomass must be too high by a factor 100. Chitin, ergosterol, protein, and laccase all show a consistent increase during the growth of biomass, but differences of values between studies (i.e., species and substrates) are of an extent requiring custom calibration of such methods. It does not appear feasible to give conversion factors of general validity for these parameters. Measurement of CO_2 is simple and sensitive, but it would seem of validity only during the active growth phase, as the output is generally reported to decline during the stationary phase.

III. Modeling

A. Importance of Process Modeling

Despite their long history, the design of SSF systems is still more an art than a technology. This is in sharp contrast to submerged fermentations, where mathematical modeling is generally recognized as a valuable tool for rational design, optimization, and process control. As outlined in the introduction, SSF processes are hampered by i.a. mass and heat transport limitations. These may cause heterogeneity (gradients in temperature and concentrations of nutrients, metabolites, water and consequently biomass), as well as overheating and desiccation of the entire bioreactor. The heterogeneity combined with the fact that the fermentation industry generally applies batch processes, seriously complicates trial-and-error-based experimental studies, process design, and scaleup. Mathematical models can improve our understanding of the complex, nonlinear behavior of SSF systems. Therefore, they are very useful tools to design experiments, interpret experimental results, and design full-scale plants. Physical problems encountered in SSF are very similar to those encountered in, for example, heterogeneous catalysis and drying, where significant advances have been made since the introduction of mechanistic mathematical models. The analogy makes it relatively easy to set up models for the physical

Table 3. "Biomass" per gram[a] of fermented SSF substrate. Columns represent ranges of data for each method, during various stages of growth. Rows represent equivalent data of several methods, insofar available from literature sources

CO_2 production ([10]Log moles/h)	By weight (mg)	Converted from hyphal length (mg)	Chitin (glucosamine) (mg)	Ergosterol (μg)	Protein (mg)	Colony-forming units ([10]log N/g)	Laccase (U)	Fluorescein diacetate esterase (OD/45 min)	Capacitance (pF)	Pressure drop (mm H_2O mm^{-1})
−4.3 A (8)	100 S (4)	6030 S (10)[b]		3000 S (2)						
−4.5 A (9)	50 S (2)	3015 A (10)[b]	2.4 S(6)	1500 A (2)	32 S (1)	7 S (7)	4 S (6)	1000 S (5)	100 A (10)	0.8 S (9)
−4.7 S (8)	50 A (4)	70 S (7)		1400 S (2)						
−4.9 S (9)	40 S (8)	50 S (2)	0.5 S (1)	300 S (7)	1 S (5)					
−5.6 A (4)	20 S (9)	20 A (2)	0.3 S (5)	250 A (2)				200 A (5)	50 S (10)	
−6.0 S (4)	20 A (2;8)	3 A (7)	0.1 A (5)	100 S (6)	0.05 A (5)	3 A (7)	0.5 S (3)			0.1 A(9)
	10 A (9)			20 A (7)			0.1 A (3)			

A = active growth phase; S = stationary growth phase. [a] Per plate in study (7). [b] Converted from hyphal length using conversion factor according to (7). Key to references: 1 Wissler et al. (1983); 2 Nout et al. (1987); 3 Wiegant et al. (1992); 4 Larroche et al. (1986); 5 Boyle and Kropp (1992); 6 Matcham et al. (1985); 7 Schnürer (1993); 8 Narahara et al. (1982); 9 Auria et al. (1993); 10 Davey et al. (1991).

phenomena in SSF processes. The bottlenecks are modeling of growth and production kinetics and the experimental validation of the models. The first bottleneck is related to the difficulties associated with biomass measurements in SSF.

B. Solid-State Fermentation Models

Mathematical models for SSF can be classified as follows:

1. Branching models describe the development of fungal mycelia as the combined result of elongation at the hyphal tips and branching.
2. Black-box kinetic models describe experimentally observed growth curves without any attempt to take cause-effect relations into account.
3. Particle-level models are mechanistic counterparts of black-box kinetic models: they predict the experimentally observed growth kinetics from intrinsic growth and maintenance kinetics of fungal biomass and space and transport limitations on or in substrate particles.
4. Reactor-level models predict the macroscopic behavior of an entire bioreactor based on a combination of reaction kinetics and conservation laws.

1. Branching Models

Prosser and Trinci (1979) successfully described the development of young mycelia with a branching model based on linear extension at the hyphal tips with constant rate and branching at a fixed interval. Microscopic observations were used to validate this phenomenological model. This branching model can simulate the initial development of fungal mycelia, but its usefulness for design of solid-state fermentors is very limited: the continuing exponential increase in biomass predicted by the model is not in agreement with the decreasing growth rate that is commonly observed.

Viniegra-Gonzalez et al. (1993) simulated this decreasing growth rate by extending the branching model with a distribution function for the number of hyphal tips that has stopped growing. This model allows accurate fits of the decreasing biomass production rate observed experimentally. However, the gradual decrease in number of growing tips is related solely to the age of the

mycelium, not to conditions outside the cells, as is attempted in the models in category (3). Therefore, this model essentially has no added value compared to the black-box kinetic models in category (2).

2. Black-Box Kinetic Models

The logistic law (Koch 1973) is frequently used to describe symmetric sigmoidal growth curves (Okazaki et al. 1980; Saucedo-Castaneda et al. 1990; Sargantanis et al. 1993; Szewczyk and Myszka 1994). A modified form of this law can be used to describe skewed sigmoidal curves (Richards 1959):

$$r_x = \mu_{max} \cdot \left[1 - \left(X/X_{max} \right)^n \right] X,$$

where r_x is the biomass production rate $(kg\,h^{-1})$, μ_{max} is the maximum specific growth rate (h^{-1}), i.e., the limit of the specific growth rate as the amount of biomass approaches zero, X is the amount of biomass present (kg), X_{max} is the maximum amount of biomass that can develop in the system (kg), and n is an exponent.

Several other empirical relations are available to describe sigmoidal growth curves (Zwietering et al. 1990; Pitt 1993). The choice should be based on the accuracy of the fit on experimental data, taking into account the number of experimental observations and model parameters (Zwietering et al. 1990).

Empirical models for substrate consumption or metabolite production, such at Pirt's linear growth law (Pirt 1965), have been used by several authors, either in predictive models (Okazaki et al. 1980; Sato et al. 1983; Saucedo-Castaneda et al. 1990; Sargantanis et al. 1993; Szewczyk and Myszka 1994; De Reu 1995) or for indirect (online) measurement of biomass (Nishio et al. 1979; Sugama and Okazaki 1979; Narahara et al. 1982; Kim et al. 1985). The latter application has been discussed in Section C.1.

No general guidelines for the best biomass indicator or the best kinetic model can be derived from the literature. Cell dry weight, glucosamine, protein, and amount of carbon dioxide produced or oxygen consumed have all been used in combination with black-box kinetic models. The empirical relation needed to describe the growth kinetics may depend entirely on the biomass indicator chosen. For example, the glucosamine development observed in experiments with pre-grown micro-colonies of *Rhizopus oligosporus* on defined agar media could be described accurately with the logistic law, but a combination of exponential growth and decay as proposed by Herbert (1958) was required to describe the biomass dry matter development (Fig. 1; A. Rinzema et al., unpubl.). Apparently, the amount of cell dry matter decreased as the result of differentiation or lysis, while the chitin skeleton remained unaffected. A similar decrease in cell dry matter has been reported by Molin et al. (1993) for *Trichoderma* sp. Although a connection to the decreasing metabolic activity observed in reactor-level modeling studies (Sargantanis et al. 1993; De Reu 1995) is plausible, this remains to be demonstrated experimentally.

The reason for the gradual decrease in biomass production rate associated with sigmoidal growth curves may be the age of the mycelium or extracellular limitations. For submerged cultures of bacteria, yeasts, and dispersed mycelium, the effect of extracellular limitations is usually described with black-box kinetic models such as the Monod model. For solid-substrate cultures, the onset of the gradual decrease in specific growth rate occurs at average nutrient concentrations far above Monod-type saturation constants reported for submerged cultures. Black-box kinetic models can be used to link the specific growth rate to average nutrient concentrations, but the model parameters will probably reflect physical characteristics of the solid substrate instead of intrinsic kinetic characteristics of the fungus. Particle-level models attempt to link the decreasing growth rate to local nutrient exhaustion in the immediate vicinity of the hyphae, or to space limitations.

3. Particle-Level Models

The goal of particle-level models is to predict the occurrence of the sigmoidal growth curves observed for most solid-state cultures from intrinsic fungal growth and maintenance kinetics and space and transport limitations on or in substrate particles. The primary benefit of these models is that they improve our understanding of SSF processes. A potential benefit is in substrate design: these models allow us to evaluate, for example, the effects of particle grinding, addition of nutrients to natural solid substrates, or the optimization of liquid media in combination with inert solid supports.

Table 4a summarizes the essential characteristics of two particle-level modeling studies supported by experimental work (Mitchell et al. 1991; Molin et al. 1993). Three other studies in this category (Edelstein and Segel 1983; Georgiou and Shuler 1986; Rajagopalan and Modak 1995b) are not discussed.

Table 4a. Particle-level solid-state fermentation models

System	Biomass indicator	Principle of the model	Fit parameters[a]	Reference
R. oligosporus on membrane covered gel with starch, overculture	Dry weight	Monod kinetics for glucose uptake and growth, uptake rate limited above a certain biomass level, diffusion of glucoamylase and glucose in the gel	q_{max}, ID_{ga}, X_0	Mitchell et al. (1991)
R. oligosporus and *T. reesei* on potato dextrose agar	Dry weight	Monod kinetics for glucose uptake and growth, maintenance has priority over growth, decay when uptake does not cover maintenance requirements, uptake rate limited above a certain biomass level, diffusion of glucose in the gel, diffusion of biomass over the surface	q_{max}, K_s, X_{smax}, m_s, $Y_{x/s}$, k_d, ID_x, X_0[b]	Molin et al. (1993)

[a] q_{max}: maximum specific glucose uptake rate (mass/mass of biomass/s), ID_{ga}: diffusion coefficient for glucoamylase in agar gel (m^2/s), K_s: Monod-constant (mass of substrate/area), X_{smax}: maximal amount of biomass in contact with the substrate surface (mass of biomass/area), $Y_{x/s}$: yield of biomass on substrate (mass/mass), ID_x: the authors replaced the radial growth rate of the colony with a diffusion coefficient for biomass (m^2/s), other symbols: see Table 4b. Symbols used by Molin et al. (1993) have been renamed to generally accepted symbols; one parameter indicated in the paper turned out to be superfluous.

[b] Added by the authors, the paper does not state clearly how this parameter was estimated.

Table 4b. Reactor-level models for solid-state fermentation

System	Biomass indicators	Principle of the model[a]	Fit parameters[b]	Reference
A. niger, cassava meal, aerated packed bed with cooling jacket, non-homogeneous, nonisothermal	1. protein during independent determination of kinetic parameters (data from literature) 2. Temperature used as fit variable in validation study, CO_2 and average protein were measured in addition	1. Logistic law with temperature dependent parameters, maintenance 2. Empirical stoichiometry 3. Radial heat conduction, evaporation not included	λ_s, a_s, α_{wall}, X_0, $r_{co2,0}$, μ, m	Saucedo et al. (1990)
R. oligosporus, corn grit, aerated rocking drum, homogeneous, nonisothermal	1. Protein in independent determination of kinetic parameters and isothermal reactor runs 2. Temperature in non-isothermal validation experiment; CO_2 and protein not measured	1. Logistic law with temperature dependent parameters, maintenance 2. Empirical stoichiometry and elemental balances 3. Homogeneous bed, heat transfer at the wall and evaporation included	T_{crit}, k_d	Sargantanis et al. (1993)
R. oligosporus, soybeans, nearly adiabatic packed bed with gas recirculation, homogeneous, non-isothermal	1. CO_2 in independent determination of kinetic parameters 2. Temperature and CO_2 in nonisothermal validation experiment.	1. Exponential growth with temperature dependent μ, no maintenance 2. Empirical stoichiometry and elemental balances 3. Homogeneous bed, heat transfer at the wall and evaporation included	X_0, substrate composition	De Reu (1995)

[a] (1) kinetics, (2) stoichiometry, (3) transport phenomena.
[b] λ_s: heat conductivity of the solids (W/m/K), a_s: heat diffusivity of the solids (m^2/s), α_{wall}: overall heat transfer coefficient at the reactor wall (W/m^2/K), X_0: initial biomass concentration (mass/volume or mass/mass), $r_{co2,0}$: initial CO_2 production rate (mol/m^3/s), μ: specific growth rate (1/s), m: maintenance coefficient (mass/mass of biomass/s), T_{crit}: critical temperature above which the fungus starts to decay irreversibly (°C), k_d: decay rate constant (1/s).

Mitchell et al. (1991) and Molin et al. (1993) used defined agar media, which allowed the use of cell dry matter as biomass indicator. Both groups measured most parameters in independent experiments or estimated them from literature data. Nevertheless, they had to adapt several parameter values to obtain agreement of calculated and measured biomass and substrate concentrations in the validation experiments. It is not clear whether the need for fitting should be attributed to inadequacies in the models or in the parameter estimation experiments. The study of Molin et al. (1993) clearly shows a decrease in biomass dry matter after substrate depletion, which is indicated indirectly by the decreasing reaction rate in several reactor-level studies. Therefore, part of the problems encountered in model validation might be due to the inadequacy of dry matter as a biomass indicator. Dry matter cannot reflect differentiation of the biomass and allows no distinction between biomass and necromass. Mitchell et al. (1991) had problems with estimating the initial amount of (viable) biomass, which were also encountered in reactor-level modeling studies.

4. Reactor-Level Models

Besides the nutrient and space limitations discussed in the previous section, macroscopic environmental conditions are frequently limiting in solid-state cultures. Temperature limitations caused by heat production associated with the microbial activity are the primary problem of large-scale solid-state fermentations. Most solid-state fermentors are cooled by aeration. Due to the low heat capacity of dry air, the cooling action is based largely on evaporation of water from the moist solid matrix. Together with the incorporation of water in the cytoplasm of new biomass, this may cause an unfavorable water activity. Finally, in tray fermentors (nonaerated static layers) oxygen depletion or carbon dioxide accumulation may become a limiting factor.

The aim of reactor-level models is to simultaneously predict the development of biomass and the limiting macroscopic environmental conditions. This requires integration of kinetic and stoichiometric models and conservation laws, taking into account the relevant physical transport phenomena and phase equilibria in the fermentor. Effects of limiting macroscopic environmental conditions are incorporated in the kinetic model. Reactor-level models can thus be used to evaluate the effect of the scale of operation (heat losses through reactor walls), solid matrix characteristics (water sorption), or aeration regime (air flow rate, temperature, humidity).

Table 4b gives the essentials of three reactor-level studies with experimental data (Saucedo-Castaneda et al. 1990; Sargantanis et al. 1993; De Reu 1995). Black-box kinetic models were used in these studies, and physical (transport) phenomena were taken into account on reactor level but not on particle level. Rajagopalan and Modak (1994, 1995a) incorporated particle-level models in a reactor-level model, but their papers contain no experimental validation and are therefore not discussed here. For the same reason, a model with black-box kinetics (Raghava Rao et al. 1993) is not discussed.

Natural substrates were used in the three studies in Table 4b, with the concomitant use of indirect biomass indicators. Saucedo-Castaneda et al. (1990) and Sargantanis et al. (1993) used protein as biomass indicator during the independent determination of kinetic parameters, but O_2 or CO_2 and temperature in the validation experiments. De Reu (1995) used CO_2 to determine kinetic parameters in independent experiments, and CO_2 combined with temperature to validate the reactor model.

None of the reactor-level models predicted the nonisothermal behavior of the fermentors using only independently determined parameter values. At least two parameters were (re-)estimated from the validation experiments in order to improve the agreement between calculated and measured values for the biomass indicator (Table 4b). Saucedo-Castaneda et al. (1990) used the most complicated fermentation system – a nonhomogeneous packed bed – and, concomitantly, the most complicated mathematical model with the highest number of parameters. Independent measurements were available for only a limited number of parameters. Missing parameters for physical transport phenomena, kinetics, and stoichiometry were estimated from the validation

experiment by fitting of model predictions on temperature measurements at various radial positions in their packed bed. Furthermore, independently measured kinetic parameters were reestimated. A disquieting fact is that the very significant effect of temperature on specific growth rate found in independent isothermal experiments was rejected by the fitting algorithm. After the fit, the model calculated the CO_2 production and the mean protein content of the bed with acceptable accuracy. Considering the number and nature of the parameters used in the fit, this proves neither the validity of the model, nor the suitability of protein as a biomass indicator for use with this fermentation process and model. Sargantanis et al. (1993) and De Reu (1995) did not revert to such extensive fitting procedures. The former authors estimated two parameters for decay kinetics from the validation experiment, in order to approximate the observed temperature decrease in their homogeneous rocking-drum fermentor. They did not alter the model or parameter values to correct the overestimation of the initial temperature rise, which may have been caused by an overestimation of the initial amount of biomass or to incorrect estimates of physical properties. De Reu (1995) obtained a good agreement between measured and calculated CO_2 production and temperature during the heating phase of a homogeneous packed-bed fermentor (Fig. 2), using the initial biomass concentration as a fit parameter. Furthermore, De Reu assumed that lipids were used as substrate (the reaction enthalpy was calculated through elemental balances based on this assumption). This assumption was based on analyses of changes in

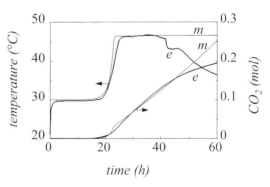

Fig. 2. Comparison of measured and predicted temperature and CO_2 development in a packed-bed reactor in which gradients were suppressed by off-gas recirculation, during cultivation of *R. oligosporus* on soybean (*e* experimental, *m* model). (After De Reu 1995)

soybean composition published previously. De Reu's model did not predict the decrease in temperature and carbon dioxide production rate observed later in the fermentation, because exponential growth with temperature as sole limitation was assumed. The observed decrease could not be simulated by incorporation of a logistic model, which suggests that decay or differentiation affected the kinetics, as observed by Sargantanis et al. (1993) and Molin et al. (1993). Neither Sargantanis and coworkers, nor De Reu measured biomass dry weight or a cell component during the nonisothermal validation experiments. Therefore, it is impossible to judge whether incorrect predictions of biomass development or incomplete kinetic models were the cause of deviations between model predictions and experiment. Both studies indicate that improved measurement of the inoculum density, and decay or other physiological changes in fungal biomass during prolonged fermentations require more attention, as can also be deduced from the particle-level modeling studies. We expect that the biomass indicators used in the studies discussed above are not sufficient to validate models that take into account the metabolic state of the biomass. This will require more specific indicators of viable biomass. Ergosterol (Newell 1992; Scheu and Parkinson 1994) and esterase activity (Boyle and Kropp 1992; Chand et al. 1994) may offer interesting possibilities.

5. Pitfalls

One aspect that is underestimated in available modeling studies is the effect that changes in physiology with concomitant changes in stoichiometry may have on parameter estimation experiments and reactor models. A striking example is the anaerobic metabolism observed by Rinzema et al. (1996) during cultivation of *Rhizopus oligosporus* on agar with mineral glucose medium. These authors observed a continuous increase in respiration quotient in experiments with several millimetres thick agar layers, long before O_2 depletion in the gas phase occurred (Fig. 3). The increase in respiration quotient was not observed in an experiment with an extremely thin agar layer. The ratio of accumulated CO_2 and biomass dry weight also increased steadily in the flasks with thick agar layers. This indicates anaerobic metabolism, which was confirmed in an experiment

under a 100% nitrogen atmosphere. Anaerobic metabolism will seriously affect specific growth rates estimated from CO_2 evolution data, and considerably complicate the modeling of bioreactors. It is therefore advisable to measure O_2, CO_2, and a biomass component during independent parameter estimation experiments, in order to obtain reliable kinetic and stoichiometric parameters.

Another problem that hampers rigorous validation of several reactor-level studies is the relatively high heat losses in small-scale reactors. Saucedo-Castaneda et al. (1990) eliminated the axial temperature gradients that are dominant in larger packed-bed fermentors (Ghildyal et al. 1994) because radial conduction was dominant in their small-diameter, noninsulated column. This is probably also the reason why they could neglect water evaporation, which is generally accepted as the most important heat removal mechanism in SSF. As indicated, the validation procedure of Sargantanis et al. (1993) may also have been hampered by inaccurate estimates of physical properties. These authors neglected the heat conduction resistance in the layer of solids and the heat capacity of the reactor itself – which is not allowed in reactors of the size used. The nearly adiabatic reactor used by De Reu (1995) eliminates conductive heat losses to a large extent and is therefore a better physical model of large-scale SSF equipment. These authors on purpose eliminated axial temperature gradients. The inclusion of these gradients in an adiabatic system presents an interesting experimental challenge.

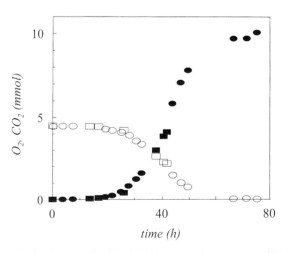

Fig. 3. Carbon dioxide (■, ●) and residual oxygen (□, ○) in two gas-tight serum flasks during cultivation of *R. oligosporus* on agar with mineral glucose medium. (A. Rinzema et al. unpubl.)

IV. Conclusions

An adequate choice of method for biomass estimations is determined by several criteria, including:

- *Why do we require a biomass estimation?* do we want to design a system, or is a measurement and control system the purpose? are we interested in total biomass, growing, metabolically active, or reproductive biomass or even in necromass?
- *What kind of substrate is used?* can it be separated from the fungus; or are we interested in the total product?
- *What kind of fermentation system (reactor) is employed?*
- *Which approaches can be applied*: direct versus indirect, on- or off-line?

Fungal ecologists investigating litter decomposition and fermentation microbiologists are able to share numerous approaches for estimating fungal biomass.

This chapter shows that there is a range of options to suit various scenarios. Up till now there is no approach which will fit in all cases, so a best choice needs to be made. In general, a direct quantitative estimation of biomass weight per gram of substrate is the most difficult to achieve. However, in many cases, such a direct approach is not an essential requirement and indirect approaches may yield useful data.

Mathematical models can play a valuable role in describing and predicting fungal growth and metabolism in fermentation reactors. Several attempts have been made to develop such models; there is still considerable scope for improvement of experimental designs and validation approaches.

Exciting challenges lie ahead!

Acknowledgment. The authors wish to thank Dr. Johan Schnürer, Swedish Agricultural University, Uppsala, Sweden, for critically reading the manuscript.

References

Ariffin R, Apostolopoulos C, Graffham A, MacDougall D, Owens JD (1994) Assessment of hyphal binding in tempe. Lett Appl Microbiol 18:32–34

Arima K, Uozumi T (1967) A new method for estimation of the mycelial weight in koji. Agric Biol Chem 31:119–123

Auria R, Morales M, Villegas E, Revah S (1993) Influence of mold growth on the pressure drop in aerated solid state fermentors. Biotechnol Bioeng 41:1007–1013

Bajracharya R, Mudgett RE (1980) Effects of controlled gas environments in solid-substrate fermentations of rice. Biotechnol Bioeng 22:2219–2235

Beeftink HH, Van der Heijden RTJM, Heijnen JJ (1990) Maintenance requirements: energy supply from simultaneous endogenous respiration and substrate consumption. FEMS Microbiol Ecol 73:203–210

Betts WB (1995) The potential of dielectrophoresis for the real-time detection of microorganisms in foods. Trends Food Sci Technol 6:51–58

Borjesson TS, Stollman UM, Schnürer JL (1993) Off-odorous compounds produced by molds on oatmeal agar – identification and relation to other growth characteristics. J Agric Food Chem 41:2104–2111

Boyle CD, Kropp BR (1992) Development and comparison of methods for measuring growth of filamentous fungi on wood. Can J Microbiol 338:1053–1060

Bull AT, Trinci APJ (1977) The physiology and metabolic control of fungal growth. Adv Microb Physiol 15:1–84

Cannel E, Moo-Young M (1980) Solid-state fermentation systems. Proc Biochem 15:2–7, 24–28

Castillo MR, Gutierrez-Correa M, Linden JC, Tengerdy RP (1994) Mixed culture solid substrate fermentation for cellulolytic enzyme production. Biotechnol Lett 16:967–972

Chand S, Lusunzi I, Veal DA, Williams LR, Karuso P (1994) Rapid screening of the antimicrobial activity of extracts and natural products. J Antibiot (Tokyo) 47:1295–1300

Cheetham JL, Bazin MJ, Markham P, Lynch JM (1995) A method utilizing mycelial fragments to estimate the relative biomass densities of fungal species in mixed culture. J Microbiol Methods 21:113–122

Cochet N, Tyagi RD, Ghose TK, Lebeault JM (1984) ATP measurement for cellulase production control. Biotechnol Lett 6:155–160

Cook PE, Campbell-Platt G (1994) *Aspergillus* and fermented foods. In: Powell KA, Renwick A, Peberdy JF (eds) The genus *Aspergillus*: from taxonomy and genetics to industrial application. Plenum Press, New York, pp 171–188

Cook PE, Owens JD, Campbell-Platt G (1991) Fungal growth during rice tape fermentation. Lett Appl Microbiol 13:123–125

Davey CL, Penaloza W, Kell DB, Hedger JN (1991) Real-time monitoring of the accretion of *Rhizopus oligosporus* biomass during the solid-substrate tempe fermentation. World J Microbiol Biotechnol 7:248–259

De Reu JC (1995) Solid-state fermentation of soya beans to tempe: process innovations and product characteristics. PhD thesis, Agricultural University Wageningen, The Netherlands

De Reu JC, Zwietering MH, Rombouts FM, Nout MJR (1993) Temperature control in solid substrate fermentation through discontinuous rotation. Appl Microbiol Biotechnol 40:261–265

De Reu JC, Griffiths AM, Rombouts FM, Nout MJR (1995) Effect of oxygen and carbon dioxide on germination and growth of *Rhizopus oligosporus* on model media and soya beans. Appl Microbiol Biotechnol 43:908–913

De Reu JC, Linssen VAJM, Rombouts FM, Nout MJR (1996) Consistency, polysaccharidase activities and non-starch polysaccharides content of soya beans during the tempe fermentation. J Sci Food Agric (in press)

De Ruiter GA, Hoopman T, Van der Lugt AW, Notermans SHW, Nout MJR (1992a) Immunochemical detection of Mucorales species in foods. In: Samson RA, Hocking AD, Pitt JI, King AD (eds) Modern methods in food mycology. Proceedings of the 2nd international workshop on the standardization of methods for the mycological examination of foods. 20–24 Aug, Baarn, The Netherlands. Centraal Bureau voor Schimmelcultures. Elsevier, Amsterdam, pp 221–227

De Ruiter GA, Van Bruggen-Van der Lugt AW, Nout MJR, Middelhoven WJ, Soentoro PSS, Notermans SHW, Rombouts FM (1992b) Formation of antigenic extracellular polysaccharides by selected strains of Mucor spp., Rhizopus spp., Rhizomucor spp., Absidia corymbifera and Syncephalastrum racemosum. Antonie van Leeuwenhoek 62:189–199

Deschamps F, Huet MC (1984) β-Glucosidase production in agitated solid fermentation, study of its properties. Biotechnol Lett 6:451–456

Desgranges C, Durand A (1990) Effect of pCO2 on growth, conidiation, and enzyme production in solid-state culture on Aspergillus niger and Trichoderma viride TS. Enzyme Microb Technol 12:546–551

Desgranges C, Georges M, Vergoignan C, Durand A (1991a) Biomass estimation in solid-state fermentation. I. Manual biochemical methods. Appl Microbiol Biotechnol 35:200–205

Desgranges C, Georges M, Vergoignan C, Durand A (1991b) Biomass estimation in solid-state fermentation. II. On-line measurements. Appl Microbiol Biotechnol 35:206–209

Desgranges C, Vergoignan C, Lereec A, Riba G, Durand A (1993) Use of solid state fermentation to produce Beauveria bassiana for the biological control of European corn borer. Biotechnol Adv 11:577–587

Dighton J, Kooistra M (1993) Measurement of proliferation and biomass of fungal hyphae and roots. Geoderma 56:317–330

Durand A, De la Broise D, Blachere H (1988) Laboratory-scale bioreactor for solid state processes. J Biotechnol 8:59–66

Durand A, Chereau D (1988) A new pilot reactor for solid state fermentation: application to the protein enrichment of sugar beet pulp. Biotechnol Bioeng 31:476–486

Edelstein L, Segel LA (1983) Growth and metabolism of mycelial fungi. J Theor Biol 104:187–210

Essers AJA, Witjes CMJW, Schurink EW, Nout MJR (1994) Role of fungi in cyanogens removal during solid substrate fermentation of cassava. Biotechnol Lett 16:755–758

Georgiou G, Shuler ML (1986) A computer model for the growth and differentiation of a fungal colony on solid substrate. Biotechnol Bioeng 28:405–416

Gervais P, Bensoussan M, Grajek W (1988) Water activity and water content: comparative effects on the growth of Penicillium roqueforti on solid substrate. Appl Microbiol Biotechnol 27:389–392

Ghildyal NP, Gowthaman MK, Rao KSMSR, Karanth NG (1994) Interaction of transport resistances with biochemical reaction in packed-bed solid-state fermentors – effect of temperature gradients. Enzyme Microb Technol 16:253–257

Gibson AM, Baranyi J, Pitt JI, Eyles MJ, Roberts TA (1994) Predicting fungal growth: the effect of water activity on Aspergillus flavus and related species. Int J Food Microbiol 23:419–431

Gordon SH, Greene RV, Freer SN, James C (1990) Measurement of protein biomass by Fourier transform infrared-photoacoustic spectroscopy. Biotechnol Appl Biochem 12:1–10

Grajek W, Gervais P (1987) Effect of the sugar beet pulp water activity on the solid-state culture of Trichoderma viride TS. Appl Microbiol Biotechnol 26:537–541

Gutierrez-Rojas M, Cordova J, Auria R, Revah S, Favela-Torres E (1995) Citric acid and polyols production by Aspergillus niger at high glucose concentration in solid state fermentation on inert support. Biotechnol Lett 17:219–224

Herbert D (1958) Some principles of continuous culture. In: Tunevall G (ed) Recent progress in microbiology. Almqvist and Wiksell, Stockholm, pp 381–396

Hong K, Tanner RD, Malaney GW, Danzo BJ (1989) Protein enrichment during baker's yeast fermentation on a semi-solid substrate in an air-fluidized bed reactor. Bioproc Eng 4:209–215

Hrubant GR, Rhodes RA, Orton WL (1989) Continuous solid-substrate fermentation of feedlot waste with grain. Biol Wastes 28:277–291

Huang SY, Wang HH, Wei C-J, Malaney GW, Tanner RD (1984) Kinetic responses of the koji solid state fermentation process. In: Wiseman A (ed) Topics in enzyme and fermentation technology, vol 10. Ellis Horwood, Chichester, pp 88–108

Hyeong-Cheol Y, Nemoto Y, Homma T, Matsuoka H, Yamada S, Sumita O, Takatori K, Kurata H (1995) Rapid viability assessment of spores of several fungi by an ionic intensified fluorescein diacetate method. Curr Microbiol 30:173–176

Ito K (1993) Studies on the mycelial growth of Aspergillus oryzae on rice grain. Seibutsu Kogaku Kaishi J Soc Ferment Bioeng 71:115–127

Jarvis B, Seiler DAL, Ould AJL, Williams AP (1983) Observations on the enumeration of moulds in food and feedingstuffs. J Appl Bacteriol 55:325–336

Kim JH, Hosobuchi M, Kishimoto M, Seki T, Yoshida T, Taguchi H, Ryu DDY (1985) Cellulase production by a solid-state culture system. Biotechnol Bioeng 27:1445–1450

Koch AL (1973) The kinetics of mycelial growth. J Gen Microbiol 89:209–216

Kronenberg HJ, Hang YD (1985) A puncture testing method for monitoring solid substrate fermentation. J Food Sci 50:539–540

Kumar PKR, Lonsane BK (1987) Gibberellic acid by solid-state fermentation: consistent and improved yields. Biotechnol Bioeng 30:267–271

Larroche C, Desfarges C, Gros JB (1986) Spore production of Penicillium roqueforti by simulated solid-state fermentation. Biotechnol Lett 8:453–456

Le Page C, Louvel L, Mary NM, Valtre C, Theilleux J (1988) Visualization and measurement of metabolizing biomass using a tetrazolium reduction method. In: Proc 8th Int Biotechnol Symp, Paris, 17–22 July. Société Francaise de Microbiologie, Paris, poster D43, p 281

Lhomme B, Roux JC (1992) Biomass production, carbon and oxygen consumption by Rhizopus arrhizus grown in submerged cultures on thin liquid films or immobilized on fibrous and particulate materials. Appl Microbiol Biotechnol 37:37–43

344

M.J.R. Nout et al.

Lonsane BK, Ghildyal NP, Budiatman S, Ramakrishna SV (1985) Engineering aspects of solid state fermentation. Enzyme Microb Technol 7:258–265

Matcham SE, Jordan BR, Wood DA (1985) Estimation of fungal biomass in a solid substrate by three independent methods. Appl Microbiol Biotechnol 21:108–112

Mitchell DA, Greenfield PF, Doelle HW (1990) Mode of growth of *Rhizopus oligosporus* on a model substrate in solid state fermentation. World J Microbiol Biotechnol 6:201–208

Mitchell DA, Do DD, Greenfield PF, Doelle HW (1991) A semimechanistic mathematical model for growth of *Rhizopus oligosporus* in a model solid-state fermentation system. Biotechnol Bioeng 38:353–362

Molimard P, Vassal L, Bouvier I, Spinnler HE (1995) Growth of *Penicillium camemberti* and *Geotrichum candidum* in pure and mixed cultures on experimental mold ripened cheese of camembert-type. Lait 75:3–16

Molin P, Gervais P, Lemiere JP, Davet T (1992) Direction of hyphal growth: a relevant parameter in the development of filamentous fungi. Res Microbiol 143:777–784

Molin P, Gervais P, Lemiere JP (1993) A computer model based on reaction diffusion equations for the growth of filamentous fungi on solid substrate. Biotechnol Prog 9:385–393

Narahara H, Koyama Y, Yoshida T, Pichangkura S, Ueda R, Taguchi H (1982) Growth and enzyme production in a solid-state culture of *Aspergillus oryzae*. J Ferment Technol 60:311–319

Newell SY (1992) Estimating fungal biomass and productivity in decomposing litter. In: Carroll GC, Wicklow DT (eds) The fungal community; its organization and role in the ecosystem. Dekker, New York, pp 521–561

Newell SY, Statzell-Tallman A (1982) Factors for conversion of fungal biovolume values to biomass, carbon and nitrogen: variation with mycelial ages, growth conditions, and strains of fungi from a salt marsh. Oikos 39:261–268

Newell SY, Arsuffi TL, Fallon RD (1988) Fundamental procedures for determining ergosterol content of decaying plant material by liquid chromatography. Appl Environ Microbiol 54:1876–1879

Nishio N, Tai K, Nagai S (1979) Hydrolase production by *Aspergillus niger* in solid-state cultivation. Eur J Appl Microbiol 8:263–270

Nout MJR (1984) Influence of sample size and analytical procedure on the variance of surface mould plate counts of maize kernels. Chem Mikrobiol Technol Lebensm 8:133–136

Nout MJR (1995) Useful role of fungi in food processing. In: Samson RA, Hoekstra E, Frisvad JC, Filtenborg O (eds) Introduction to food-borne fungi. Centraal Bureau voor Schimmelcultures, Baarn, pp 295–303

Nout MJR, Rombouts FM (1990) Recent developments in tempe research. J Appl Bacteriol 69:609–633

Nout MJR, Bonants-Van Laarhoven TMG, De Dreu R, Gerats IAGM (1985) The influence of some process variables and storage conditions on the quality and shelf-life of soybean tempeh. Antonie van Leeuwenhoek 51:532–534

Nout MJR, Bonants-Van Laarhoven TMG, De Jongh P, De Koster PG (1987) Ergosterol content of *Rhizopus oligosporus* NRRL 5905 grown in liquid and solid substrates. Appl Microbiol Biotechnol 26:456–461

Okazaki N, Sugama S, Tanaka T (1980) Mathematical model for surface culture of Koji mold. J Ferment Technol 58:471–476

Owens JD, Konirova L, Thomas DS (1992) Causes of conductance change in yeast cultures. J Appl Bacteriol 72:32–38

Ozawa S, Amada K, Sato K (1993) Semi-continuous production of enzymes by a novel solid-state fermentation system by using urethan foam carriers. In: Abstract books, vol 2, 6th European congress on biotechnology, Firenze, 13–17 June 1993. European Federation of Biotechnology. Finito di Stampare dalla Tipografia TAF s.r.l., Borgo Stella 21, 50124 Firenze, Italy, abstract TU119

Packer HL, Thomas CR (1990) Morphological measurements on filamentous microorganisms by fully automatic image analysis. Biotechnol Bioeng 35:870–881

Penaloza W, Davey CL, Hedger JN, Kell DB (1992) Physiological studies on the solid-state quinoa tempe fermentation, using on-line measurements of fungal biomass production. J Sci Food Agric 59:227–235

Pirt SJ (1965) The maintenance energy of bacteria in growing cultures. Proc R Soc Lond 163 B:224–231

Pitt RE (1993) A descriptive model of mold growth and aflatoxin formation as affected by environmental conditions. J Food Prot 56:139–146

Prosser JI, Trinci APJ (1979) A model for hyphal growth and branching. J Gen Microbiol 111:153–164

Raghava Rao KSMS, Gowthaman MK, Ghildyal NP, Karanth NG (1993) A mathematical model for solid state fermentation in tray bioreactors. Bioproc Eng 8:255–262

Rajagopalan S, Modak JM (1994) Heat and mass transfer simulation studies for solid-state fermentation processes. Chem Eng Sci 49:2187–2193

Rajagopalan S, Modak JM (1995a) Modeling of heat and mass transfer for solid state fermentation process in tray bioreactor. Bioproc Eng 13:161–169

Rajagopalan S, Modak JM (1995b) Evaluation of relative growth limitation due to depletion of glucose and oxygen during fungal growth on a spherical solid particle. Chem Eng Sci 50:803–811

Ramana Murthy MV, Thakur MS, Karanth NG (1993) Monitoring of biomass in solid state fermentation using light reflectance. Biosens Bioelectron 8:59–63

Richards FJ (1959) A flexible growth function for empirical use. J Exp Bot 10:290–300

Ride JP, Drysdale RB (1971) A rapid method for the chemical estimation of filamentous fungi in plant tissue. Physiol Plant Pathol 2:7–15

Rinzema A, De Reu JC, Oostra J, Nagel FJI, Nijhuis GJA, Scheepers AA, Nout MJR, Tramper J (1996) Models for solid-state cultivation of *Rhizopus oligosporus*. In: Roussos S, Lonsane BK, Raimbault M, Viniegra-Gonzalez G (eds) Advances in solid state fermentation. Kluwer, Dordrecht (in press)

Rodriguez Leon JA, Sastre L, Echevarria J, Delgado G, Bechstedt W (1988) A mathematical approach for the estimation of biomass production rate in solid-state fermentation. Acta Biotechnol 8:307–310

Sargantanis J, Karim MN, Murphy VG, Ryoo D, Tengerdy RP (1993) Effect of operating conditions on solid substrate fermentation. Biotechnol Bioeng 42:149–158

Sato K, Nagatani M, Nakamura K, Sato S (1983) Growth estimation of *Candida lipolytica* from oxygen uptake in a solid-state culture with forced aeration. J Ferment Technol 61:623–629

Saucedo-Castaneda G, Gutierrez-Rojas M, Bacquet G, Raimbault M, Viniegra-Gonzalez G (1990) Heat-transfer simulation in solid-state fermentation. Biotechnol Bioeng 35:802–808

Scheu S, Parkinson D (1994) Changes in bacterial and fungal biomass C, bacterial and fungal biovolume and ergosterol content after drying, remoistening and incubation of different layers of cool temperate forest soils. Soil Biol Biochem 26:1515–1525

Schnürer J (1993) Comparison of methods for estimating the biomass of three food-borne fungi with different growth patterns. Appl Environ Microbiol 59:552–555

Schnürer J, Rosswall T (1982) Fluorescein diacetate hydrolysis as a measure of total microbial activity in soil and litter. Appl Environ Microbiol 43:1255–1261

Seitz LM, Sauer DB, Burroughs R, Mohr HE, Hubbard JD (1979) Ergosterol as a measure of fungal growth. Phytopathology 69:1202–1203

Silman RW (1980) Enzyme formation during solid-substrate fermentation in rotating vessels. Biotechnol Bioeng 22:411–420

Silman RW (1988) Continuous estimation of viscosity, fluid density, surface tension, and heat output during fermentations. Biotechnol Tech 2:221–226

Silman RW, Black LT, Norris K (1983) Assay of solid-substrate fermentation by means of reflectance infrared analysis. Biotechnol Bioeng 25:603–607

Soccol C, Rodriguez Leon J, Marin B, Roussos S, Raimbault M (1993) Growth kinetics of Rhizopus arrhizus in solid state fermentation of treated cassava. Biotechnol Tech 7:563–568

Sugama S, Okazaki N (1979) Growth estimation of Aspergillus oryzae cultured on solid media. J Ferment Technol 57:408–412

Szewczyk KW, Myszka L (1994) The effect of temperature on the growth of A. niger in solid state fermentation. Bioproc Eng 10:123–126

Trinci APJ (1969) A kinetic study of the growth of Aspergillus nidulans and other fungi. J Gen Microbiol 57:11–24

Trinci APJ (1971) Influence of the width of the periferal growth zone on the radial growth rate of fungal colonies on solid media. J Gen Microbiol 67:325–344

Viniegra-Gonzales G, Saucedo-Castaneda G, Lopez-Isunza F, Favela-Torres E (1993) Symmetric branching model for the kinetics of mycelial growth. Biotechnol Bioeng 42:1–10

Whipps JM, Lewis DH (1980) Methodology of a chitin assay. Trans Br Mycol Soc 74:417–418

Wicklow DT, Dowd PF, Gloer JB (1994) Antiinsectan effects of Aspergillus metabolites. In: Powell KA, Renwick A, Peberdy JF (eds) The genus Aspergillus: from taxonomy and genetics to industrial application. Plenum Press, New York, pp 93–114

Wiegant WM (1991) A simple method to estimate the biomass of thermophilic fungi in composts. Biotechnol Tech 5:421–426

Wiegant WM, Wery J, Buitenhuis ET, De Bont JAM (1992) Growth-promoting effect of thermophilic fungi on the mycelium of the edible mushroom Agaricus bisporus. Appl Environ Microbiol 58:2654–2659

Wissler MD, Tengerdy RP, Murphy VG (1983) Biomass measurement in solid-state fermentations using 15N mass spectrometry. Dev Ind Microbiol 24:527–538

Xue MJ, Liu DM, Zhang HX, Qi HY, Lei ZF (1992) A pilot process for solid state fermentation from sugar beet for the production of microbial protein. J Ferment Bioeng 73:203–205

Yang SS, Jang HD, Liew CM, Du Preez JC (1993) Protein enrichment of sweet potato residue by solid-state cultivation with mono- and co-cultures of amylolytic fungi. World J Microbiol Biotechnol 9:258–264

Zhu Y, Smits JP, Knol W, Bol J (1994) A novel solid-state fermentation system using polyurethane foam as inert carrier. Biotechnol Lett 16:643–648

Zwietering MH, Jongenburger I, Rombouts FM, Van 't Riet K (1990) Modeling of the bacterial growth curve. Appl Environ Microbiol 56:1875–1881

21 Enzymatic Conversion of Plant Biomass

R.L. SINSABAUGH and M.A. LIPTAK

CONTENTS

I. Introduction

The use of fungi in food processing predates human civilization. By that timeline, the use of fungi for modifying lignocellulose is a recent innovation. The oldest processes are probably composting and retting, in which mixed microbial communities partially decompose plant material to generate soil amendments and fiber.

In natural environments, filamentous fungi are the dominant decomposers of plant fiber. This dominance is due to more than the capacity to produce the proper enzymes. Their hyphal growth habit, their ability to secrete enzymes freely into the environment, particularly from growing tips, and their ability to translocate nutrients through extensive mycelia, give them ecological advantages over bacteria (Rayner and Boddy 1988; Dix and Webster 1995).

Considerable effort has been expended over the past three decades to unravel the biochemistry of plant fiber degradation; progress is regularly reviewed (Dekker 1985; Eriksson and Wood 1985;

Biology Department, University of Toledo, Toledo, Ohio 43606, USA

Marsden and Gray 1986; Kirk and Farrell 1987; Eriksson et al. 1990; Higuchi 1990). Much of the effort has concentrated on model systems such as the cellulases of *Trichoderma* and the ligninases of *Phanerochaete*. Recently, interest has extended to the fungi of extreme environments. The enzymatic decomposition of plant fiber is complex and synergistic: individual enzymes operate as components of multienzyme systems to effect the degradation of specific polymers, systems of enzymes interact to degrade the matrix of polymers that comprise plant cell walls, and microbial consortia create higher-order synergisms through interactions among their enzyme systems.

While the biochemistry of fiber degradation has been investigated primarily from an applied perspective, ecological researchers have contributed much to our understanding of fungi at the physiological and community levels (Ljungdahl and Eriksson 1985; Rayner and Boddy 1988; Dix and Webster 1995). We can crudely partition the commercial applications of fungi along the same levels, into processes that utilize abiotic enzymes and those that involve the growth and activity of viable organisms. Since abiotic processes require the cultivation of enzyme production strains, it is clear that successful applications are dependent on both ecological and biochemical insight.

In this chapter, we provide overviews of the major enzyme systems involved in the degradation of lignocellulose and their application or potential application in industrial processes.

II. Ligninases

A. Lignin Structure and Biodegradation

Lignin is the most recalcitrant, biotically synthesized organic molecule, second only to cellulose in biospheric abundance, accounting for about a quarter of terrestrial primary production. In

The Mycota IV
Environmental and Microbial Relationships
Wicklow/Söderström (Eds.)
© Springer-Verlag Berlin Heidelberg 1997

plants, it provides rigidity to secondary cell walls and shields structural polysaccharides from microbial attack. In a broader ecological context, lignin abundance affects rates of plant litter decomposition and humus formation (Dix and Webster 1995). While many fungi and bacteria are capable of modifying the structure of lignin to varying degrees, the organisms primarily responsible for its decomposition are filamentous basidiomycetes (Kirk and Farrell 1987; Rayner and Boddy 1988). Because the depolymerization of brown-colored lignin whitens plant fiber, these fungi are commonly referred to as white rot fungi (Eriksson et al. 1990).

Lignin is a polymer of phenylpropanoid units, principally coniferyl, sinapyl, and p-hydroxycinnamyl alcohols, which are joined through an oxidative free radical process (Higuchi 1990). The most frequent linkage is an ether bond between the second carbon (C_β) of a glycerol moiety and an aromatic ring of another monomer (Fig.

1). Hemicelluloses are bound to lignin through ester linkages. Several molecular attributes account for lignin's recalcitrance: large size, high degree of branching, diversity of monomers, diversity of linkages, and stability of C-C and ether bonds. These attributes militate the fact that ligninolytic agents are nonspecific and oxidative.

The primary function of ligninases is to gain access to shielded polysaccharides. Even white rot fungi are unable to use lignin as a sole carbon source. In soil and litter, ligninolytic activity is typically induced by low N availability. Conversely, the addition of N to decomposing plant litter depresses lignin decomposition (Fog 1988).

The biochemistry of lignin degradation has been difficult to unravel, and much remains unclear (Kirk and Farrell 1987; Eriksson et al. 1990; Blanchette 1991). A model for the process exists only for white rot fungi, based heavily on studies of *Phanerochaete chrysosporium*. The first ligninolytic enzyme isolated was lignin peroxidase

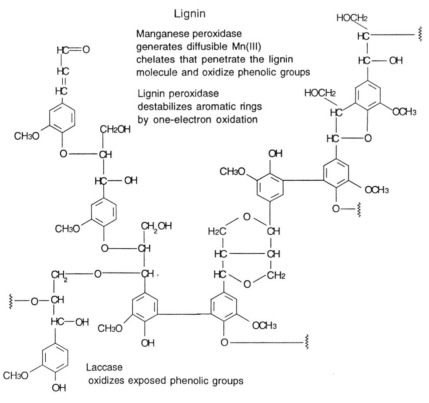

Fig 1. Lignin structure and decomposition. The heterogeneity of lignin molecules and the stability of the linkages require ligninolytic agents that are oxidative and nonspecific. At present, only a few ligninolytic enzymes are known. Lignin peroxidase is able to withdraw single electrons directly from aromatic rings; the resulting destabilization can lead to cleavage of intermonomer linkages such as the C_α-C_β bond of glycerol moieties. Manganese peroxidase generates chelated Mn(III) and perhaps other diffusible oxidants that can penetrate lignin molecules and oxidize phenolic moieties, creating reactive phenoxy radicals. Laccases are oxidases that act on phenolic moieties through the single electron oxidations

(Lip) (Glenn et al. 1983; Tien and Kirk 1983). The enzyme embraces a heme iron prosthetic group which is oxidized by H_2O_2 and reduced by performing two single electron oxidations of donor substrates. LiP is the strongest peroxidase known, able to pull single electrons from aromatic rings. The most frequent outcome of this destabilization is cleavage of the C_α-C_β bond of the glycerol moiety, which releases a peripheral monomer and a two carbon (C_β-C_γ) glycoaldehyde fragment.

Another component of the *Phanerochaete* ligninolytic system is manganese peroxidase (MnP), first characterized by Kuwahara et al. (1984). Unlike LiP, manganese peroxidase can generate small, diffusible oxidants capable of penetrating the cell wall matrix. Like LiP, the heme prosthetic group is oxidized by H_2O_2. The enzyme returns to its ground state by oxidizing Mn(II) to Mn(III), which chelates with organic acids like glycolate or oxalate to form a diffusible oxidant capable of attacking phenolic (but not nonphenolic) moieties. Some ligninolysis occurs through cleavage of C_α-aryl bonds, and quinones are created that participate in secondary reactions. It has recently been suggested that MnP may participate in the oxidation of nonphenolic moieties by peroxidizing unsaturated lipids which are stronger oxidants than Mn(III) (Bao et al. 1994).

Both LiP and MnP require a source of H_2O_2 to sustain activity. Thus, extracellular oxidases that reduce O_2 to H_2O_2 are also elements of the ligninolytic system. Glucose oxidase, cellobiose oxidase, and aryl alcohol oxidases are examples. Another enzyme found in white rot fungi is glyoxal oxidase, which uses small aldehydes like glycoaldehyde, a cleavage product of LiP, as electron donors, creating the potential for a catalytic cycle (Hammel et al. 1994).

Peroxidases are not the only type of enzyme capable of attacking lignin. Laccases (EC 1.10.3.2) are Cu-bearing oxidases that catalyze the single electron oxidations of phenolic molecules. After accepting four electrons, the enzymes regenerate by reducing molecular oxygen to water. Like MnP, laccases can attack the phenolic moieties of lignin, generating reactive phenoxy radicals.

Neither LiP nor MnP is universally produced by white rot fungi, thus neither appears to be a necessary element of a ligninolytic system. The lignin-degrading mechanisms of other fungal groups are even more obscure. Brown rot fungi are basidiomycetes that preferentially degrade cellulose; they are of considerable importance in plant fiber decomposition and humus formation (Dix and Webster 1995). They modify lignin through oxidation and demethylation reactions but do not substantially decompose it. In high moisture environments, ascomycetes and deuteromycetes, collectively described as soft rot fungi, are prominent decomposers (Eriksson et al. 1990; Dix and Webster 1995). They attack cellulose preferentially and may have lignin-degrading capabilities, though the mechanisms are unknown. Studies of *Aspergillus* and other model deuteromycetes indicate that they do not possess LiP or MnP (Duarte and Costa-Ferreira 1994).

B. Biotechnical Applications

Because it is the most recalcitrant component of plant fiber, lignin degradation is the limiting process in the conversion of wood and other phytomass into commercial products. Traditionally, the delignification of plant fiber is accomplished by various combinations of mechanical, chemical, and thermal treatment which are energy-intensive and create problematic waste streams. Potentially, these disadvantages could be mitigated by substituting biological or enzymic processing. However, there are major impediments. First, lignin is effectively degraded only under aerobic conditions and only by white rot fungi, restricting biological delignification to a rather small (by industrial standards) niche defined by axes of pH, temperature, moisture, nutrient availability, and atmosphere. These restrictions might be circumvented by using an abiotic enzyme approach, but the enzymology of ligninolysis is complex and only coarsely understood.

At present several potential applications of white rot fungi or ligninolytic enzymes are being explored (Eriksson et al. 1990; Ghosh and Singh 1993). One is delignification of chipped wood as a pretreatment for pulping. A labile carbon source like malt must be mixed with the chips to sustain fungal growth. Unfortunately, ligninolysis is also accompanied by cellulolysis. This problem can be avoided by using engineered strains that lack cellulolytic capabilities, but these strains grow more slowly and break down less lignin.

Ligninases might also be applied later in the pulping process as bleaching agents, reducing the need for chlorine or other chemical oxidants. The practical difficulties include modifying the temperature and pH of pulp liquors to those suitable for enzymatic activity and developing systems for enzyme production and recovery (Messner and Srebotnik 1994; Reid and Paice 1994). More fundamental, however, is lack of understanding of the enzymatic process and the optimal suite of enzymes required. Depending on the conditions, ligninolytic enzymes are capable of polymerizing as well as depolymerizing soluble lignin fragments.

Ligninases might also be used to treat waste bleach liquors by breaking down degraded lignin fragments. This would be especially beneficial if the bleaching process involved chlorine because of concerns over the environmental impact of chlorinated aromatic compounds. The problems here are similar to those associated with the use of ligninases for bleaching: what is the optimal mix of enzymes; and how to produce and recycle them? A related application is decompostion of waste lignosulfonates from the kraft-pulping process. Specific products like vanillin may be marketable, but at present there are no breakdown products with sufficent commercial demand to make them economical.

Delignification of wood, straw, and bagasse (the residual fiber from sugar cane processing) increases their value as livestock feed (Eriksson et al. 1990; Ghosh and Singh 1993). White rot fungi could partly replace the physicochemical process of steam explosion and alkaline delignification, with the added benefit of increasing protein content. Pilot trials have been successful but scaleup creates need for aeration and temperature/moisture control systems that make the economics unfavorable. The process is also susceptible to invasion by undesired microbes.

III. Cellulases

A. Cellulose Structure and Biodegradation

As the principal structural component of plants, cellulose is the most abundant organic molecule in the biosphere, accounting for nearly half of the annual primary production (Eriksson et al. 1990). The rigidity and crystalline nature of cellulose microfibrils account for both its functional role in plant cell walls and its resistance to microbial decomposition. Cellulose is less recalcitrant than lignin; a broader spectrum of fungi and bacteria are

capable of attacking it, but, like lignin, the number of organisms capable of fully decomposing the polymer in its native crystalline state is comparatively small (Marsden and Gray 1986).

Cellulose is a linear polymer of glucose residues linked by β-1,4-glycosyl bonds (Fig. 2). The beta configuration of these linkages allows reinforcing hydrogen bonds to form between adjacent residues conferring rigidity to the molecule. Microfibrils, the basic structural unit of plant cell walls, are formed through alignment and hydrogen bonding of cellulose molecules. The structure of microfibrils and their arrangement in plant cell walls vary (Eriksson et al. 1990). These patterns lead to differences in degree of crystallinity, accessible surface area, pore size distributions, moisture-holding capacities, and other properties that determine the susceptibility of the cellulose to enzymatic attack (Marsden and Gray 1986). In addition, the microfibrils are physically intercalated with hemicelluloses, which, in turn, are covalently linked with lignin.

The literature on fungi and cellulose decomposition is extensive, and the subject of frequent reviews (e.g., Eriksson and Wood 1985; Ljungdahl and Eriksson 1985; Marsden and Gray 1986; Eriksson et al. 1990) beginning with Waksman and Skinner (1926). Biochemical investigations of cellulose degradation began in the 1940s. In 1950, Reese et al. proposed the first mechanistic model of cellulolysis, based on the recognition that enzymes capable of hydrolyzing soluble cellulose derivatives were not generally effective at decomposing native cellulose. Their two-step $C1$-C_x model remained the paradigm for cellulolysis for 20 years. C_1 was the designation for enzymes capable of decomposing native cellulose into short, noncrystalline molecules; C_x designated enzymes that hydrolyzed amorphous cellulose into assimilable products.

Over the succeeding decades, much of the effort toward unraveling the biochemistry of cellulose decomposition focused on a few fungal models: *Trichoderma*, *Sporotrichum* (*Phanerochaete*), *Aspergillus*, *Penicilium*. As enzymes were isolated and characterized, it became clear that cellulolysis was more complex than the two-step C_1-C_x model (Eriksson et al. 1990). Three classes of hydrolytic enzymes are required (Fig. 2): exo-1,4-β-glucanases (cellobiohydrolases) (EC 3.2.1.91), which are capable of binding to crystalline domains and hydrolyzing cellobiose or glucose from the nonreducing ends of cellulose

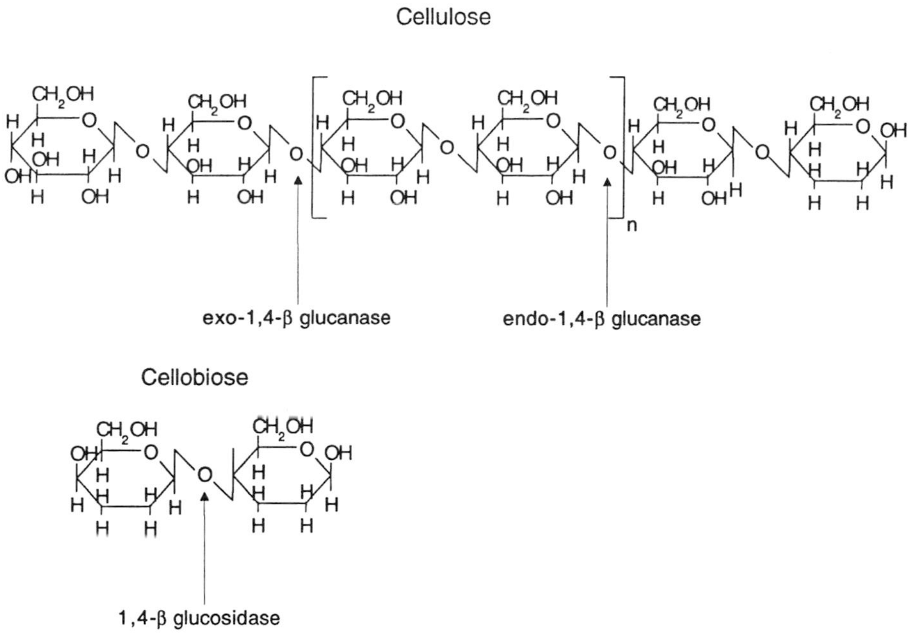

Cellulose

Fig 2. Cellulose structure and decomposition. Three classes of hydrolytic enzymes are necessary for cellulose degradation. Exo-1,4-β-glucanases hydrolyze cellobiose or glucose from the nonreducing end of cellulose. Endo-1,4-β- glucanases cleave glucosidic linkages at random spots along the polymer chain. 1,4-β glucosidases hydrolyze cellobiose into glucose

molecules; endo-1,4-β-glucanases (EC 3.2.1.4), which randomly cleave glucosidic linkages along noncrystalline domains; and 1,4-β-glucosidases (EC 3.2.1.21), which release glucose from celloligosaccharides and aryl-β-glucosides. For white rot fungi, the decomposition of native cellulose is the result of synergistic interaction between exoglucanase and endoglucanase; neither enzyme alone is able to effect significant breakdown. Other mechanisms may operate for brown rot fungi and bacteria (see review by Eriksson et al. 1990). The oxidative enzymes cellobiose:quinone oxidoreductase (cellobiose dehydrogenase) and cellobiose oxidase also contribute to cellulose degradation; the former reduces quinones by oxidizing cellobiose to cellobiolactone, the latter oxidizes cellobiose to cellobionic acid using molecular oxygen.

B. Biotechnical Applications

Interest in lignocellulose as a renewable source of fuel and chemical feedstocks can be traced to the oil crises of the 1970s. Since that time, considerable progress has been made in understanding the structure and biodegradation of these materials.

However, the major problems of pretreatment, hydrolysis, enzyme recovery, and pentose fermentation remain (Eriksson et al. 1990).

Pretreatments disrupt the ordered structure of plant cell walls, allowing for enzyme access. In the case of cellulose, enzymatic saccharification is limited by crystallinity and low specific surface area in addition to the shielding provided by lignin and hemicellulose (Ghosh and Singh 1993). A variety of pretreatments are possible: mechanical grinding or milling, radiation, alkaline hydrolysis, acid hydrolysis, oxidation, and steam explosion (Ghosh and Singh 1993). All are effective in increasing surface area, breaking some lignin linkages, and reducing cellulose crystallinity, but costly relative to the commercial value of the products generated.

After pretreatments expose the cellulose and hemicellulose, these are susceptible to enzymatic hydrolysis. The major problems at the industrial scale are production of an optimal mix of enzymes and recovery of those enzymes for reuse (Eriksson et al. 1990). At present, cellulases are expensive to produce, though the potential exists for creating hyperproducing strains through genetic engineering. For fuel production, thermophilic anaerobic bacteria may be more economical than fungi, since they can directly ferment cellulose to ethanol at temperatures up to 70 °C.

In abiotic saccharification systems, enzyme recovery can be a limiting constraint; a large fraction of the applied enzyme becomes adsorbed to the substrate. Enzymes bound to undigested particles can be recycled by either releasing them from the substrate by raising the pH with phosphate or directly mixing hydrolyzed residue with fresh incoming material (Eriksson et al. 1990).

In the absence of a single-stage cellulose to ethanol fermentation, the final step in a biomass to fuel process would be fermentation of enzymatically released saccharides. This step is complicated by the presence of D-xylose and L-arabinose, pentose components of hemicellulose that comprise up to 30% of the carbohydrates of plant material. Yeasts are very efficient at converting hexoses into ethanol, but not pentoses (Mishra and Singh 1993). The economics of ethanol production would be improved if hexoses and pentoses were fermented concurrently. Some filamentous fungi can ferment both glucose and xylose from lignocellulosic substrates, but the process involves alternating aerobic and anaerobic regimes. Bacteria, e.g., *Bacillus*, *Aeromonas*, and *Klebsiella*, are more efficient, but have far lower tolerances for ethanol than fungi.

An alternative use for lignocellulosic wastes is production of microbial protein for animal feed. Since the early 1900s, the yeast *Candida utilis* and the fungus *Paecilomyces varoti* have been cultured on the sugars and acetate present in waste liquors from the kraft-pulping process. In the 1970s, new processes for treating lignocellulosic wastes using *Sporotrichum pulverulentum* and *Chaetomium cellulolyticum* were developed, but are not in commercial use (Eriksson et al. 1990).

IV. Hemicellulases

A. Hemicelluose Structure and Biodegradation

Hemicelluloses link lignin with cellulose microfibrils. In wood, hemicellulose accounts for 20–30% of cell wall mass (Eriksson et al. 1990). Compared with lignin and cellulose, hemicellulose is labile. Its low degree of polymerization, heterogeneous composition, and amorphous structure render it susceptible to attack by many saprotrophic microorganisms. In general, the same microbes that degrade cellulose also degrade hemicellulose. As is the case for other cell wall components, the principal impediment to decomposition is access; thus fungi have the advantage, at least in the initial stages of cell wall breakdown.

Hemicelluloses are heteropolysaccharides; the diversity of monomers and linkages may exceed that of lignin, although their comparatively small size makes them less recalcitrant. Hemicelluloses can be branched or unbranched with individual molecules composed of a subset of D-xylose, D-mannose, D-galactose, L-arabinose, D-glucuronic acic, and D-glucose (Figs. 3, 4). The largest constituents of hemicellulose are xylan and mannan. Xylan is a polymer of β-1,4-linked xyloses with β-1,2-linked side chains of 4-O-methylglucuronic acid and β-1,3-linked side chains of arabinose. The methylglucuronic acid may be further modifed by acetylation. In hardwoods, the degree of polymerization is 150–200 and in softwoods about 70–130 (Viikari et al. 1994). Mannans are polymers of glucose and mannose, in random order, connected by β-1,4-glycosidic linkages. Galactose side chains branch from main chain mannose and glucose units via α-1,6 bonds. In addition, 20–30% of mannose and glucose units are acetylated at carbon 2 or 3.

The most extensively studied hemicellulase systems are those of the soft rot fungi *Aspergillus* and *Trichoderma*. The complexity of hemicellulose mandates participation of many enzymes for complete decomposition (see reviews by Biely 1985; Dekker 1985; Wong et al. 1988; Eriksson et al. 1990; Viikari et al. 1993; Duarte and Costa-Ferreira 1994). Endo-1,4-β-xylanases (EC 3.2.1.8) and endo-1,4-β-mannases (EC 3.2.1.78) depolymerize the main chains (Figs. 3, 4). The resulting oligosaccharides are further hydrolyzed by 1,4-β-xylosidases (EC 3.2.1.37), 1,4-β-D-mannosidases (EC 3.2.1.25), and 1,4-β-glucosidases (EC 3.2.1.21). Branch chain cleavage requires α-L-arabinosidases (EC 3.2.1.55), α-glucuronidases (EC 3.2.1) and α-galactosidases (EC 3.2.1.22). Acetyl substitutents are hydolyzed by acetyl xylan esterase and acetyl galactoglucomannan esterases.

B. Biotechnical Applications

Fungal xylanases have potential application in a number of phytomass conversion processes, including wood pulping, production of animal feed, and conversion of lignocellulose into feedstocks

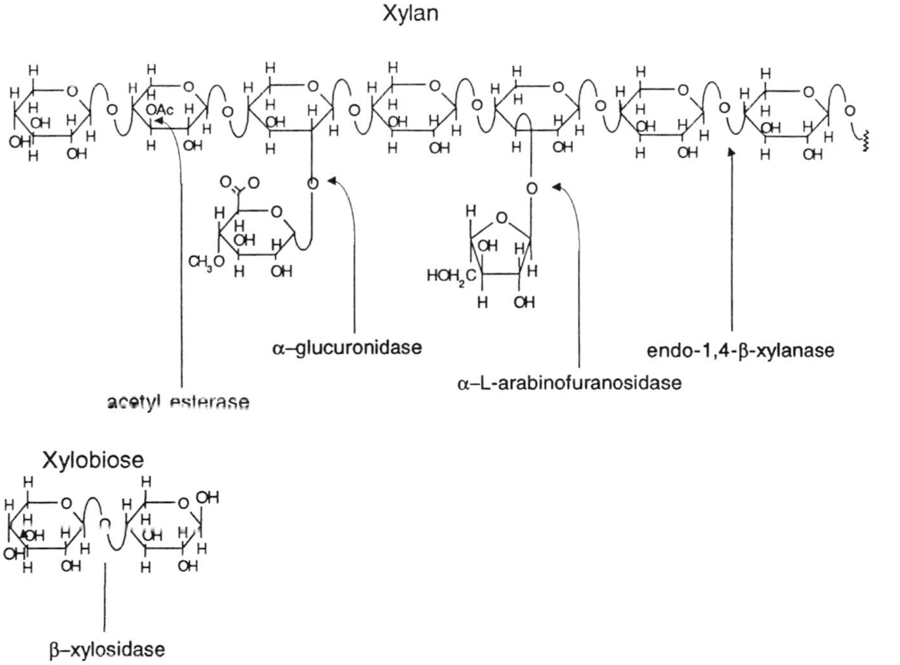

Xylan

α–glucuronidase

acetyl esterase

α–L-arabinofuranosidase

endo-1,4-β-xylanase

Xylobiose

β–xylosidase

Fig 3. Xylan structure and decomposition. Endo-1,4-β-xylanases hydrolyze linkages along the polymer chain. 1,4-β-xylosidases hydrolyze the resulting oligosaccharides to xylose. Side chains are cleaved by α-L arabinosidases, α-glucuronidases, and α-galactosidases. Acetyl substituents are hydrolyzed by acetyl xylan esterases

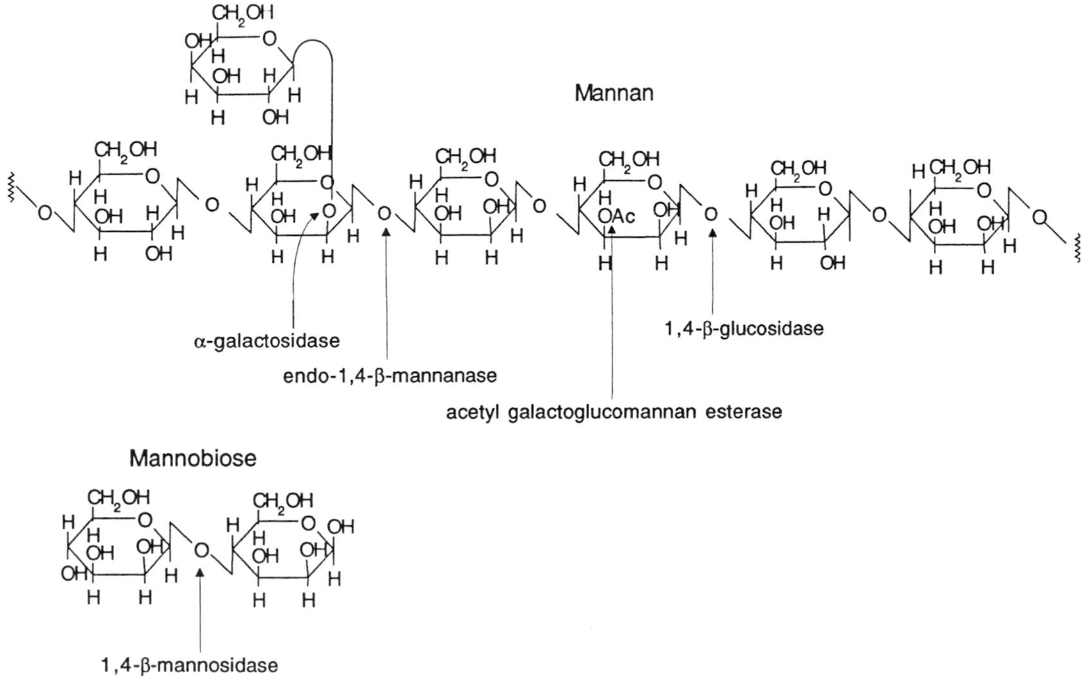

Mannan

α-galactosidase

endo-1,4-β-mannanase

acetyl galactoglucomannan esterase

1,4-β-glucosidase

Mannobiose

1,4-β-mannosidase

Fig 4. Galactoglucomannan structure and decomposition. Endo-1,4-β-mannanases hydrolyze glucosidic mannan linkages at random spots within the polymer chain. 1,4-β-D-mannosidases hydrolyze the resulting oligosaccharides to mannose. Galactose substituents are cleaved by α-galactosidases, and glucose units in the chain are cleaved by 1,4-β-glucosidases. Acetyl substituents are hydrolyzed by acetyl galactoglucomannan esterases

and fuels (Li et al. 1993). Xylan derivatives may even be useful as biodegradable plastics (Glasser et al. 1995). However, the most extensive commercial use of hemicellulases has been pulp bleaching. Unbleached kraft pulps are brown-colored due to lignin, which is degraded and removed through a series of bleaching and alkali extraction steps. In general, the combination of oxidants applied varies with pulp type and target brightness, but traditionally, bleaching involved the application of elemental chlorine, and chlorine dioxide. Growing environmental concerns over the high concentration of chlorinated organic compounds in bleaching effluents has led to increased substitution of ozone and oxygen (Daneault et al. 1994).

In 1986, Viikari et al. proposed using xylanase for bleaching kraft pulp. After further studies, this process was first commercially applied in Finland in 1988 and has since been widely adopted in Europe and North America. Enzymatic pretreatment improves final brightness and reduces the consumption of bleaching agents. For softwood pulps, the average reduction in chlorine consumption is 25% (Viikari et al. 1994). As the pulping industry moves increasingly toward totally chlorine-free bleaching, enzymatic pretreatment remains an economical method for increasing final brightness.

The most effective commercial delignifying enzymes are endo-β-1,4-xylanases. Early studies focused on *Trichoderma reesei* as a source, but the search for more efficient production strains has expanded considerably. The organisms used for commercial production of xylanases include *Trichoderma reesei*, *Thermomyces lanuginosus*, *Aureobasidium pullulans*, and *Streptomyces lividans* (Viikari et al. 1993). Because high pH and high temperature optima are desirable properties (current practice requires reducing the pH of pulp liquor to 5–7 and cooling it to 40–50°C), most recent research effort has turned to bacteria. *Bacillus* produces alkaline active enzymes with pH optima circa 9 (Rättö et al. 1992; Nakamura et al. 1993; Lundgren et al. 1994). *Thermotoga* produces the most thermostable xylanases known, with a half-life of 90 min at 95°C (Simpson et al. 1991).

Xylanase may promote bleaching through several mechanisms. The principal one appears to be hydrolysis of reprecipitated xylan. The heating of kraft pulp under alkaline conditions dissolves up to half the xylan and hydrolyzes its side chains. Upon cooling, this xylan reprecipitates in crystalline form on the surface of microfibrils, which physically restricts the extraction of degraded lignin from the pulp fiber. Glucomannans are also dissolved, but are less stable, and tend to be completely hydrolyzed, thus additions of endo-β-1,4-mannanase only slightly improve delignification even in softwood, where glucomannan is the main hemicelluose. Side-chain-cleaving enzymes hydrolyze the ester linkages that attach hemicellulose to lignin, but these are generally broken during kraft treatment in any case.

The major treatment variables for xylanase bleaching are pH, temperature, enzyme dosage, mixing, and reaction time (Viikari et al. 1994). Bacterial xylanases operate at pH 6–9, fungal enzymes at pH 4–6. Temperature optima range from 35–60°C. Optimal dosage is 2–5 IU per g dry pulp. With efficient mixing, most of the benefit is realized with about 1 h of contact. Extending contact may improve pulp properties, or degrade them depending on the quantity of cellulase in the enzyme preparation. Cellulase activity progressively reduces the viscosity and mechanical strength of the pulp. For this reason, some research has been directed at cloning xylanase genes into noncellulolytic hosts.

V. Pectinase

A. Pectin Structure and Biodegradation

Pectin is found in the primary cell walls and middle lamellae of plants, where it functions as an adhesive. Pectin is a linear polymer of α-1,4-linked galacturonic acid residues, up to 70% of which are methoxylated, with neutral sugar side branches. Size, charge, and degree of substitution vary. The charges on the nonmethoxylated carboxylic acids are neutralized by Ca^{2+}, Na^+ or K^+. In plants, the side branches link pectin to proteins, hemicelluloses, and cellulose.

Pectin degradation is catalyzed by three types of enzymes (Fig. 5; Sakai et al. 1993). Polygalacturonases (PG) are hydrolytic enzymes; endo-1,4-α-polygalacturonase (EC 3.2.1.15) randomly cleaves linkages along the interior; exo-1,4-α-polygalacturonase (EC 3.2.1.67) releases galacturonate from nonreducing ends. Pectin lyases (PL) cleave 1,4-α-glycosidic linkages using a *trans*-elimination mechanism. Like the polygalacturonases, pectin lyases come in two varieties:

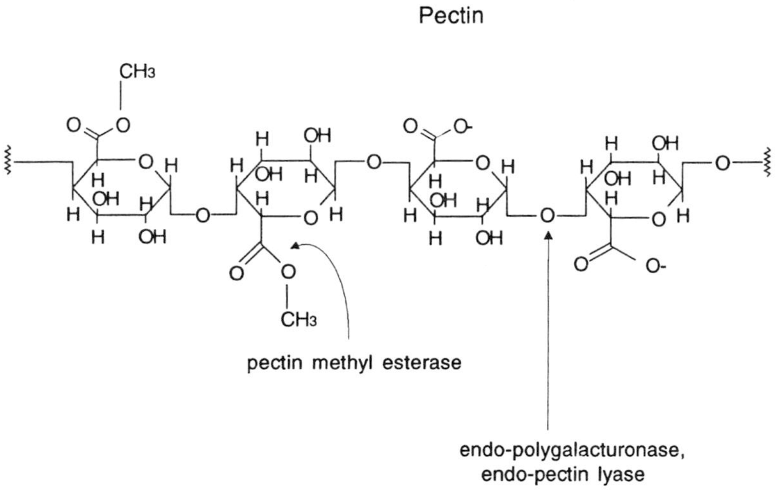

Pectin

pectin methyl esterase

endo-polygalacturonase,
endo-pectin lyase

Fig 5. Pectin structure and decomposition. Pectin esterases hydrolyze methoxy side groups, releasing methanol. Polygalacturonate can be depolymerized by polygalacturonases and pectin lyases. Endo-1,4-α-polygalacturonase randomly cleaves linkages along the interior; exo-1,4-α-polygalacturonase releases galacturonate from nonreducing ends. Pectin lyases (PL) cleave 1,4-α-glycosidic linkages by *trans*-elimination. Endo-PLs catalyze random cleavages and exo-PLs remove residues from nonreducing ends

endo-acting enzymes (EC 4.2.2.2) that catalyze random cleavages and exo-acting enzymes (EC 4.2.2.9) that remove residues from nonreducing ends. Finally, there are pectin esterases (PE; EC 3.1.1.11) that mediate hydrolysis of methoxy side groups.

Because it is more accessible, pectin is labile compared to lignocellulose. Polygalacturonases are produced by many ascomycetes and deuteromycetes. Pectin lyases and esterases are found in fewer taxa, frequently those classified as plant pathogens.

B. Biotechnical Applications

Pectinolytic fungi or their enzymes are used commercially to clarify raw fruit juices, process plant fiber, and release pectin from fruit by-products (Sakai et al. 1993). In its native state, pectin is generally insoluble, but in fruit, partial degradation of pectin is part of the ripening process. Consequently, the raw juices of many fruits, e.g., apples, cherries, and raspberries, contain pectin, which renders them viscous and turbid. The enzymatic treatment of fruit juices to improve handling, ease filtration, and increase clarity began in the 1930s and is now an established procedure (e.g., Chang et al. 1994). Commercial enzyme preparations containing mixtures of PG, PGL, and PE come from *Aspergillus*. There are at least three culture methods in use but details are proprietary. These preparations have some disadvantages. The hydrolysis of methoxy esters releases methanol into the juice and deesterified pectins tend to precipitate. Thus, there is interest in preparations containing only PL.

Pectin degradation is also part of the production process for textile fibers, like jute, hemp, and flax. A fermentation process called retting uses bacteria, e.g., *Clostridium*, *Bacillus*, and fungi, e.g., *Aspergillus*, *Penicillium*, *Cladosporium*) to decompose pectins, releasing the fibers.

Apple pomace and citrus peels are by-products of the fruit-processing industry. Pectin extracted from these materials is marketed as a gelling agent for jams and jellies and as a binder for pharmaceuticals and cosmetics. The extraction process involves cooking the fruit products in weak acid, separating the extract, and precipitating the pectin with salts or organic solvents; the precipitated pectin is then dried and ground. For mandarin orange peels, this chemical extraction process is too harsh, which has led to the development of an alternative enzymatic process. Peels are inoculated with *Trichosporon penicillatum*, and incubated for 24 h at 30 °C, releasing nearly all the pectin (Sakai et al. 1993).

VI. Conclusions

Plant fiber is a renewable resource. The bioconversion of lignocellulose to fuel, feed, or precursors for commercial syntheses may reduce consumption of fossil fuels and reduce waste streams. However, except for limited applications of fungal pectinases and xylanases, the conversion of lignocellulose into marketable products using enzyme technology remains a long-term goal rather than a practical reality. The proximate limitations are economic, but these economic constraints largely reflect our still fragmentary understanding of the biochemistry of plant fiber decomposition and the ecology of the decomposers. At the biochemical level, a mechanistic understanding of the apparently diverse ligninolytic systems of fungi and bacteria may be a prerequisite to the development of a practical delignification process. At the ecological level, the exploration of microbial diversity, especially in extreme environments, may yield new strains with commercial potential. As additional biochemical and ecological information accumulates, the possibilities for reduced costs through the engineering of transgenic strains expand.

References

Bao W, Fukushima Y, Jensen KA, Moen MA, Hammel KE (1994) Oxidative degradation of non-phenolic lignin during lipid peroxidation by fungal manganese peroxidase. FEBS Lett 354:297–300

Biely P (1985) Microbial xylanolytic systems. Trends Biotechnol 3:286–290

Blanchette RA (1991) Delignification by wood-decay fungi. Annu Rev Phytopathol 29:381–398

Chang T, Siddiq M, Sinha NK, Cash JN (1994). Plum juice quality affected by enzyme treatment and fining. J Food Sci 59:1065–1069

Daneault C, Luduc C, Valade JL (1994) The use of xylanases in kraft pulp bleaching: a review. Tappi J 77:125–131

Dekker RFH (1985) Biodegradation of the hemicelluloses. In: Higuchi T (ed) Biosynthesis and biodegradation of wood components. Academic Press, Tokyo, pp 503–533

Dix NJ, Webster J (1995) Fungal ecology. Chapman and Hall, London, 549pp

Duarte JC, Costa-Ferreira M (1994) Aspergilli and lignocellulosics: enzymology and biotechnological applications. FEMS Microbiol Rev 13:377–386

Eriksson K-E, Wood TM (1985) Biodegradation of cellulose. In: Higuchi T (ed) Biosynthesis and biodegradation of wood components. Academic Press, New York, pp 469–503

Eriksson K-EL, Blanchette RA, Ander P (1990) Microbial and enzymatic degradation of wood components. Springer, Berlin Heidelberg New York, 407pp

Fog K (1988) The effect of added nitrogen on the rate of decomposition of organic matter. Biol Rev 63:433–462

Ghosh P, Singh A (1993) Physicochemical and biological treatments for enzymatic/microbial conversion of lignocellulosic biomass. Adv Appl Microbiol 39:295–333

Glasser WG, Ravindran G, Jain RK, Samaranayake G, Todd J (1995) Comparative enzyme biodegradability of xylan, cellulose, and starch derivatives. Biotechnol Prog 11:552–557

Glenn JK, Morgan MA, Mayfield MB, Kuwahara M, Gold MH (1983) An extracellular H_2O_2-requiring enzyme preparation involved in lignin biodegradation by the white-rot basidiomycete *Phanerochaete chrysosporium*. Biochem Biophys Res Commun 114:1077–1083

Hammel KE, Mozuch MD, Jensen KA, Kersten PJ (1994) H_2O_2 recycling during oxidation of the arylglycerol β-aryl ether lignin structure by lignin peroxidase and glyoxal oxidase. Biochemistry 33:13349–13354

Higuchi T (1990) Lignin biochemistry: biosynthesis and biodegradation. Wood Sci Technol 24:23–63

Kirk TK, Farrell RL (1987) Enzymatic "combustion": the microbial degradation of lignin. Annu Rev Microbiol 41:465–505

Kuwahara M, Glenn JK, Morgan MA, Gold MH (1984) Separation and characterization of two extracellular H_2O_2-dependent oxidases from ligninolytic cultures of *Phanerochaete chrysosporium*. FEBS Lett 169:247–250

Li X, Zhang Z, Dean JFD, Eriksson K-EL, Ljundahl LG (1993) Purification and characterization of a new xylanase (APX-II) from the fungus *Aureobasidium pullulans* Y-2311-1. Appl Environ Microbiol 59:3212–3218

Ljungdahl LG, Eriksson K-E (1985) Ecology of microbial cellulose degradation. Adv Microbial Ecol 8:237–299

Lundgren KR, Bergkvist L, Hogman S, Joves H, Eriksson G, Bartfai T, van der Laan J, Rosenberg E, Shoham Y (1994) TCF mill trial on softwood pulp with korsnas thermostable and alkaline stable xylanase T6. FEMS Microbiol Rev 13:365–368

Marsden WL, Gray PP (1986) Enzymatic hydrolysis of cellulose in lignocellulosic materials. CRC Crit Rev Biotechnol 3:235–276

Messner K, Srebotnik E (1994) Biopulping: an overview of developments in an environmentally safe paper-making technology. FEMS Microbiol Rev 13:351–364

Mishra P, Singh A (1993) Microbial pentose utilization. Adv Appl Microbiol 39:91–152

Nakamura S, Wakabayashi K, Nakai R, Aono R, Horikoshi K (1993) Production of alkaline xylanase by a newly isolated alkaliphilic *Bacillus* sp. strain 41M-1. World J Microbial Biotechnol 9:221–224

Rättö M, Poutanen K, Viikari L (1992) Production of xylanolytic enzymes by an alkalitolerant *Bacillus circulans* strain. Appl Microbiol Biotechnol 37:470–473

Rayner ADM, Boddy L (1988) Fungal decomposition of wood. Its biology and ecology. Wiley, Chichester, 587pp

Reese ET, Siu RGH, Levinson HS (1950) The biological degradation of soluble cellulose derivatives and its relationship to the mechanism of cellulose hydrolysis. J Bacteriol 59:485–497

Reid ID, Paice MG (1994) Biological bleaching of Kraft pulps by white-rot fungi and their enzymes. FEMS Microbiol Rev 13:369–376

Sakai T, Sakamoto T, Hallaert J, Vandamme EJ (1993) Pectin, pectinase, and protopectinase: production, properties, and applications. Adv Appl Microbiol 39:213–294

Simpson HD, Haufler UR, Daniel RM (1991) An extremely thermostable xylanase from the thermophilic eubacterium *Thermotoga*. Biochem J 277:413–417

Tien M, Kirk TK (1983) Lignin-degrading enzyme from the hymenomycete *Phanerochaete chrysosporium* burds. Science 221:661–663

Viikari L, Ranua M, Kantelinen A, Linko M, Sundquist J (1986) Bleaching with enzymes. Proceedings of the 3rd international conference on biotechnology in the pulp and paper industry, STFI, Stockholm, pp 67–69

Viikari L, Tenkanen M, Buchert J, Rättö M, Bailey M, Siika-aho M, Linko M (1993) Hemicellulases for industrial applications. In: Saddler JN (ed) Bioconversion of forest and agriculatural plant residues. CAB International, Wallingford, pp 131–182

Viikari L, Kantelinen A, Sundquist J, Linki M (1994) Xylanases in bleaching: from an idea to the industry. FEMS Microbiol Rev 13:335–350

Waksman SA, Skinner CE (1926) Microorganisms concerned in the decomposition of celluloses in soil. J Bacteriol 12:57–84

Wong KKY, Tan LUL, Saddler JN (1988) Multiplicity of β-1,4-xylanase in microorganisms: functions and applications. Microbiol Rev 52:305–317

Generic Index

Subject Index

Druck: STRAUSS OFFSETDRUCK, MÖRLENBACH
Verarbeitung: SCHÄFFER, GRÜNSTADT